Glacial deposits provide a long-term record of climate and sea level changes on Earth. Detailed study of sedimentary rocks deposited during and immediately after glacial episodes is paramount to accurate paleoclimatic reconstructions and for our understanding of global climatic and eustatic changes. This book presents new information and interpretations of the ancient glacial record, looking in particular at the Late Proterozoic and Late Paleozoic eras. The influence of global tectonics on the origins and distribution of ice masses and the character of glacial deposits through geologic time is emphasised. Sequence stratigraphic techniques are applied to glaciogenic successions, and explanations are put forward for possible low-latitude glaciation during the Late Proterozoic era and the association of carbonate deposits with glaciogenic rocks. Early interglacial conditions, represented by dark-grey mudrocks and ice keel scour features, are discussed. These studies, from key workers in International Geological Correlation Program Project 260, will aid the understanding of Earth's climatic history.

Earth's Glacial Record

World and regional geology series
Series Editors: M. R. A. Thomson and J. A. Reinemund

This series comprises monographs and reference books on studies of world and regional geology. Topics will include comprehensive studies of key geological regions and the results of some recent IGCP projects.

Geological Evolution of Antarctica   M. R. A. Thomson, J. A. Crame & J. W. Thomson (eds.)

Permo-Triassic Events in the Eastern Tethys   W. C. Sweet, Yang Zunyi, J. M. Dickins & Yin Hongfu (eds.)

The Jurassic of the Circum-Pacific   G. E. G. Westermann (ed.)

Global Geological Record of Lake Basins vol. 1   E. Gierlowski-Kordesch & K. Kelts (eds.)

Earth's Glacial Record   M. Deynoux, J. M. G. Miller, E. W. Domack, N. Eyles, I. J. Fairchild & G. M. Young (eds.)

Paleocommunities: A Case Study from the Silurian and Lower Devonian   A. J. Boucot & J. D. Lawson (eds.)

Earth's Glacial Record

EDITED BY

M. DEYNOUX, J. M. G. MILLER,
E. W. DOMACK, N. EYLES,
I. J. FAIRCHILD, G. M. YOUNG

International Geological Correlation Project 260:
Earth's Glacial Record

PUBLISHED BY THE PRESS SYNDICATE OF THE UNIVERSITY OF CAMBRIDGE
The Pitt Building, Trumpington Street, Cambridge, United Kingdom

CAMBRIDGE UNIVERSITY PRESS
The Edinburgh Building, Cambridge CB2 2RU, UK
40 West 20th Street, New York NY 10011–4211, USA
477 Williamstown Road, Port Melbourne, VIC 3207, Australia
Ruiz de Alarcón 13, 28014 Madrid, Spain
Dock House, The Waterfront, Cape Town 8001, South Africa

http://www.cambridge.org

First published 1994
First paperback edition 2003

A catalogue record for this book is available from the British Library

Library of Congress cataloguing-in-publication data

Earth's glacial record / edited by M. Deynoux . . . [et al.].
    p. cm.
'International Geological Correlation Project 260: Earth's Glacial
Record.'
ISBN 0 521 42022 9 hardback
1. Glacial epoch. I. Deynoux, Max. II. International Geological
Correlation Project 260: Earth's Glacial Record.
QE697.E17 1994
551.7′92–dc20
93–29826 CIP

ISBN 0 521 42022 9 hardback
ISBN 0 521 54803 9 paperback

Contents

Contributors

L. BRADBY
School of Earth Sciences, University of Birmingham, Edgbaston, Birmingham B15 2TT, UK

J.R. CANUTO
Instituto de Geosciencias, Universidade de Sao Paulo, CP 20.889,01498-970, Sao Paulo, Brazil

A.D.M. CHRISTIE
Anglo-American Corporation, PO Box 561, Bethel 2310, South Africa

N. CHRISTIE-BLICK
Department of Geological Sciences and Lamont-Doherty Geological Observatory of Columbia University, Palisades, New York 10964, USA

D.I. COLE
Geological Survey, PO Box 572, Belleville 7535, South Africa

J.W. COLLINSON
Department of Geological Sciences, Ohio State University, Columbus, Ohio 43210, USA

A.R. CROSSING
Wonga Station, via Broken Hill, New South Wales, 2880 Australia

M. DEYNOUX
Centre C.N.R.S. de Geochimie de la Surface, Institut de Geologie, 1 rue Blessig, 67084 Strasbourg, France

E.W. DOMACK
Department of Geology, Hamilton College, Clinton, New York 13323, USA

P.R. DOS SANTOS
Instituto de Geosciencias, Universidade de Sao Paulo, CP 20.889,01498-970, Sao Paulo, Brazil

J.A. DOWDESWELL
Scott Polar Research Institute, University of Cambridge, Cambridge CB2 1ER, UK

N. EYLES
Department of Geology, Scarborough Campus, University of Toronto, 1265 Military Trail, Scarborough, Ontario M1C 1A4, Canada

I.J. FAIRCHILD
School of Earth Sciences, University of Birmingham, Edgbaston, Birmingham B15 2TT, UK

A.B. FRANÇA
PETROBRAS/NEXPAR, Rua Padre, Camargo, 285 Curitiba, Pr 80060, Brazil

GAO ZHENJIA
Xinjian Institute and Bureau of Geology, Urumgi, P.R. China

M.R. GIPP
Department of Geology, Scarborough Campus, University of Toronto, Scarborough, Ontario M1C 1A4, Canada

V.A. GOSTIN
Department of Geology and Geophysics, University of Adelaide, GPO Box 498, Adelaide, South Australia 5001, Australia.

LI HUAIKUN
Tianjin Institute of Geology and Mineral Resources, 8th Road, Dazhigu, Tianjin, P.R. China

LI YUZHEN
Ningxia Institute of Geology and Mineral Resources, New City Region of Yinchuan City 750021, Ningxia, P.R. China

P.K. LINK
Department of Geology, Idaho State University, Pocatello, Idaho 83209, USA

LU SONGNIAN
Tianjin Institute of Geology and Mineral Resources, 8th Road, Dazhigu, Tianjin, P.R. China

J.M.G. MILLER
Department of Geology, Vanderbilt University, Nashville, Tennessee 37235, USA

M.F. MILLER
Department of Geology, Vanderbilt University, Nashville, Tennessee 37235, USA

J.N. PROUST
Universite de Lille, U.F.R. des Sciences de la Terre, Unite de Recherche C.N.R.S. 'Tectonique et Sedimentation', 59655 Villeneuve d'Ascq, France

QI RUI ZHANG
Institute of Geology, Academia Sinica, PO Box 634, Beijing, P.R. China

A.C. ROCHA-CAMPOS
Instituto de Geosciencias, Universidade de Sao Paulo, CP 20.889,01498-970, Sao Paulo, Brazil

B. SPIRO
NERC Isotope Geosciences Laboratory, Kingsley Dunham Centre, Keyworth, Nottingham NG12 5GG, UK

J.N.J. VISSER
Department of Geology, University of the Orange Free State, PO Box 339, Bloemfontein 9300, South Africa

V. VON BRUNN
Department of Geology, University of Natal, PO Box 375, Pietermaritzburg, 3200 South Africa

G.E. WILLIAMS
Department of Geology and Geophysics, University of Adelaide, GPO
Box 498, Adelaide, South Australia 5001, Australia

C.M.T. WOODWORTH-LYNAS
Centre for Cold Ocean Resources Engineering, Memorial University of
Newfoundland, St John's, Newfoundland A18 3X5, Canada

G.M. YOUNG
Department of Geology, University of Western Ontario, London,
Ontario N6A 5B7, Canada

ZHENG ZHAOCHANG
Ningxia Institute of Geology and Mineral Resources, New City Region
of Yinchuan City 750021, Ningxia, P.R. China

Preface

An IGCP contribution

This volume represents the final contribution of the International Geological Correlation Program Project 260 'Earth's Glacial Record'. The project was active from 1987 until 1991. It succeeded IGCP Project 24 'Quaternary Glaciations in the Northern Hemisphere', and Project 38 'Pre-Pleistocene Tillites'. Thus, its goal was to promote further research on glaciations, whatever their age and location, and to encourage geologists working in Modern or Quaternary sequences and those working on Pre-Pleistocene rocks to share their experience and different approaches. In this way we hoped to emphasize and illuminate global aspects of glacial phenomena in addition to facies description and regional problems. Participants in the project were encouraged to consider the tectonic versus climatic control on sedimentation in glacially influenced basins, the nature of the feedback between plate positioning, tectonics, and climate, and the paleoenvironmental significance and distribution of specific rock types (e.g. black shales, carbonates, iron formations, etc.) during and immediately after glacial periods.

These ambitious objectives were addressed by three subgroups that were in charge of, respectively, geodynamic setting, paleomagnetic reconstructions, and significance of specific rock types. Yearly meetings and field trips were held in various countries (10 field trips in Canada, Brazil, USA, UK, Mali), covering a vast variety of structural terranes (active and passive margins, intracratonic basins) from Proterozoic up to Modern in age. This clearly demonstrated the interest in and necessity for such a comprehensive approach among geologists. We do not pretend that all objectives were perfectly met, but aspects have been clarified and progressively more communications on global climatic or tectonic implications were presented during the successive meetings. This awareness of the importance of a global approach is certainly one of the most significant results of the project.

A second outcome of the project is a definite improvement in communication among workers on ancient and modern deposits leading to more sophisticated facies interpretation. Such interaction, as well as the idea of submitting a common IGCP project, began at the Till Mauretania 83 Symposium (Deynoux, 1985a), when Quaternary and 'paleo' geologists examined West African Late Proterozoic glacial deposits. The final decision to submit the project, which was encouraged and supported in particular by J.C. Crowell and N.M. Chumakov, was made after an informal meeting during the 1986 International Sedimentological Congress in Canberra.

Content of the volume

The present volume complements the largely descriptive compilation made by Hambrey and Harland (1981) for IGCP 38. It is less encyclopedic and more interpretive. It reflects results of research on the IGCP 260 project themes outlined above.

One of the most puzzling climatic phenomena in the Earth's history is the possibility of glaciations at low latitudes, inferred from low paleomagnetic inclinations in many Late Proterozoic (Neoproterozoic) strata containing glacial deposits. In response to this paradox, a glacial origin for most alleged Late Proterozoic 'tillites' was denied by Schermerhorn (1974) in an excellent and somewhat provocative paper. Among his arguments was the occurrence of diamictites in active tectonic settings, implying that many diamictites may be explained better as deposited by debris-flow processes or in association with mountain glaciations, and the association of these diamictites with rocks generally indicative of warm climate, such as carbonate or iron deposits. Such an extreme position has in many cases been undermined by field evidence (Hambrey and Harland, 1981). However, Schermerhorn's paper forced reappraisal of arguments in terms of plate tectonic activity and reinforcement of those arguments which concern the glaciogenic origin of diamictite-bearing facies associations. IGCP Project 260 extended these themes by investigating the effect of tectonic setting upon the distribution and type of sedimentary facies in proven glaciogenic successions. The first nine papers of this volume address these themes.

N. Eyles and G.M. Young (p. 1) give an overview of glaciations in Earth's history and focus on the role of plate tectonic processes in the production and preservation of glaciogenic deposits. They demonstrate that strong uplift in active collisional margins or on the flanks of basins undergoing regional extension, and the resultant enhanced weathering that causes drawdown of atmospheric

CO_2, provide first-order controls on glaciations. Many points that they discuss are fleshed out in papers later in the volume. The paper by P.K. Link, J.M.G. Miller, and N. Christie-Blick (p. 29) may serve as an illustration of the Eyles and Young hypothesis on the origin of glaciations. They present well-documented examples of Late Proterozoic glacial-marine sedimentation along the margin of differentially subsiding basins that developed during an episode of rifting in western North America. They provide careful descriptions of diamictite-bearing facies associations and emphasize the common occurrence of relatively deep-water facies in most of the Late Proterozoic sequences of inferred glacial origin. In contrast, J.M.G. Miller (p. 47) illustrates the merits of Schermerhorn's arguments in an excellent example of Late Proterozoic diamictite-bearing facies associations in a continental rift system in eastern North America. She proposes that diamictites were deposited in a lacustrine environment under the influence of a local alpine glaciation.

The next two papers show the influence of local tectonics upon glacial sedimentation patterns during the Late Paleozoic. V. Von Brunn (p. 60) proposes a Permo-Carboniferous model in which glacial-marine deposits, including a large amount of diamictite, were formed within and on the flanks of a subsiding trough which developed over a failed rift. Depositional architecture appears largely controlled by pre-existing topography and glacially related sea-level changes. A.B. França (p. 70) presents an overview of the stratigraphy and hydrocarbon potential of the Carboniferous-Permian Itararé Group in the whole Brazilian Paraná Basin. The Itarare Group forms a continuous and thick record of a temperate glacial-marine environment in which the distribution and thickness of sedimentary units were affected by structural lineaments. This contribution is original because it concerns an economic aspect of the glacial sequences. According to structural setting and proximity of source rocks, the Itarare Group constitutes an excellent model for petroleum exploration.

Difficulty determining the origin of massive diamictites was frequently discussed at IGCP 260 meetings, J.N.J. Visser (p. 83) addresses this problem using four examples from the Permo-Carboniferous Dwyka Formation in South Africa. He proposes well-defined criteria, based on clast fabric and facies context, for the absolute identification of mechanisms of deposition of these diamictites in a glacial marine environment. Harking back to Schermerhorn (1974), Lu Songnian and Gao Zhenjia (p. 95) questioned the origin of two superposed Late Proterozoic diamictite-bearing formations in West China. Both formations were previously interpreted as glacial. Using stratigraphic and sedimentologic arguments, the authors demonstrate that the 'lower diamictite' corresponds to non-glacial debris-flows, the 'upper diamictite' to continental glacial deposits. Zheng Zhaochang and Li Yuzhen (p. 101) report Late Proterozoic glaciogenic successions in northwestern China, which consist of massive to bedded diamictite layers deposited subglacially or as subaqueous debris flows, overlain by transgressive post-glacial, thinly bedded siltstones and shales. They emphasize the importance of regional and local facies context in inferring the glaciogenic origin of the diamictite.

Lastly, the paper by M.R. Gipp (p. 109) provides sedimentological models for the large-scale architecture of glaciated shelf and slope systems in Late Cenozoic active (Gulf of Alaska) and passive (Nova Scotia) margins. Processes of deposition appear identical on both active and passive margins, but the gross depositional architecture of glacial marine deposits differs depending on the preservation potential of sediments, which is controlled by tectonics and relative sea-level changes.

Recent years have seen the introduction of the exciting new concepts of sequence stratigraphy. Such ideas are particularly relevant to glaciogenic successions because of the associated rapid and large-scale sea-level changes. However, sequence stratigraphy remains scarcely used in glacial rock successions owing probably to local effects related to the common occurrence of glacigenic deposits in tectonically active areas. J.N. Proust and M. Deynoux (p. 121) propose a sequence stratigraphic model based on the definition of a depositional genetic unit and its evolution through space and time in the marine to continental transitional zone of an intracratonic glacially influenced basin. Their model is developed from detailed field analysis of Late Proterozoic glacially related deposits on the West African platform. These genetic units and their different development in a stacking pattern lead to the definition of different orders of stacked sequences that are interpreted in terms of short-term climatically (glacially) controlled and long-term tectonically driven baselevel fluctuation cycles.

As shown by the papers quoted above, geologists have strong arguments which confirm the glacial origin of several Late Proterozoic successions. However, this does not solve the problem of low latitude glaciations (Chumakov and Elston, 1989). Hypotheses such as fast-moving plates (Crowell, 1983) or global glacial climate (Harland, 1964) have been proposed but are difficult to support on paleomagnetic and geologic grounds. Astronomical causes have also been proposed (Williams, 1975, Sheldon, 1984), and in this volume G.E. Williams (p. 146) again addresses this problem which 'challenges conventional views on the nature of the geomagnetic field, climatic zonation, and the earth's planetary dynamics in Late Proterozoic time'. Williams gives new evidence (paleomagnetic and time-series analysis of tidalites, and paleoclimatic interpretation of periglacial structures) supporting his previous hypothesis of a large obliquity of the ecliptic ($> 54°$) leading to a reverse climatic zonation and marked seasonality.

Although still used as an argument against glaciation, the co-occurrence of carbonate rocks and glacial deposits is common. However, there are many facets of the association. Skeletal carbonates are common in high latitude seas today but they do not have exact equivalents in Proterozoic rocks. As reviewed by Fairchild (1992) the extensive ice sheets of Late Proterozoic times appear to have encroached onto previously warm carbonate-forming environments which returned following glaciation. Glacial deposits are carbonate-rich primarily because of the incorporation of detrital carbonate. Subglacial redistribution of detrital carbonate by dissolution and reprecipitation by stress-related melting–freezing processes was proposed for some Late Proterozoic terrestrial tillites (Deynoux, 1985b). The discovery of marine recrystallization of

Late Proterozoic glacially transported rock flour in Svalbard (Fairchild *et al.*, 1989) gave new insight to the problem. Following this discovery, Crossing and Gostin (p. 165) investigated examples of diamictites in the Adelaide geosyncline of south Australia and found geochemical evidence from the composition of the matrix for deposition in a sea diluted by meltwater. Additionally, correlation between high Fe and reduced δ^{13} C values in the matrix suggests that bacterial activity accompanied the diagenesis of the rock flour. In order to gain a better understanding of the chemical processes involved when carbonate rock flours interact with fluids in glacial systems, investigations of carbonate-rich Quaternary glacial systems have been started, and I.J. Fairchild, L. Bradby, and B. Spiro (p. 176) report their preliminary conclusions. Although they stress the importance of postglacial processes in controlling lithification of the sediments, evidence is also found for the precipitation of calcite in the matrix of a refrozen meltout till. Since this article was prepared they have found similar material in unfrozen meltout till and basal ice.

The succession of glacial deposits in Late Paleozoic sequences across much of Gondwana is punctuated by a sharp contract between diamictites and overlying late- or post-glacial dark- to black-colored mudstones (Domack *et al.*, 1992). Such a sharp contact has also been described in intracratonic Late Proterozoic and Late Ordovician glacial sequences on the West African platform (see Deynoux and Trompette in Hambrey and Harland, 1981). The diamictite–black mudstone transition is important in that it is widely believed to represent the end of glacial climates within the regional extent of the basins in which it is found. Hence these rocks preserve a record of global warming associated with an apparently rapid 'glacial' to 'interglacial' transition. A similar hypothesis was also proposed for certain Proterozoic carbonate horizons capping glacial deposits (e.g. Williams, 1979). Depositional mechanisms for the dark to black mudstones are varied. J.N.J. Visser (p. 193) describes the conditions that prevailed in a shallow to moderately deep Late Carboniferous foreland basin (Dwyka glaciations of the Karoo Basin) after self-destructive collapse of a marine ice sheet resulting from a relative sea-level rise. The syn- to post-glacial dark to black mudrocks, which overlie the glaciogenic deposits, were deposited by suspension settling of mud and mud turbidites. Basin tectonics, oceanic circulation, and climate controlled the organic-rich black mud deposition. D.I. Cole and A.D.M. Christie (p. 204) also describe Early Permian black mudrocks overlying diamictites (debris and turbidity flows) deposited by a retreating tidewater glacier during the final phase of the Dwyka glaciation. The black mudrocks are the product of pelagic mud settling proximal to the ice front where freshwater plumes mixed with basinal saline water. They account for a sudden increase in organic productivity, indicating that the rapid termination of the Dwyka glaciation was accompanied by a sharp rise in temperature. M.F. Miller and J.W. Collinson (p. 215) describe the processes and environments that characterize the filling of a large Lower Permian post-glacial inland sea in Antarctica. Deposition in relatively shallow water was dominated by turbidity currents carrying fine-grained sediments in channel-overbank systems. The glacial environment allows the definition of a model in which a fine-grained turbidite system is paradoxically fed by a coarse-grained braided stream of the outwash plain.

Recent years have seen increasing interest in the sedimentary record of ice scours on continental shelves and lakes. Such structures and associated diamict deposits are widespread on Pleistocene shelves but are rarely known from the rock record. The paper by A.C. Rocha-Campos, P.R. dos Santos, and J.R. Canuto (p. 234) describes Early Permian iceberg scours. These scours are encountered on bedding planes of rhythmites inferred to be varves in a relatively deep freshwater body. C.M.T. Woodworth-Lynas and J.A. Dowdeswell (p. 241) argue that many striated surfaces, and associated diamictites found in ancient successions may have been produced by marine (or lacustrine) ice keel scour and report a modern analogue from the Greenland shelf. The need for a more critical examination of ancient glacial striated surfaces and associated facies is clearly indicated. Qui Rui Zhang (p. 260) describes periglacial indicators such as iceberg scours and dropstones, ice wedge casts, and glaciotectonic structures. These structures indicate that, instead of an abrupt erosional unconformity, the glaciogenic deposits of the Nanhua Ice Age (Late Proterozoic) of South China are locally conformably underlain by rocks that mark a progressive climatic transition from warm to cold.

Significance

The results of IGCP Project 260 are very relevant to current concerns about global change. As more details of the present climate and the climate of the Pleistocene are revealed, the interplay of ocean, atmosphere, biosphere and lithosphere with the internal heat engine of the Earth is being revealed as something of great complexity. We are still far from a complete understanding of the Earth's climatic system but glaciogenic deposits provide critical palaeoclimatic data. In spite of recent suggestions that some tillites/diamictites may be the ejecta of large planetesimal impacts (Rampino, 1992; Oberbeck *et al.*, 1993), extensive research and field studies demonstrate that the vast majority of documented ancient glaciogenic deposits correctly record periods of cold climate during Earth history. The discovery of mechanisms governing the appearance and disappearance of glaciers on Earth is paramount to the understanding of long-term climatic change, and the only long-term record of climatic change is the geologic record. The rock record of glaciation has been catalogued (Hambrey and Harland, 1981) but still has not been perfectly described. Since the 1983 Mauritanian meeting (Deynoux, 1985a), good facies descriptions of ancient glacial deposits have been proposed, and lately sequence stratigraphic concepts have developed. Now glacial sedimentology has joined the mainstream of sedimentology. In the near future, because most ancient glacial evidence is preserved in marine sequences, we must achieve a better understanding of the workings of glacially related marine basins.

References

Chumakov, N.M. & Elston, D.P. (1989). The paradox of late Proterozoic glaciation at low latitude. *Episodes*, **12**, 115–19.

Crowell, J.C. (1983). Ice ages recorded on Gondwana continents. *Transactions of the Geological Society of South Africa*, **85**, 238–61.

Deynoux, M., (ed.) (1985a). Glacial record. Proceedings of the Till Mauretania Symposium, Nouakchott-Atar, 4–15 January 1983, *Palaeogeography Palaeoclimatology Palaeoecology*, Special Issue, **51**, 451pp.

Deynoux, M. (1985b). Terrestrial or waterlain glacial diamictites? Three case studies of the Late Precambrian and Late Ordovician glacial drifts in West Africa. *Palaeogeography Palaeoclimatology Palaeoecology*, **51**, 97–141.

Domack, E.W., Burkley, L.A., Domack, C.R. & Banks, M.R. (1992). Facies analysis of glacial marine pebbly mudstones in the Tasmania Basin: implications for regional paleoclimates during the Late Paleozoic. In *Gondwana 8: Assembly, evolution and dispersal*, eds., Findlay, R.H., Banks, M.R., Veevers, J.J., Unrug, R., Rotterdam: Balkema, 471–84.

Fairchild, I.J., Hambrey, M.J., Jefferson, T.H. & Spiro, B. (1989). Late Proterozoic glacial carbonates in NE Spitsbergen: new insights into the carbonate–tillite association. *Geological Magazine*, **126**, 469–90.

Fairchild, I.J. (1993). Balmy shores and icy wastes: the paradox of carbonates associated with glacial deposits in Neoproterozoic times. *Sedimentary Review*, **1**, 1–6.

Hambrey, M.J. & Harland, W.B., (eds.) (1981). Earth's Pre-Pleistocene glacial record. Cambridge: Cambridge University Press, 1004pp.

Harland, W.B. (1964). Critical evidence for a great Infracambrian glaciation. *Geological Rundschau*, **54**, 45–61.

Oberbeck, V.R., Marshall, J.R., Aggarvad, H. (1993). Impacts, tillites, and the breakup of Gondwanaland. *Journal of Geology*, **101**, 1–19.

Rampino, M.R. (1992). Ancient 'glacial' deposits are ejecta of large impacts: the ice age paradox explained. *EOS, Transactions of the American Geophysical Union*, **73**, p. 99.

Schermerhorn, L.J.G. (1974). Late Precambrian mixtites: glacial and/or non-glacial. *American Journal of Science*, **274**, 673–824.

Sheldon, R.P. (1984). Ice-ring origin of the Earth's atmosphere and hydrosphere and Late Proterozoic-Cambrian phosphogenesis. *Geological Survey of India Special Publication*, **17**, 17–21.

Williams, G.E., (1975). Late Precambrian glacial climate and the Earth's obliquity. *Geological Magazine*, **112**, 441–65.

Williams, G.E. (1979). Sedimentology, stable-isotope geochemistry and palaeoenvironments of dolostones capping Late Precambrian glacial sequences in Australia. *Journal of the Geological Society of Australia*, **26**, 377–86.

Strasbourg, July 1993   M. Deynoux (Strasbourg,
                                              France)
                                     J.M.G. Miller (Nashville, USA)
                                     E.W. Domack (Clinton, USA)
                                     N. Eyles (Toronto, Canada)
                                     I.J. Fairchild (Birmingham, UK)
                                     G.M. Young (London, Canada)

Acknowledgements

This volume reflects only a small part of the contributions by several active members of the IGCP 260 Project. The following colleagues organized our annual meetings and field trips, and also specific symposia during international congresses: N. Eyles (Canada, 1987, 1991), C.J.S. Alvaranga, J.R. Canuto, M. Deynoux, A.C. Rocha-Campos, P.R. Santos, R. Trompette (Brazil, 1988), E.W. Domack, C.H. and N. Eyles, J.M.G. Miller, B. Molnia (USA, 1989), M.J. Hambrey, A.M. McCabe, A.C.M. Moncrief (UK, 1990), M. Deynoux, C.S. Diawara, N.D. Keita, S. Keita, J.N. Proust (Mali, 1991).

We would also like to acknowledge our colleagues who acted as critical and constructional reviewers of the submitted papers: C.H. Eyles (Toronto, Canada), M.R. Gipp (Toronto, Canada), V. Gostin (Adelaide, Australia), P. Herrington (London, UK), L.A. Krissek (Columbus, USA), O. Lopez-Gamundi (Buenos Aires, Argentina), M.F. Miller (Nashville, USA), H.T. Ore (Pocatello, USA), J. Sarfati (Montpellier, France), W.W. Simpkins (Ames, USA), B. Spiro (Keyworth, UK), M.E. Tucker (Durham, UK), T. Warman (Toronto, Canada) C. Woodworth-Lynas (St Johns, Canada).

1 Geodynamic controls on glaciation in Earth history

NICHOLAS EYLES and GRANT M. YOUNG

Abstract

The geological record of glaciation is sporadic. It begins with poor and fragmentary evidence from Archean rocks but there is unequivocal evidence of glaciation in the Paleoproterozoic of North America, Scandinavia and possibly South Africa, western Australia and India. There follows a long Mesoproterozoic non-glacial period, between about 2.0 and 1.0 Ga with no well-constrained evidence of glaciation but there is a return to sporadic glacial conditions from about 800 Ma to the Cambrian. Neoproterozoic glaciations were very widespread and evidence is preserved on all the present day continents. Some palaeomagnetic evidence suggests that Neoproterozoic glaciation may have taken place at low paleolatitudes but new data supports earlier concerns with regard to rapid plate motions and low latitude 'overprinting' in the Cambrian. 'Warm' climate strata in many Neoproterozoic glacial successions appear to be detrital in origin.

In Phanerozoic times, glaciation is reported from Ordovician successions in Africa, possibly Brazil and Arabia; evidence of Silurian and Devonian glaciation is largely limited to South America. The most significant Phanerozoic glaciation took place in the Permo-Carboniferous, between 350 and 250 Ma, across a large area of the Gondwanan supercontinent. There is no direct geologic record of Mesozoic glaciation but small ice masses may have developed in the interiors of landmasses at high latitudes (*e.g.* Antarctica, Siberia). Small-scale fourth-order cycles of sea-level change recorded on several carbonate and siliciclastic shelves at this time are unlikely to be of glacio-eustatic origin.

The earliest Late Tertiary glaciation is recorded from Antarctica about 36 Ma; glaciation in the northern hemisphere was initiated at about 6 Ma with large continental ice sheets developing after about 3 Ma.

Global tectonic cycles of supercontinent amalgamation and dispersal are a first-order control on glaciation in Earth history. These cycles control the long-term composition of the Earth's atmosphere and have created alternating 'icehouse' and 'greenhouse' climates through geologic time. Plate tectonic activity results in amalgamation and dispersal of continental crust, which in turn controls the amount of continental freeboard. The elevation and planetary distribution of continental crust also affects such things as albedo and oceanic circulation, both of which play important roles in controlling surface temperatures. Crustal thickening in both extensional and collisional setting may have played a role in initiating 'adiabatic' glaciations. Active tectonism initiates large sediment fluxes from the continents, the weathering of which may draw down atmospheric CO_2 allowing the growth of larger ice masses controlled in part, by relatively weak 'Milankovitch' astronomical rhythms.

Recent work identifies the importance of palaeogeography in further controlling planetary temperatures and the tendency for very rapid global change involving abrupt changes in ocean–atmosphere coupling.

Introduction

The major questions that need to be answered in relation to Earth's glacial record are as follows:

(1) Why does glaciation occur? Is there a single cause or a series of separate or combined factors that can bring about cold climatic conditions on the planet?
(2) Why is the record of glaciation sporadic?
(3) What is the explanation for glaciation allegedly extending into marine basins at apparent equatorial latitudes in the Neoproterozoic?
(4) Why are many Neoproterozoic glacial deposits associated with other sedimentary rocks that appear to suggest warm climatic conditions?

The geologic record, although imperfect and incomplete, remains the only primary source of information concerning the long-term climatic history of the planet. In this paper we present a brief review of the Earth's glacial record and attempt to place it in the context of the geodynamic evolution of the planet. In doing so we hope to provide tentative answers to some of the questions posed above. We would emphasize that space does not permit a detailed treatment of all glacial occurrences. An exhaustive descriptive summary of Earth's glacial record can be found in Hambrey and Harland (1981). The more recent interpretive summaries include those by

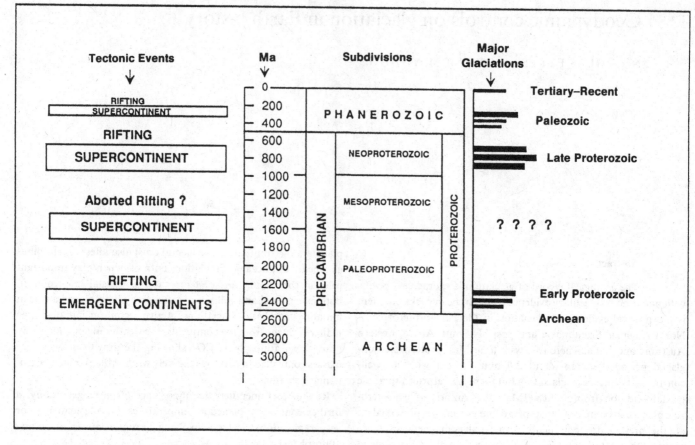

Fig. 1.1. Distribution of major glaciations throughout geologic time (subdivisions after Plumb, 1991). Note the association between some periods of supercontinentality and some glaciations. See text for discussion.

Crowell (1982), Deynoux (1985b), Veevers (1990), Worsley and Nance (1989), Young (1991a), Frakes *et al.* (1992) and Chumakov (1992). A detailed treatment of the relationship between glaciation and geodynamic setting can be found in Eyles (1993); Socci (1992) provides a thorough overview of likely global geochemical controls.

Any attempt to depict the distribution of glaciogenic rocks in Earth history is frustrated by a lack of precise and reliable geochronological data from the largely sedimentary successions in which they occur. A second problem is inadequate discrimination between diamictites formed by direct glacial activity (tillites) and those formed by other processes. As pointed out by Schermerhorn (1974) this problem is compounded by the fact that many glaciogenic successions were deposited in tectonically active settings and had complex histories sometimes involving both glacial processes and deposition by sediment gravity flows.

Figure 1.1 shows the distribution of glaciogenic rocks based on radiometric age dating of the past 2500 Ma. The age of most ancient glacial deposits is very poorly constrained and Figure 1.1 probably underestimates the frequency and duration of glaciation in Earth history.

Late Archean–Paleoproterozoic glaciations

The preserved record of glaciation is confined to the last half of geologic history when four major glacial eras (Chumakov, 1985) occurred over a period of about 2.5 Ga (Fig. 1.1). The dearth of evidence of glaciation in Archean times is puzzling because of the widely held belief that the sun's luminosity was much less (about 70% of the present value) during the early part of Earth history (Sagan and Mullen, 1972; Kasting and Toon, 1989; Gilliland, 1989; Caldeira and Kasting, 1992). It has been estimated (Kasting *et al.*, 1984) that under the present atmosphere, such a weakly radiant sun would have resulted in surface temperatures on Earth about 30 °C lower than at present. At such temperatures liquid water could not have existed on the surface of the planet but preservation of abundant waterlaid Archean rocks clearly indicates that this was not the case. The simplest resolution of what has been called the 'faint young sun paradox' (Kasting, 1987) is to invoke an enhanced greenhouse effect due to a much higher CO_2 content in the early atmosphere (Hart, 1978). If this interpretation is correct, then the paleoclimatic controls in the early part of Earth history differed greatly from those today. If it is accepted that solar luminosity was much weaker than at present, then cooling could easily have been induced by drawdown of atmospheric CO_2.

The Archean Witwatersrand Supergroup of South Africa, deposited between 3000 and 2700 Ma (de Wit et al., 1992), contains diamictites and other rocks that have been interpreted as products of glaciation (Wiebols, 1955; Harland, 1981) but a non-glacial origin is also possible (Kingsley, 1984; Tainton and Meyer, 1990). Von Brunn and Gold (1993) report glacialastic strata from the Pongola Sequence of Southern Africa dated between 3100 and 2800 Ma. Dropstones in Archean rocks associated with the Stillwater Complex in western USA have also been ascribed to glacial transport (Page, 1981). Apart from these small occurrences the Archean record is barren of glaciogenic deposits.

In the Early Proterozoic (Paleoproterozoic in the new terminology of Plumb (1991)) there is evidence of glaciation in several regions of North America, and in Karelia, South Africa, Australia, India and Scandinavia (e.g. Okajangas, 1988). Early Proterozoic glacial deposits are widespread but most again are poorly dated. The most clearly exposed deposits are those of the Gowganda Formation in Ontario, Canada dated to about 2300 Ma (Young and Nesbitt, 1985; Figs. 1.2, 1.3). Other, poorly constrained deposits occur in the Northwest Territories (Padlei Fm; older than 2.1 Ga; Patterson and Heaman, 1991), several occurrences in Michigan (2100–2000 Ma) and in the Black Hills (2560–1620 Ma). Deposits of similar age occur in southern Africa (Griquatown Basin; c. 2300 Ma), India (Bijawar Group, c. 1815 Ma) and Australia (Meteorite Bore Member, c. 2500–2000 Ma). Diamictites, some of which may be glaciogenic (Ojakangas, 1988), occur in the Baltic shield and in east-central and northern former USSR.

The most widespread and most convincing glaciogenic deposits of this era are found in North America. The Gowganda Formation, which forms part of the Huronian Supergroup deposited between about 2.5 and 2.2 Ga (Krogh et al., 1984), is perhaps the best known (Miall, 1985; Young and Nesbitt, 1985) but similar rocks have also been described (Young, 1970) from Michigan, Chibougamau in northern Quebec, the Hurwitz Group in the Canadian Northwest Territories and from Wyoming (Fig. 1.2). In some Paleoproterozoic successions there is evidence of multiple glaciations. For example, the Gowganda Formation (Fig. 1.3) is the youngest of three diamictite-bearing formations, all of which may be glaciogenic. Thus there is evidence in the Huronian Supergroup (and also possibly in the Snowy Pass Supergroup of Wyoming; Houston et al., 1981) of three major glacial intervals. Geochemical and sedimentological data (Fig. 1.3) suggest that glacial episodes in the Huronian Supergroup were separated by periods of intense chemical weathering (Nesbitt and Young, 1982).

In both areas, the tectonic setting has been interpreted in terms of a rifting continental margin (Karlstrom et al., 1983; Zolnai et al., 1984; Young and Nesbitt, 1985; Goodwin, 1991). Emergence of large areas of continental crust in the late Archean–Early Proterozoic (Taylor and McLennan, 1985) may have been a critical factor in reduction of global temperatures to the point where glaciation

could occur. The quantitatively most important weathering reaction on the surface of the planet (Nesbitt and Young, 1984) is that between carbonic acid ($CO_2 + H_2O$) and labile minerals of the continental crust (particularly plagioclase felspars). This reaction at present accounts for about 80% of the drawdown of CO_2 from the atmosphere (Houghton and Woodwell, 1989). Any tectonic process leading to emergence and subaerial exposure of significant regions of crustal material should therefore result in increased rates of drawdown of atmospheric CO_2. Under the conditions of weak solar luminosity inferred for the early part of Earth history, such a reduction in CO_2 may have caused a sufficiently strong 'antigreenhouse' effect to bring about the first glaciations.

How can we explain the 'anomalous' association of glacial deposits with sedimentary rocks that are highly weathered? This has been ascribed by Young (1991) to a negative feedback mechanism. Glacial cover inhibits surface weathering and results in increased albedo which contributes to a general lowering of surface temperatures. All of these factors contribute to a decrease in chemical weathering, so that CO_2, which is constantly being supplied from volcanic sources, can once more build up, leading eventually to the destruction of the ice sheets and initiation of a 'warm' period. Such 'warm' periods could only have occurred when concentrations of CO_2 were sufficiently high to counteract the inferred weak sun. This interpretation is in keeping with the strong chemical depletion observed in Huronian formations between the glacial units (Nesbitt and Young, 1982) and elsewhere in Archean and Early Proterozoic sedimentary rocks (Reimer, 1986; Eriksson et al., 1990). Thus negative feedback may provide an explanation for multiple glaciations separated, at least in the early part of Earth history, by sedimentary rocks carrying evidence of intense chemical weathering.

Mesoproterozoic non-glacial epoch

Perhaps even more puzzling than the glacial periods themselves is the apparent absence of glaciation on the planet between about 2.0 and 1.0 Ga (Fig. 1.1). Chumakov and Krasil'-nikov (1992) report possible Riphean glacial deposits from the Jena area of the former Soviet Union but their age is uncertain. Tectonic syntheses such as that of Hoffman (1989) suggest that the period between about 2.0 Ga and 1.8 Ga was a time of growth and aggregation of continents (at least in the Laurentian shield). A supercontinental configuration has been proposed by many (Windley, 1977; Piper, 1978; Hoffman, 1989) for the ensuing period. According to ideas outlined by Worsley et al. (1984) and Nance et al. (1988) such a supercontinental configuration should have provided suitable conditions for glaciation by the mechanism of drawdown of atmospheric CO_2 as a result of enhanced weathering reactions on the high standing supercontinent. Hoffman (1989), noting widespread magmatic activity (including unusually abundant anorthosites and rapakivi granites) during the period from about 1.8 to 1.3 Ga, proposed a unique stage in the thermal evolution of the planet, invoking a 'mantle superswell' that caused

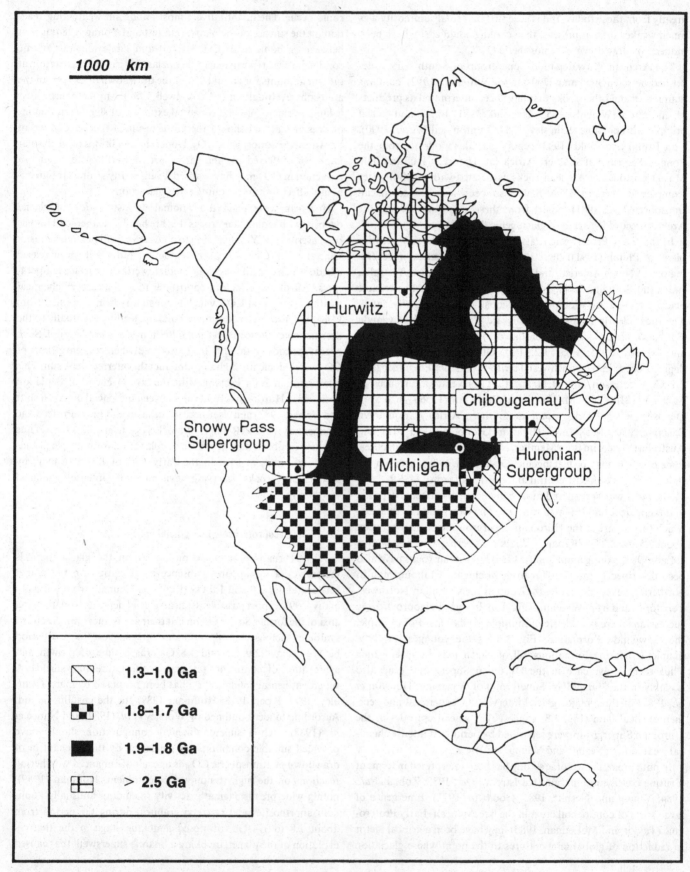

Fig. 1.2. Sketch map to show the location of Early Proterozoic glaciogenic rocks in North America in relation to major tectonic provinces (after Hoffman, 1988) the ages of which are shown at bottom left. Note that most occurrences are at or close to the margins of Archean cratons.

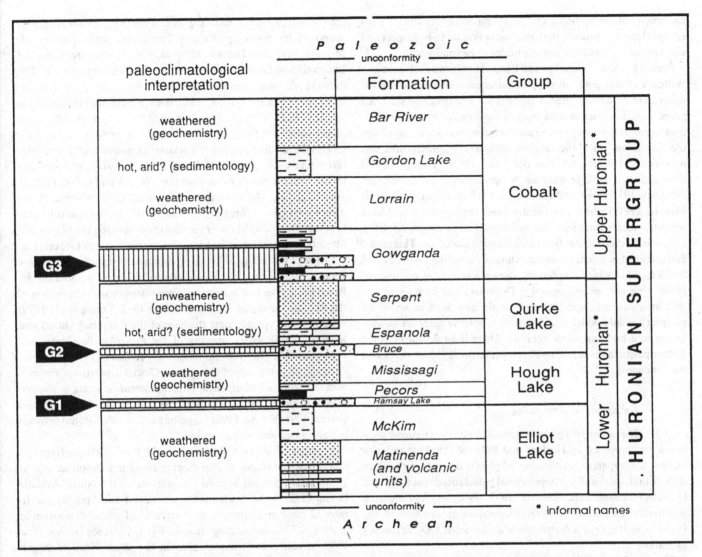

Fig. 1.3. Schematic representation of the stratigraphic succession of the Paleoproterozoic Huronian Supergroup (maximum thickness about 12 km) together with inferred paleoclimatic changes. Note the three major glacial periods (G1–3) separated by warm and/or humid periods of intense chemical weathering. These fluctuations are attributed to changes in atmospheric CO_2 content, ascribed to negative feedback (see text). Abbreviations in the stratigraphic column are as follows: fine dots, sandstone; large dots, diamictites; dashes, siltstones; black, mudstones; horizontal bricks, limestones; inclined bricks, dolostones; vertical hatching, volcanic units in the basal part of the Huronian. See text for discussion.

elevated temperatures beneath the first extensive supercontinent. Outgassing of CO_2 during this period of extensive magmatism and aborted continental fragmentation (Emslie, 1978) may have outstripped drawdown by weathering, causing the Earth to enter a long period of greenhouse-induced warm climate.

Neoproterozoic glaciations

Overview

Late Proterozoic glaciogenic deposits are known from all the continents. They provide evidence of the most widespread and long-ranging glaciation on Earth, extending from about 1.0 Ga to just before the Cambrian. Most regions display evidence of several glaciations, preceded, separated and followed by periods of relati-vely warm climate. This alternation of glacial and 'warm' climatic conditions, on a scale of tens of millions of years, is similar to that observed in the Paleoproterozoic, and could likewise have resulted from negative feedback, related to weathering reactions on a high-standing supercontinent (Young, 1991a, b). One of the most puzzling aspects of Late Proterozoic history is that glaciers appear to have descended to sea level in low paleolatitudes (Chumakov and Elston, 1989). Many explanations for these anomalous occurrences have been proposed. Williams (1975) suggested that the obliquity of the ecliptic was much higher than today and argued that this would have led to preferential glaciation at low latitudes as inferred from paleomagnetic study of Late Proterozoic glaciogenic sediments in South Australia (Embleton and Williams, 1986). Williams also suggested that this model provides an explanation for evidence of seasonality of low latitudes. Others (Stupavsky *et al.*, 1982; Crowell, 1982) have considered the paleomagnetic data to be suspect and some of the reported low paleolatitudes to be spurious.

Recent work by Chumakov and Elston (1989) and Embleton and Williams (1986) tends to support an equatorial positioning for many Late Proterozoic glacial deposits but new data also show the probable importance of high latitude positioning during the Neoproterozoic followed by very rapid plate motions to low latitudes in the Cambrian when the original remanent magnetization was overprinted (Meert and Van der Voo, 1992). Young (1991a) proposed that a high standing Neoproterozoic supercontinent (Windley, 1977; Piper, 1978; Khramov, 1983; Hoffman, 1989, 1991; Moores, 1991; Dalziel, 1991) centred over the equator would have provided ideal conditions for withdrawal of atmospheric CO_2, reducing the greenhouse effect and leading to glaciation. This model finds support in recent theoretical climate modelling (Worsley and Kidder, 1991) which emphasizes the critical role of continental positioning and paleogeography. Deynoux (1985b) pointed out that an emergent supercontinent would also lead to increased planetary albedo which could provide a positive feedback contributing to a build-up of ice volumes. There is an urgent need for further paleomagnetic studies of Neoproterozoic glacial strata and related deposits.

Timing and tectonic setting

The notion of a globally correlative Neoproterozoic glaciation was proposed by Harland and Bidgood (1959) in order to explain the apparent relationship of glacial deposits with warm climate indicators and low depositional paleolatitudes as suggested by paleomagnetic data. Roberts (1971, 1976) suggested that a world-wide episode of dolomite deposition at the end of the Proterozoic triggered a drawdown of atmospheric CO_2, favouring glaciation.

The association between diamictites of supposedly glacial origin and warm climatic indicators such as dolostones led Schermerhorn (1974, 1977) to question the glaciogenic nature of many diamictites and to emphasize their active tectonic setting. Subsequent research (reported in Hambrey and Harland, 1981), stimulated in part by Schermerhorn's controversial review, has confirmed the glaciogenic nature of many Neoproterozoic diamictite-bearing successions. In addition, however, work also stresses the importance of active tectonic settings and the deposition of non-glacial debris flows and thick turbidite successions (N. Eyles, 1990). Schermerhorn (1977) also emphasized the role of regional tectonics as a means of depleting Late Proterozoic atmospheric CO_2 levels thereby providing a background to Late Proterozoic glaciation.

The notion of a widespread, global late Proterozoic glaciation has been weakened by new tectonic models (Hoffman, 1991) and by new data regarding the nature of allegedly 'warm' strata interbedded within glaciclastic successions (Fairchild, 1993). A key factor in the origin and timing of Neoproterozoic glaciation is the history of accretion and fragmentation of the Neoproterozoic supercontinent. In turn, the structural setting is central to understanding the

sedimentology of the resulting glacial deposits. Hoffman (1991) proposed that break-up of a Late Proterozoic supercontinent (cf. Moores, 1991 and Dalziel, 1991) resulted in the 'break out' of Laurentia and fan-like rotation of the other component parts. The initial break-up occurred along the Proto-Pacific margin of Laurentia after 750 Ma and is recorded by thick passive margin deposits in western North America, Australia and China. An early episode of plate tectonic activity is indicated by the presence of a thick succession of fine-grained siliciclastic deposits and carbonates (Wernecke Supergroup of Delaney, 1981) that were deformed and thrust onto the North American craton (Clark and Cook, 1992) at about 1.2 Ga. The Neoproterozoic break-up in western North America (Hayhook 'orogeny' of Young *et al.*, 1979) appears to have begun at about 750 Ma near the transition between the Mackenzie Mountains Supergroup and the glaciogenic Rapitan Group (Fig. 1.4). The exact timing of break-up has remained elusive but it now appears that there were two episodes of rifting as pointed out by Ross (1991). The first may be contemporaneous with deposition of the Rapitan Group at 750 Ma (Stewart, 1972; Young *et al.*, 1979) with a second episode near the base of the Cambrian (Bond and Kominz, 1984). Early opening of the Proto-Pacific Ocean is a prerequisite of the tectonic model of Hoffman (1991) for Pan-African amalgamation of Gondwana. Glacial deposits are a prominent (correlative?) component of Neoproterozoic strata in western North America (e.g. Eisbacher, 1985; Ross, 1991; Young and Gostin, 1991; Young, 1992) suggesting a close relationship between rifting and glaciation (see below).

Younger rifting (< 650 Ma) along the Proto-Atlantic (Iapetus) margin of Laurentia is also associated with widespread glacial strata now scattered around the margins of the North Atlantic Ocean in Scotland, Scandinavia, Greenland and Spitsbergen. In most of these areas there is clear evidence of glacial deposition in extensional tectonic settings (e.g. C. Eyles, 1988; see below). New U-Pb zircon dates from granites in the Blue Ridge of North Carolina and Tennessee, however, have given ages of 740 to 760 Ma (Su *et al.*, 1992). The significance of these dates is that the granites are thought to be closely associated with the Mount Rogers volcanic centre in Virginia and could be evidence of an early episode of rifting of Laurentia from the Neoproterozoic supercontinent. A relationship between rifting and the preservation of glaciogenic strata of the Mount Rogers volcanic centre is examined by Miller (this volume).

In contrast, glaciogenic deposits of the Pan-African belt now preserved around the margins of the North Atlantic Ocean are young (they post-date *c.* 550 Ma) and were deposited in an active plate margin setting. Diamictites form a prominent stratigraphic component of basin fills in eastern North America, North Africa and Europe and are closely associated with voluminous volcanic sediments; local glaciation of volcanic cordillera is indicated in many areas (e.g. Socci and Smith, 1987; N. Eyles and C. Eyles, 1989; N. Eyles, 1990). Elsewhere, in North Africa, glaciation covered a huge (2×10^6 km²) foreland region of the West African platform (Taoudeni Basin; Deynoux and Trompette, 1976; Deynoux, 1985a,b) and is described elsewhere in this volume.

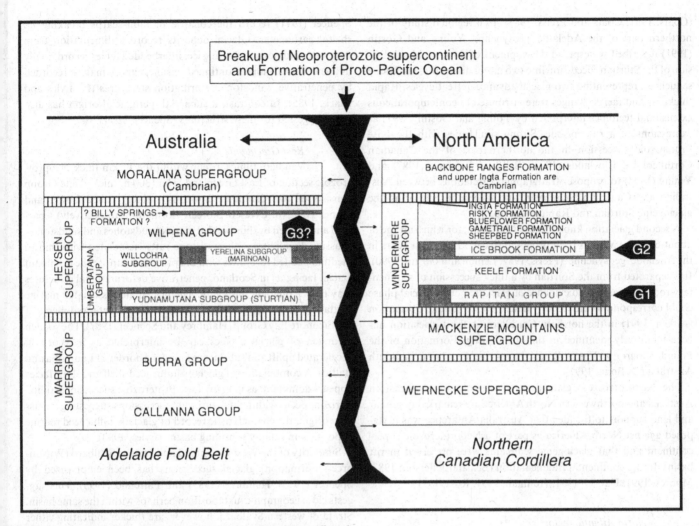

Fig. 1.4. Comparison of Late Proterozoic stratigraphy of South Australia and the northern Canadian Cordillera. Glacial episodes are shown by the grey ornament and are designated G1–3. Black ornament in central part of the figure represents oceanic crust formed by the break-up of the inferred Neoproterozoic Supercontinent. New paleontological data after MacNaughton *et al.*, (1992).

The importance of extensional rift basins

Most Neoproterozoic glacial deposits accumulated as glacially influenced marine strata along rifted continental margins or interiors; phases of crustal extension preceded disintegration of the Neoproterozoic supercontinent starting at around 750 Ma along its western (Pacific) margin. Given that strong, localized uplifts of *many* kilometres can occur along the boundary faults or rifted terrains, a glacial source area and sediment repository are thus provided. Dolomites that underlie many glaciclastic strata may record clastic-starved sedimentation in restricted warm-water basins prior to onset of uplift on the basin margins. Dolomitic strata that occur within several glaciclastic successions are detrital in origin reflecting the uplift and glacial erosion of underlying dolomites (Fairchild, 1993). Strong uplift and the development of local

ice masses adjacent to rapidly subsiding rifted basins could also provide the uniformitarian key to the occurrence of Late Proterozoic glacial deposits in low to middle paleolatitudes (Schermerhorn, 1977). However, whilst uplift may initiate glaciation on elevated massifs, the deposition of *glaciomarine* sediments indicates cold temperatures at sea level. It is the case however, that paleomagnetic work on which models of low latitude glacial deposition are based is increasingly suspect (e.g. Meert and Van der Voo, 1992).

South Australia/northern Canadian Cordillera

Glaciogenic rocks figure prominently in the Neoproterozoic stratigraphy of southeastern Australia and the northern Canadian Cordillera (Fig. 1.4). The Sturtian glaciogenic succession (*c.* 740 Ma) unconformably overlies rocks of the Burra Group

(Coats, 1981; Coats and Preiss, 1987). In a regional study of the northern part of the Adelaide 'geosyncline' Young and Gostin (1991) described widespread development of a four-fold subdivision of the Sturtian succession into two major diamictite–mudstone sequences, representing two glacial advance–retreat cycles. Rapid thickness and facies changes were attributed to contemporaneous extensional tectonics interpreted by Young and Gostin (1991) as expressions of a rift episode. Strong similarities with the Late Proterozoic succession in the northern part of the Canadian Cordillera led Rowlands (1973), Bell and Jefferson (1987) and Young (1992) to propose stratigraphic correlations between Australian and Canadian Neoproterozoic successions including the glaciogenic Sturtian and Rapitan Group.

A second glaciation known as the Marinoan, for which an age of about 680–690 Ma has been proposed (Coats, 1981) also ocurs in the Adelaide 'geosyncline' (Preiss, 1987; Lemon and Gostin, 1990). It is separated from the Sturtian by a thick succession of sedimentary rocks containing no evidence of glaciation. This glacial phase could correspond to the recently described Ice Brooke Formation (Aitken, 1991) in the northern Cordillera. A third glaciation has been tentatively identified in the Billy Springs Formation of the Pound Subgroup (Fig. 1.4) just below the Cambrian in South Australia (Di Bona, 1991).

The Neoproterozoic glaciogenic successions of southeastern Australia and northwestern North America are remarkably similar and lend support to the idea that Australia/Antarctica was juxtaposed against North America as part of a Neoproterozoic supercontinent and that glaciogenic sediments were preserved in rift basins during continental fragmentation (Bell and Jefferson, 1987; Moores, 1991; Dalziel, 1991; Hoffman, 1991; Ross, 1991).

North Atlantic sector

Late Proterozoic glacial deposits accumulated under extensional crustal regimes around the rifted margins of several cratonic blocks.

Scotland

In Scotland, the Port Askaig Tillite of the Dalradian Supergroup, was first described by Thompson (1871). The succession, about 750 m thick, is seen in strike-parallel outcrops from southwest Ireland to northeast Scotland (700 km) and consists of a relatively simple alternation of massive diamictites conformable with sandstones, minor conglomerates and laminated siltstones, some of tidal origin. Spencer (1971, 1985) emphasized repeated cycles of sedimentation below a grounded ice sheet; Eyles (1988) argued that the geometry and sedimentology of the deposit were inconsistent with subglacially deposited facies and argued for a subaqueous glaciomarine origin. She highlighted the structural setting of the Port Askaig Tillite at the base of a deepening-upward succession of turbidites and basinal facies (Argyll Group; Anderton, 1982) deposited in fault-bounded blocks and basins on the northern margins of what was to become the Iapetus Ocean. Large olistostromes composed of dolomite (e.g. the Great Breccia;

Spencer, 1971) record the collapse of fault-scarps breaking up shelfal carbonates. Glacial deposits record sedimentation from meltwater plumes and floating ice; interbedded facies record tidally influenced conditions. Continued seismic activity may be recorded by penetrative, sandstone deformation structures (N. Eyles and Clark, 1985). In contrast, a subaerial, periglacial origin has also been suggested for these structures (Spencer, 1971).

East Greenland

Vendian 'tillites' occur within the 17 km thick Neoproterozoic section of East Greenland. The 1300 m thick Tillite Group consists of two diamictite horizons variably called Lower and Upper Tillite (or Ulvescö, Storeelv Formations respectively) separated and capped by thin-bedded marine mudstones and sandstones. Massive diamictites with sandstone rafts, associated with thin to medium bedded turbidites with dropstones comprise the dominant glacial facies; as in Scotland, penetrative deformational structures may record ongoing seismicity. The glacial and associated marine strata rest on predominantly dolomitic shallow water sediments of the Eleonore Bay Group (Hambrey and Spencer, 1987). The sudden incursion of glacial clastics can be interpreted as a record of accelerated uplift and subsidence along the border of Laurentia and Baltica. A combination of glacially influenced shelf and slope facies appears dominant; as in many Late Proterozoic successions 'tillite' horizons occur within thick turbidite successions suggesting a mass flow origin; the characteristic record of glacially influenced marine deposition in rapidly subsiding basins (Eyles, 1993).

Similarity of the West Greenland and eastern Svalbard (Wilsonbreen Formation) glacial successions has been emphasized by Spencer (1975), Hambrey (1983) and Fairchild (1993). This suggests close geographic juxtaposition perhaps within the same basin. Strata of western Svalbard, however, are thicker indicating either accelerated subsidence or deposition in another basin. Western Svalbard was juxtaposed against central and eastern Svalbard by later strike-slip movement.

Scandinavia

The importance of block faulting during early rifting of the Iapetus Ocean is strongly apparent in the Neoproterozoic glacial deposits on the west-facing margin of Baltica.

In Finnmark, sedimentation occurred on the western rifted flank of Baltica, adjacent to the Timanian Aulacogen which separated Baltica from the Barents Craton to the north. Thick (450 m) glacial deposits of the Vestertana Group (Smalfjord, Nyborg, Mortensnes Formations; Edwards, 1975) began to accumulate in the Gaissa Basin following late Sturtian uplift of shallow water stromatolitic dolomites (Porsanger, Grasdal Formations); uplift of at least 1500 m can be inferred in the eastern boundary of the Gaissa Basin but nearly 8 km of strata were removed in other areas (Gayer and Rice, 1989). It is highly significant that coeval strata in closely adjacent basins to the west and north (Baltoscandian miogeocline and Timanian Aulacogen) are represented by alluvial fan, fluvial facies and shallow marine facies; these contain diamictites formerly regarded as glacial (Siedlecka and Roberts, 1972) but now interpreted as alluvial fan debris flows (Laird, 1972, Gayer and Rice, 1989). These findings are especially pertinent to the 'tillites' of the Vestertana Group.

Diamictites of the Smalfjord Formation are rich in dolomite clasts and record downslope slumping and sediment gravity flow of coarse-grained sand and gravel facies. Diamictites are interbedded with shallow marine, storm-influenced facies. A fan-delta depositional setting is suggested by Warman (personal communication, 1991). The so-called 'interglacial' Nyborg Formation comprises a thick succession of turbidites probably recording enhanced fault-controlled subsidence. Toward the top, it is interbedded with massive diamictites of the Mortensnes Formation which record glaciomarine sedimentation primarily from suspended sediment plumes and icebergs, with secondary mass flow. Initial extension and rifting of the continental margin is marked by dykes dated at 640 Ma (Beckinsale et al., 1976). These are part of a wider network along the margin of Baltoscandia interpreted as syn-Iapetus rifting intrusions and are correlative with turbidites of the Nyborg Formation.

Broadly similar 'fan-delta' settings as identified in the Vestertana Group of the Gaissa Basin occur in other Norwegian basins along the former Baltoscandian continental margin (e.g. Tietzsch-Tyles, 1989). In central and southern Scandinavia five main basins (Risbäck, Hedmark, Valdres, Engerdalen, Tossasfjallet) record sedimentation in extensional settings along the Baltoscandian margin. Nystuen (1987) recognized a common succession in the western Baltoscandian basins of lowermost diamictites overlain by laminated mudstones with decreasing quantities of ice-rafted debris upward in section. This succession is most easily interpreted as a basinal deep-water assemblage recording the downslope slumping of primary glacial debris and syndepositional subsidence.

The structural setting and sedimentology of the glacial deposits in Scotland, Greenland, Spitsbergen and Scandinavia point to the importance of uplift during regional extension along the paleo-Atlantic continental margin of Laurentia. At about 750 Ma the North Atlantic region of the Neoproterozoic supercontinent was characterized by the widespread development of inter- and intra-cratonic basins peripheral to an early Iapetus Ocean. Geophysical models show that large-scale extension and thinning of the crust, whether by simple or pure shear, is associated with topographic 'arching' on the margins of the extended terrain (Lister et al., 1991). This may create a glacial source area and a depositional repository (see below; Late Cenozoic Glaciations). Many areas undergoing extension show broad topographic domes separated from the active rift by a fault-bounded topographic escarpment. The modern Red Sea rift and associated updomed Arabian Shield is one example; the Death Valley area east of the Sierra Nevada also shows a marked topographic updoming; the lowest (−86 m asl) and highest (4419 m) points in mainland USA occur in this area (Wernicke, 1985).

Strong localized uplift of rift shoulder flanks and associated crustal blocks resulted in widely-dispersed glacial centers, across a very broad range of paleolatitudes, adjacent to rapidly subsiding Neoproterozoic rift basins (cf. Yeo, 1981; Young, 1989). Most of the diamictites preserved in these basins are not true tillites but glacially influenced marine deposits recording the release of large volumes of mud and coarser debris to the marine environment. Schermerhorn (1975) argued that tectonic differentiation involving complementary uplift and subsidence set the scene for deposition of many Neoproterozoic 'glacial' successions. Schermerhorn's arguments were largely ignored by geologists who felt that he had largely written off all Neoproterozoic tillites in favour of non-glacial, tectonically generated mass flows. This is not so, for as Schermerhorn (1975) concluded, 'though the Late Precambrian mixtite formations were laid down in obvious tectonic settings, among them exist glacial deposits or glacially influenced sediments, and one task of future unprejudiced studies will be to determine the proportion of glacial components and the origin and extent of glacial activity within the frame of mixtite deposition'. Because of local tectonic controls, individual diamictite horizons have a limited potential for lithostratigraphic correlations (cf. Chumakov, 1985).

Discussion

Strong tectonic uplift in extensional regimes may explain the widespread and seemingly paradoxical association of diamictites and carbonates in Neoproterozoic strata. Fairchild (1993) showed that carbonates below 'glacial' strata in Greenland, Scotland and Spitsbergen record an upward change from limestone to dolostone facies consistent with upward shallowing. The shallowing upward carbonate successions may record strong tectonic uplift accompanying the initiation of an extensional tectonic regime. This uplift is recorded by progressive unroofing of carbonates and associated exposure of basement lithologies and by abundant clastic carbonate within overlying 'glacial' strata.

Carbonates within diamictite successions are overwhelmingly of clastic origin (most are dolarenites) and record the glacial erosion of uplifted fault-bounded carbonate massifs and rift shoulders. Fairchild and Spiro (1990) and Fairchild et al. (1989) also report primary lacustrine precipitates from Spitsbergen comparable to those of 'arid' saline lakes in present-day Antarctica. Cap dolostones are common in Australia, West Africa and North America (Williams, 1979; Deynoux, 1985a). Fairchild (1993) stressed the lack of detailed data and could not discriminate between the effects of diagenesis or primary deposition. The most important variable appears to be the availability of large amounts of reactive carbonate rock flour produced by glacial erosion.

In summary, the widespread development of clastic-starved, carbonate platforms in restricted Neoproterozoic rift basins may have been followed by the shedding of glacial and clastic carbonate debris from uplifted crustal blocks. The ultimate cause of global cooling may, however, be the drawdown of atmospheric CO_2 by enhanced weathering of the uplifted supercontinent. This may have taken place several times in response to negative feedback as described earlier for the Paleoproterozoic (see above).

Another outstanding problem that may be resolvable by reference to tectonic setting is the 'enigma' of low paleolatitude glaciation first recognized by Harland and Bidgood (1959) and supported by paleomagnetic data from other workers (e.g. Chumakov and Elston, 1989). Glacial deposition in low equatorial paleolatitudes has been used to argue for global refrigeration (Harland, 1975) and for the use of tillites as global chronostratigraphic marker horizons (Chumakov, 1985). The stratigraphic association of tillites and carbonates was cited as evidence of short-lived cold conditions in low latitudes (paleoclimatic crises).

Piper (1982) argued that supposed primary magnetization showing alleged low paleolatitudes is the result of secondary overprinting and this was clearly demonstrated for the Port Askaig Formation (see above) by Stupavsky et al. (1982). Concerns with the quality of paleomagnetic data were also voiced by Crowell (1982) and have been borne out by new data (Meert and Van der Voo, 1992). Schermerhorn (1974) had earlier argued that many supposed tillites were non-glacial mass flows and it is true that a wide variety of diamictites deposited under very different tectonic settings have been lumped together as Late Proterozoic tillites (see above). The most compelling evidence for cold climates at low latitude is forthcoming from the Marinoan succession of the Adelaide Geosyncline in South Australia (c. 650 Ma). Fine-grained tidal laminites (Elatina rhythmite, member of the Elatina Formation; Williams, 1989) record low inclinations consistent with deposition in a paleolatitude belt between 20 °N and 12 °S of the paleoequator (Schmidt et al., 1991). Coeval strata host permafrost structures recording mean annual temperatures of less than -4 °C.

Embleton and Williams (1986), whilst recognizing that reasonable doubt remains regarding many Neoproterozoic paleomagnetic data, suggested three hypotheses to account for cold climates in low latitudes. The first (global glaciation) was rejected; a 'frozen-over Earth' requires an unrealistic increase in solar luminosity to thaw (e.g. Bahcall and Ulrich, 1988), but changes in atmospheric composition (increasing CO_2) were not considered. A second argument proposes that the present-day situation of an axial geocentric dipole model for the Earth's magnetic field is invalid for the Late Proterozoic. The third hypothesis argues for a considerably increased obiquity of the ecliptic such that areas at low latitudes would receive less solar radiation during an annual cycle than areas at high latitudes. Both these arguments are reliant on more but better geophysical and paleomagnetic data combined with rigorous analysis of the geodynamic setting of supposed glacials.

If the available paleomagnetic data stand the test of time and it is shown that many Neoproterozoic glaciogenic successions were definitely deposited in low paleolatitudes then these data must be accommodated within an appropriate depositional model. One possibility is that, given an atmosphere richer in CO_2 than at present, a high-standing supercontinent that straddled the equator (e.g. Hoffman, 1991) where both rainfall and temperatures are high (Gyllenhaal et al., 1991), would have been particularly susceptible to chemical weathering. Resultant drawdown of atmospheric CO_2, together with a slightly weak sun, could have brought about glaciation even at low latitudes (Worsley and Kidder, 1991). Glaciation could eventually have been terminated by build-up of CO_2 due to negative feedback (Young, 1991a). The resultant sediments would reflect the tropical location of many of the depocentres. In this way, the 'anomalous' occurrence of warm climate indicators such as dolomites between glacial formations can be explained. However, as noted above, dolomites within glaciogenic formations are largely clastic and reflect the cannibalization of earlier-deposited carbonate units during uplift of the shoulders of rifts in which many Neoproterozoic glaciogenic successions are preserved (Schermerhorn, 1974; Yeo, 1981; Young, 1989).

The answer to the enigma of low latitude glaciation lies in further detailed facies studies of Neoproterozoic strata to determine their precise depositional and tectonic setting, paleoclimatic modelling using general circulation models and better age dating to determine correlative or diachronous glaciation. However, before the theory of low latitude glaciation can be accepted, a first and critical requirement is the need for more paleomagnetic investigations of remanence acquisition in Neoproterozoic sediments and the clear demonstration that low latitude magnetization is primary. Geochemical studies, to determine the degree of weathering undergone by 'interglacial' clastic units, could also contribute to resolution of this problem.

Early and Late Paleozoic glaciations

Caputo and Crowell (1985) argued that the pattern of Early and Late Paleozoic glaciation of Gondwana could be explained by the migration of the supercontinent across the South Pole. The Late Ordovician south pole was sited over northern Africa and there is a well-defined record of glaciation across North Africa and Saudi Arabia that identifies an ice sheet similar in size to that of the present day Antarctic ice sheet (Beuf et al., 1971; Deynoux, 1985a,b; Vaslet, 1990). Glacio-eustatic sea-level changes, at about 440 Ma, have been held responsible for a series of well-defined extinction events in marine fauna (Brenchley, 1989; Fig. 1.5).

A major glacial episode at c. 440 Ma (Figs. 1.4, 1.6), is recorded in Late Ordovician strata (predominantly Ashgillian) in West Africa (Tamadjert Formation of the Sahara), in Morocco (Tinduf Basin) and in west-central Saudi Arabia, all areas at polar latitudes at this time (Vaslet, 1990). Less well-dated and poorly understood deposits occur in Scotland (Macduff tillite), Ireland (Maumtransa Formation), Normandy (Tillite de Feugeurolles) and Spain and Portugal. Late Ordovician deposits that may record small ice centres in eastern North America are represented by the Halifax Formation of Nova Scotia and Gander Bay tillites of Newfoundland. The sedimentology of these deposits is not well known and is probably non-glacial (Long, 1991). Other well-constrained Ordovician deposits occur in South Africa (Pakhuis Formation).

From the Late Ordovician to the Early Silurian the centre of glaciation moved from northern Africa to southwestern South America. The Cancaniri Formation of Argentina, Bolivia and Peru and Iapo and Furnas Formations of the Parana Basin, Brazil and the Trombetas Group of the Amazonas Basin may record glacially influenced deposition in middle to high latitudes as a result of polar wander (Fig. 1.7) but these deposits are not well-studied as yet (Grahn and Caputo, 1992). From the mid-Silurian to the early Late

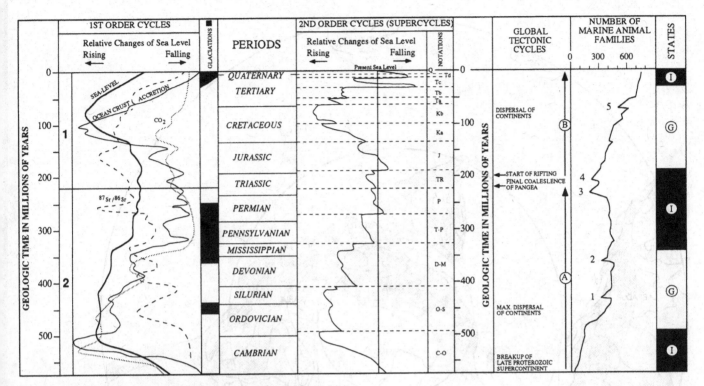

Fig. 1.5. Earth's Phanerozoic glacial record with record of sea level (after Vail *et al.*, 1977), atmospheric CO_2 (after Berner, 1990), ratio of $^{87}SR/^{86}Sr$ in marine waters (after Koepnick *et al.*, 1988), rate of ocean crust accretion (after Gaffin, 1987), global tectonic cycles (after Worsley *et al.*, 1986) and marine animal extinctions (1–5 after Raup and Sepkovsky, 1982) and climate 'states' (G, greenhouse; I, icehouse) as proposed by Fischer (1984).

Devonian no record of glaciation is known. Caputo (1985) suggested that the South Pole was located over the Pacific Ocean at this time where no substrate was available to support an ice sheet (see below) but the absence of cold climate marine deposits is still unexplained. Renewed Late Devonian glaciation is well documented in three large intracratonic basins in Brazil (Solimoes, Amazonas and Paranaiba basins; Caputo, 1985; Caputo and Crowell, 1985) and in Bolivia (Diaz *et al.*, 1993).

By the Early Carboniferous (*c.* 350 Ma) glacial strata were beginning to accumulate in sub-Andean basins of Bolivia, Argentina and Paraguay. By the mid-Carboniferous glaciation had spread to Antarctica, Australia, southern Africa, the Indian Subcontinent, Asia and the Arabian Peninsula (Figs. 1.6, 1.8). During the Late Carboniferous glacial culmination (*c.* 300 Ma) a very large area of the Gondwana land mass was experiencing glacial conditions. The thickest glacial deposits of Permo-Carboniferous age are the Dwyka Formation (1000 m thick) in the Karoo Basin in southern Africa, the Itarare Group of the Parana Basin, Brazil (1400 m) and the Lyons Formation (2500 m) of the Carnarvon Basin in eastern Australia. Dropstones in Australia show that some glacial ice (or at least seasonally cold conditions) persisted into the early Late Permian (*c.* 255 Ma). Powell and Veevers (1987) argued that strong Namurian uplift in southern South America and in eastern and central Australia was the trigger for the extensive

Permo-Carboniferous glacial culmination during the Late Westphalian at *c.* 300 Ma (Fig. 1.8). Elevation of continental areas, either by tectonothermal thickening of the continental crust or by direct continental collisions has also been ascribed as a precursor to Quaternary glaciations (Ruddiman *et al.*, 1989). The influence of global geochemical cycles on Late Palaeozoic glaciations has been modelled by Crowley and Baum (1992) who assessed the role of atmospheric CO_2 levels and changes in paleogeography.

South America

Late Palaeozoic glaciated basins of southern South America are of economic importance (e.g. Franca and Potter, 1991). In Bolivia, glacial deposits host the Bermejo, Palmer and Santa Cruz oil fields as do correlative strata in Argentina (e.g. Duran, Leva and Madrejones oil fields). In Brazil, the Parana Basin has subcommercial gas shows in the glaciogenic Itararé Group; significant coal deposits occur in early postglacial strata. Figure 1.8 is a compilation of age data from Gondwanan glaciated basins. The early onset of glaciomarine sedimentation in the early Visean in Argentina and Paraguay can be related to Hercynian orogenesis along the proto-Andean margin (Herve *et al.*, 1987; Lopez Gamundi, 1987; Goualez Bonorino, 1992). Away from the active plate margin, glacial sediments began to accumulate on the Brazilian craton during the latest Westphalian (*c.* 300 Ma). The Parana Basin is the largest

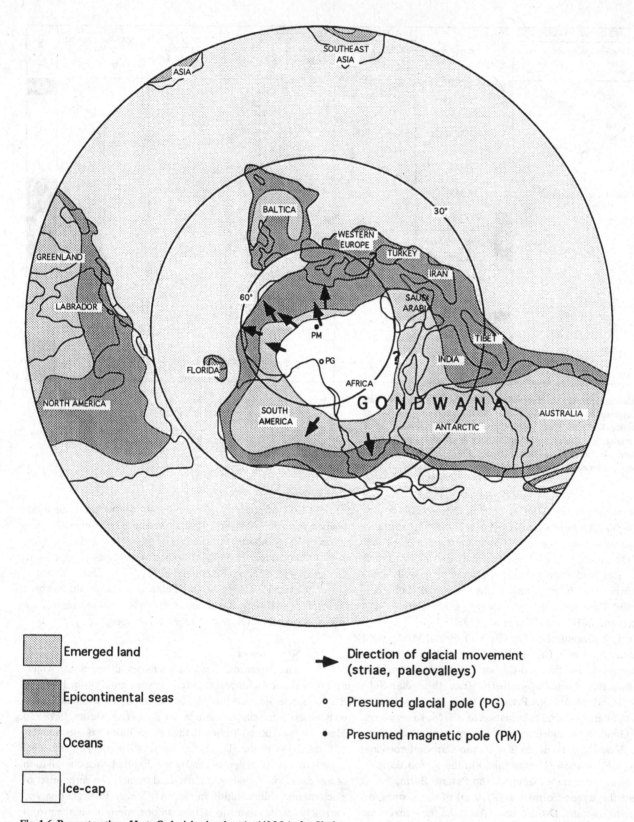

Fig. 1.6. Reconstruction of Late Ordovician ice sheet (*c.* 440 Ma) after Vaslet (1990).

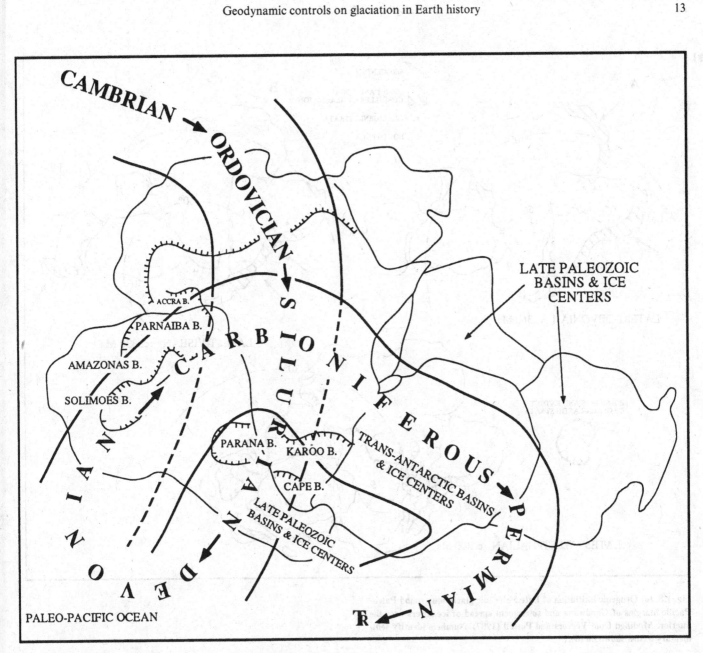

Fig. 1.7. Migration of Paleozoic glacial centres across Gondwana (after Caputo and Crowell, 1985).

(1 600 000 km²) intracratonic basin in southern South America and contains a thick (1400 m) Late Palaeozoic glacial succession. A new pollen biostratigraphic zonation indicates glacial deposition between the latest Westphalian (300 Ma) to Early Kungurian (260 Ma). Three formations (Lagoa Azul, Campo Mourao and Taciba) can be recognized in the basin, each recording a renewed phase of basin subsidence as a result of rifting along steeply dipping basement structures (C. Eyles et al., 1993). These structures, principally Late Proterozoic fold belts and basement lineaments, were produced during the major collisional orogenies that were responsible for cratonization of the Late Proterozoic Afro-Brazilian supercontinent (Murphy and Nance, 1991). Expansion of the Parana Basin during deposition of the Itarare Group was controlled by faulting

and reactivation of these structures (Eyles and Eyles, 1993) as has been documented in other basins (e.g. Gibbs, 1984; Klein and Hsui, 1987; Leighton et al., 1990; Daly et al., 1991).

Subsidence in the Parana Basin began when 'far-field' tectonic stresses from the active western margin of Gondwana were communicated into the previously passive craton. 'Far field' reactivation of intracontinental basement structures by rifting results from distant collisional processes at an active continental margin (e.g. Jorgensen and Bosworth, 1989; Daly et al., 1991). Peripheral orogeny around the paleo-Pacific margins of the Gondwana supercontinent was responsible not only for initiating Cordillera-style glaciation along the plate margin but because of regional uplift, crustal thickening and complementary intracontinental rifting, allowed the growth of

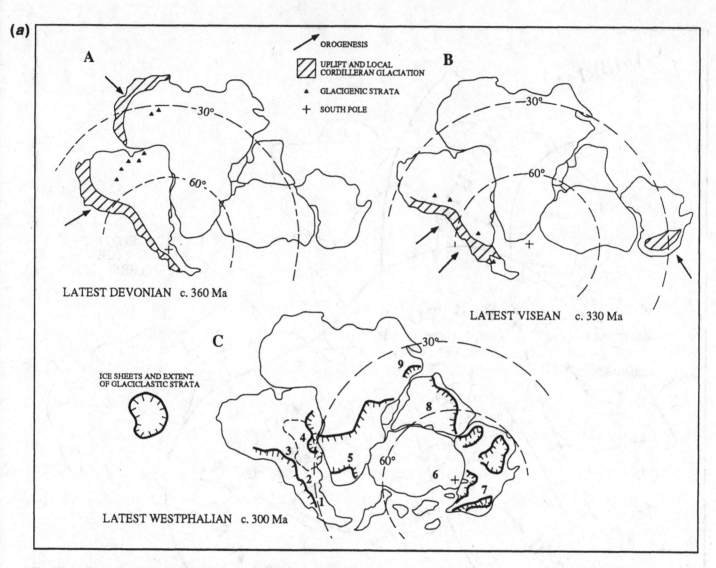

Fig. 1.8. (*a*) Orogenic intitiation of Late Paleozoic glaciation around Paleo-Pacific margins of Gondwana and subsequent spread of ice covers over the interior. Modified from Veevers and Powell (1987). Numbers identify sedimentary basins shown on (*b*).

ice caps in the interior and preservation of a depositional record. The same tectonic model involving intracratonic extension can be applied to the Southern African and Australian and Peninsula Indian sectors of Gondwana (Fig. 1.9).

South Africa

Stratigraphic similarities between Permo-Carboniferous records of the Parana of Brazil and Karoo Basin of southern Africa suggest a common structural control. Both basins share early Permian postglacial shales containing *Mesosaurus* (Visser, 1991; Franca and Potter, 1991). Oelofsen (1987) proposed a central seaway allowing deposition of the Irati shale in Brazil and the Whitehill Formation in southern Africa during regional postglacial transgression. It is likely that this seaway existed much earlier

during deposition of the glacial Itararé Group and Dwyka Formation. Visser (1987, 1989, 1990) has examined the paleogeography of the Karoo Basin during Dwyka glaciation and identifies a northern landmass (the Cargonian Highlands) on which glacioterrestrial and near coastal glaciomarine deposits are preserved and a southern deeper water glaciomarine basin. This part of the basin contains over 1000 m of glacial strata and suggests that the southern margin of the Cargonian Highlands lies along faulted Proterozoic basement structures as in the Parana. Progressive subsidence of the southern Karoo Basin, south of the Cargonian Highland fault, results in the characteristic 'overstepping' of successively younger stratigraphic units over older. This is the same stratigraphic relationship as seen in the Parana Basin where repeated subsidence and stratigraphic overstep results in the classic 'steer's head' infill architecture.

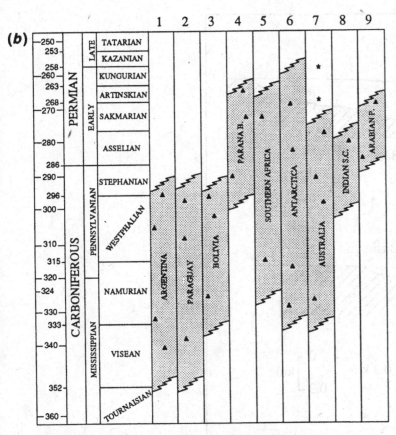

(*b*) **Timing of the glacial record across Gondwana (from Eyles *et al.* 1993). Better age dating of the record is required; significant episodes of non-glaciation, glacial erosion and restricted mountain glaciation are not shown.**

Australia

In eastern Australia, glacial deposition is alleged to have begun in the Late Carboniferous (Early Namurian) within the Tamworth forearc basin (Veevers and Powell, 1987). The 'glacial' succession, however, is predominantly of mass flow origin up to 1500 m thick (Lindsay, 1966) and contains abundant acid and intermediate volcanic debris. Massive diamictites are still in some cases interpreted as tillites (e.g. McKelvey, 1981) but ash fall tuffs and coarse fluvial sediments suggest a depositional setting proximal to a volcanic arc where mass flows can be expected. The same glaciogenic/volcanigenic association is clearly evident in the large foreland basin collectively referred to as the Sydney-Bowen Basin (McPhie, 1987). A volcanic arc lay to the east and volcanism was widespread during the latest Carboniferous and Early Permian. Glaciated highlands centred at latitude 50 °S supported extensive valley glaciers on the westward side of the basin; Herbert (1981) describes eastward-draining, 'paleovalleys', up to 500 m deep, 3 km wide and 100 km in length, thought to be glacially excavated. These are filled with fluvial conglomerates (Tallong and Yadboro conglomerates; Herbert, 1981) together with minor diamictites, up to 250 m thick. These basins are similar in scale to the 'paleofjords' reported from the Kaokoveld coastal margin of northwestern Namibia and Angola (Martin, 1981).

Away from the active eastern coastal margin of Australia, Late Paleozoic glaciated basins comprise many interior basins. Basins are floored by glacial sediments of lower Permian age. These basins however, are systematically related to structures in underlying strata and involve substantial rifting coincident with the onset of glaciation (Wopfner, 1981; Fig. 1.9). Marine glacial conditions are recorded in the Denman, Ackaringa, Troubridge and Denmark basins. Glaciomarine sediments are absent in the Pedirka Basin and a limited marine incursion occurred in the Cooper Basin (Wopfner, 1981). Fully marine conditions obtained at the northern, seaward reaches of the Canning and Carnarvon basins with brackish water to the south.

The age difference between the Late Carboniferous glacial successions of the active eastern Australian margin and the Permian successions of the intracratonic basins to the west may be a function of delayed transmission of far-field stresses to the interior. The infills of the Arckaringa, Officer, Troubridge, Pedirka, Denman basins are dominated by mass flow deposits (graded facies, slumped horizons, debris flow diamictites and conglomerates) with evidence of direct glacial activity restricted to the basin margins. Terrestrial, subglacial facies were probably widely developed on surrounding highlands but are not preserved (e.g. Youngs, 1975). Cross-sections through the relatively well-studied Cooper Basin show that succes-

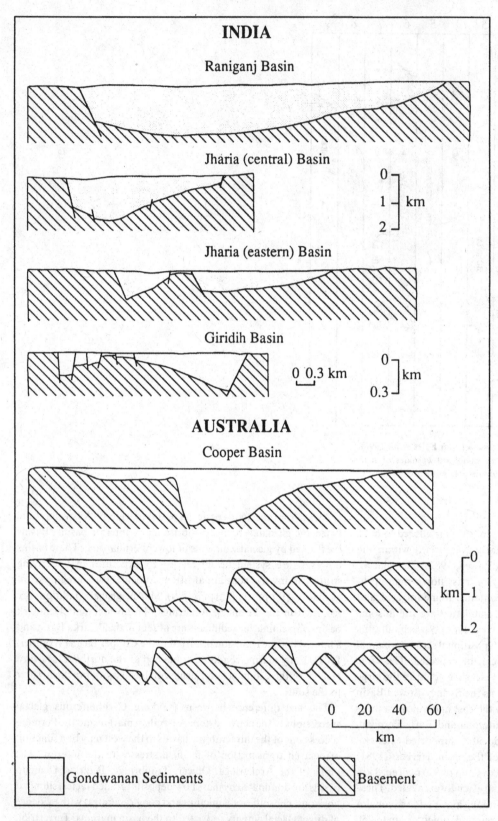

Fig. 1.9. Representative, simplified cross-sections through Gondwanan intra-cratonic glaciated basins showing structural control by basement faults. Glaciclastic strata occur at the base of many basins recording intimate relationship between glaciation, rifting and preservation of a glacial record (From Eyles, 1993).

sively younger glacial sediments overstep underlying units. This is typical of basins undergoing progressive subsidence and expansion by faulting of the basin margins. The similarity of the central and southern Australian basinfills to the Parana Basin of southern America and Karoo Basin of southern Africa invites further study.

Antarctica

In Antarctica, widespread Permian glacial sediments occur at the base of a thick Permian to Jurassic succession (the Victoria Group) and rest unconformably on a widespread erosion surface (Barrett, 1982). The glaciogenic deposits are widely scattered along the Transantarctic Mountains from Victoria Land to Dronning Maud Land (Miller and Waugh, 1991). Considerable thickness variations along the outcrop belt are evident; massive diamictites up to several hundreds of metres thick are reported in the Ellsworth Mountains (Whiteout Tillite). These thicknesses, the presence of interbedded mudstones, and transitional upper formational contacts with mudstones containing ice-rafted debris provide unambiguous evidence of a marine setting and rapidly subsiding basins (Matsch and Ojakangas, 1991). Dalziel et al. (1987) suggested the presence of several discrete sedimentary basins on continental crust undergoing crustal extension and rifting.

Southeast Asia

Correlative Late Paleozoic glaciomarine diamictites can be traced as an outcrop belt 200 km wide over 1000 km from Malaya (Singa Formation), Thailand (Phuket Series) to Burma (Mergus, Martaban series; Stauffer and Mantijit, 1981; Stauffer and Peng, 1986). The age of the diamictites is poorly constrained (Waterhouse, 1982). Facies are dominated by laminated (turbiditic) muds, silts and sandstones, and diamictites that contain a well-developed Cruziana ichofacies assemblage typical of glaciated continental shelf areas dominated by muddy substrates (N. Eyles et al., 1992).

The pre-drift Late Paleozoic paleogeography of northeast Gondwana is not well known. Older reconstructions of Gondwanaland depicted a large area of oceanic crust (Paleotethys) northward of Australo-India and Eurasia. In contrast, Lin and Watts (1988) and Smith (1988) argue from paleomagnetic and paleobiographic evidence that a 'Cathaysian' landmass (south China block, Thai–Malay terrane, West Burma terrane, Indochina terrane) occupied this large triangular area and did not become separated from Gondwana proper until the middle Permian. The location of the glacial ice centres feeding mud and ice-rafted debris to the 'Cathaysian' shelf is not well constrained; a Himalayan source area seems likely.

Arabian Peninsula

Late Paleozoic glacial deposits have been argued to occur in Yemen, Saudia Arabia and Oman. In Oman, the glaciogenic Al Khlata Formation (up to 750 m thick) was deposited at about 50 °S paleolatitude on the northern margin of Gondwana. The rifted epicratonic basin (800 km × 200 km) was bounded northwards by a highland (Oman Mountains) and possibly extended into Saudi Arabia and onto the Indian Plate; Arabia and India later separated along the axis of the basin. The onset of glaciation coincided with renewed subsidence and glaciogenic sediments are underlain by Lower Paleozoic sandstones and Proterozoic evaporites; Levell et al. (1988) identified a predominantly glaciolacustrine setting with possible marine influence recorded by marine algae and acritarchs. Large glaciofluvial delta bodies pass downbasin into rhythmically laminated and massive silty clays, associated diamictites are interpreted as 'rain-out' facies. Around the basin margin, diamictites rest erosively on, and seal, underlying deltaic bodies and are thought to be true tillites. The glaciogenic sequence is overlain by an extensive shale unit (Rahab Shale) which may be marine. The Oman deposits are of particular interest in that they contain significant oil deposits (more than 3.5 billion bbl of oil discovered to date). Oil migrated from Late Precambrian algal source rocks and the main prospective belt occurs in coarse-grained glaciodeltaic facies preserved around the former basin margin.

The extent (and even age) of Late Paleozoic glaciation in southern Saudi Arabia and coterminous Yemen is still controversial (e.g. Hadley and Schmidt 1975; Kruck and Thiele, 1983; McClure et al., 1988) though the presence of striated and shaped clasts suggests glacial influence. Sedimentological details of sections exposed through the Wajid Sandstone are few. The overall tectonic-stratigraphic picture of Late Paleozoic glaciation in Arabia (McClure et al., 1988), Yemen (Kruck and Thiele, 1983) and Oman (Levell et al., 1988) suggests local intracratonic subsidence in a 'passive' tectonic setting.

India and Pakistan

An active tectonic setting is indicated for deposits in northwestern India, Pakistan and Nepal. Reconstruction of northern Gondwana (deWit et al., 1988) places these areas in close proximity. The glaciogenic record reflects active collisional processes where the Indian Subcontinent abutted against Laurasia resulting in upthrusting of the Himalayas and closure of the paleo-Tethys Ocean. The 'Blaini Tillites' of the Lesser Himalaya, India, the Rangit Pebble-Slate and Lachi Formation of Nepal and the eastern Himalaya, are correlative with diamictites of the Salt Range in Pakistan (Tobra Formation). Diamictites occur at the base of the Gondwana sequence and marine fossils are correlative with those of the intracratonic glacial basins of peninsular India. Most of the Himalayan diamictites are interbedded with and contain pyroclastic strata; basaltic flows are also known. Acharyya (1975) argued that the diamictites are laterally persistent, tectonically generated submarine mudflows deposited close to the rising Himalayan Mountain front (see also Alavi, 1991). Kemp (1975) reports Cymatiosphaera acritarchs indicative of marine conditions.

The widely separated, linear Gondwana basins in peninsular India contain significant coal reserves (98% of India's reserves) and their subsurface geology is relatively well understood (Chowdhury et al., 1975). They are narrow intracratonic rifts and half-grabens that show on one side a steeply dipping boundary fault parallel to the basin axis resulting from the reactivation of east–west and northwest–southeast trending Proterozoic structures (Ghosh,

1975). The other margin of these basins typically shows an unconformable 'overstepping' of strata as a result of repeated subsidence. Systematic drilling shows that the glaciogenic Talchir Formation forms the basal unit within the basins with a more restricted, commonly fault-bounded, subsurface extent compared to later non-glaciogenic sediment (Fig. 1.9). Basu and Shrivastava (1981) showed that the basins are 'cradled' by Late Proterozoic and Archean fold belts between cratonic blocks and ascribed their rifted origin to Hercynian orogenesis.

Discussion

The preservation of Late Paleozoic glacial sediments across Gondwana from 350 to 250 Ma may be the result of widely dispersed ice centres adjacent to rift basins. The growth of these ice masses, like those of the Proterozoic, may be due in part to atmospheric CO_2 depletion (Crowley and Baum, 1992), related to the supercontinental configuration, but was also influenced by the positioning of Gondwana as it drifted across the south polar latitudes (Caputo and Crowell, 1985) and as tectonic differentiation occurred between the rifted basins and its uplifted margins (Veevers and Powell, 1987). Glacial sediments comprise the 'first flushing' into the newly developed rift basins and are associated with clear evidence of syndepositional tectonics. The 'overdeepened' character of many of these basins is due more to rifting than glacial erosion *per se*; the latter has probably been overemphasized (see Wopfner, 1981, p. 187). The inherited Late Proterozoic structure of Gondwana is essentially that of cratonic blocks that record the closure of internal oceans and complex collisional histories during supercontinent amalgamation. This fundamental structure dictated the subsequent structural history of Gondwana during widespread Late Paleozoic rifting and crustal extension. Extension appears to have been in response to the propagation of 'far-field' stresses into the plate interior directed from the active southwestern (South American), northern (Laurussian) and southeastern (Australian) margins. There are excellent prospects for modelling the dynamics of these interior basins in terms of global tectonics and megascale depositional sequences (e.g. Leighton and Kolata, 1990; Sloss, 1988, 1990). Palaeoceanographic studios (e.g. Domack *et al.*, 1993) offer great potential for understanding climate change. The Permo-Carboniferous glaciations are significant because of the marked glacio-eustatic changes in sea-level that resulted and which are recorded in non-glacial basins (e.g. Crowell, 1978; Klein and Willard, 1989; Klein and Kupperman, 1992).

Mesozoic glaciation

Direct geological evidence of glaciers during the Triassic, Jurassic and Cretaceous has not been found though ice-rafted horizons in Jurassic and Cretaceous strata of Siberia, and in Cretaceous strata of Australia, indicate seasonally cold conditions at sea-level (Epshteyn, 1978; Frakes and Francis, 1988). This raises the important question as to whether significant glacier ice cover may have developed at this time in continental interiors at high latitudes (e.g. Antarctica, Siberia). Epshteyn (1978) describes pebbly glaciomarine clays of Jurassic age in Siberia. These strata contain glendonites typically formed in polar oceans at *c.* 0 °C. Subsequently, Brandt (1986) identified cycles of eustatic sea-level variation, of between several hundred thousand years to 2 million years, in early Jurassic strata (Sinemurian–Pliensbachian) of southern Germany. These were related to the waxing and waning of an extensive ice sheet in northern Siberia. However, no direct glacial geologic evidence exists for such an ice sheet and it is noted that Epshteyn (1978, p. 54) used 'glaciomarine' to include 'material of the beach and shallow-water area, brought in by shore ice floes' and nowhere identified glacially ice-rafted debris.

Frakes and Francis (1988) report Cretaceous ice-rafted horizons in Australia and concluded that ice-rafted deposits exist for every period of the Phanerozoic except the Triassic. Again it is not clear whether Cretaceous rafted debris was emplaced by icebergs; rafting by seasonal ice is probable. Nonetheless an absence of widespread glacially ice-rafted debris during the Mesozoic is not a conclusive argument for the absence of continental ice sheets because the latter are unlikely to have extended to sea-level. The existence of significant ice cover over the Siberian and Antarctica land masses cannot be ruled out since both these areas lay at high paleolatitudes. Barron (1989) stressed that the geologic record of warm Cretaceous paleoclimates is derived from shallow marine sequences and does not exclude the presence of ice in continental interiors at high latitudes. Oxygen isotope studies, using concepts of Late Cenozoic relationships between $^{16}O/^{18}O$, rule out any significant Cretaceous ice cover but the results are arguable (Barron, 1989). Polar ice caps may have little influence on global isotope ratios if they were small, located in areas of low topographic relief, or were fed by precipitation derived from closely adjacent warm oceans. Barron (1989) identified two key aspects that had been largely ignored by previous work on Cretaceous climates, namely the well-established presence of a significant seasonal cycle in continental interiors and uncertainty regarding the nature of polar climates. Paleobotanic investigations suggest cool temperate conditions in high paleolatitudes during the mid-Cretaceous (Spicer and Parrish, 1986; see also Benton, 1991). These data are important because continental land masses at high latitudes that are surrounded by warm oceans satisfy the one fundamental requirement for ice sheet growth, namely relatively mild, wet winters and cool summers.

The pattern of past sea-level change can provide proxy evidence of the existence of ice sheets and thus has considerable bearing on the possible existence of ice during the Mesozoic. The growth of large ice sheets results in marked glacio-eustatic changes of absolute sea-level. Alternating sea-level drawdown and rise generates globally synchronous unconformities that define depositional sequences on continental margins. These glacially induced changes in sea-level, of up to 150 m during Late Cenozoic glaciations, are equivalent to fourth order sea-level cycles typically on the order of a few tens or hundred of thousands of years duration. Good examples are the well-known depositional cycles of transgression and regression evident in Pennsylvanian coal-bearing strata of the eastern United States ('cyclotherms'; Weller, 1930) that record changes in the volume of ice sheets across Gondwanaland (Crowell, 1978; Klein and Willard, 1989). Well-defined fourth order sea-level cycles have been identified from various strata deposited in the Cretaceous Interior Seaway of North America (Walker and Eyles, 1988; Plint, 1991) and in the Jurassic of southern Germany (Brandt, 1986; though see above). In Middle and Late Triassic strata in the Dolomite region of northern Italy, several workers have identified fourth and fifth order eustatic sea-level cycles (10 k and 20 k respectively) from platformal limestones and invoked a glacial control (Goldhammer et al., 1987; Masetti et al., 1991). It is worth emphasizing that small-scale subdivision of the Mesozoic timescale is not possible, many sea-level changes may not be cyclic and that Milankovitch-driven changes in global precipitation or ground-water storage may be responsible. It is doubtful whether any Mesozoic ice masses were ever large enough to influence global sea-levels.

In conclusion, the possibility of Mesozoic ice caps at high paleolatitudes remains open; there are no data that flatly contradict the presence of ice but proxy sea-level data suggesting substantial ice covers during this period need further evaluation.

Late Cenozoic glaciations

The role of tectonic uplift

The 'classic' glaciations of the Quaternary are relatively recent phenomena of the last 3.5 Ma but they are the result of Cenozoic climatic changes that began at around 60 Ma. Tectonic uplift of key middle latitude zones has been identified as a major trigger (Ruddiman et al., 1989; Kutzbach et al., 1989; Behrendt and Cooper, 1991) though the data are questioned by some (Molnar and England, 1990; Wilch et al., 1993). Ruddiman et al. (1989) summarized geological evidence showing an increase in uplift rates and absolute elevation during the Late Cenozoic of areas such as the Tibetan Plateau, Himalayas, and the western Cordillera of North and South America (Fig. 1.10). Computer models show that lowered atmospheric heating rates, and in turn global temperatures, result in longer winters and cooler, moister summers. A greater snow cover increases albedos thereby creating favourable conditions for ice sheet growth. A further corollary is that increased weathering of mineral suites released from uplifted areas would take up more atmospheric CO_2 and provide an anti-greenhouse effect. Thus, accelerated Late Cenozoic uplift results in lowered temperatures and an increased predisposition to Milankovitch-forced glacial/interglacial climate cycles.

Molnar and England (1990) argue, on the other hand, that the evidence of accelerated Late Cenozoic uplift has been exaggerated. Their thesis is that a large part of the evidence used to argue for Late Cenozoic uplift, mostly paleobotanical and geomorphological data, can also be the result of global climate changes that were happening independent of tectonic uplift. They argue that such climatic change would result in increased mechanical weathering of mountain ranges, giving the appearance of recent uplift. Similarly, paleobotanical studies that argue uplift of warm floras into cooler

climate zones can also be simply interpreted as the result of independent climate change.

The earliest record of Neogene glaciation is from Antarctica at c. 36 Ma (Barrett et al., 1989; Zachos et al., 1992) where strong uplift of the western Antarctic rift margin (Transantarctic Mountains) has been implicated in the initiation, advance and retreat and thermal regime of the ice sheet (Stump and Fitzgerald, 1992). Behrendt and Cooper (1991) speculated that the relationship between Cenozoic tectonism and the waxing and waning of the Antarctic Ice Sheet may have started as early as the early Oligocene (60 Ma). It should be noted in addition, that the West Antarctic rift system was initiated much earlier, during Jurassic rifting of Africa from Antarctica such that possible *Cretaceous* Antarctic sheet(s) may also have been adiabatically controlled (see above). The role of tectonic forcing in Antarctica is subject to debate (see Wilch et al., 1993).

The Cenozoic tectonic uplift of the West Antarctic rift shoulder after 60 Ma occurred while Australia was moving away from the continent and a strong circum-Antarctic current was established (Kennett, 1982). The present day 'polar' characteristics of the Antarctic Ice Sheet (i.e. dry-based, little meltwater production and a 'clastic-starved' continental shelf) were only established as a result of renewed uplift during the Late Pliocene (Barrett et al., 1992). This transition coincides with the formation of major Northern Hemispheric ice sheets. The earliest input of glacial sediment to the Ross Sea occurred at about 36 Ma when temperate wet-based glaciers reached the Victoria Land rift basin adjacent to the Cenozoic West Antarctic rift shoulder (Transantarctic Mountains; Barrett et al., 1989).

The best exposed record of Late Cenozoic glaciation is that preserved around the Gulf of Alaska (Yakataga Formation) where more than 5 km of glacially influenced marine strata has been uplifted as a result of plate margin convergence (C. Eyles and Lagoe, 1990; C. Eyles et al., 1991; N. Eyles et al., 1992). Local glaciation started at c. 6 Ma primarily in response to strong orogenic uplift of coastal mountains (e.g. Mt. St. Elias: > 5 km elevation). Large volumes of glaciclastic sediment were flushed into a rapidly subsiding forearc basin; in contrast, only a vestigial glacial record has been preserved on the adjacent continent (C. Eyles and N. Eyles, 1989).

Lagoe et al. (1993) summarize the marine and onshore record of the northern Pacific Ocean and demonstrate initial tidewater glaciation between 6.7 and 5.0 Ma, followed by a mid-Pliocene warm interval from about 4.2 to 3.0 Ma. A major increase in the severity of glaciation occurs between 3.5 and 2.4 Ma when continental ice sheets began to grow in North America and Europe; a coeval Late Miocene–Pleistocene glacial history appears to be recorded in the southern hemisphere. Mudie and Helgason (1983) report evidence for climatic cooling in Iceland as early as 9.8 Ma; the onset of ice rafting may have begun as early as 10 Ma in Baffin Bay (Anstey, 1992).

The first extensive influx of ice-rafted debris into the North Atlantic occurred at 2.4 Ma, when the amplitude of the $\delta^{18}O$ signal in the deep oceans exceeded 10% for the first time, though small ice

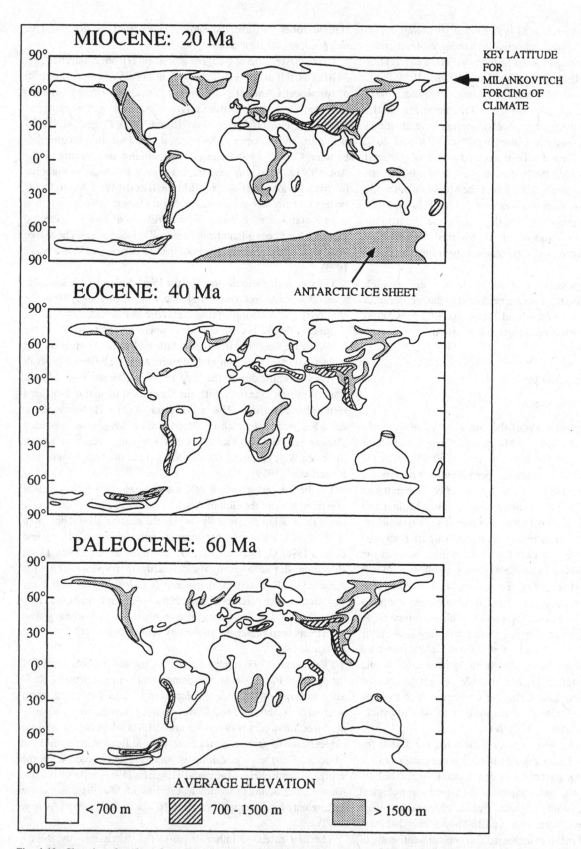

Fig. 1.10. Changing elevation of continental surfaces after 60 Ma as a precondition for Late Cenozoic glaciation. After Barron (1985). The biggest tectonic influences on global climate are uplift of the Tibetan Plateau (Raymo and Ruddiman, 1992), and continued passive margin uplift around the North Atlantic Ocean (N. Eyles, 1993).

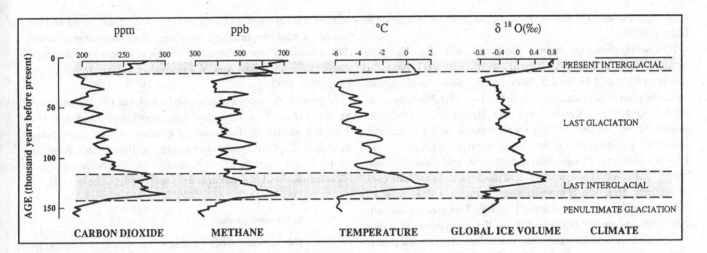

Fig. 1.11. Global change over the last 150 Ka as determined by study of the Vostok ice core from Antarctica (after Lorius *et al.*, 1990). An overall astronomical control is apparent (Raymo *et al.*, 1989) but the rapid and synchronous linkage between ice sheets, oceans and atmosphere suggests abrupt reorganizations of the Earth's climate system (Broecker and Denton, 1989). Did such rapid global change characterize pre-Quaternary glaciations and, if so, how would we recognize them?

sheets began to develop at about 3 Ma (Raymo *et al.*, 1989). Sequence stratigraphy (Haq *et al.*, 1988) identifies an accompanying sea-level lowstand around 2.4 Ma. Classic Late Pleistocene (< 2 Ma) glaciations are driven by relatively weak 'Milankovitch' astronomical forcing reinforced by changes in atmospheric CO_2 levels (Martinson *et al.*, 1987; Lorius, *et al.*, 1990; Fig. 1.11). Climate cycles with periodicities of 100, 41 and 23 Ka are linked to orbital cycles of eccentricity, obliquity and precession respectively. The volume changes of the early ice sheets were driven by a dominant 41 ka cycle of orbital obliquity with 100 ka cycle becoming established after 1.6 Ma. At the same time the decay and growth phases of the northern hemisphere ice sheets (dominantly the Laurentide Ice Sheet) became markedly asymmetric with deglaciation being much more rapid than ice sheet growth (Fig. 1.11). Prior to 2.4 Ma ice covers may have formed but did not reach tidewater. Glaciation in Iceland for example, commenced about 3 Ma but did not reach sea-level until 2 Ma (Shackleton *et al.*, 1984). This timing is of particular tectonic significance given that uplift of the Reykjanes Ridge occurred at around 2.5 Ma in response to a change in spreading direction and rate along most of the North Atlantic spreading system (Vogt and Tucholke, 1989).

The role of ocean currents

The abrupt 'switch' from glacial to interglacial conditions and the synchronous global changes from northern to southern hemisphere (e.g. Fig. 1.11) clearly cannot be explained by reference to relatively weak changes in seasonal solar insolation produced by astronomical rhythms. The major problem, as discussed by Broecker and Denton (1989) is that weak Milankovitch insolation variations are most effective at about latitute 60 °N but have to generate a *global* signal. A further problem is that the deep marine sedimentary record of the growth and decay of Late Cenozoic ice sheets reveals a strong 100 Ka rhythm which is at odds with the fact that variation in incoming solar radiation that can be attributed to the 100 Ka eccentricity cycle is extremely small. Liu (1992) argues that the 100 Ka climate signal is the result of variation in the frequency of the obliquity cycle. Broecker and Denton (1989) argue that the climatic 'jumps' that are evident from the Late Cenozoic glacial record are the result of dramatic changes in ocean circulation and heat delivery resulting from changes in salinity of surface waters (Duplessy *et al.*, 1992). These changes are complex, rapid and recorded by marine microfossils in oceanic sediment and ice cores from Greenland and Antarctica (Lorius *et al.*, 1990; Johnsen *et al.*, 1992). This work is a reminder of the complexity and interlinkage of controls on glaciation on planet Earth and emphasizes the difficulty of isolating these controls in those pre-Cenozoic glaciated basins containing no paleontological record. Regardless of the precise atmospheric and oceanic controls acting to determine the growth cycles of Late Cenozoic ice sheets, it is very unlikely that Late Cenozoic glaciations and the evolution of modern man (Molnar, 1990) would have occurred without significant Late Tertiary tectonic uplift in the mid-latitude belt (Ruddiman, *et al.*, 1989; Raymo and Ruddiman, 1992).

Conclusions

At the commencement of this brief review we posed some questions as to the nature of controls on glaciation in Earth history. Finding the answers to these and other questions will occupy geologists for many years to come but some provisional responses can be made. Arguably the most important conclusion is that there is no single causative factor that controls the timing of glaciation in Earth history.

New work shows increasingly the significance of global plate tectonic cycles as a central organizing principle underlying the stratigraphical and biological history of the planet, and increasingly, the planet's climate history (e.g. Fischer, 1984; Worsley et al., 1984; Leighton and Kolata, 1990; Des Marais et al., 1992). As emhasized by Veevers (1990), there appears to be a tectonic supercycle of about 400 million years duration (at least for the Phanerozoic) from a single continent (Pangea) and ocean (Panthalassa), having an atmosphere deficient in CO_2 ('icehouse' climate), to multiple dispersed continents and oceans in a CO_2-enriched atmosphere ('greenhouse' climate; Fig. 1.5). This general 'tectonic cycle' climate model is supported by the occurrence of major Neoproterozoic and Late Paleozoic glaciations at times of icehouse climate but is weakened because Late Ordovician and Quaternary glaciations occur under greenhouse conditions (Fig. 1.5). Early Paleozoic glaciation occurred at a time when global CO_2 levels may have been as much as 16 times *greater* than at present (Berner, 1990) but was restricted to the south polar region (Deynoux, 1985a,b; Vaslet, 1990). Nonetheless, the importance of atmospheric CO_2 as a major constraint on climate (Chamberlin, 1899) is supported by a large body of data (Raymo, 1991; Raymo and Ruddiman, 1992).

Within the general framework of global tectonic cycles and global 'climate states', combinations of factors such as continental positioning, tectonic uplift and increased elevation of continental surfaces, and enhanced weathering results in atmospheric CO_2 drawdown thereby providing a background for regional glaciation. Atmospheric CO_2 is also implicated as a control on the formation of Paleoproterozoic glaciations and in creating long Mesoproterozoic and Archean *non-glacial* intervals.

Tectonic uplift and differentiation of uplifted source areas from subsiding basins provides the structural framework for the initiation of adiabatic glaciers and for accommodating glaciogenic sediments; the same tectonism produces a large flux of crustal sediment which when weathered consumes and draws down atmospheric CO_2 to a level where large regional ice centres might form. Once tectonism and reduced levels of atmospheric CO_2 have created the necessary conditions for expanded glaciation, relatively weak 'Milankovitch' astronomical forcing may then come into play and control fine-scale temporal fluctuations of ice masses. This is well demonstrated for Late Pleistocene glaciations (e.g. Martinson et al., 1987) but the stability of the Milankovitch rhythms during the Phanerozoic has been questioned (e.g. Laskar, 1989; Berger and Loutre, 1989). New work stresses, however, that very rapid climatic reorganizations from glacial to interglacial states is dependent on correspondingly abrupt changes in ocean circulation (Broecker and Denton, 1989). This work provides an interesting link with the climate modelling of Worsley and Kidder (1991) who identify the importance of continental positioning and paleogeography as a further control on the CO_2 composition of the atmosphere.

The rapid drift rates now identified for several Neoproterozoic continents (e.g. Hoffman, 1989, 1991; Meert and Van der Voo, 1992) give a mechanism for overprinting an original high latitude remanent magnetism with a secondary low latitude record and thus provide a ready explanation for alleged 'low latitude glaciations.' The supposedly 'warm climate' dolomitic interbeds associated with many Neoproterozoic glacial successions appear to be detrital (Fairchild, 1993).

In general, the record of glaciation on planet Earth supports the model of alternating 'ice-house' and 'greenhouse' conditions (Fig. 1.5); the overriding, fundamental control on long-term climate evolution therefore appears to be global tectonic cycles. Resolving the finer-scale controls will be dependent on new geochronological, plate tectonic and geochemical data and geodynamic models.

Acknowledgements

The authors wish to thank the Natural Science and Engineering Research Council of Canada for supporting their research on the stratigraphic record of glaciation. We are especially grateful to Max Deynoux and Julia Miller for shouldering responsibility for IGCP 260 and providing a forum for numerous fruitful discussions and field conferences on glaciogenic rocks. We thank Victor Gostin, Max Deynoux and Michael Gipp for their comments on the manuscript.

References

Acharyya, S.K. (1975). Tectonic framework of sedimentation of the Gondwana of the Eastern Himalayas, India. In *Gondwana Geology*, ed. K.S.W. Campbell, Australian National University Press, pp. 663–74.

Aitken, J.D. (1991). Two late Proterozoic glaciations, MacKenzie Mountains, northwestern Canada. *Geology*, **19**, 445–8.

Alavi, M. (1991). Sedimentary and structural characteristics of the paleo-Tethys remnants in northeastern Iran. *Geological Society of America Bulletin*, **103**, 938–92.

Anderton, R. (1982). Dalradian deposition and the late Precambrian–Cambrian history of the N. Atlantic region: a review of the early evolution of the Iapetus Ocean. *Journal of the Geological Society of London*, **139**, 421–31.

Anstey, C.E. (1992). Biostratigraphic and palaeoenvironmental interpretation of Late Middle Miocene through Early Pleistocene dinocyst, achritarch and other algal palynomorph assemblages from Ocean Drilling Program Site 645, Baffin Bay, Unpublished M.Sc thesis, University of Toronto, 212 pp.

Bahcall, J.N. & Ulrich, R.K. (1988). Solar models, neutrino experiments and helioseismology. *Reviews of Modern Physics*, **60**, 297–372.

Barrett, P.J. (1982). History of the Ross Sea region during the deposition of the Beacon Supergroup 400–180 million years ago. *Journal of the Royal Society of New Zealand*, **11**, 447–58.

Barrett, P.J., Hambrey, M.J., Harwood, D.M., Pyne, A.R. & Webb, P.N. (1989). Synthesis. In *Antarctic Cenozoic History from the CIROS-1 Drillhole, McMurdo Sound*, ed. P.J. Barrett, Department of Scientific and Industrial Research (DSIR), 245, 241–51.

Barrett, P.J., Adams, C.J., McIntosh, W.C., Swisher, C.C., III & Wilson, G.S. (1992). Geochronological evidence supporting Antarctic deglaciation three million years ago. *Nature*, **359**, 816–18.

Barron, E.J. (1989). Studies of cretaceous climate. In *Understanding Climate Change*, eds. A. Berger, R.E. Dickinson & J.W. Kidson, Washington, American Geophysical Union, International Union of Geodesy and Geophysics (IUGG), 52, 149–157.

Basu, T.N. & Shrivastava, B.B.P. (1981). Structure and tectonics of Gondwana basins of Peninsula India. In *Gondwana 5*, eds. M.M. Cresswell & P. Vella, Rotterdam, A.A. Balkema, 177–83.

Beckinsale, R.D., Reading, H.G. & Rex, D.C. (1976). Potassium-argon ages for basic dykes from East Finnmark: stratigraphic and structural implications. *Scottish Journal of Geology*, **12**, 51–65.

Behrendt, J.C. & Cooper, A. (1991). Evidence of rapid Cenozoic uplift of the shoulder escarpment of the Cenozoic West Antarctic rift system and a speculation on possible climate forcing. *Geology*, **19**, 360–3.

Bell, R.T. & Jefferson, C.W. (1987). An hypothesis for an Australian–Canadian connection in the Late Proterozoic and the birth of the Pacific Ocean. In *Proceedings, Pacific Rim Congress, '87*. Parkville, Victoria, Australian Institute of Mining and Metallurgy, pp. 39–50.

Benton, M.J. (1991). Polar dinosaurs and ancient climates. *Trends in Ecology and Evolution*, **6**, 28–30.

Berger, A. & Loutre, M.F. (1989). Pre-Quaternary Milankovitch frequencies. *Nature*, **332**, 133.

Berner, R.A. (1990). Atmospheric carbon dioxide over Phanerozoic time. *Science*, **249**, 1382–6.

Beuf, S., Biju-Duval, B., de Charpal, O., Rognon, P., Gariel, O. & Bennacef, A. (1971). *Les gres du Paleozoique inferieur au Sahara*. Paris: Editions Technip, 464 p.

Bond, G.C. & Koimnz, M.A. (1984). Construction of tectonic subsidence curves for the early Paleozoic miogeocline, southern Canadian Rocky Mountains. Implications for subsidence mechanisms, age of break-up, and crustal thinning. *Geological Society of America Bulletin*, **95**, 155–73.

Brandt, K. (1986). Glacioeustatic cycles in the Early Jurassic? *Neues Jahrbuch fur Geologie und Palaontologie, Monatshefte*, **5**, 257–74.

Brenchley, P.J. (1989). The Late Ordovician extinction. In *Mass Extinctions*, ed. S.K. Donovan, New York, Columbia University Press, pp. 104–32.

Broecker, W.S. & Denton, G.H. (1989). The role of ocean–atmosphere reorganizations in glacial cycles. *Geochemica Cosmochemica Acta*, **53**, 2465–501.

Caldeira, K. & Kasting, J.F. (1992). Susceptibility of the Earth to irreversible glaciation caused by carbon dioxide clouds. *Nature*, **359**, 226–28.

Caputo, M.V. (1985). Late Devonian glaciation in South America. *Palaeogeography, Palaeoclimatology, Palaeoecology*, **51**, 291–317.

Caputo, M.V. & Crowell, J.C. (1985). Migration of glacial centers across Gondwana during Paleozoic Era. *Geological Society of America Bulletin*, **96**, 1020–36.

Chamberlin, T.C. (1899). An attempt to frame a working hypothesis of the cause of glacial periods on an atmospheric basis. *Journal of Geology*, **7**, 545–84.

Chowdhury, M.K.R., Laskar, B. & Mitra, N.D. (1975). Tectonic control of Lower Gondwana sedimentation in peninsular India. In *Gondwana Geology*, ed. K.S.W. Campbell, Canberra, Australian National University Press, pp. 675–680.

Chumakov, N.M. (1985). Glacial events of the past and their geological significance. *Palaeogeography, Palaeoclimatology, Palaeoecology*, **51**, 319–46.

Chumakov, N.M. (1992). The problems of old glaciations (Pre-Pleistocene glaciogeology in the USSR). *Geology Review*, **1**, 1–209.

Chumakov, N.M. & Elston, D.P. (1989). The paradox of Late Proterozoic glaciations at low latitudes. *Episodes*, **12**, 115–1120.

Chumakov, N.M. & Krasil'nikov, S.S. (1992). Lithology of Riphean tilloids: Urinsk uplift area, Lena region. *Lithology and Mineral Resources*, **26**, 249–64.

Clark, E.A. & Cook, F.A. (1992). Crustal-scale ramp in a Middle Proterozoic orogen, Northwest Territories, Canada. *Canadian Journal of Earth Sciences*, **29**, 142–57.

Coats, R.P. (1981). Late Proterozoic (Adelaidean) tillites of the Adelaide Geosyncline. In *Earth's Pre-Pleistocene Glacial Record*, ed. M.J. Hambrey and W.B. Harland, Cambridge, Cambridge University Press, pp. 537–48.

Coats, R.P. & Preiss, W.V. (1987). Stratigraphy of the Umberatana Group. In *The Adelaide Geosyncline; Late Proterozoic Stratigraph, Sedimentation, Palaeontology and Tectonics*, ed. W.V. Preiss, Geological Survey of South Australia, **53**, 125–209.

Crowell, J.C. (1978). Gondwana glaciation, cyclotherms, continental positioning and climate change. *American Journal of Science*, **278**, 1345–72.

Crowell, J.C. (1982). Continental glaciation through geologic time. In *Climate in Earth History*. Washington, D.C., National Academy Press, pp. 77–82.

Crowley, T.J. & Baum, S.K. (1992). Modeling late Paleozoic glaciation. *Geology*, **20**, 507–10.

Daly, M.C., Lawrence, S.R., Kimun'a, D. & Binga, M. (1991). Late Palaeozoic deformation in central Africa: A result of distant collision? *Nature*, **350**, 605–7.

Dalziel, I.W.D. (1991). Pacific margins of Laurentia and East Antarctica–Australia as a conjugate rift pair: Evidence and implications for an Eocambrian supercontinent. *Geology*, **19**, 598–601.

Dalziel, I.W.D., Garrett, S.W., Grunow, A.M., Pankhurst, R.J., Storey, B.C. & Vennum, W.R. (1987). The Ellsworth-Whitmore Mountains crustal block: Its role in the tectonic evolution of West Antarctica. In *Gondwana 6*, ed. G.D. McKenzie. American Geophysical Union, Geophysical Monograph, **41**, 173–82.

de Wit, M., Jeffery, M., Bergh, H. & Nicolaysen, L. (1988). Geological map of sectors of Gondwana. Tulsa, Oklahoma, American Association of Petroleum Geologists, Scale 1:10 000 000.

de Wit, M.J., Roering, C., Hart, R.J., Armstrong, R.A., de Ronde, C.E., Green, R.W.E., Tredoux, M., Peberdy, E. & Hart, R.A. (1992). Formation of an Archaean continent. *Nature*, **357**, 553–56.

Des Marais, D.J., Strauss, H., Summons, R.E. & Hayes, J.M. (1992). Carbon isotope evidence for the stepwise oxidation of the Proterozoic environment: *Nature*, **359**, 605–9.

Deynoux, M. (1985a). Terrestrial or waterlain glacial diamictites? Three case studies from the Late Precambrian and Late Ordovician glacial drifts in S. Africa. *Palaeogeography, Palaeoclimatology, Palaeoecology*, **51**, 97–142.

Deynoux, M. (1985b). Les glaciations du Sahara. *la Recherche*, **16**, 986–97.

Deynoux, M. & Trompette, R. (1976). Discussion: Late Precambrian mixtites: glacial and/or nonglacial? Dealing especially with the mixtites of West Africa. *American Journal of Science*, **276**, 1302–15.

Diaz E., Isaaccson, P.E. & Sablock, P.E. (1993). Late Paleozoic latitudinal shift of Gondwana: stratigraphy/sedimentology and biogeography evidence from Bolivia. Document des Laboratoires Géologiques de Lyon, **25**, 119–38.

DiBona, P.A. (1991). A previously unrecognized Late Proterozoic glacial succession: Upper Wilpena Group, northern Flinders Ranges, South Australia. *Geological Survey of South Australia Quarterly Geological Notes*, **117**, 2–9.

Domack, E.W., Burkley, L.A., Domack, C.R. & Banks M.R. (1993). Facies analysis of glacial marine pebbly mudstones in the Tasmania Basin: implications for regional paleoclimates during the Paleozoic. In *Gondwana 8*, eds. R.H. Finlay, R. Unrug, M.R. Banks, J.J. Veevers. Rotterdam: Balkema, 471–84.

Duplessy, J.C., Labeyrie, L., Arnold, M., Paterne, M., Duprat, J. & van Weering, T.C.E. (1992). Changes in surface salinity of the North Atlantic Ocean during the last deglaciation. *Nature*, **358**, 485–488.

Edwards, M.B. (1975). Glacial retreat sedimentation in the Smalfjord Formation, Late Precambrian, North Norway. *Sedimentology*, **22**, 75–94.

Eisbacher, G.H. (1985). Late Proterozoic rifting, glacial sedimentation and sedimentary cycles in the light of Windermere deposition, Western Canada. *Palaeogeography, Palaeoclimatology, Palaeoecology*, **51**, 231–54.

Embleton, B.J.J. & Williams, G.E. (1986). Low palaeolatitude of deposition for late Precambrian periglacial varvites in South Australia: implications for palaeoclimatology. *Earth and Planetary Science Letters*, **79**, 419–30.

Emslie, R.F. (1978). Anorthosite massifs, rapakivi granites, and Late Proterozoic rifting of North America. *Precambrian Research*, 7, 61–98.

Epshteyn, O.G. (1978). Mesozoic–Cenozoic climates of Northern Asia and glacial-marine deposits. *International Geology Review*, **20**, 49–58.

Ericksson, P.G., Twist, D., Snyman, C.P. & Burger, L. (1990). The geochemistry of the Silverton Shale Formation, Transvaal Sequence. *South African Journal of Geology*, **93**, 454–62.

Eyles, C.H. (1988). Glacially and tidally influenced shallow marine sedimentation of the Late Precambrian Port Askaig Formation, Scotland. *Palaeogeography, Palaeoclimatology, Palaeoecology*, **68**, 1–25.

Eyles, C.H. & Eyles, N. (1989). The Late Cenozoic White River 'Tillites' of southern Alaska: Subaerial slope and fan delta deposits in a strike-slip setting. *Geological Society of America Bulletin*, **101**, 1091–102.

Eyles, C.H., Eyles, N. & Lagoe, M.B. (1991). The Yakataga Formation; A late Miocene to Pleistocene record of temperate glacial marine sedimentation in the Gulf of Alaska. In *Glacial Marine Sedimentation; Paleoclimatic Significance*, ed. J.B. Anderson & G.M. Ashley. Boulder, Colorado, Geological Society of America, **261**, 159–80.

Eyles, C.H. & Lagoe, M.B. (1990). Sedimentation patterns and facies geometries on a temperate glacially-influenced continental shelf: the Yakataga Formation, Middleton Island, Alaska. In *Glaciomarine Environments: Processes and Sediments*. Ed. J.A. Dowdeswell & J.D. Scourse, *Geological Society of London Special Publicaton*, **53**, 363–386.

Eyles, C.H., Eyles, N. & França, A.B. (1993). Late Palaeozoic glaciation in an active intracratonic basin. Itararé Group, Parana Basin, Brazil. *Sedimentology*, **40**, 1–25.

Eyles, N. (1990). Late Precambrian 'tillites' of the Avalonian–Cadomian belt; marine debris flows in an active tectonic setting. *Palaeogeography, Palaeoclimatology, Palaeoecology*, **79**, 73–98.

Eyles, N. (1993). Earth's glacial record and its tectonic setting. *Earth Science Reviews*, **35**, 1–248.

Eyles, N. & Clark, B.M. (1985). Gravity induced soft-sediment deformation structures in glaciomarine sequences of the Late Proterozoic Port Askaig Formation, Scotland. *Sedimentology*, **32**, 784–814.

Eyles, N. & Eyles, C.H. (1989). Glacially influenced deep marine sedimentation of the Late Precambrian Gaskiers Formation, Newfoundland, Canada: *Sedimentology*, **36**, 601–20.

Eyles, N., Lagoe, M.B. & Vossler, S. (1992). Ichnology of a glacially influenced continental shelf and slope; the Late Cenozoic Gulf of Alaska (Yakataga Formation). *Palaeogeography, Palaeoclimatology, Palaeoecology*, **92** (in press).

Eyles, N. & Eyles, C.H. (1993). Glacial geologic confirmation of an intraplate bounding crossing the Parana Basin of Brazil. *Geology*, **21**, 459–62.

Fairchild, I.J. (1993). Balmy shores and icy wastes: the paradox of carbonates associated with glacial deposits in Neoproterozoic times. *Sedimentology Review*, **1**, 1–6.

Fairchild, I.J., Hambrey, M.J., Spiro, B. & Jefferson, T.H. (1989). Late Proterozoic glacial carbonates in northeast Spitsbergen: new insights into the carbonate–tillite association. *Geological Magazine*, **126**, 469–90.

Fairchild, I.J. & Spiro, B. (1990). Carbonate minerals in glacial sediments: geochemical clues to palaeoenvironment. In *Glaciomarine*

Environments: Processes and Sediments, eds. J.A. Dowdeswell & J.D. Scourse, *Geological Society of London Special Publication*, **53**, 201–16.

Fischer, A.G. (1984). The two Phanerozoic supercycles. In *Catastrophes in Earth History; The New Uniformitarianism*, eds. W.A. Berggren & J.A. Van Couvering, Princeton, N.J., Princeton University Press, pp. 129–50.

Frakes, L.A. & Francis, J.E. (1988). A guide to Phanerozoic cold polar climates from high latitude ice-rafting in the Cretaceous. *Nature*, **333**, 547–9.

Frakes, L.A., Francis, J.E. & Syktus, J.I. (1992). *Climate modes of the Phanerozoic*. Cambridge: Cambridge University Press, 274 pp.

Franca, A.B. & Potter, P.E. (1991). Stratigraphy and reservoir potential of glacial deposits of the Itarare Group (Carboniferous–Permian), Parana Basin, Brazil. *American Association of Petroleum Geologists Bulletin*, **75**, 62–85.

Gaffin, S. (1987). Ridge volume dependence on sea floor generation rate and inversion using long term sea-level change. *American Journal of Science*, **287**, 596–611.

Gayer, R.A. & Rice, A.H.N. (1989). Palaeogeographic reconstruction of the pre-to syn-Iapetus rifting sediments in the Caledonides of Finnmark, N. Norway. In *The Caledonide Geology of Scandinavia*, ed. R.A. Gayer, London, Graham and Trotman, pp. 127–42.

Ghosh, P.K. (1975). The environment of coal formation in the peninsular Gondwana basins of India. In *Gondwana Geology*, ed. K.S.W. Campbell, Canberra, Australian National University Press, pp. 221–31.

Gibbs, A.D. (1984). Structural evolution of extensional basin margins. *Journal of the Geological Society of London*, **141**, 609–20.

Gilliland, R.L. (1989). Solar evolution. *Palaeogeography, Palaeoclimatology, Palaeoecology* (Global and planetary change section), **75**, 35–55.

Goldhammer, R.K., Dunn, P.A. & Hardie, L.A. (1987). High frequency glacio-eustatic sea-level oscillations with Milankovitch characteristics recorded in Middle Triassic platform carbonates in northern Italy. *American Journal of Science*, **287**, 853–92.

Gonzalez Bonorino, G. (1992). Carboniferous glaciation in Gondwana. Evidence for grounded marine ice and continental glaciation in southwestern Argentina. *Palaeogeography Palaeoclimatology Palaeoecology*, **91**, 363–75.

Goodwin, A.M. (1991). *Precambrian Geology*. London, Academic Press. 666 pp.

Grahn, Y. & Caputo, M.V. (1992). Early Silurian Glaciations in Brazil. *Palaeogeography Palaeoclimatology Palaeoecology*, **99**, 9–15.

Hadley, D.G. & Schmidt, D.L. (1975). Non-glacial origin for conglomerate beds in the Wajid Sandstone of Saudi Arabia. In *Gondwana Geology*, ed. K.S.W. Campbell, Canberra, Australian National University Press, pp. 357–71.

Hambrey, M.J. (1983). Correlation of Late Proterozoic tillites in the North Atlantic region and Europe. *Geological Magazine*, **120**, 209–32.

Hambrey, M.J. & Harland, W.B., eds. (1981). *Earth's Pre-Pleistocene Glacial Record*. Cambridge, Cambridge University Press, 1004 pp.

Hambrey, M.J. & Spencer, A.M. (1987). Late Precambrian glaciation of central East Greenland. *Meddelelser Gronland Geoscience*, **19**, 50 pp.

Haq, B.U., Hardenbol, J. & Vail, P.R. (1988). Mesozoic and Cenozoic chronostratigraphies and eustatic cycles. In *Sea-level changes: an integrated approach*, eds. C.S. Vilgus et al. Tulsa, Oklahoma, Society of Economic Palaeontologists and Mineralogists Special Publication, **42**, 71–108.

Harland, W.B. (1981). The Late Archaean(?) Witwatersrand conglomerate, South Africa. In *Earth's Pre-Pleistocene Glacial Record*, eds. M.J. Hambrey & W.B. Harland, Cambridge, Cambridge University Press, pp. 185–7.

Harland, W.B. & Bidgood, D.E.T. (1959). Palaeomagnetism in some Norwegian sparagmites and the Late pre-Cambrian Ice Age. *Nature*, **184**, 1860–2.

Harland, W.B. & Herod, K.N. (1975). Glaciations through time. In *Ice Ages: Ancient and Modern*, eds. A.E. Wright & F. Moseley, Liverpool, Seel House Press, pp. 189–216.

Hart, M.H. (1978). The evolution of the atmosphere of the Earth. *Icarus*, **33**, 23–39.

Herbert, C. (1981). Late Palaeozoic glacigenic sediments of the southern Sydney Basin, New South Wales. In *Earth's Pre-Pleistocene Glacial Record*, eds. M.J. Hambrey & W.B. Harland, Cambridge, Cambridge University Press, pp. 488–491.

Herve, F., Godoy, E., Parada, M.A., Ramos, V., Rapela, C., Mpodozis, C. & Davidson, J. (1987). A general view on the Chilean–Argentine Andes, with emphasis on their early history. In *Circum-Pacific Orogenic Belts and Evolution of the Pacific Ocean Basin*, eds. J.W.H. Monger & J. Francheteau, Washington D.C., American Geophysical Union, **18**, 97–113.

Hoffman, P.F. (1988). United plates of America, the birth of a craton: Early Proterozoic assembly and growth of Laurentia. *Annual Reviews of Earth and Planetary Science*, **16**, 543–603.

Hoffman, P.F. (1989). Speculations on Laurentia's first gigayear (2.0 to 1.0 Ga). *Geology*, **17**, 135–8.

Hoffman, P.F. (1991). Did the break-out of Laurentia turn Gondwanaland inside-out? *Science*, **252**, 1409–12.

Houghton, R.A. & Woodwell, M.G. (1989). Global climatic change. *Scientific American*, **260**, 36–44.

Houston, R.S., Lanthier, L.R., Karstrom, K.K. & Sylvester, G.G. (1981). Late Proterozoic diamictite of southern Wyoming. In *Earth's Pre-Pleistocene Glacial Record*. Eds. M.J. Hambrey & W.B. Harland, Cambridge, Cambridge University Press, pp. 795–799.

Johnsen, S.J., Clausen, H.B., Dansgaard, W., Fuhrer, K. Gundestrup, N., Hammer, C.U., Iversen, P., Jouzel, J., Stauffer, B., and Steffensen, J.P. (1992). Irregular glacial interstadials recorded in a new Greenland ice core: *Nature*, **359**, 311–13.

Jorgensen, G.J. & Bosworth, W. (1989). Gravity modeling in the Central African Rift System, Sudan: rift geometrics and tectonic significance. *Journal of African Earth Sciences*, **8**, 283–306.

Karlstrom, K.K., Flurkey, A.J. & Houston, R.S. (1983). Stratigraphy and depositional setting of Proterozoic rocks of southeastern Wyoming: record of an Early Proterozoic Atlantic-type cratonic margin. *Geological Society of America Bulletin*, **94**, 1287–94.

Kasting, J.F. (1987). Theoretical constraints on oxygen and carbon dioxide concentrations in the Precambrian atmosphere. *Precambrian Research*, **34**, 205–29.

Kasting, J.F. & Toon, O.B. (1989). Climate evolution on the terrestrial planets. In *Origin And Evolution Of Planetary And Satellite Atmospheres*, eds, S.K. Atneya, J.B. Pollack & M.S. Matthews, Tucson, University of Arizona Press, pp. 423–49.

Kasting, J.F., Pollack, J.B. & Ackerman, T.P. (1984). Response of Earth's surface temperature to increases in solar flux and implications for loss of water from Venus. *Icarus*, **57**, 335–55.

Kemp, E.M. (1975). The Palynology of Late Palaeozoic Glacial Deposits. In *Gondwana 3*, ed. K.S.W. Campbell, Canberra, Australian National University Press, pp. 397–413.

Kennett, J.P. (1982). *Marine Geology*. Englewood Cliffs, N.J., Prentice-Hall, 812 pp.

Khramov, A.N. (1983). Global reconstruction of the position of ancient cratons during late Precambrian. In *Paleomagnetism of Upper Precambrian of U.S.S.R.*, ed, A.N. Khramov, Leningrad, VNIGRI, pp. 127–37.

Kingsley, C.S. (1984). Dagbreek fan-delta: an alluvial placer to prodelta sequence in the Proterozoic Welkom goldfield, Witwatersrand, South Africa. In *Sedimentology of Gravels and Conglomerates*, eds. E.H. Koster & R.J. Steel, Canadian Society of Petroleum Geologists Memoir, **10**, 321–330.

Klein, G. de V. & Hsui, A.T. (1987). Origin of cratonic basins. *Geology*, **19**, 330–42.

Klein, G. de V. & Willard, D.A. (1989). Origin of the Pennsylvanian coal-bearing cyclothems of North America. *Geology*, **17**, 152–5.

Klein, G. de V. & Kupperman, J.B. (1992). Pennsylvanian cyclothems: Methods of distinguishing tectonically induced changes in sea-level from climatically induced changes. *Geological Society of America Bulletin*, **104**, 166–75.

Koepnick, R.B., Denison, R.E. & Dahl, D.A. (1988). The Cenozoic seawater 87Sr/86Sr curve: data review and implications for correlation of marine strata. *Paleoceanography*, **3**, 743–56.

Krogh, T.E., Davis, D.W. & Covbn, F. (1984). Precise U-Pb zircon and baddeleyite ages for the Sudbury area. In *The Geology and Ore Deposits of the Sudbury Structure*, eds. E.G. Pye, A.J. Naldrell & P.E. Giblin, Ontario Geological Survey, Special Volume 1, 431–446.

Kruck, W. & Thiele, J. (1983). Late Palaeozoic Glacial Deposits in the Yemen Arab Republic. *Geologische Jahrbuch Reihe B*, **46**, 3–29.

Kutzbach, J.E., Guetter, P.J., Ruddiman, W.F. & Prell, W.L. (1989). Sensitivity of climate to Late Cenozoic uplift in southern Asia and the American West: numerical experiments. *Journal of Geophysical Research*, **94**, 18,393–407.

Lagoe, M.B., Eyles, C.H., Eyles, N. & Hale, C. (1993). Dating the onset of Late Cenozoic glaciation in the north Pacific Ocean. *Geological Society of America Bulletin*, (in press).

Laird, M.G. (1972). Stratigraphy and sedimentology of the Laksefjord Group, Finnmark. *Norges Geologiske Undersokelse*, **278**, 13–40.

Laskar, J. (1989). A numerical experiment on the chaotic behaviour of the solar system. *Nature*, **338**, 237–238.

Leighton, M.W., Kolata, D.R., Oltz, D.F. & Eidel, J.J., eds. (1990). *Interior Cratonic Basins*. American Association of Petroleum Geologists Memoir 51, 819 pp.

Leighton, M.W. & Kolata, D.R. (1990). Selected interior cratonic basins and their place in the scheme of global tectonic: a synthesis. In *Interior Cratonic Basins*, eds. M.W. Leighton, D.R. Kolata, D.F. Oltz & J.J. Eidel, American Association of Petroleum Geologists Memoir 51, pp. 729–799.

Lemon, N.M. & Gostin, V.A. (1990). Glacigenic sediments of the Late Proterozoic Elatina Formation and equivalents, Adelaide Geosyncline, South Australia. In: *The Evolution of a Late Precambrian-Early Palaeozoic Rift Complex: the Adelaide Geosyncline*, eds. J.B. Jago & P.S. Moore, Geological Society of Australia, **16**, 149–63.

Levell, B.K., Braakman, J.H. & Rutlen, K.W. (1988). Oil-bearing sediments of Gondwana glaciation in Oman. *American Association of Petroleum Geologists Bulletin*, **72**, 775–96.

Lin, J. & Watts, D.R. (1988). Paleomagnetic constraints on Himalayan Tibetan tectonic evolution. In *Tectonic Evolution of the Himalayas and Tibet*, eds. R.M. Shackleton, J.F. Dewey & B.F. Windley, Royal Society of London, pp. 172–188.

Lindsay, J.F. (1966). Carboniferous subaqueous mass-movement in the Manning-Macleay Basin, Kempsey, New South Wales. *Journal of Sedimentary Petrology*, **36**, 719–32.

Lister, G.S., Etheridge, M.A. & Symonds, P.A. (1991). Detachment models for the formation of passive continental margins. *Tectonics*, **10**, 1038–64.

Liu, H.-S. (1992). Frequency variations of the Earth's obliquity and the 100-kyr ice-age cycles. *Nature*, **358**, 397–9.

Long, D.G.F. (1991). A non-glacial origin for the Ordovician (Middle Caradocian) Cosquer Formation, Vesyarc'h, Crozon Peninsula, Brittany, France. *Geological Journal*, **26**, 279–94.

Lopez Gamundi, O.R. (1987). Depositional models for the glaciomarine sequences of Andean Late Paleozoic basins of Argentina. *Sedimentary Geology*, **52**, 109–26.

Lorius, C., Jouzel, J., Raynaud, D., Hansen, J. & Le Treut, H.E. (1990). The ice-core record: climate sensitivity and future greenhouse warming. *Nature*, **347**, 139–45.

MacNaughton, R.B., Robert, B. & Narbonne, G.M. (1992). Ichnology of a Neoproterozic-Early Cambrian siliciclastic succession, Mackenzie Mountains, N.W.T., Canada. Canadian Paleontology Conference, Program and Abstracts, **2**, 17–18.

Martin, H. (1981). The Late Paleozoic Dwyka Group of the South Kalahari Basin in Namibia and Botswana and the subglacial valleys of the Kaokoveld in Nambia. In *Earth's Pre-Pleistocene Glacial Record*, eds. M.J. Hambrey & W.B. Harland, Cambridge, Cambridge University Press, pp. 61–5.

Martinson, D.G., Pisias, N.G., Hays, J.D., Imbrie, J., Moore, T.C. & Shackleton, N.J. (1987). Age dating and the orbital theory of the Ice Ages: development of a high-resolution 0 to 300,000 year chronostratigraphy. *Quaternary Research*, **27**, 1–29.

Masetti, D., Neri, C. & Bosellini, A. (1991). Deep-water asymmetric cycles and progradation of carbonate platforms governed by high frequency eustatic oscillations (Triassic of the dolomites, Italy). *Geology*, **19**, 336–9.

Matsch, C.L. & Ojakangas, R.W. (1991). Comparisons in depositional style of 'polar' and 'temperate' glacial ice: Late Paleozoic Whiteout Conglomerate (West Antarctica) and Late Proterozoic Mineral Fork Formation (Utah). In *Glacial Marine Sedimentation: Paleoclimatic Significance*, eds. J.B. Anderson & G.M. Ashley, Geological Society of America Special Paper, **261**, 191–206.

McClure, H.A., Hussey, E. & Kaill, I. (1988). Permian–Carboniferous glacial deposits in southern Saudi Arabia. *Geology Jahrbuch*, **68**, 3–31.

McKelvey, B.C. (1981). Carboniferous tillites in the New England area of New South Wales. In *Earth's Pre-Pleistocene Glacial Record*, eds. M.J. Hambrey & W.B. Harland, Cambridge: Cambridge University Press, pp. 476–479.

McPhie, J. (1987). Andean analogue for Late Carboniferous volcanic arc and arc flank environments of the western New England Orogen, New South Wales, Australia. *Tectonophysics*, **138**, 269–88.

Meert, J.G. & Van der Voo, R. (1992). Evidence for a high paleolatitude of North America in the latest Precambrian and rapid drift during the Cambrian. *Eos*, **73**, 150 (abstract supplement).

Miall, A.D. (1985). Sedimentation on an early Proterozoic continental margin under glacial influence: the Gowganda Formation (Huronian), Elliot Lake area, Ontario, Canada. *Sedimentology*, **32**, 763–88.

Miller, J.M.G. & Waugh, B. (1991). Permo-Carboniferous glacial sedimentation in the central Transantarctic Mountains and its palaeotectonic significance. In *Geological Evolution of Antarctica*, eds. M.R.A. Thomson, J.A. Crame & J.W. Thompson, Cambridge: Cambridge University Press, pp. 205–208.

Miller, R.M. & Burger, A.J. (1983). U-Pb zircon age of the early Damara Naaumpoort Formation. In *Evolution of the Damara Orogen of South West Africa/Namibia*, ed. R.M. Miller, Geological Society of South Africa, Special Publication 11, 267–72.

Molnar, P. (1990). The rise of mountain ranges and the evolution of humans: A causal relation? *Irish Journal of Earth Sciences*, **10**, 199–207.

Molnar, P. & England, P. (1990). Late Cenozoic uplift of mountain ranges and global climate change: chicken or egg? *Nature*, **346**, 29–34.

Moores, E.M. (1991). Southwest U.S.–East Antarctic (SWEAT) connection: a hypothesis. *Geology*, **19**, 425–8.

Murphy, J.B. & Nance, R.D. (1991). Supercontinent model for the contrasting character of Late Proterozoic orogenic belts. *Geology*, **19**, 469–72.

Nance, R.D., Worsley, T.R. & Moody, J.B. (1988). The supercontinent cycle. *Scientific American*, **259**, 72–9.

Nesbitt, H.W. & Young, G.M. (1982). Early Proterozoic climates and plate motions inferred from major element chemistry of lutites. *Nature*, **299**, 715–717.

Nesbitt, H.W. & Young, G.M. (1984). Prediction of some weathering trends of plutonic and volcanic rocks based on thermodynamic and kinetic considerations. *Geochimica Cosmochimica Acta*, **48**, 1523–34.

Nystuen, J.P. (1987). Synthesis of the tectonic and sedimentological evolution of the late Proterozoic – early Cambrian Hedmark Basin, the Caledonian Thrust Belt, southern Norway. *Norsk Geologisk Tidskrift*, **67**, 395–418.

Oelofsen, B.W. (1987). The biostratigraphy and fossils of the Whitehill and Irati Shale Formations of the Karoo and Parana Basins. In *Gondwana 6*, ed. G.D. McKenzie, American Geophysical Union, Geophysical Monograph, pp. 131–138.

Ojakangas, R.W. (1988). Glaciation: an uncommon 'mega-event' as a key to intracontinental and intercontinental correlation of Early Proterozoic basin fill, North American and Baltic Cratons. In *New Perspectives in Basin Analysis*, eds. K.L. Kleinspehn & C. Paola, New York, Springer-Verlag, 431–44.

Page, N.J. (1981). The Precambrian diamictite below the base of the Stillwater Complex, Montana. In *Earth's Pre-Pleistocene Glacial Record*, eds. M.J. Hambrey & W.B. Harland, Cambridge: Cambridge University Press, pp. 821–5.

Patterson, J.G. & Heaman, L.M. (1991). New geochronologic limits on the depositional age of the Humitz Group, Trans-Hudson Hinterland, Canada. *Geology*, **19**, 1137–40.

Piper, J.D.A. (1978). *Paleomagnetism and the Continental Crust*. Milton Keynes, Open University Press, 434 pp.

Piper, J.D.A. (1981). Palaeomagnetic study of the (Late Precambrian) West Greenland Kimberlife-Lamprophyre suite: definition of the Hadrynian Track. *Physics of the Earth and Planetary Interiors*, **27**, 164–86.

Plint, A.G. (1991). High frequency relative sea-level oscillations in Upper Cretaceous shelf clastics of the Alberta foreland basin: Possible evidence for a glacio-eustatic control? In *Sedimentation, Tectonics and Eustasy*, ed. D.I.M. MacDonald, International Association of Sedimentologists Special Publication, **12**, 409–28.

Plumb, K.A. (1991). New Precambrian time scale. *Episodes*, **14**, 139–40.

Powell, C.M. & Veevers, J.J. (1987). Namurian uplift in Australia and South America triggered the main Gondwanan glaciation. *Nature*, **326**, 177–9.

Preiss, W.V., ed. (1987). *The Adelaide Geosyncline-Late Proterozoic Stratigraphy: Sedimentation, Palaeontology and Tectonics*. Geological Survey of South Australia, Bulletin, **53**, 438 pp.

Raup, D.M. Sepkovsky, J.J. (1982). Mass extinctions in the marine fossil record. *Science*, **215**, 1501–3.

Raymo, M.E. (1991). Geochemical evidence supporting T.C. Chamberlin's theory of glaciation. *Geology*, **19**, 344–348.

Raymo, M.E., Ruddiman, W.F., Backman, J., Clement, B.M. & Martinson, D.G. (1989). Late Pliocene variation in northern hemisphere ice sheets and north Atlantic deep water circulation. *Paleoceanography*, **4**, 413–46.

Raymo, M.E. & Ruddiman, W.F. (1992). Tectonic forcing of late Cenozoic climate. *Nature*, **359**, 117–22.

Reimer, T.O. (1986). Alumina-rich rocks from the Early Precambrian of the Kaapvaal Craton as indicators of paleosols and as products of other decompositional reactions. *Precambrian Research*, **32**, 155–79.

Roberts, J.D. (1971). Later Precambrian glaciation: an anti-greenhouse effect? *Nature*, **234**, 216–17.

Roberts, J.D. (1976). Late Precambrian dolomites, Vendian glaciation, and synchroneity of Vendian glaciations. *Journal of Geology*, **84**, 47–63.

Rocha-Campos, A.C. & Rosler, O. (1978). *Late Paleozoic Faunal and Floral Successions in the Parana Basin, Southeastern Brazil.* Boletim IG, Instituto de Geociencias, Universidade de Sao Paulo, Brazil, **9**, 1–16.

Ross, G.M. (1991). Tectonic setting of the Windermere Supergroup revisted. *Geology*, **19**, 1125–8.

Rowlands, N.J. (1973). The Adelaidean System of South Australia: a review of its sedimentation, tectonics and copper occurrences. In *Belt Symposium*. Moscow, University of Idaho and Idaho Bureau of Mines and Geology, pp. 80–112.

Ruddiman, W.F., Prell, W.L. & Raymo, M.E. (1989). Late Cenozoic uplift in Southern Asia and the American West: rationale for general circulation modeling experiments. *Journal of Geophysical Research*, **94**, 18,379–91.

Sagan, C. & Mullen, G. (1972). Earth and Mars: evolution of atmospheres and surface temperatures. *Science*, **177**, 52–6.

Schermerhorn, L.J.G. (1974). Late Precambrian mixtites: Glacial and/or non-glacial. *American Journal of Science*, 673–824.

Schermerhorn, L.J.G. (1975). Tectonic framework of Late Precambrian supposed glacials. In *Ice Ages: Ancient and Modern*, eds. A.E. Wright & F. Moseley, Geological Journal Special Issue, **6**, 241–74.

Schermerhorn, L.J.G. (1977). Late Precambrian dolomites, Vendian glaciation and synchroneity of Vendian glaciations: a discussion. *Journal of Geology*, **85**, 247–250.

Schmidt, P.W., Williams, G.E. & Embleton, B.J.J. (1991). Low palaeolatitude of Late Proterozoic glaciation: early timing of remanence in haematite of the Elatina Formation, South Australia. *Earth and Planetary Science Letters*, **105**, 355–67.

Shackleton, N.J., Backman, J., Zimmerman, H., Kent, D.V., Hall, M.A., Roberts, D.G. *et al.* (1984). Oxygen isotope calibration of the onset of ice-rafting and history of glaciation in the North Atlantic region. *Nature*, **307**, 620–3.

Siedlecka, A. & Roberts, D. (1972). A late Precambrian tilloid from Varangerhalvoya: evidence of both glaciation and subaqueous mass movement. *Norsk Geologiske Tiddskrifte*, **52**, 135–41.

Sloss, L.L. (1988). Conclusions. In *Sedimentary Cover: North American Craton*, ed. L.L. Sloss, Geological Society of America, Geology of North America Series D-2, pp. 25–51.

Sloss, L.L. (1990). Epilog. In *Interior Cratonic Basins*, eds. M.W. Leighton, D.R. Kolata, D.F. Oltz & J.J. Eidel, American Association of Petroleum Geologists Memoir 51, pp. 799–805.

Smith, A.B. (1988). Late Paleozoic biogeography of East Asia and paleontological constraints on plate tectonic reconstructions. In *Tectonic Evolution of the Himalayas and Tibet*, eds. R.M. Shackleton, J.F. Deewey & B.F. Windley, Royal Society of London, pp. 189–226.

Socci, A.D. (1992). Climate, glaciation and deglaciation, controls, pathways, feedbacks, rates and frequencies. *Modern Geology*, **16**, 279–316.

Socci, A.D. & Smith, G.W. (1987). Evolution of the Boston Basin: A sedimentological perspective. In *Basin Forming Mechanisms*, eds. C. Beaumont & A.J. Tankard, Canadian Society of Petroleum Geologists Memoir, **12**, 87–99.

Spencer, A.M. (1971). Late Precambrian glaciation in Scotland. *Memoir Geological Society of London*, **6**, 1–48.

Spencer, A.M. (1975). Late Precambrian glaciation in the North Atlantic region. In *Ice Ages: Ancient and Modern*, eds. A.E. Wright & F. Moseley, Liverpool, Seel House Press, **6**, 217–240.

Spencer, A.M. (1985). Mechanisms and environments of deposition of late Precambrian geosynclinal tillites: Scotland and East Greenland. *Palaeogeography, Palaeoclimatology, Palaeoecology*, **51**, 143–58.

Spicer, R.A. & Parrish, J.T. (1986). Paleobotanical evidence for cool north polar climates in middle Cretaceous (Albian–Cenomanian) time. *Geology*, **14**, 703–6.

Stanistreet, I.G., Kukla, P.H. & Henry, G. (1991). Sedimentary basinal responses to a Late Precambrian Wilson cycle: the Damara orogen and Nama foreland, Namibia. *Journal of Africa Earth Sciences*, **13**, 141–56.

Stauffer, P.H. & Mantajit, N. (1981). Late Palaeozoic tilloids of Malaya, Thailand and Burma. In *Earth's Pre-Pleistocene Glacial Record*, eds. M.J. Hambrey & W.B. Harland, Cambridge: Cambridge University Press, pp. 331–5.

Stauffer, P.H. & Peng, L.C. (1986). Late Paleozoic glacial marine facies in Southeast Asia and its implications. *Geological Society of Malaysia Bulletin*, **20**, 363–97.

Stewart, J.H. (1972). Initial deposits in the Cordilleran Geosyncline: evidence of a Late Precambrian (<850 M.y) continental separation. *Geological Society of America Bulletin*, **83**, 1345–60.

Stump, E. & Fitzgerald, P.G. (1992). Episodic uplift of the Transantarctic Mountains. *Geology*, **20**, 161–4.

Stupavsky, M., Symons, D.T.A. & Gravenor, C.P. (1982). Evidence for metamorphic remagnetisation of upper Precambrian tillite in the Dalradian Supergroup of Scotland. *Transactions of the Royal Society of Edinburgh, Earth Sciences*, **73**, 59–65.

Su, Q., Fullagar, P.D., Goldberg, S.A. & Martin, M.W. (1992). U-Pb zircon and Sm-Nd whole-rock ages of the plutonic rocks of the Crossnone Complex in the southern Appalachian Blue Ridge: implications for the timing of Late Proterozoic rifting. *Geological Society of America, Abstracts with Programs*, **24**, A178.

Tainton, S. & Meyer, F.M. (1990). The stratigraphy and sedimentology of the Promise Formation of the Witwatersrand Supergroup in the western Transvaal. *South African Journal of Geology*, **93**, 103–17.

Taylor, S.R. & McLennan, S.M. (1985). *The Continental Crust: Its Composition and Evolution*. Oxford, Blackwell Scientific Publications, 312 pp.

Thompson, J. (1871). On the occurrence of pebbles and boulders of granite in schistose rocks in Islay. 40th annual meeting, British Association, Liverpool Transactions, 88 pp.

Tietzsch-Tyles, D. (1989). Evidence of intracratonic Finnmarkian orogeny in central Norway. In *The Caledonide Geology of Scandinavia*, ed. R.A. Gayer, London, Graham and Trotman, pp. 47–62.

Vail, P.R., Mitchum, R.M. & Thompson, S. (1977). Seismic stratigraphy and global changes in sea-level. *American Association of Petroleum Geologists Memoir*, **26**, 83–97.

Van der Voo, R. (1988). Paleozoic palaeogeography of North America, Gondwana and intervening displaced terranes: comparisons of paleomagnetism with paleoclimatology and biogeographic patterns. *Geological Society of America Bulletin*, **100**, 311–24.

Vaslet, D. (1990). Upper Ordovician glacial deposits in Saudi Arabia. *Episodes*, **13**, 147–61.

Veevers, J.J. (1990). Tectono-climatic supercycle in the billion-year plate tectonic eon: Permian Pangean icehouse alternates with Cretaceous dispersed-continents greenhouse. *Sedimentary Geology*, **68**, 1–16.

Veevers, J.J. & Powell, C.M. (1987). Late Paleozoic glacial episodes in Gondwanaland reflected in transgressive-regressive depositional sequences in Euramerica. *Geological Society of America Bulletin*, **98**, 475–87.

Visser, J.N.J. (1987). Influence of topography on the Permo-Carboniferous glaciation in the Karoo Basin and adjoining areas, southern Africa. In *Gondwana 6*, ed. G.D. McKenzie, American Geophysical Union, Geophysical Monograph, **41**, 123–129.

Visser, J.N.J. (1989). Episodic Palaeozoic glaciation in the Cape-Karoo Basin, South Africa. In *Glacier Fluctuations and Climatic Change*, ed. J. Oerlemans, Boston, Kluwer, pp. 1–12.

Visser, J.N.J. (1900). The age of the Late Palaeozoic glacigene rocks in southern Africa. *South African Journal of Geology*, **93**, 366–75.

Visser, J.N.J. (1991). The paleoclimatic setting of the Late Paleozoic marine ice sheet in the Karoo Basin of southern Africa. In *Glacial Marine*

Sedimentation: Paleoclimatic Significance, eds. J.B. Anderson & G.M. Ashley, Geological Society of America, Special Paper 261, 181–90.

Vogt, P.R. & Tucholke, B.E. (1989). North Atlantic Ocean basin: aspects of geologic structure and evolution. In *The Geology of North America: An Overview*, eds. A.W. Bally & A.R. Palmer, Geological Society of America, pp. 53–80.

Von Brunn, V. & Gold, D.J.C. (1993). Diamictite in the Archean Pongola Sequence of southern Africa. *Journal of African Earth Sciences*, **16**, 367–74.

Walker, R.G. & Eyles, C.H. (1988). Geometry and facies of stacked shallow-marine sandier upward sequences dissected by erosion surface, Cardium Formation, Willesden Green, Alberta. *American Association of Petroleum Geologists Bulletin*, **72**, 1469–94.

Waterhouse, J.B. (1982). An early Permian cool-water fauna from pebbly mudstones in south Thailand. *Geological Magazine*, **119**, 337–55.

Weller, J.M. (1930). Cyclic sedimentation of the Pennsylvanian Period and its significance. *Journal of Geology*, **38**, 97–135.

Wernicke, B. (1985). Uniform-sense normal simple shear of the continental lithosphere. *Canadian Journal of Earth Sciences*, **22**, 108–25.

Wiebols, J.H. (1955). A suggested glacial origin for the Witwatersrand conglomerates. *Transactions Geological Society of South Africa*, **48**, 367–87.

Wilch, T.I., Lux, D.R., Denton, G.H. & McIntosh, W.C. (1993). Minimal Pliocene-Pleistocene uplift of the dry valleys sector of the Transantarctic Mountains: a key parameter in ice-sheet reconstructions. *Geology*, **21**, 841–4.

Williams, G.E. (1975). Late Precambrian glacial climate and the Earth's obliquity. *Geological Magazine*, **112**, 441–65.

Williams, G.E. (1979). Sedimentology, stable isotope geochemistry and palaeoenvironment of dolostones capping late Precambrian glacial sequences in Australia. *Journal of the Geological Society of Australia*, **26**, 377–86.

Williams, G.E. (1989). Late Precambrian tidal rhythmites in South Australia and the history of the Earth's rotation. *Journal of the Geological Society of London*, **146**, 97–111.

Windley, B.F. (1977). *The Evolving Continents*. New York, Wiley, 385 pp.

Wopfner, H. (1981). Development of Permian intracratonic basins in Australia. In *Gondwana 5*, eds. M.M. Creswell & P. Vella, Rotterdam, A.A. Balkema, pp. 185–190.

Worsley, T.R., Nance, D. & Moody, J.B. (1984). Global tectonics and eustacy for the past 2 billion years. *Marine Geology*, **58**, 373–400.

Worsley, T.R., Nance, R.D. & Moody, J.B. (1986). Tectonic cycles and the history of the Earth's biogeochemical and paleoceanographic record. *Palaeoceanography*, **1**, 233–63.

Worsley, T.R. & Nance, R.D. (1989). Carbon redox and climate controls through Earth history: a speculative reconstruction. *Palaeogeography, Palaeoclimatology, Palaeoecology* (Global and Planetary Change Section), **75**, 259–82.

Worsley, T.R. & Kidder, D.L. (1991). First-order coupling of paleogeography and CO_2, with global surface temperature and its latitudinal contrast. *Geology*, **19**, 1161–4.

Yeo, G.M. (1981). The Late Proterozoic Rapitan glaciation in the northern Cordillera. In *Proterozoic Basins of Canada*, ed. F.H.A. Campbell, Geological Survey of Canada, pp. 81–10, 25–46.

Young, G.M. (1970). An extensive Early Proterozoic glaciation in North America. *Palaeogeography, Palaeoclimatology, Palaeoecology*, **7**, 85–101.

Young, G.M. (1989). Glaciation and tectonics. *Episodes*, **12**, 117.

Young, G.M. (1991a). The geologic record of glaciation: relevance to the climatic history of Earth. *Geoscience Canada*, **18**, 100–8.

Young, G.M. (1991b). Glaciation in the context of global change. *Geological Society of America, Abstracts with Programs*, **23**, A239.

Young, G.M. (1992). Late Proterozoic stratigraphy and the Canada–Australia connection. *Geology*, **20**, 215–18.

Young, G.M. & Gostin, V.A. (1991). Late Proterozoic (Sturtian) succession of the North Flinders Basin, South Australia: an example of temperate glaciation in an active rift setting. In *Glacial Marine Sedimentation: Paleoclimatic Significance*, eds. J.B. Anderson & G.M. Ashley, Geological Society of America Special Paper, **261**, 207–23.

Young, G.M. & Nesbitt, H.W. (1985). The Gowganda Formation in the southern part of the Huronian outcrop belt, Ontario, Canada: Stratigraphy, depositional environments and regional tectonic significance. *Precambrian Research*, **29**, 265–301.

Youngs, B.C. (1975). The geology and hydrocarbon potential of the Pedirka Basin. Department of Mines, Geological Survey of South Australia, Report of Investigations, 44.

Zachos, J., Breza, J.R. & Wise, S.W. (1992). Early Oligocene ice-sheet expansion on Antarctica: stable isotope and sedimentological evidence from Kerquelen Plateau, southern Indian Ocean. *Geology*, **20**, 569–73.

Zolnai, A.I., Price, R.A. & Helmstaedt, H. (1984). Regional cross-section of the Southern Province adjacent to Lake Huron, Ontario: Implications for the tectonic significance of the Murray Fault Zone. *Canadian Journal of Earth Sciences*, **21**, 447–56.

2 Glacial-marine facies in a continental rift environment: Neoproterozoic rocks of the western United States Cordillera

PAUL K. LINK, JULIA M.G. MILLER and NICHOLAS CHRISTIE-BLICK

Abstract

Diamictite-bearing strata present in Neoproterozoic successions in the Death Valley area of southeastern California, northern Utah, and southeastern Idaho, USA record glacial-marine sedimentation. Though no radiometric dates have been obtained from these rocks, lithostratigraphic correlations suggest that they were deposited during the Sturtian (Rapitan) glacial epoch. These strata were deposited within and along the margins of a number of differentially subsiding basins that are inferred to have developed between 780 and 730 Ma, during an episode of rifting that preceded the formation of a passive continental margin in latest Neoproterozoic or early Cambrian time.

Three diamictite-bearing associations of sedimentary facies are recognized. The *massive diamictite association* is interpreted to represent rain-out and/or redeposition by sediment gravity flow, of debris derived from a partially floating or disintegrating ice sheet, perhaps near its grounding line, in the absence of bottom currents. Two bedded diamictite-bearing facies associations are recognized. The *stratified diamictite and graded sandstone association* includes heterogeneous diamictite, graded sandstone, bedded conglomerate and grit, as well as fine-grained strata including rhythmite and dropstone-bearing laminite. It is interpreted to include subwave base sediment gravity flow deposits, in part redistributed by thermo-haline currents, with a variable component of ice-rafting. Generally this facies represents more ice-distal or basinal paleogeographic settings than the *massive diamictite association*. The *diamictite and laminated sandstone association* contains heterogeneous diamictite and distinctive, texturally mature, massive to parallel-laminated sandstone beds. It is interpreted to represent ice-proximal deposits which form in the presence of subglacial or glacial-fluvial meltwater. Two non-glacial facies associations are recognized: the *carbonate, shale and sandstone association* (mainly marine), and the *cross-bedded sandstone association* (fluvial).

The general scarcity of shallow water and terrestrial facies suggests that deep-water facies had significantly enhanced preservation potential. This is due to both syn-rift fault-related subsidence and post-rift thermal subsidence, the combination of which provided basinal depositional sites. Isostatic loading by ice further depressed the sedimentary basins, and post-glacial isostatic rebound may have served to bring any shallow water or terrestrial facies which were deposited into uplifted sites where they were subject to erosion. The abundance of relatively deep-water facies in many Neoproterozoic glacial-marine successions worldwide may be due to their deposition in rifted tectonic settings in which net subsidence kept the sediment–water interface below the level of storm wavebase.

Introduction

A distinctive combination of diamictite-bearing sedimentary strata with mafic lavas and intrusive rocks is present within Neoproterozoic (1000 Ma to the base of the Cambrian (Plumb, 1991); 'Late Proterozoic' in United States Geological Survey terminology) successions of the western United States (Fig. 2.1). These rocks, which are generally included within the Windermere Supergroup, contain evidence of glacial-marine sedimentation. They correlate with similar rocks in Canada to show that Neoproterozoic glacial sediments are distributed over a north–south distance of approximately 3600 km in the North American Cordillera.

The stratigraphic continuity throughout the North American Cordillera of Neoproterozoic diamictite and volcanic rocks overlain by a thick succession of siltstone and quartzite was recognized by Crittenden and others (1972) and Stewart (1972), who interpreted the rocks as having accumulated in a rift to passive-margin setting. Recent studies of the early Paleozoic subsidence history have shown that thermal subsidence of the Paleozoic margin did not begin until between 600 and 545 Ma, at least 130 m.y. after the ~780–~730 Ma rifting event associated with deposition of the diamictites (Armin and Mayer, 1983; Bond *et al.*, 1983, 1985; Bond and Kominz, 1984; Christie-Blick and Levy, 1989; Levy and Christie-Blick, 1991*b*). It thus appears that two continental rift events, of which we are dealing with the first, affected western North America during the Neoproterozoic (Ross, 1991; Link *et al.*, 1993). We infer that subsidence during deposition of the diamictites (730–780 Ma) was driven at least in part by post-rift lithospheric cooling,

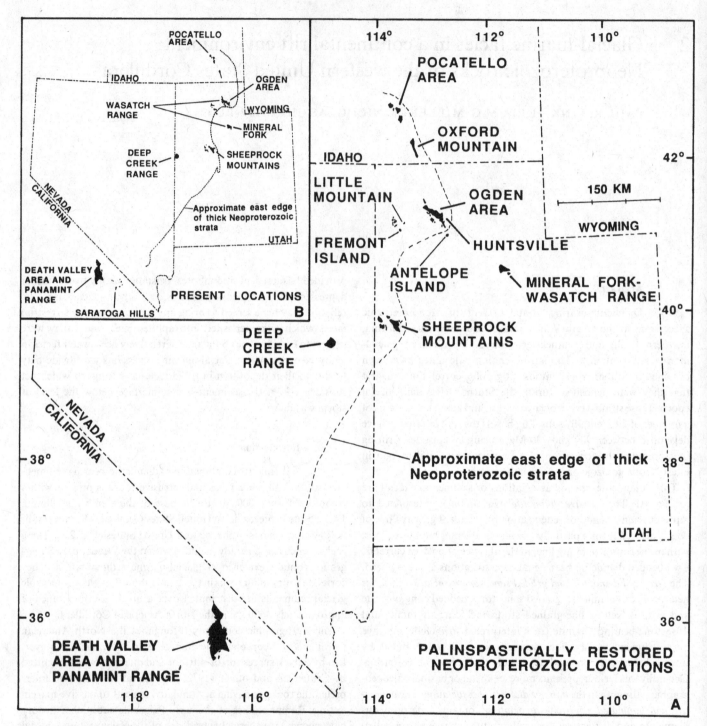

Fig. 2.1. (*a*) Palinspastically restored locations of Neoproterozoic diamictite-bearing successions of the western United States (after Levy and Christie-Blick, 1989, Figure 3). The palinspastically restored outcrop areas in some cases include Neoproterozoic rocks stratigraphically above the diamictite-bearing units, and locally Mesoproterozoic rocks stratigraphically below the diamictites. (*b*) Present locations of Neoproterozoic diamictite-bearing successions of the western United States (after Levy and Christie-Blick, 1989, Figure 2). The outcrop areas in some cases include Neoproterozoic rocks stratigraphically above the diamictite-bearing units.

but we do not know whether this was in an intracratonic setting (Levy and Christie-Blick, 1991a) or one associated with a developing passive margin that was itself rifted again during latest Proterozoic to earliest Cambrian time (e.g. Link, 1984; Miller, 1987; Ross, 1991).

The evidence for rifting at the time of diamictite accumulation is for the most part indirect. The diamictite-bearing strata contain coarse-grained and mineralogically immature sedimentary rocks which display abrupt facies and thickness changes. This suggests syndepositional faulting and proximity to margins of fault-bounded basins that formed during crustal thinning (Stewart, 1972; Stewart and Suczek, 1977; Ross, 1991). Mafic igneous rocks associated with the diamictites in both the United States and in Canada have the chemistry of tholeiitic continental rift basalts (Stewart, 1972; Harper and Link, 1986; Roots and Parrish, 1988). In northeast Washington and the Canadian Cordillera these basalts have been broadly dated within the range 780 to 730 Ma (Devlin and others, 1985, 1988; Jefferson and Parrish, 1989). Obviously, confident statements about age and chronocorrelation await availability of better geochronologic data.

The age assignment of 780 to 730 Ma places these glacigenic rocks within the second of three clusters of ages for Neoproterozoic glacial rocks cited by Hambrey and Harland (1981; 1985), Harland (1983) and Crowell (1983). This cluster, of Late Riphean age, represents the Sturtian (Rapitan) glaciation using the terminology of Harland and others (1990). Its precise age is not well known; recent summary papers place it at ~725 Ma (Knoll, 1991; Knoll and Walter, 1992).

Beginning in Late Devonian time, the United States Cordillera was subject at different times to both crustal shortening and extension, and especially in the Mesozoic to Late Cenozoic, was affected by widespread magmatic activity. The details of this deformational history are controversial, especially in the more complexly deformed areas of east-central Nevada and eastern California, and estimates of the amounts and directions of shortening and extension are subject to considerable uncertainties. The most recent attempt at a palinspastic reconstruction of the Neoproterozoic rocks was published by Levy and Christie-Blick (1989), and this reconstruction forms the basis of Fig. 2.1a.

The purposes of this paper are (1) to characterize and interpret the principal glacial facies associations that are present in Neoproterozoic rocks of Idaho, Utah, and California, with emphasis on regional variations within a continental rift setting; (2) to show how the facies associations vary with paleogeographic position; and (3) to suggest reasons why most of the glacigenic strata preserved here were deposited in relatively deep-water settings.

Important regional discussions of these strata include Stewart (1972), Eisbacher (1985) and Young (1992). Sedimentological details have been published for virtually all of the key localities. Key references include: for Utah, Hintze (1913), Blackwelder (1932), Varney (1976), Ojakangas and Matsch (1980), Blick (1981), Christie-Blick (1982a, b, 1983, 1985), Crittenden et al. (1971, 1983), Rodgers (1984), Christie-Blick and Link (1988), and Matsch and Ojakangas (1991); for Idaho, Ludlum (1942), Crittenden et al.

(1971), Link (1981, 1983, 1986, 1987), Link and LeFebre (1983), Christie-Blick and Levy (1989), and Link and Smith (1992); and for southeastern California, Hazzard (1939), Troxel (1966), Wright et al. (1976), Troxel et al. (1977), Miller et al. (1981, 1988) and Miller (1985). Correlative rocks not discussed in detail here include the Windermere Supergroup of northeastern Washington and adjacent British Columbia (Aalto, 1971; Eisbacher, 1985), the Rapitan Group of northwestern Canada (Yeo, 1981; Eisbacher, 1981, 1985) and the upper Tindir Group of eastern Alaska (Young, 1982).

Evidence for overall glacial origin

Evidence for glacigene sedimentation includes: (1) presence of thick successions of massive or crudely bedded diamictite containing both extrabasinal and intrabasinal clasts; (2) presence locally of striated and/or faceted stones in diamictite; (3) dropstones and isolated sediment pods in laminated fine-grained sedimentary rocks; (4) a striated and grooved pavement overlain by diamictite at one locality (Mineral Fork) in the central Wasatch Range, Utah; and (5) the lateral persistence of correlative diamictite-bearing strata throughout the Cordillera from eastern Alaska to eastern California (Stewart, 1972; Eisbacher, 1985).

We infer that much of the preserved rock accumulated in a marine setting, though at times and in some places close to the grounding line of the ice sheet or in a subglacial setting, and rarely in a glacial-fluvial environment. A marine rather than lacustrine setting is inferred on the basis of (1) the continuity along strike for hundreds of kilometres of successions dominated by diamictite; (2) the lack of definitive clastic varves (annual silt-clay couplets) in fine-grained laminated rocks, suggesting the presence of saline water in which flocculation suppressed the tendency for varves to develop; and (3) the detection of an apparent post-glacial eustatic sea level rise in Idaho and Utah, marked by laminated shale overlying post-glacial carbonate and sandstone at the top of the succession.

Facies associations

Three diamictite-bearing facies associations are recognized in the Neoproterozoic rocks of the western USA: *Massive diamictite*, *stratified diamictite and graded sandstone* (sandstones are generally texturally immature), and *diamictite and laminated sandstone* (sandstones are generally texturally mature). Two non-diamictite bearing associations are recognized: *cross-bedded sandstone*, and *carbonate, shale and sandstone*. The occurrences of these facies associations and their percentages in seven generalized stratigraphic sections are indicated on Fig. 2.2.

The *massive diamictite association* (Fig. 2.3) contains generally thick (up to a few hundred metres), sheetlike, massive diamictite with very few stratified interbeds. Indistinct bedding may be defined by clast-concentrated and clast-free layers, but the rocks are everywhere matrix-supported. Inverse and normal grading are occasionally visible. Stratified sandstone or argillite interbeds or lenses are rarely present, and may show normal grading or channelled bases. Stone clots and clast imbrication (Fig. 2.3b) are

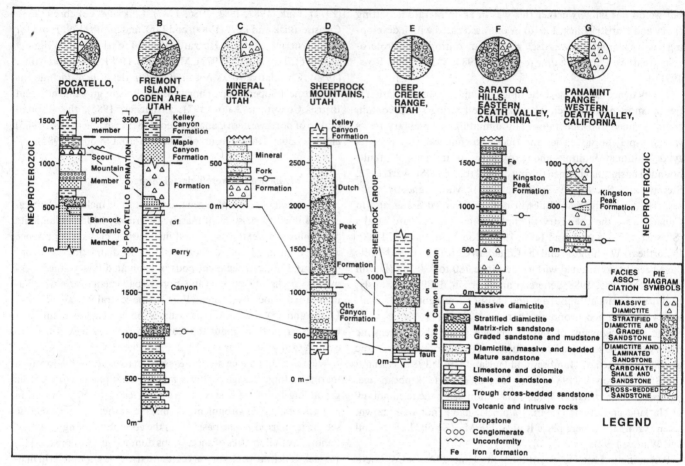

Fig. 2.2. Generalized stratigraphic columns for Neoproterozoic diamictites of the western United States. Pie diagrams above show approximate percentages of facies associations in each column. Sources for data as follows: Column A, Pocatello (Link, 1983; Link and LeFebre, 1983); Column B, Fremont Island section in lower part, Ogden area above break in section symbol (Blick, 1979; Crittenden *et al.*, 1983); Column C, Mineral Fork (Christie-Blick, 1983); Column D, Sheeprock Mountains (Christie-Blick, 1982b); Column E, Deep Creek Mountains (Rodgers, 1984); Column F, Eastern Death Valley (Miller *et al.*, 1988); Column G, Western Death Valley (Miller, 1985).

Fig. 2.3. *Massive diamictite association.* (a) Massive cleaved boulder diamictite, Scout Mountain Member, Pocatello Formation; southeast of Pocatello (Fig. 2.2, Column A, 1200 m). (b) Imbrication in cobble clasts in massive diamictite, Scout Mountain Member, Pocatello Formation, south of Oxford Mountain. Clasts are quartzites and vesicular volcanic rocks.

Fig. 2.4. *Stratified diamictite and graded sandstone association.* (a) Graded siltstone to shale beds below lower diamictite, Scout Mountain Member, southeast of Pocatello; (Fig. 2.2, Column A, 400 m). (b) Thick bedded volcanogenic diamictite with mainly clasts of Bannock Volcanic Member, Scout Mountain Member, Pocatello Formation (Fig. 2.2, Column A, 550 m). (c) Very thick bedded clast-supported cobble to boulder conglomerate, Scout Mountain Member, Pocatello Formation, southeast of Pocatello (Fig. 2.2, Column A, 850 m). (d) Graded sandstone turbidite bed with siltstone rip-ups near top. Sperry Wash, east of Saratoga Hills, Kingston Peak Formation unit 3 (Fig. 2.2, Column F, 1100 m).

present locally. The association is present in all locations except the Sheeprock Mountains and Deep Creek Range of Utah (Figs. 2.1, 2.2), and makes up 7–47% of the generalized sections.

The *stratified diamictite and graded sandstone association* (Fig. 2.4) contains medium- to thick-bedded matrix-supported diamictites (Fig. 2.4b), clast-supported conglomerates in both laterally continuous beds (Fig. 2.4c) and small pods, graded sandstones, and fine-grained rhythmites and laminites (Fig. 2.4a,d), locally with dropstones. Diamictite beds in this association may be tabular to lenticular, 10 cm to 50 m thick; they are commonly massive but may show silty wisps, diffuse bedding due to variable clast concentrations, or lenticular interbeds of sandstone and shale. Conglomerates are generally massive and disorganized (Figure 2.4c). Sandstones are generally texturally and mineralogically immature. They may be massive or show normal grading, parallel laminae, rare ripple cross-lamination, cross-bedding, dish structures and rip-up clasts. Fine-grained rocks may be massive, rhythmically bedded, or parallel-laminated. Dropstones exist in places. In some Death Valley sections, iron formation (hematitic laminated mudstone) is present. The association is present in all of the localities except Mineral Fork, making up 16–80% of the generalized sections.

The *diamictite and laminated sandstone association* (Fig. 2.5)

Fig. 2.5. *Diamictite and laminated sandstone association.* (*a*) **Dropstone in laminated siltstone, Mineral Fork Formation (Fig. 2.2, Column C, 350 m).** (*b*) **Two metre boulder of granite in diamictite, possible dropstone, upper Mineral Fork Formation (Fig. 2.2, Column C, 700 m).** (*c*) **Conglomerate lense filling an** inferred subglacial tunnel, surrounded by bedded diamictite, Mineral Fork Formation (Fig. 2.2, Column C, 450 m). (*d*) Beds of laminated, texturally mature fine-grained sandstone surrounded by cobble to boulder diamictite, Mineral Fork Formation (Fig. 2.2, Column C, 500 m).

contains stratified diamictite (Fig. 2.5*a*,*d*), with contorted bodies of fine-grained sediment and lenses of pebble to boulder conglomerate (Fig. 2.5*c*), distinctive evenly interstratified black laminated mudstone and medium to very fine-grained sandstone containing sparse dropstones (Fig. 2.5*a*), and texturally mature parallel-laminated medium and coarse-grained sandstone (Fig. 2.5*d*). The *diamictite and laminated sandstone association* makes up the bulk (80%) of the Mineral Fork Formation, and is present in all other localities, making up 16–38% of those sections. It is especially important in the Sheeprock Mountains and Deep Creek Range of Utah.

The *carbonate, shale and sandstone association* (Fig. 2.6) contains a variety of fine- to coarse-grained strata, lacking an obvious imprint of glacial sedimentation. Carbonates are generally laminated (Fig. 2.6*a*), from one to several tens of metres thick, and include limestones and dolomites. Mudrocks are variously laminated, cross-laminated, and graded. Fine-, medium-, and rarely coarse-grained sandstones (Fig. 2.6*c*,*d*) are included in the association. The association is present in all locations except Mineral Fork, making up 16–50% of the sections.

The *cross-bedded sandstone association* (Fig. 2.7) contains

medium- to large-scale trough cross-bedded medium- to coarse-grained sandstone, locally pebbly, but generally texturally submature to mature. The association is only recognized near the top of the glacial succession in the Panamint Range, Death Valley area, where it is unconformably overlain by massive diamictite.

Review of stratigraphy

Diamictite-bearing successions are exposed discontinuously in the Death Valley region of eastern California, in western and northern Utah, and in southeastern Idaho (Fig. 2.1). All of the outcrops are within the Mesozoic to early Cenozoic Cordilleran thrust and fold belt, and in most cases the rocks were displaced by thrusting in excess of 100 km in an approximately eastward direction from their sites of deposition (Fig. 2.1a; Levy and Christie-Blick, 1989). The least-thrusted strata are present in the Mineral Fork Formation at Mineral Fork of Big Cottonwood Canyon in the central Wasatch Range of northern Utah. These rocks are parautochthonous with respect to the continental interior of North America, and displaced no more than a few kilometres.

Fig. 2.6. *Carbonate, shale and sandstone association.* (*a*) Parallel-laminated Sourdough Limestone Member, Kingston Peak Formation, Panamint Range (Fig. 2.2, Column G, 650 m). (From Miller, 1985, Figure 14.) (*b*) Breccia of clasts of laminated carbonate (now dolomite), unconformably above upper diamictite, Scout Mountain Member, Pocatello Formation, southeast of Pocatello (Fig. 2.2, Column A, 1150 m). (*c*) Low-angle planar cross beds interpreted to represent beach (foreshore) environment. Hammer handle rests on bedding plane, with foresets above the hammer head; upper Scout Mountain Member, at Little Gap, 20 km southeast of Pocatello (Fig. 2.2, Column A, 1250 m). (*d*) Inverse and normal grading in granule conglomerate, lower part of South Park Member, Kingston Peak Formation, Panamint Range (Fig. 2.2, Column G, 700 m).

Most of the outcrops are located also within the late Cenozoic Basin and Range extensional province. Depending on their position with respect to the eastern boundary of this province, the rocks were displaced by extensional faulting in an approximately west-north-west direction by as little as a few kilometres to as much as many tens of kilometres. Rocks of the Mineral Fork area were least affected by this deformation, and those in the Death Valley region most affected. In the latter area, the amount of extension is so large that palinspastic reconstruction (Fig. 2.1*a*) superposes many of the outcrop shapes evident on the present-day map (Fig. 2.1*b*). Uncertainties in reconstruction associated with local structural details make the interpretation of paleogeography in the Death Valley region problematic.

Fig. 2.7. *Cross-bedded sandstone association.* Trough cross-bedding in upper South Park Member, Kingston Peak Formation, Panamint Range (Fig. 2.2, Column G, 900 m). (From Miller, 1985, Figure 16.)

Death Valley region of southeastern California,
(Kingston Peak Formation)

In the structurally complex Death Valley region of south-eastern California (Fig. 2.1; Fig. 2.2, columns F and G), diamictite is present within the Kingston Peak Formation, the upper formation of the Pahrump Group. The Kingston Peak Formation ranges in thickness from 40 to over 2000 m and shows a different internal stratigraphy in different parts of the Death Valley area (Miller *et al.*, 1981, 1988; Miller, 1985). Syndepositional tectonism is indicated by abrupt thickness changes, lateral variability in stratigraphy, inferred buried faults, and large allochthonous blocks within the sedimentary section.

The Kingston Peak Formation is overlain by the Noonday Dolomite or by correlative argillite, sandstone and limestone of the Ibex Formation (Wright and Troxel, 1984). In places in the Panamint Range, basal Noonday Dolomite strata are interbedded with the upper Kingston Peak diamictite (Miller, 1987). East of Death Valley the Kingston Peak–Noonday contact is conformable in southern sections but angularly unconformable elsewhere (Troxel *et al.*, 1987; Miller *et al.*, 1988), demonstrating post-depositional tectonism.

In the Panamint Range (Fig. 2.1*b*; Fig. 2.2, column G) a lower diamictite unit (Surprise Member), over 400 m thick, belongs to the *massive diamictite association.* It is dominantly structureless, with sparse bedding and rare sandstone interbeds. It locally contains at least 50 m of, as yet undated, pillow basalt. It overlies either laminated shale with diamictite and sandstone interbeds of lower parts of the Kingston Peak Formation (*stratified diamictite and graded sandstone* and *diamictite and laminated sandstone* associations) or the subjacent Beck Spring Dolomite. The thinly laminated Sourdough Limestone Member (*carbonate, shale and sandstone association*, Fig. 2.6*a*), deposited in relatively deep water during a flooding event, abruptly overlies massive diamictite of the Surprise Member. The upper, South Park Member (up to 300 m thick) contains (1) an upward-fining sequence of interbedded graded sandstone, siltstone and limestone (*carbonate, shale and sandstone association*, strata in Fig. 2.6*d* are near the base), abruptly overlain by (2) trough cross-bedded pebbly sandstone (*cross-bedded sandstone association*, Fig. 2.7), and (3) unconformably overlying massive diamictite (*massive diamictite association*) 0 to 190 m thick. The erosive base, cutting into and through fluvial sandstone, and the irregular shape of the upper diamictite in the Panamint Range make it distinct from other massive diamictites in these Cordilleran successions. Miller (1987) interprets this diamictite as a terrestrial lodgement till.

In the eastern Death Valley area, the Kingston Peak Formation comprises a northern and a southern facies interpreted by some to have been deposited on opposite sides of a fault-bounded basin or graben (Wright, 1974; Wright and Troxel, 1984; Wright *et al.*, 1976; Troxel, 1966, 1982; Walker *et al.*, 1986; Miller *et al.*, 1988). Column F of Fig. 2.2 represents the northern facies, in the Saratoga Hills. The lower part of the formation contains siltstone and sandstone with rare thin carbonate interbeds (*carbonate, shale and sandstone*

association) (Tucker, 1986; Miller *et al.*, 1988). The middle member contains the *massive diamictite association*, 50 to 370 m thick, with minor sandstone and argillite with isolated stones. The upper member thickens southward from 60 to about 2000 m. It contains *stratified diamictite and graded sandstone association* (laminated siltstone with dropstones, sandstone turbidites (Fig. 2.4*d*), stratified diamictite, conglomerate, breccia beds, and some hematitic laminated mudstone). In eastern parts of the Death Valley region within the upper member, there is a southward facies change from thin conglomerate and breccia, to thick (over 1000 m) interbedded siltstone and graded sandstone. The upper member locally contains a thick section of undated volcanic flows, breccia, and tuff in the eastern Kingston Range (east of the Saratoga Hills, Fig. 2.1*b*). The southern facies, deposited on the southern side of the basin, contains argillite with isolated stones, graded sandstone, stratified diamictite, conglomerate and some breccia (Basse, 1978).

The Kingston Peak Formation in the Panamint Range is interpreted to record two ice advances, with associated sea level variations (Miller, 1985). However, correlation between Kingston Peak sections is uncertain (Miller *et al.*, 1988). The thick diamictite unit of the eastern Death Valley region may correlate with either diamictite unit in the Panamint Range.

Central Wasatch Range and Antelope Island, northern
Utah (Mineral Fork Formation)

The Mineral Fork Formation crops out discontinuously from Antelope Island in the Great Salt Lake to the southern part of the Wasatch Range (Christie-Blick, 1983). Bounded above and below by unconformities, the formation ranges in thickness from a few to about 800 m. At its type locality in Mineral Fork of Big Cottonwood Canyon, central Wasatch Range (column C of Fig. 2.2; Fig. 2.5), it occupies two glacially eroded valleys, one about 400 m deep and the other as much as 900 m deep (Crittenden *et al.*, 1952; Crittenden, 1965*a,b*; Christie-Blick, 1983; Christie-Blick and Link, 1988; Christie-Blick and Levy, 1989). The lower contact is locally striated and characterized by features interpreted as roches moutonnées (Ojakangas and Matsch, 1980; Christie-Blick, 1982*a*; 1983). Overall, the Mineral Fork Formation is composed of bedded and locally massive diamictite (Fig. 2.5*b,d*), containing contorted bodies of fine-grained sediment and lenses of pebble to boulder conglomerate (Fig. 2.5*c*), with lesser proportions of sandstone and siltstone (Ojakangas and Matsch, 1980; Christie-Blick, 1983). The diamictite contains rare sandy or silty inhomogeneities or pebble layers (Fig. 2.5*d*); in places it is intimately interstratified with sandstone and siltstone (Fig. 2.5*d*). Sandstone, forming laterally discontinuous and in places channelized units, ranges from massive to laminated, and is rarely cross-stratified. Nearly 15% of the formation is composed of rhythmically interstratified graded sandstone and finely laminated siltstone and sandstone, locally with dropstones (Fig. 2.5*a*).

The Mineral Fork Formation represents the type example of the *diamictite and laminated sandstone* facies association. In places, where the diamictite is many tens of metres thick, there is an obvious gradation into the *massive diamictite association*. In other places, where sandstones are graded and texturally immature, the association grades into *stratified diamictite and graded sandstone*.

Our preferred interpretation for the Mineral Fork Formation (Christie-Blick, 1983; Christie-Blick and Link, 1988; Christie-Blick and Levy, 1989) is that it represents sedimentation in a temperate glacial-marine environment, close to the grounding line of a partially buoyant ice sheet. The best evidence for grounding is syndepositional deformation and abundant well-stratified sandstone interpreted as subaqueous outwash (Fig. 2.5*d*). The bulk of the diamictite appears to have accumulated from a combination of ice-rafting and resedimentation by sediment gravity flow and downslope sliding. The depositional environment contained abundant meltwater and powerful water flow, suggesting it was close to the ice margin (Fig. 2.5*c*). The lack of major internal erosion surfaces indicates deposition during a single overall glacial retreat, although with many minor oscillations of the ice margin.

Ojakangas and Matsch (1980) and Matsch and Ojakangas (1991) have interpreted the Mineral Fork Formation in its type area as representing an overall transition from glacial-terrestrial (lower two-thirds) to glacial-marine conditions (the remainder), and suggest that the inferred relative sea-level rise of more than 800 m required by this interpretation was due largely to long-term subsidence of the sedimentary basin. The main arguments against the terrestrial interpretation for the lower part are (1) the amount of subsidence required is implausibly large for a platformal succession in which the total thickness of the entire younger Neoproterozoic is in most places less than 1000 m; (2) the fact that the sandstones are largely massive to parallel-laminated and not cross-stratified, as would be expected in a terrestrial setting; (3) the absence of clastic varves in fine-grained sediments, a feature that might be expected if any of the succession was lacustrine; (4) the presence of microfossils of probable marine origin in both marine and 'terrestrial' units (Knoll *et al.*, 1981); and (5) the fact that rhythmically stratified siltstone and sandstone with dropstones, and clearly of subaqueous aspect, can be mapped laterally down depositional dip into supposedly terrestrial outwash.

Fremont Island, Little Mountain and Ogden area, northern Utah (formation of Perry Canyon)

The formation of Perry Canyon, exposed for 25 km along strike in the Willard thrust plate, contains both massive and stratified diamictite interbedded with laminated shale which locally contains dropstones and till clots (column B of Fig. 2.2) (Blick, 1979; Crittenden *et al.*, 1983). Basaltic tuff and pillow basalt are present locally. In the thickest exposed section, on Fremont Island in the Great Salt Lake (Fig. 2.1*a*), two glacial stratigraphic units are separated by 1000 m of non-glacial deposits (*carbonate, shale and sandstone association*). On Fremont Island the lower glacial horizon contains mainly slaty argillite and sandy turbidites, with dispersed sand particles and isolated dropstones (pebbles to boulders up to 2.5 m in diameter; *stratified diamictite and graded sandstone association*). It rests gradationally upon an upward-fining succession of

coarse- to fine-grained turbidite sandstones and shales. In the Ogden area the lower horizon is primarily stratified diamictite. An upper interval of the *massive diamictite association* (up to 600 m thick), containing mainly massive diamictite, with minor graded sandstone, conglomerate and laminated shale, is intercalated at Little Mountain with undated basaltic volcanic rocks, including pillow basalt (Crittenden *et al.*, 1983; Christie-Blick, 1985; Harper and Link, 1986). The upper contact of diamictite is abrupt to graywacke and siltstone (*carbonate, shale and sandstone association*) where exposed at Perry Canyon northeast of Ogden.

Sheeprock Mountains and Deep Creek Range, western Utah (Sheeprock Group and Horse Canyon Formation)

Thick successions of diamictite are present in the Sheeprock Group of the Sheeprock Mountains (column D of Fig. 2.2) and Horse Canyon Formation of the Deep Creek Range, western Utah (column E of Fig. 2.2) (Blick, 1979; Christie-Blick, 1982*b*; Crittenden *et al.*, 1983; Rodgers, 1984). In the Sheeprock Group, stratified diamictite is present in the middle part of the Otts Canyon Formation, where it overlies several hundred metres of black slaty argillite and underlies with interfingering contact as much as 1000 m of quartzite with minor conglomerate, siltstone and diabase sills. The overlying Dutch Peak Formation is composed primarily of diamictite, conglomerate, sandstone and siltstone. Several hundred metres of quartzite are present in the upper part of the formation in the western part of the Sheeprock Mountains, where it passes eastward into sandy diamictite containing tabular to lenticular bodies of quartzite. Throughout the Sheeprock Group, the diamictite tends to be compositionally and texturally heterogeneous, and for the most part corresponds to the *stratified diamictite and graded sandstone association*. In the upper part of the Dutch Peak Formation, diamictite is associated with mature quartzite that is predominantly parallel-laminated. These rocks are assigned to the *diamictite and laminated sandstone association*.

Diamictite crops out in two units in the much-deformed Horse Canyon Formation of the Deep Creek Range, separated by about 200 m of quartzite (column E of Fig. 2.2), and these two units are thought to correlate with the two diamictite units of the Sheeprock Mountains (Christie-Blick, 1983; Rodgers, 1984). The diamictite is heterogeneous, associated with a range of non-diamictic clastic facies, and is assigned to the *stratified diamictite and graded sandstone facies association*.

Interpretation of sedimentary facies in the Sheeprock Mountains is problematic. Crittenden and co-workers (1983) inferred deposition near the margin of a locally subsiding marine embayment, and interpreted the quartzites as shallow marine. Although stratification in the diamictite of the upper Dutch Peak Formation resembles that of the Mineral Fork Formation, there is no evidence in the Sheeprocks that the ice was ever grounded. The associated quartzite in both the Otts Canyon and Dutch Peak Formations is similar lithologically and in terms of included structures to that in the Mineral Fork Formation, but is much thicker. We tentatively propose the quartzite in the Sheeprock Mountains may have accumulated in a glacial-fluvial to braid-delta setting, and that the overall stratigraphic relations are consistent with two large-scale shoaling-upward successions, one corresponding approximately with the Otts Canyon Formation and the other with the Dutch Peak Formation.

If this interpretation is correct, and the succession in the Deep Creek Range is similarly interpreted, it implies that towards the end of the glaciation, fluvial conditions may have persisted much farther west than was thought by Crittenden and colleagues (1983). The absence of fluvial facies in the Mineral Fork Formation, paleogeographically far to the east, suggests that fluvial sediments may once have been deposited but were subsequently removed by erosion.

Southeastern Idaho (Pocatello Formation)

In the Pocatello Formation of southeastern Idaho (column A of Fig. 2.2), diamictite and intercalated basaltic volcanic rocks are present for 100 km north–south along strike, from the Pocatello area south to Oxford Mountain, near the Utah border (Fig. 2.1a). The Bannock Volcanic Member (200–450 m thick) contains greenstone, agglomerate, pillow breccia and pillow lava, plus a body of metadiabase or metagabbro. The chemistry of mafic rocks from the formation of Perry Canyon and the Bannock Volcanic Member indicates that they are tholeiitic; their trace element signatures resemble continental rift basalts (Harper and Link, 1986). They have not yet been successfully dated radiometrically.

Within the Scout Mountain Member, a lower interval of *stratified diamictite and graded sandstone association* contains medium- to thick-bedded diamictite interbedded with graded sandstone (Fig. 2.4a,b). A prominent thick-bedded clast-supported cobble conglomerate (Fig. 2.4c) is present in the middle of the member and is interpreted as the outwash of a subglacial meltwater stream which filled a submarine channel, below wave base. An upper interval of *massive diamictite association* (Fig. 2.3a,b) is present above the conglomerate. A thin cap carbonate lies directly on the upper diamictite and is unconformably overlain by a sedimentary breccia (Fig. 2.6b) containing angular clasts of the carbonate. This breccia is succeeded by an upward fining, shallow marine, transgressive succession of the *carbonate, shale and sandstone association* containing texturally mature sandstone, flaser-bedded siltstone and laminated limestone. Fine-grained strata at the top of the Scout Mountain Member south of Pocatello pass southward 15 km along the outcrop trend into low-angle tabular cross-bedded sandstone interpreted to represent a beach environment (Fig. 2.6c). Laminated shale of the upper member of the Pocatello Formation, which correlates with the Kelley Canyon Formation in Utah, caps the succession, suggesting integration of basins for at least 400 kilometres north–south.

The main part of the Scout Mountain Member is interpreted to represent subwave base deposition by sediment gravity flows and rainout from disintegrating ice. The stratigraphic relations above the upper diamictite are interpreted by one of us (PKL) to indicate

deposition of laminated carbonate, possibly in deep water, following retreat of the ice sheet, followed by isostatic uplift caused by removal of the ice load which produced a local unconformity. The succeeding upward fining transgressive succession may be the playa to subwave base deposits produced by the post-glacial eustatic rise in sea level. A partly lacustrine setting cannot be disproven. This explanation suggests that the ice sheet was grounded and produced isostatic depression within tens of kilometres of the Pocatello area. The presence of local source areas, inferred from abrupt changes in clast composition, also suggests that grounded ice was close by (Link, 1983, 1986).

Interpretation of facies associations

Massive diamictite facies association: ice-proximal nonreworked deposits

Massive diamictite indicates ice-proximal deposition in the absence of sorting by meltwater or sediment gravity flow. Interpretation of massive diamictite within a glacial-marine setting has a contentious history. Two interpretations have recurred over the past 10 years: (1) the massive diamictite represents grounding line deposits, deposited during glacial retreat in the absence of significant volumes of meltwater, and thus probably by a polar ice sheet; or (2) the massive diamictite represents ice-rafted deposits which accumulate during very rapid sedimentation by suspension fallout, during disintegration of an ice shelf.

The first interpretation, that the diamictite is a grounding line deposit, implies that its lateral extent is caused by lateral movement of the grounding line. This interpretation is based on the observation that in the Antarctic, till near the grounding line of polar ice sheets is massive and structureless. These deposits have been called 'basal till' or 'transitional glacial-marine sediments' (Anderson et al., 1983, 1991). 'Basal till' implies deposition directly from grounded ice when the ice sheet extended over and was grounded on the continental shelf. 'Transitional glacial-marine sediments' describes deposits in a zone transitional between that of a fully grounded ice sheet and a buoyant ice shelf. Deposition occurs in a narrow zone adjacent to the ice shelf grounding line where basal debris melts out of the ice and circulation is sluggish (Anderson et al., 1983, 1991). Several papers by the present authors (Link, 1983; Christie-Blick, 1983, 1985; Miller, 1985) advocated this interpretation for massive diamictite within the Neoproterozoic successions of the United States Cordillera. Some recent syntheses of Paleozoic and Neoproterozoic diamictites also advocate a grounding line origin for *massive diamictite* (Moncrieff and Hambrey, 1990; Matsch and Ojakangas, 1991). The absence of stratified interbeds in a grounding line setting indicates an absence of meltwater, and so implies either very rapid deposition or dry-based glacier or polar conditions. Polar conditions raise paleogeographic problems and are not compatible with the evidence for presence of abundant meltwater in the *diamictite and laminated sandstone association*.

The second interpretation, that of rapid rain-out below icebergs or a disintegrating ice shelf in a temperate or subpolar regime, has been advocated for several pre-Pleistocene examples of thick *massive diamictite*. In the Permo-Carboniferous Dwyka Formation of the Karoo basin of southern Africa, Visser (1983,*a,b*, 1989, 1991) interprets massive clast-poor diamictite as laid down very rapidly at the end of the glaciation as sea-level rose and the ice sheet became decoupled and self-destructive. Once an ice sheet starts to disintegrate, caused perhaps by eustatic rise in sea level, the process proceeds rapidly, itself triggering further sea-level rise (Anderson and Thomas, 1991). In the 5 km thick Miocene to Pleistocene Yakataga Formation in the Gulf of Alaska (Armentrout, 1983; C. Eyles *et al.*, 1985, 1991; C. Eyles, 1987; N. Eyles and Lagoe, 1989; C. Eyles and Lagoe, 1990), thick rain-out diamictites are inferred to represent a mixture of coarse debris from icebergs and muds from suspended sediment plumes. In the Neoproterozoic (Sturtian) succession of South Australia, Young and Gostin (1988, 1989, 1990, 1991) interpret massive diamictite to have been deposited by rain-out during times when abundant floating ice may have clogged surface waters in a narrow rift basin, precluding development of marine bottom currents. Link and Gostin (1981) had previously interpreted some of the same massive diamictites as subaqueous basal tillite formed near the grounding line of an ice sheet.

Under either hypothesis for its origin, *massive diamictite* implies proximity to ice that is laden with basal debris, and requires deposition in an environment free of bottom currents. The two inferred mechanisms of deposition are not mutually exclusive, and we list them both as possible origins.

Facies context of the *massive diamictite* is critical, especially in regards to presence of the *diamictite and laminated sandstone association*, containing meltwater deposits, the *cross-bedded sandstone association*, implying fluvial deposition, and carbonate rocks of the *carbonate, shale, and sandstone association*. The carbonate cap which abruptly overlies massive diamictite in the Pocatello and Panamint Range sections indicates that terrigenous sedimentation terminated very rapidly. This could be caused by decoupling and rapid disintegration of an ice sheet or shelf due to sea-level rise, suggesting a rain-out origin for the *massive diamictite association*. The stratigraphic position of the upper diamictite in the Panamint Range, unconformably occupying erosional troughs in cross-bedded sandstone, suggests an origin as terrestrial lodgement till (Miller, 1987).

Stratified diamictite and graded sandstone association: sediment gravity flow deposits

Stratified diamictite and graded sandstone is the most voluminous diamictite-bearing facies association present in the Neoproterozoic glacial strata of the western United States. The association includes both very coarse-grained strata deposited proximal to the ice sheet and fine-grained strata with a minor ice-rafted component, deposited in ice-distal settings. This association grades into essentially non-glacial deposits of the *carbonate, shale and sandstone association*. It is distinguished from the *diamictite and laminated sandstone association* mainly by the textural maturity

(less mature) and sedimentary structures of the associated sandstones (indicating sediment gravity flow rather than tractive currents). However, in cases where sandstones are uniformly immature, or where exposure is limited, as in drill core, the facies associations may become indistinguishable.

The *stratified diamictite and graded sandstone association* is interpreted as reworked glacial-marine sediment formed under conditions where meltwater or thermo-haline bottom currents were present, and deposited below wave base by sediment gravity flows, locally with a component of ice-rafted detritus. It probably includes both residual and compound glacial-marine sediment as classified in the Antarctic (Anderson *et al.*, 1983), that is, unsorted gravel and sand with minor silt and clay, and mud with some ice-rafted debris. Both of these sediment types were deposited on the continental shelf after the ice retreated. However, in the western USA this association displays a much greater influence of redeposition and downslope movement than sediments on the Antarctic shelf. The tectonic setting of these Neoproterozoic successions in actively subsiding rift basins could explain this difference.

Diamictite and laminated sandstone association: ice-proximal deposits with abundant meltwater

The *diamictite and laminated sandstone association* is distinguished by the presence of well-sorted sandstone that is locally parallel-laminated, interbedded with heterogeneous facies including diamictite and conglomerate. It is interpreted as an ice-proximal facies in which meltwater was a major depositional factor, generally deposited during the final stages of the glaciation (Christie-Blick, 1983). Texturally mature sandstone is present in most of the outcrop areas, suggesting glacial-fluvial input.

Diamictite and laminated sandstone of the Mineral Fork Formation is interpreted to represent sediments deposited in a narrow glaciated trough by a combination of sediment gravity flow, downslope sliding, and possible ice-contact deformation. Lenses and pods of boulder conglomerate suggest deposits of powerful water flows close to the ice margin. Distinctive evenly interstratified black laminated mudstone and medium to very fine-grained sandstone containing sparse dropstones are interpreted as a combination of distal turbidites and cyclopsams (sandstone laminae lacking ripples and derived by rain-out from buoyant plumes; Mackiewicz *et al.*, 1984; Powell, 1988). Meltwater entered this environment as high-concentration sheet underflows and sediment gravity flows that emanated from subglacial tunnels. The lack of significant sedimentation from waning flows under lower flow regime conditions is consistent with the generation of buoyant plumes where fresh water emerged below sea level from tunnels in the ice.

In the Sheeprock Mountains, diamictite of the *diamictite and laminated sandstone association* is inferred to have been deposited by a combination of ice-rafting and sediment gravity flow. Associated texturally mature quartzites are interpreted as glacial-fluvial deposits. Stratigraphic relations here illustrate the transitional boundary between *diamictite and laminated sandstone* and *stratified diamictite and graded sandstone* associations.

Carbonate, shale and sandstone association: non-glacial marine deposits

Carbonate is present in small amounts in most locations. A thin (∼1 m) laminated dolomite is present immediately above massive diamictite in the upper part of the Scout Mountain Member, Pocatello Formation (column A of Fig. 2.2). Laminated limestone, up to 45 m thick, is also present above, between or below diamictite units in most Kingston Peak sections (columns F and G of Fig. 2.2; Fig. 2.6a), and is present between the two diamictite horizons at Fremont Island and the Odgen area (column B of Fig. 2.2). These carbonates may represent condensed sections deposited in deep water following eustatic sea-level rise and removal of the clastic sediment supply, at the termination of a glacial epoch (Miller, 1985; Tucker, 1986; Domack, 1988; Christie-Blick and Levy, 1989). Several recent studies of the carbonate–glacial-marine association (Fairchild et al., 1989; Fairchild and Spiro, 1990; Fairchild, 1993) suggest that such precipitation must be accompanied by significant warming of water temperatures if comminuted carbonate rock flour is not present in the diamictites themselves. Such rock flour may have been present in the Mineral Fork and Kingston Peak formations, but probably not in the Pocatello Formation, where carbonate clasts are not present in diamictite.

Terrigenous detrital rocks in the *carbonate, shale and sandstone association* represent a spectrum of marine environments, both above and below wave base. Laminated mudrocks included in this association are deposits from suspension settling, with some reworking by bottom currents. In the Kingston Peak Formation, sandstones included in this association were deposited by turbidites. The upper part of the Pocatello Formation, above any evidence of glacial influence, contains an upward-fining succession of sandstone to shale. The succession lies above an unconformity in the upper Scout Mountain Member, interpreted to be due to post-glacial isostatic rebound. The succession may contain fluvial or playa deposits at the base (Fig. 2.6b), but mainly consists of shallow marine facies (Fig. 2.6c; Link, 1983).

Cross-bedded sandstone association: fluvial deposits

Cross-bedded sandstone is an important lithofacies near the top of the Kingston Peak Formation in the Panamint Range (column G of Fig. 2.2). This pebbly, coarse-grained, trough cross-bedded sandstone (Fig. 2.7) is interpreted as a braided stream deposit, possibly glacial outwash.

Discussion: facies distribution, subsidence mechanisms, unconformities and preservation potential in rift settings

The distribution of facies associations with respect to paleogeographic and glacier margin settings is shown in cartoon form in Fig. 2.8. The diagram is not meant to show palinspastic position. The diagram shows a time of maximum glaciation but during early stages of glacial retreat, as sediment is being deposited rapidly.

The following discussion attempts to integrate the proportion of various facies associations represented at the seven localities (Fig. 2.2) with the depositional models of Fig. 2.8. An ice-proximal eroded trough, similar to the Mineral Fork setting, is suggested on the right side of Fig. 2.8 and accumulated mainly the *diamictite and laminated sandstone association*. The Sheeprock Mountains area contains similar facies to Mineral Fork, but in a very different paleogeographic setting, well out into the rifted basins. Ice sheets there may have scoured an intra-rift high area and were close to the mouth of a major river, which delivered texturally mature quartzose sediment (center right of diagram). Several areas including Pocatello, Ogden area, and Death Valley are represented by the ice-proximal and ice-distal deposits suggested at the middle front of the diagram. These successions contain *massive diamictite* deposited from rain-out and *stratified diamictite and graded sandstone* deposited in deeper water toward the center of the basin. The Pocatello area, with glacial erosion of young basaltic shield volcanoes as well as glacial transport of debris eroded from distant source areas, is shown in the middle left of Fig. 2.8. Fig. 2.8 illustrates the essential difference in the depositional setting of the *stratified diamictite and graded sandstone association* versus the *diamictite and laminated sandstone association*. The former represents mainly sediment gravity flow below wave base while the latter represents glacio-fluvial and braid-delta processes.

Subsidence mechanisms

The largest part of the preserved Neoproterozoic glacial successions of the western USA contains basinal strata. Shallow water facies were probably deposited, but subsequently eroded. In general, sediments deposited below wave base are preferentially preserved from subsequent erosion (Bjorlykke, 1985), but more importantly this preservation demonstrates that during deposition of the glacial successions the balance of subsidence rates and sea-level change produced net subsidence to depths below wave base. Net subsidence of depositional sites is ascribed to a combination of fault-controlled and thermally induced mechanisms. Subsidiary components of glacio-isostatic and sediment loading were also present.

Generally in rift settings vertical crustal movements have greater amplitude than succeeding relative sea-level change and thick successions may be preserved. In many Phanerozoic examples these successions are terrestrial or lacustrine, but such facies are sparse in the Neoproterozoic of the United States Cordillera. One reason for this may be that subsidence of the 780–730 Ma basins may have begun during a time of relatively high sea-level, therefore prior to glaciation and to input of large volumes of terrigenous sediment. In four locations (Pocatello, Ogden, Sheeprock Mountains and locally in the Death Valley area), the lowest part of the diamictite-bearing strata contains deep-water facies. Glacial-eustatic drawdown may have subsequently allowed coarse-grained sediment to prograde into the rifting basins.

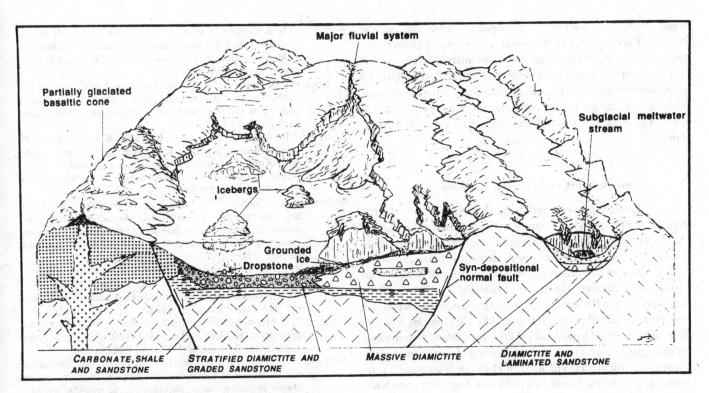

Fig. 2.8. Block diagram showing glacial-marine environments within a tectonic regime of continental rifting. This spectrum of facies associations is represented and preserved in the Neoproterozoic strata of the western United States. Locations of deposition of the *massive diamictite*, *stratified diamictite and graded sandstone*, *diamictite and laminated sandstone*, and *carbonate, shale and sandstone* associations are shown. No geographic context or scale is implied. See text for discussion. (Diagram drawn by H.T. Ore, Idaho State University.)

Relation of facies to tectonic position

A few clear relationships exist between position on the palinspastic map (Fig. 2.1a) and the observed facies associations.

The palinspastically restored Mineral Fork locality of the central Wasatch Range lies significantly closer to the continent than all other outcrops. In this locality, tectonic subsidence caused by rifting was less significant than in more basinal areas to the west. The thin and ice-proximal facies association there was deposited instead in a basin scoured and deepened by glacial ice (Fig. 2.8, right side).

In three general areas (Pocatello, Ogden, and Death Valley) abundant deep-water strata, mainly of the *stratified diamictite and graded sandstone association* suggest an important component of rift-related subsidence. Both thermal- and fault-related subsidence may have contributed to the importance of deep-water deposition in these rift basins. The tectonic subsidence, related to crustal extension, may have been accentuated by isostatic depression by the ice sheet, as suggested in Fig. 2.8.

The Sheeprock Mountains and Deep Creek Range plot near the geographic center of exposed parts of the Neoproterozoic strata, and contain inferred fluvial to braid-delta deposits of the *diamictite*

and laminated sandstone association. This suggests that an intra-rift high was present, and brings into question more simplistic paleogeographic models which show facies becoming deeper to the west (Crittenden *et al.*, 1983).

Generation of unconformities

The unconformable upper contact of the Mineral Fork Formation, and perhaps the unconformity in the Pocatello Formation that separates the carbonate cap of the upper diamictite from transgressive deposits of the upper part of the Scout Mountain Member, may have formed during isostatic rebound. Studies of post-Pleistocene rebound in Greenland suggest that generally the timing of post-glacial rebound lags the glacial-eustatic rise, but in the vicinity of the ice sheet the rate of rebound is more rapid than the rate of eustatic sea-level rise on timescales of thousands of years (Boulton and others, 1982). A comparable scenario, followed by a major eustatic rise, may explain unconformities near the top of the Pocatello Formation and above the Mineral Fork Formation. The presence of these unconformities implies that these localities lay proximal to the ice sheet.

A second mechanism for development of unconformities and

removal of shallow-water strata, especially applicable to the Mineral Fork locality, is Neoproterozoic sea-level change that exceeded net tectonic subsidence during the interval ~700 to ~600 Ma. Lack of accommodation space at Mineral Fork relative to more western locations (Pocatello, Ogden) may have been caused by less rapid tectonic subsidence subsequent to the 780–730 Ma rift event. This is consistent with the more cratonal location of Mineral Fork.

Geologic setting of the Rapitan glaciation

Limited paleomagnetic evidence suggests that North America maintained a low to medium latitude position throughout the Neoproterozoic (Chumakov and Elston, 1989; Hoffman, 1991; Elston *et al.*, 1993). The association of laminated marine carbonate with glacial-marine strata suggests rapid warming of ocean water temperatures between glacial stades (Fairchild, 1993). Furthermore, evidence of the Rapitan (Sturtian) glaciation is preserved on six continents (Hambrey and Harland, 1981, 1985) making a purely high latitude glacial hypothesis unlikely. In addition, tectonic and paleomagnetic evidence is lacking for rapid continental drift through polar latitudes. The Rapitan glaciation is associated with rifting of at least the North American and Australian continents (Eisbacher, 1985; Young, 1988, 1992) and it is reasonable that uplift of rift-margin blocks favored accumulation of glacial ice. The association of glaciation with rifting recurs in Neoproterozoic successions, and preserved facies are similar to those described here, biased toward deep-water deposits. We suggest that rift-related subsidence is the primary cause for this preferential preservation of deep-water glacial marine deposits.

Acknowledgements

This manuscript is a synthesis of research begun in 1975 by 'Crowell's Trolls', at University of California, Santa Barbara under the supervision of J.C. Crowell and supported by National Science Foundation Grant EAR 77–06008. Subsequently, Link's research in Neoproterozoic stratigraphy has been supported by grants from the Idaho State University Faculty Research Committee, the Kackley Family Endowment, and the Idaho Geological Survey. Link's contribution to this paper was facilitated by sabbatical leave at the Department of Geological Sciences, Cambridge University. Christie-Blick's research has been supported by National Science Foundation Grant EAR85–17923, and American Chemical Society, Petroleum Research Fund Grants PRF 16042–G2 and PRF 19989–AC2. We acknowledge reviews by H.T. Ore and N. Eyles.

References

Aalto, K.R. (1971). Glacial marine sedimentation and stratigraphy of the Toby Conglomerate (Upper Proterozoic), Southeastern British Columbia, northwestern Idaho and northeastern Washington. *Canadian Journal of Earth Sciences*, **8**, 753–87.

Anderson, J.B. & Domack, E.W. (1991). Foreword. In *Glacial Marine Sedimentation: Paleoclimatic Significance*, eds. J.B. Anderson & G.M. Ashley, Boulder, Colorado, Geological Society of America Special Paper 261, v–viii.

Anderson, J.B. & Thomas, M.A. (1991). Marine ice-sheet decoupling as a mechanism for rapid, episodic sea-level change: the record of such events and their influence on sedimentation. *Sedimentary Geology*, **70**, 87–104.

Anderson, J.B., Brake, C., Domack, E., Myers, N. & Wright, R. (1983). Development of a polar glacial-marine sedimentation model from Antarctic Quaternary deposits and glaciological information. In *Glacial-marine Sedimentation*, ed. B.F. Molnia, New York, Plenum, pp. 233–64.

Anderson, J.B., Kennedy, D.S., Smith, M.J. & Domack, E.W. (1991). Sedimentary facies associated with Antarctica's floating ice masses. In *Glacial Marine Sedimentation: Paleoclimatic Significance*, eds. J.B. Anderson & G.M. Ashley, Boulder, Colorado, Geological Society of America Special Paper 261, 1–25.

Armentrout, J.M. (1983). Glacial lithofacies of the Neogene Yakataga Formation, Robinson Mountains, southern Alaska Coast Range, Alaska. In *Glacial-marine Sedimentation*, ed. B.F. Molnia, New York, Plenum, pp. 629–65.

Armin, R.A. & Mayer, L. (1983). Subsidence analysis of the Cordilleran miogeocline: implications for timing of late Proterozoic rifting and amount of extension. *Geology*, **11**, 702–5.

Ashley, G.M., Boothroyd, J.C. & Borns, H.W., Jr. (1991). Sedimentology of late Pleistocene (Laurentide) deglacial-phase deposits, eastern Maine: an example of a temperate marine grounded ice-sheet margin. In *Glacial Marine Sedimentation: Paleoclimatic Significance*, eds. J.B. Anderson & G.M. Ashley, Boulder, Colorado, Geological Society of America Special Paper 261, 107–125.

Basse, R.A. (1978). Stratigraphy, sedimentology and depositional setting of the late Precambrian Pahrump Group, Silurian Hills, California. M.Sc. Thesis, Stanford University, Stanford, California, 79 pp.

Bjorlykke, K. (1985). Glaciations, preservation of their sedimentary record and sea-level changes: a discussion based on the Late Precambrian and Lower Palaeozoic sequence in Norway. *Palaeogeography, Palaeoclimatology, Palaeoecology*, **51**, 197–207.

Blackwelder, E. (1932). An ancient glacial formation in Utah. *Journal of Geology*, **40**, 289–304.

Blick, N. (1981). Late Precambrian glaciation in Utah. In *Earth's Pre-Pleistocene Glacial Record*, eds. M.J. Hambrey & W.B. Harland, Cambridge, Cambridge University Press, pp. 740–4.

Blick, N.H. (1979). Stratigraphic, structural and paleogeographic interpretation of the Upper Proterozoic glaciogenic rocks in the Sevier orogenic belt, northwestern Utah. Ph.D. dissertation, University of California, Santa Barbara, California, 633 pp.

Bond, G.C. & Kominz, M.A. (1984). Construction of tectonic subsidence curves for the early Paleozoic miogeocline, southern Canadian Rocky Mountains: implications for subsidence mechanisms, age of breakup, and crustal thinning. *Geological Society of America Bulletin*, **95**, 155–73.

Bond, G.C., Kominz, M.A. & Devlin, W.J. (1983). Thermal subsidence and eustasy in the Lower Paleozoic miogeocline of western North America. *Nature*, **306**, 775–9.

Bond, G.C., Christie-Blick, N., Kominz, M.A. & Devlin, W.J. (1985). An Early Cambrian rift to post-rift transition in the Cordillera of western North America. *Nature*, **316**, 742–5.

Boulton, G.S., Baldwin, C.T., Peacock, J.D., McCabe, A.M., Miller, G., Jarvis, J. *et al.* (1982). A glacioisostatic facies model and amino acid stratigraphy for late Quaternary events in Spitsbergen and the Arctic. *Nature*, **298**, 437–41.

Christie-Blick, N. (1982a). Upper Precambrian (Eocambrian) Mineral Fork Tillite of Utah: a continental glacial and glaciomarine sequence: discussion. *Geological Society of America Bulletin*, **93**, 184–7.

Christie-Blick, N. (1982*b*). Upper Proterozoic and Lower Cambrian rocks of the Sheeprock Mountains, Utah: regional correlation and significance. *Geological Society of America Bulletin*, **93**, 735–50.

Christie-Blick, N. (1983). Glacial-marine and subglacial sedimentation, Upper Proterozoic Mineral Fork Formation, Utah. In *Glacial-marine sedimentation*, ed. B.F. Molnia, New York, Plenum, pp. 703–76.

Christie-Blick, N. (1985). Upper Proterozoic glacial-marine and subglacial deposits at Little Mountain, Utah. *Brigham Young University Geology Studies*, **32**, part 1, 9–18.

Christie-Blick, N. & Levy, M., eds. (1989). *Late Proterozoic and Cambrian Tectonics, Sedimentation, and Record of Metazoan Radiation in the Western United States*. Field Trip Guidebook T331, 28th International Geological Congress, American Geophysical Union, Washington D.C. 113 pp.

Christie-Blick, N. & Link, P.K. (1988). Glacial-marine sedimentation, Mineral Fork Formation (Late Proterozoic) Utah. In *Geological Society of America Annual Meeting Field Trip Guidebook 1988*, ed. G.S. Holden, Golden, Colorado, Professional Contributions, Colorado School of Mines, **12**, 259–74.

Chumakov, N.M. & Elston, D.P. (1989). The paradox of Late Proterozoic glaciations at low latitudes. *Episodes*, **12**, 115–20.

Cowan, E.A. & Powell, R.D. (1991). Ice-proximal sediment accumulation rates in a temperate glacial fjord, southeastern Alaska. In *Glacial Marine Sedimentation; Paleoclimatic Significance*, eds. J.B. Anderson & G.M. Ashley, Boulder, Colorado, Geological Society of America Special Paper, 261, 61–73.

Crittenden, M.D. Jr. (1965*a*). Geology of the Mount Aire quadrangle, Salt Lake County, Utah. U.S. Geological Survey, Geologic Quadrangle Map, GQ-379, scale 1:24000.

Crittenden, M.D. Jr. (1965*b*). Geology of the Dromedary Peak quadrangle, Utah. U.S. Geological Survey Geologic Quadrangle Map, GQ-378, scale 1:24000.

Crittenden, M.D., Sharp, B.J. & Calkins, F.C. (1952). Geology of the Wasatch Mountains east of Salt Lake City, Parleys Canyon to Traverse Range. *Guidebook to the Geology of Utah*, ed. R.E. Marsell, Utah Geological Society, **8**, 1–37.

Crittenden, M.D. Jr, Schaeffer, F.E., Trimble, D.E., & Woodward, L.A. (1971). Nomenclature and correlation of some upper Precambrian and basal Cambrian sequences in western Utah and southeastern Idaho. *Geological Society of America Bulletin*, **82**, 581–602.

Crittenden, M.D. Jr., Stewart, J.H. & Wallace, C.A. (1972). Regional correlation of upper Precambrian strata in western North America. *24th International Geological Congress*, Montreal, section 1, 334–341.

Crittenden, M.D. Jr., Christie-Blick, N. & Link, P.K. (1983). Evidence for two pulses of glaciation during the Late Proterozoic in northern Utah and southeastern Idaho. *Geological Society of America Bulletin*, **94**, 437–50.

Crowell, J.C. (1983). The recognition of ancient glaciations. In *Proterozoic Geology: Selected Papers from an International Proterozoic Symposium*, eds. L.G. Medaris Jr., C.W. Byers, D.M. Mickelson & W.C. Shanks, Boulder, Colorado, Geological Society of America Memoir, 161, 289–97.

Devlin, W.J. & Bond, G.C. (1988). The initiation of the early Paleozoic Cordilleran miogeocline: evidence from the uppermost Proterozoic-Lower Cambrian Hamill Group of southeastern British Columbia. *Canadian Journal of Earth Sciences*, **25**, 1–19.

Devlin, W.J., Bond, G.C. & Brueckner, H.K. (1985). An assessment of the age and tectonic setting of volcanics near the base of the Windermere Supergroup in northeastern Washington: implications for latest Proterozoic–earliest Cambrian continental separation. *Canadian Journal of Earth Sciences*, **22**, 829–837.

Devlin, W.J., Brueckner, H.K. & Bond, G.C. (1988). New isotopic data and a preliminary age for volcanics near the base of the Windermere Supergroup, northeastern Washington, USA. *Canadian Journal of Earth Sciences*, **25**, 1906–11.

Deynoux, M. (1985). Terrestrial or waterlain diamictites? Three case studies from the Late Precambrian and Late Ordovician glacial drifts in West Africa. *Palaeogeography, Palaeoclimatology, Palaeoecology*, **51**, 97–114.

Domack, E.G. (1988). Biogenic facies in the Antarctic glacimarine environment: basis for a polar glacimarine summary. *Palaeogeography, Palaeoclimatology, Palaeoecology*, **63**, 357–72.

Dreimanis, A. (1983). Quaternary glacial deposits: Implications for the interpretation of Proterozoic glacial deposits. In *Proterozoic Geology: Selected Papers from an International Proterozoic Symposium*, eds. L.G. Medaris Jr., C.W. Byers, D.M. Mickelson & W.C. Shanks, Boulder, Colorado, Geological Society of America Memoir, 161, 299–308.

Edwards, M. (1986). Glacial environments. In *Sedimentary Environments and Facies*, 2nd edition, ed. H.G. Reading, Oxford, Blackwell Scientific Publications, pp. 445–70.

Eisbacher, G.H. (1981). Sedimentary tectonics and glacial record in the Windermere Supergroup, Mackenzie Mountains, northwestern Canada. Geological Survey of Canada Paper 80-27, 40 pp.

Eisbacher, G.H. (1985). Late Proterozoic rifting, glacial sedimentation, and sedimentary cycles in the light of Windermere deposition, western Canada. *Palaeogeography, Palaeoclimatology, Palaeoecology*, **51**, 231–54.

Elston, D.P., Link, P. K., Winston, D. & Horodyski, R.J. (1993). Correlations of Middle and Late Proterozoic successions. In *Precambrian: Conterminous U.S.*, eds. J.C. Reed, Jr., M.E. Bickford, R.S. Houston, P.K. Link, D.W. Rankin, P.K. Simms, & W.R. Van Schmus, Boulder, Colorado, Geological Society of America, The Geology of North America, C-2, 468–87.

Eyles, C.H. (1987). Glacially influenced submarine channel sedimentation in the Yakataga Formation, Middleton Island, Alaska. *Journal of Sedimentary Petrology*, **57**, 1004–17.

Eyles, C.H. & Eyles, N. (1983). Sedimentation in a large lake; a reinterpretation of the Late Pleistocene stratigraphy at Scarborough Bluffs, Ontario, Canada. *Geology*, **11**, 146–52.

Eyles, C.H., Eyles, N. & Miall, A.D. (1985). Models of glaciomarine sedimentation and their application to the interpretation of ancient glacial sequences. *Palaeogeography, Palaeoclimatology, Palaeoecology*, **51**, 15–84.

Eyles, C.H. & Lagoe, M.B. (1990). Sedimentation patterns and facies geometries on a temperate glacially influenced continental shelf; The Yakataga Formation, Middleton Island, Alaska. In *Glacimarine Environments: Processes and Sediments*, eds. J.A. Dowdeswell & J.D. Scourse, Geological Society of London Special Publication 53, 363–386.

Eyles, C.H., Eyles, N. & Lagoe, M.B. (1991). The Yakataga Formation; A late Miocene to Pleistocene record of temperate glacial marine sedimentation in the Gulf of Alaska. In *Glacial Marine Sedimentation; Paleoclimatic Significance*, eds. J.B. Anderson & G.M. Ashley, Boulder, Colorado, Geological Society of America Special Paper 261, 159–80.

Eyles, N. & Lagoe, M.B. (1989). Sedimentology of shell-rich deposits (coquinas) in glaciomarine facies of the Late Cenozoic Yakataga Formation, Middleton Island, Alaska. *Geological Society of America Bulletin*, **101**, 124–42.

Fairchild, I.J. (1993). Balmy shores and icy wastes: the paradox of carbonates associated with glacial deposits in Neoproterozoic times. *Sedimentology Review*, **1**, 1–16.

Fairchild, I.J. & Spiro, B. (1990). Carbonate minerals in glacial sediments: geochemical clues to palaeoenvironment. In *Glacimarine Environments: Processes and Sediments*, eds. J.A. Dowdeswell & J.D. Scourse, Geological Society of London Special Publication 53, 201–16.

Fairchild, I.J., Hambrey, M.J., Spiro, B. & Jefferson, T.H. (1989). Late Proterozoic glacial carbonates in northeast Spitsbergen: new insights into the carbonate–tillite association. *Geological Magazine*, **126**, 469–90.

Hambrey, M.J. & Harland, W.B., eds. (1981). *Earth's Pre-Pleistocene Glacial Record*. Cambridge, Cambridge University Press, 1004 pp.

Hambrey, M.J. & Harland, W.B. (1985). The Late Proterozoic glacial era. *Palaeogeography, Palaeoclimatology, Palaeoecology*, **51**, 255–72.

Harland, W.B. (1983). The Proterozoic glacial record. In *Proterozoic Geology: Selected Papers from an International Proterozoic Symposium*, eds. L.G. Medaris Jr., C.W. Byers, D.M. Mickelson & W.C. Shanks, Boulder, Colorado, Geological Society of America Memoir 161, 279–88.

Harland, W.B., Armstrong, R.L., Cox, A.V., Craig, L.A., Smith, A.G. & Smith, D.G. (1990). *A Geologic Time Scale 1989*. Cambridge, Cambridge University Press, 263 pp.

Harper, G.D. & Link, P.K. (1986). Geochemistry of Upper Proterozoic rift-related volcanics, northern Utah and southeastern Idaho. *Geology*, **14**, 864–7.

Hazzard, J.C. (1939). Possibility of pre-Cambric glaciation in southeastern California. *Pan American Geological Abstracts*, **71**, 47–8.

Hintze, F.F. Jr. (1913). A contribution to the geology of the Wasatch Mountains, Utah. *New York Academy of Sciences Annals*, **23**, 85–143.

Hoffman, P.F. (1991). Did the breakout of Laurentia turn Gondwanaland inside-out? *Science*, **252**, 1409–12.

Jefferson, C.W. & Parrish, R.R. (1989). Late Proterozoic stratigraphy, U-Pb zircon ages, and rift tectonics, Mackenzie Mountains, northwestern Canada. *Canadian Journal of Earth Sciences*, **26**, 1784–801.

Knoll, A.H. (1991). End of the Proterozoic Eon. *Scientific American*, **265**(4), 64–73.

Knoll, A.H. & Walter, M.R. (1992). Latest Proterozoic stratigraphy and Earth history. *Nature*, **356**, 673–8.

Knoll, A.H., Blick, N. & Awramik, S.M. (1981). Stratigraphic and ecologic implications of late Precambrian microfossils from Utah. *American Journal of Science*, **281**, 247–63.

Levy, M. & Christie-Blick, N. (1989). Pre-Mesozoic palinspastic reconstruction of the eastern Great Basin (western United States). *Science*, **245**, 1454–62.

Levy, M. & Christie-Blick, N. (1991a). Late Proterozoic paleogeography of the eastern Great Basin. In *Paleozoic Paleogeography of the Western United States II*, eds. J.D. Cooper & C.H. Stevens, Los Angeles, California, Pacific Section, Society of Economic Paleontologists and Mineralogists, Special Publication 67, vol. 1, pp. 371–86.

Levy, M. & Christie-Blick, N. (1991b). Tectonic subsidence of the early Paleozoic passive continental margin in eastern California and southern Nevada. *Geological Society of America Bulletin*, **103**, 1590–606.

Lindsey, K.A. & Link, P.K. (1988). Stratigraphic evidence for two episodes of continental rifting in the basal Cordilleran miogeocline from the northern Great Basin to northeastern Washington. *Geological Society of America Abstracts with Programs*, 20, 176.

Link, P.K. (1981). Upper Proterozoic diamictites in southeastern Idaho, USA. In *Earth's Pre-Pleistocene Glacial Record*, eds. M.J. Hambrey & W.B. Harland, Cambridge, Cambridge University Press, pp. 736–9.

Link, P.K. (1983). Glacial and tectonically influenced sedimentation in the upper Proterozoic Pocatello Formation, southeastern Idaho. In *Stratigraphic and Tectonic Studies in the Eastern Great Basin*, eds. D.M. Miller, V.R. Todd & K.A. Howard, Boulder, Colorado, Geological Society of America Memoir, 157, 165–82.

Link, P.K. (1984). Comment and reply on 'Subsidence analysis of the Cordilleran miogeocline: implications for timing of Late Proterozoic rifting and amount of extension'. *Geology*, **12**, 699.

Link, P.K. (1986). Tectonic model for deposition of the Late Proterozoic Pocatello Formation, southeastern Idaho. *Northwest Geology*, **15**, 1–7.

Link, P.K. (1987). The Late Proterozoic Pocatello Formation: a record of continental rifting and glacial marine sedimentation, Portneuf Narrows, southeastern Idaho. In *Rocky Mountain Section of the Geological Society of America, Centennial Field Guide vol. 2*, ed. S.S. Bues, Boulder, Colorado, Geological Society of America, pp. 139–142.

Link, P.K. & Gostin, V.A. (1981). Facies and paleogeography of Sturtian glacial strata (Late Precambrian), South Australia. *American Journal of Science*, **281**, 353–74.

Link, P.K. & LeFebre, G.B. (1983). Upper Proterozoic diamictites and volcanic rocks of the Pocatello Formation and correlative units, southeastern Idaho and northern Utah. *Utah Geological and Mineral Survey Special Studies*, **60**, (ed. K.D. Gurgel), 1–32.

Link, P.K. & Smith, L.H. (1992). Late Proterozoic and Early Cambrian stratigraphy, paleobiology, and tectonics, northern Utah and southeastern Idaho. In *Field Guide to Geologic Excursions in Utah and Adjacent Areas of Nevada, Idaho, and Wyoming*, ed. J.R. Wilson, Salt Lake City, Utah, Utah Geological Survey, Miscellaneous Publication, pp. 92–3, 461–481.

Link, P.K., Jansen, S.T., Halimdihardja, P., Lande, A., & Zahn, P. (1987). Stratigraphy of the Brigham Group (Late Proterozoic-Cambrian), Bannock, Portneuf, and Bear River Ranges, southeastern Idaho. In *The thrust belt revisited*, ed. W.R. Miller, Wyoming Geological Association Field Conference Guidebook, Casper, Wyoming, **38**, 113–148.

Link, P.K., Christie-Blick, N., Stewart, J.H., Miller, J.M.G., Devlin, W.J. & Levy, M.E. (1993). Late Proterozoic strata of the United States cordillera. In *Precambrian: Conterminous U.S.*, eds. J.C. Reed, Jr., M.E. Bickford, R.S. Houston, P.K. Link, D.W. Rankin, P.K. Simms & W.R. Van Schmus, Boulder, Colorado, Geological Society of America, The Geology of North America, C-2, 536–58.

Ludlum, J.C. (1942). Pre-Cambrian formations at Pocatello, Idaho. *Journal of Geology*, **50**, 85–95.

Mackiewicz, N.E., Powell, R.D., Carlson, P.R. & Molnia, B.F. (1984). Interlaminated ice-proximal glaciomarine sediments in Muir Inlet, Alaska. *Marine Geology*, **57**, 113–47.

Matsch, C.L. & Ojankangas, R.W. (1991). Comparisons in depositional style of 'polar' and 'temperate' glacial ice; Late Paleozoic Whiteout Conglomerate (West Antarctica) and Late Proterozoic Mineral Fork Formation (Utah). In *Glacial Marine Sedimentation; Paleoclimatic Significance*, eds. J.B. Anderson & G.M. Ashley, Boulder, Colorado, Geological Society of America Special Paper 261, 191–206.

Miller, J.M.G. (1985). Glacial and syntectonic sedimentation: The upper Proterozoic Kingston Peak Formation, southern Panamint Range, eastern California. *Geological Society of America Bulletin*, **96**, 1537–53.

Miller, J.M.G. (1987). Paleotectonic and stratigraphic implications of the Kingston Peak–Noonday contact in the Panamint Range, eastern California. *Journal of Geology*, **95**, 75–85.

Miller, J.M.G., Wright, L.A. & Troxel, B.W. (1981). The late Precambrian Kingston Peak Formation, Death Valley region, California. In *Earth's Pre-Pleistocene Glacial Record*, eds. M.J. Hambrey & W.B. Harland, Cambridge, Cambridge University Press, pp. 745–8.

Miller, J.M.G., Troxel, B.W. & Wright, L.A. (1988). Stratigraphy and paleogeography of the Proterozoic Kingston Peak Formation, Death Valley region, California. In *Geology of the Death Valley Region*, eds. J.L. Gregory & E.J. Baldwin, Los Angeles, California, South Coast Geological Society, Annual Field Trip Guidebook No. 16, 118–142.

Molnia, B.F. (1983). Subarctic glacial-marine sedimentation: a model. In *Glacial-marine Sedimentation*, ed. B.F. Molnia, New York, Plenum, 95–144.

Moncrieff, A.C.M. & Hambrey, M.J. (1990). Marginal-marine glacial sedimentation in the late Precambrian succession of East Greenland. In *Glacimarine environments: Processes and Sediments*, eds. J.A. Dowdeswell & J.D. Scourse, Geological Society of London Special Publication 53, 387–410.

Ojakangas, R.W. & Matsch, C.L. (1980). Upper Precambrian (Eocambrian) Mineral Fork Tillite of Utah: a continental glacial and glaciomarine sequence. *Geological Society of America Bulletin*, **91**, 495–501.

Plumb, K.A. (1991). New Precambrian time scale. *Episodes*, **14**, 139–40.

Powell, R.D. (1988). Processes and facies of temperate and subpolar glaciers with tidewater fronts. Boulder, Colorado, *Geological Society of America Short Course Notes*, 75–93.

Powell, R.D. (1991). Grounding-line systems as second-order controls on fluctuations of tidewater termini of temperate glaciers. In *Glacial Marine Sedimentation; Paleoclimatic Significance*, eds. J.B. Anderson & G.M. Ashley, Boulder, Colorado, Geological Society of America Special Paper 261, 75–94.

Rodgers, D.W. (1984). Stratigraphy, correlation, and depositional environments of Upper Proterozoic and Lower Cambrian rocks of the southern Deep Creek Range. In *Geology of Northwest Utah, Southern Idaho and Northeast Nevada*, eds. G.J. Kerns & R.L. Kerns Jr., Salt Lake City, Utah, Utah Geological Association Publication, 13, 79–91.

Roots, C.F. & Parrish, R.R. (1988). Age of the Mount Harper Volcanic Complex, southern Ogilvie Mountains, Yukon. In *Radiogenic Age and Isotopic Studies: Report 2*. Geological Survey of Canada Paper 88–2, pp. 29–35.

Ross, G.M. (1991). Tectonic setting of the Windermere Supergroup revisited. *Geology*, **19**, 1125–8.

Schermerhorn, L.J.G. (1974). Late Precambrian mixtites: glacial and/or nonglacial? *American Journal of Science*, **274**, 673–824.

Stewart, J.H. (1972). Initial deposits in the Cordilleran geosyncline: evidence of a Late Precambrian (<805 m.y.) continental separation. *Geological Society of America Bulletin*, **83**, 1345–60.

Stewart, J.H. (1982). Regional relations of Proterozoic Z and Lower Cambrian rocks in the western United States. In *Geology of Selected Areas in the San Bernardino Mountains, Western Mojave Desert, and Southern Great Basin*, eds. J.D. Cooper, B.W. Troxel & L.A. Wright, Guidebook for Field trip No. 9, Cordilleran Section Geological Society of America, Shoshone, California, Death Valley Publishing Company, pp. 171–186.

Stewart, J.H. (1983). Extensional tectonics in the Death Valley area, California: transport of the Panamint Range structural block 80 km northwestward. *Geology*, **11**, 153–7.

Stewart, J.H. & Suczek, C.A. (1977). Cambrian and latest Precambrian paleogeography and tectonics in the western United States. In *Paleozoic Paleogeography of the Western United States*, eds. J.H. Stewart, C.H. Stevens & A.E. Fritsche, Los Angeles, California, Society of Economic Paleontologists and Mineralogists, Pacific Section, Pacific Coast Paleogeography Symposium I, pp. 1–17.

Stewart, T.G. (1991). Glacial marine sedimentation from tidewater glaciers in the Canadian High Arctic. In *Glacial Marine Sedimentation; Paleoclimatic Significance*, eds. J.B. Anderson & G.M. Ashley, Boulder, Colorado, Geological Society of America Special Paper 261, 95–105.

Troxel, B.W. (1966). Sedimentary features of the upper Precambrian Kingston Peak Formation, Death Valley, California (abstract), Boulder, Colorado, Geological Society of America Special Paper 101, 341.

Troxel, B.W. (1982). Description of the uppermost part of the Kingston Peak Formation, Amargosa rim canyon, Death Valley region, California. In *Geology of Selected Areas in the San Bernardino Mountains, Western Mojave Desert, and Southern Great Basin, California*, eds. J.D. Cooper, B.W. Troxel & L.A. Wright, Guidebook Field Trip No. 9, Cordilleran Section, Geological Society of America, Shoshone, California, Death Valley Publishing Company, 61–70.

Troxel, B.W., Wright, L.A. & Williams, E.G. (1977). Late Precambrian history derived from the Kingston Peak Formation, Death Valley region, California. *Geological Society of America Abstracts with Programs*, **9**, 517.

Troxel, B.W., McMackin, M.A. & Calzia, J.P. (1987). Comment on 'Late Precambrian tectonism in the Kingston Range, southern California'. *Geology*, **15**, 274–5.

Tucker, M.E. (1986). Formerly aragonitic limestones associated with tillites in the late Proterozoic of Death Valley, California. *Journal of Sedimentary Petrology*, **56**, 818–30.

Varney, P.J. (1976). Depositional environment of the Mineral Fork Formation (Precambrian), Wasatch Mountains, Utah. In *Geology of the Cordilleran Hingeline*, ed. J.G. Hill, Denver, Colorado, Rocky Mountain Association of Geologists Symposium, pp. 91–102.

Visser, J.N.J. (1983a). Submarine debris flow deposits from the Upper Carboniferous Dwyka Tillite Formation in the Kalahari Basin, South Africa. *Sedimentology*, **30**, 511–23.

Visser, J.N.J. (1983b). Glacial-marine sedimentation in the Late Paleozoic Karoo Basin, southern Africa. In *Glacial-marine Sedimentation*, ed. B.F. Molnia, New York, Plenum, pp. 667–701.

Visser, J.N.J. (1989). The Permo-Carboniferous Dwyka Formation of southern Africa: deposition by a predominantly subpolar marine ice sheet. *Palaeogeography, Palaeoclimatology, Palaeoecology*, **70**, 377–91.

Visser, J.N.J. (1991). The paleoclimatic setting of the late Paleozoic marine ice sheet in the Karoo Basin of southern Africa. In *Glacial Marine Sedimentation; Paleoclimatic Significance*, eds. J.B. Anderson & G.M. Ashley, Boulder, Colorado, Geological Society of America Special Paper 261, 181–9.

Walker, J.D., Klepacki, D.W. & Burchfiel, B.C. (1986). Late Precambrian tectonism in the Kingston Range, southern California. *Geology*, **14**, 15–18.

Wernicke, B., Axen, G.J. & Snow, J.K. (1988). Basin and range extensional tectonics at the latitude of Las Vegas, Nevada. *Geological Society of America Bulletin*, **100**, 1738–57.

Wright, L.A. (1974). Geology of the southeast quarter of the Tecopa Quadrangle, San Bernardino and Inyo Counties, California. California Division of Mines and Geology, map sheet 20.

Wright, L.A. & Troxel, B.W. (1984). Geology of the northern half of the Confidence Hills 15-minute quadrangle, Death Valley region, eastern California: the area of the Amargosa chaos. California Division of Mines and Geology, map sheet 34.

Wright, L.A., Troxel, B.W., Williams, E.G., Roberts, M.T. & Diehl, P.E. (1976). Precambrian sedimentary environments of the Death Valley Region, eastern California. California Division of Mines and Geology Special Report 106, 7–15.

Yeo, G.M. (1981). The late Proterozoic Rapitan glaciation in the northern Cordillera. In *Proterozoic Basins of Canada*, ed. F.H.A. Campbell, Geological Survey of Canada Paper, 81–10, 25–46.

Young, G.M. (1982). The late Proterozoic Tindir Group, east-central Alaska: Evolution of a continental margin. *Geological Society of America Bulletin*, **93**, 759–83.

Young, G.M. (1988). Proterozoic plate tectonics, glaciation, and iron-formations. *Sedimentary Geology*, **58**, 127–44.

Young, G.M. (1992). Late Proterozoic stratigraphy and the Canada–Australia connection. *Geology*, **20**, 215–18.

Young, G.M. & Gostin, V.A. (1988). Stratigraphy and sedimentology of Sturtian glacigenic deposits in the western part of the North Flinders Basin, South Australia. *Precambrian Research*, **39**, 151–70.

Young, G.M. & Gostin, V.A. (1989). An exceptionally thick late Proterozoic (Sturtian) glacial succession in the Mount Painter area, South Australia. *Geological Society of America Bulletin*, **101**, 834–45.

Young, G.M. & Gostin, V.A. (1990). Sturtian glacial deposition in the vicinity of the Yankaninna anticline, North Flinders Basin, South Australia. *Australian Journal of Earth Sciences*, **37**, 447–58.

Young, G.M. & Gostin, V.A. (1991). Late Proterozoic (Sturtian) succession of the North Flinders Basin, South Australia: an example of temperate glaciation in an active rift setting. In *Glacial marine Sedimentation; Paleoclimatic Significance*, eds. J.B. Anderson & G.M. Ashley, Boulder, Colorado, Geological Society of America Special Paper 261, 207–22.

3 The Neoproterozoic Konnarock Formation, southwestern Virginia, USA: glaciolacustrine facies in a continental rift

JULIA M.G. MILLER

Abstract

The Konnarock Formation (new name, Rankin, 1993) comprises about 1100 m of diamictite, sandstone and argillite and crops out within a limited area in the Blue Ridge thrust belt of southwestern Virginia, USA. The formation overlies thick, bimodal volcanic rocks, approximately 760 Ma, with interbedded sandstone and conglomerate (Mount Rogers Formation; Rankin, 1993), or orthogneisses about 1 Ga. It underlies areally extensive Lower Cambrian sandstones and shales. The Konnarock Formation was deposited on the North American craton, most likely during an aborted phase of rifting which preceded continental separation and formation of the Iapetus Ocean.

Argillite, including coarse and fine laminites and massive mudstone, occurs at the base of the formation and may be up to about 650 m thick. This sequence records seasonal sedimentation in a thermally stratified, ice-contact lake. Fine laminites (varvites) were largely deposited from overflows and interflows. Turbidity currents deposited coarser laminae and graded beds. Dropstones and till pellets were dropped from floating ice. Massive and graded sandstone and massive or bedded diamictite are interbedded with the argillite and become more abundant higher in the section. They were deposited in the lake by turbidity currents, density-modified grain flows, and debris flows. Liquefied flows resedimented mixed sandstone and argillite units, and post-depositional liquefaction modified some beds.

Diamictite is most common in the upper part of the formation. Diamictite may be massive and up to 400–500 m thick (including some sandstone and argillite interbeds and lenses), bedded on a 1 to 20 cm scale, or laminated on a millimetre to centimetre scale. It records deposition through melting at the bottom of floating to partially floating ice, by rain-out, and/or by subaqueous debris flows. Massive diamictite at the top of the formation is interpreted to indicate proximity to an ice cliff or ramp near the lake margin and/or glacial advance across the lake.

The Konnarock Formation most likely reflects the influence of both glaciation and tectonism upon sedimentation. The section represents a progradational sequence which filled a glacial lake and culminated in ice advance across the region. These rocks probably accumulated in a graben within a larger rift system. Great thicknesses of argillite and diamictite imply rapid subsidence and, for the diamictite, high sediment supply. Both are consistent with contemporaneous tectonism. Because of the existence of significant topographic relief and the rarity of coeval glacial deposits in autochthonous eastern North America, the Konnarock Formation may record local, alpine glaciation.

Introduction

The Konnarock Formation (formerly the *upper* Mount Rogers Formation, Rankin, 1993), exposed in southwestern Virginia (Fig. 3.1), contains the best record of Neoproterozoic (Late Proterozoic; Plumb, 1991) glaciation in southeastern North America. It overlies both the Mount Rogers Formation (redefined by Rankin, 1993) and 1 Ga metamorphic rocks, and is situated on the North American craton. Both the Konnarock and Mount Rogers formations probably accumulated in a rift-valley prior to opening of the Iapetus Ocean (Rankin, 1975, 1976; Wehr and Glover, 1985; Schwab, 1981, 1986). The Konnarock Formation is therefore one of many Neoproterozoic glaciogenic sequences in both North America and the world which were deposited in a rift tectonic setting (Schermerhorn, 1974; Young, 1991). It is distinct in that (1) it is geographically isolated from other Neoproterozoic glaciogenic sequences, and (2) it overlies a thick accumulation of subaerial volcanic rocks (the Mount Rogers Formation; Rankin, 1970, 1993) and was deposited in continental, in contrast to marine, glacial environments.

These glaciogenic rocks of the Konnarock Formation were formerly included within the Mount Rogers Formation and referred to as the upper Mount Rogers Formation (Fig. 3.2; Rankin, 1970; Miller, 1989). Rankin (1993) introduced the name Konnarock Formation to include diamictite, sandstone and argillite at the top of this Neoproterozoic section. Thus, he redefined the Mount Rogers Formation and limited it to include only the underlying volcanic and subordinate interbedded sedimentary rocks. Rankin (1967, 1969, 1970) proposed the existence of glaciogenic rocks within the Konnarock Formation. Later Blondeau and Lowe (1972, and Blondeau, 1970), Schwab (1976, 1981) and

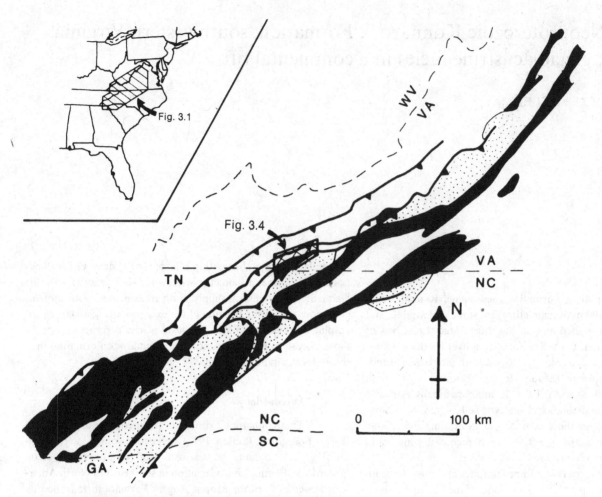

Fig. 3.1. Simplified geologic map of part of southern Appalachians, showing location of study area in southwestern Virginia. Stippled areas, older Proterozoic rocks; shaded, Neoproterozoic rocks; blank, Paleozoic and younger rocks. (Modified after Rankin et al., 1972.)

Rankin (1970)		Rankin (1993)	Lithology
MOUNT ROGERS FORMATION	UPPER	KONNAROCK FORMATION	diamictite, sandstone argillite
	MIDDLE	MOUNT ROGERS FORMATION	dominantly rhyolite, some basalt, conglomerate, argillite
	LOWER		

Fig. 3.2. Stratigraphic nomenclature, previous and revised, for the Mount Rogers and Konnarock formations.

Rexroad (1978) described facies of the formation and interpreted them to be various glacial and periglacial deposits.

Evidence for glacial deposition of the Konnarock Formation includes the presence of the following: (1) thick, generally structureless diamictite, (2) dropstones and till clasts and pellets embedded in laminite and laminated diamictite (Fig. 3.3; see also Fig. 11 in Rankin, 1993), and (3) laminites with distinct varve-like couplets. The purpose of this paper is to establish the character of glaciogenic sedimentation in the Konnarock Formation and thus to document an example of glaciolacustrine sedimentation in a continental rift.

Stratigraphic and structural setting

The Konnarock Formation (KF) comprises approximately 1100 m of diamictite, sandstone and argillite. Underlying it, the Mount Rogers Formation (MRF) is about 3000 m thick and contains thick rhyolite in addition to subordinate basalt, conglomerate, sandstone and shale (Rankin, 1970, 1993). The MRF–KF contact is conformable in places and unconformable elsewhere (Rankin, 1993). Rocks of both formations unconformably overlie the Cranberry Gneiss, a foliated plutonic rock about 1 to 1.1 Ga (Rankin, 1970; Rankin et al., 1983, 1989). The KF is disconformably overlain by conglomerate and sandstone of the basal Chilhowee Group (Rankin, 1975).

Exposures of the KF extend about 45 km along strike (northeast–southwest) within the Shady Valley thrust sheet and the Mountain City window in the northwestern part of the Blue Ridge province (Figs. 3.1, 3.4; Rankin, 1993). These rocks show lower greenschist facies metamorphism. They were transported northwestward during the Alleghanian orogeny by somewhere between about 175 and 260 km (Secor et al., 1986; Hatcher et al., 1989). Rankin (1970) suggests less than about 30 km relative displacement between the

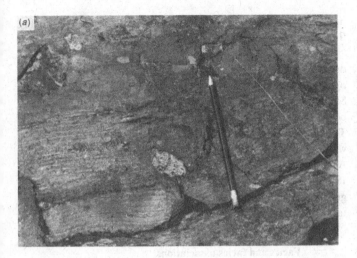

Fig. 3.3. (*a*) Dropstone in facies transitional between fine laminite and laminated diamictite. Note massive mudstone at and above pencil point. Outcrop is about 3 km east of Weaver's (Fig. 3.4) on Shady Valley thrust sheet. (Photo taken by D.W. Rankin.) (*b*) Dropstone in laminated diamictite, Trout Dale section. Arrows point to edge of stone which is 15 cm wide and composed of schist.

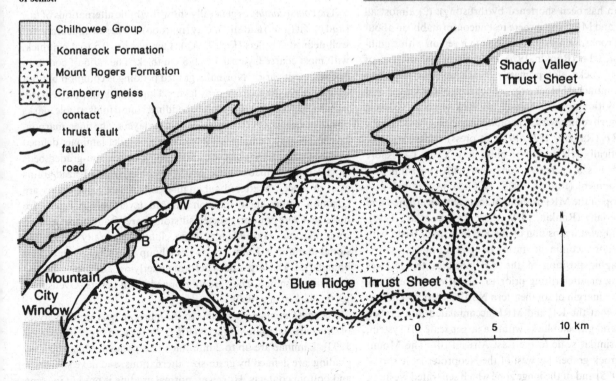

Fig. 3.4. Location map showing distribution of Konnarock Formation in different structural blocks and locations of stratigraphic columns in Figs. 3.5 and 3.6. (Modified after Rankin *et al.*, 1972; Rankin, 1993) B, Big Hill; K, Konnarock; T, Trout Dale; W, Weaver's store.

different structural blocks in which the KF is exposed, based upon the formation's consistent internal stratigraphy. Facies analysis (this study) supports such small relative motion (Figs. 3.5, 3.6).

New U-Pb zircon ages from MRF rhyolites define an age of 759 ± 2 Ma (Aleinikoff *et al.*, 1991). These data supersede an earlier estimate of approximately 820 Ma (U-Pb on zircons from three different Neoproterozoic units in the Blue Ridge; Rankin *et al.*, 1969). Trace and body fossils from the overlying lower Chilhowee Group show that it is Neoproterozoic (Vendian) to Early Cambrian (Placentian or younger) in age (Simpson and Sundberg, 1987; Walker and Driese, 1991). Thin basalt flows in the lower Chilhowee may correlate with the Catoctin Formation (570–600 Ma; Badger and Sinha, 1988; Aleinikoff *et al.*, 1991). Glaciogenic rocks of the Konnarock Formation are therefore between 760 and 570 Ma.

Evidence for coeval glaciation in others parts of autochthonous eastern North America is restricted to scanty, and in many cases controversial, exposures within neighbouring parts of the southern Appalachians (e.g., Schwab, 1981; Wehr, 1986; Thiesmeyer, 1939). Neoproterozoic glaciogenic rocks exist in Newfoundland and Massachusetts (Eyles and Eyles, 1989; Socci and Smith, 1987) within terranes accreted to North America during the Paleozoic (Williams and Hatcher, 1982).

Paleotectonic setting

Evidence pertinent to the paleotectonic setting of the Konnarock Formation is: (1) the formation's limited extent, all outcrops occur within an area about 45 by 10 km (although the latter dimension has been shortened by thrusting); (2) almost all units of the KF and MRF somewhere rest unconformably on about 1 Ga basement rocks, implying deposition in a region with significant topographic relief (Rankin, 1970, 1993); (3) modal analyses of KF sedimentary rocks plot primarily in the uplifted basement portion of the continental block provenance field of Dickinson and Suczek (1979; Withington, 1986), and (4) thick bimodal volcanic rocks, including rhyolites with a peralkaline affinity, exist in the underlying MRF (Rankin, 1975, 1976, in press). MRF volcanic rocks are genetically related to the plutonic-volcanic Crossnore Complex which has a chemistry consistent with extrusion in a tensional environment (Rankin, 1975, 1976). A welded ash-flow sheet near the top of the MRF is thought to have been erupted and emplaced subaerially (Rankin, 1970, 1993; Rankin *et al.*, 1989). These data all suggest deposition of the KF in a small basin with significant relief, on a continent, during or after rifting.

The stratigraphic position of the KF indicates that it was deposited during or after rifting prior to formation of the Early Paleozoic passive margin of southeastern North America. Thomas (1991) suggests that the KF and MRF accumulated in a graben amongst horst and graben blocks within a larger scale rift system, probably on a similar scale to the East African rift. The Mount Rogers–Konnarock graben lay west of the Neoproterozoic shoreline (Rankin, 1975) and of the hinge zone which separated western, less attenuated from eastern, more attenuated continental crust (Wehr and Glover, 1985; Schwab, 1986). The KF therefore accu-

mulated landward of the edge of the Iapetus Ocean, probably before oceanic crust began to form farther east (Thomas, 1991).

Isotopic data from volcanic rocks support existence of two distinct igneous events in the southern Appalachians, prior to opening of the Iapetus Ocean (Aleinikoff *et al.*, 1991; Badger and Sinha, 1988). Mount Rogers volcanic rocks belong to the first which occurred 700 to 760 Ma and did not lead to continental separation. The Catoctin Formation records the second event, 570 to 600 Ma, which did lead to continental separation (Badger and Sinha, 1988; Aleinikoff *et al.*, 1991; Simpson and Ericksson, 1989; Bond *et al.*, 1984). Glaciogenic rocks of the KF were probably deposited in a graben formed during the first, aborted phase of Iapetan continental rifting (Rankin, 1993).

Facies and facies associations

Three facies, each with three subdivisions, exist within the KF. Figures 3.5 and 3.6 show the facies distribution in generalised stratigraphic columns.

Argillite

The argillite facies is subdivided into: (i) coarse laminite, typically with centimeter-scale laminae, (ii) fine laminite, typically with millimeter-scale laminae, and (iii) massive mudstone. However, all variations exist; the argillite subfacies grade into one another, and argillites may show features transitional between categories (Fig. 3.7*a*).

The *coarse laminites* generally show rhythmic alternations of fine sand or silt (gray) and silt or clay (maroon), giving an appearance of well-defined couplets (Fig. 3.7*b*,*c*). Laminae are 0.5 to 4 cm thick, with most coarse layers 0.3 to 0.8 cm thick. The ratio of coarse to fine sediment varies. Normally graded bedding is common but not ubiquitous in the coarser, gray layers. Ungraded layers have sharp, parallel bed contacts. Convolute bedding, slump folds, ripple marks and load casts occur locally in the coarser layers. The finer, maroon, layers typically contain millimeter-scale parallel laminae, defined by grain size alternations without grading or by mini-graded beds (Fig. 3.7*b*). Laminae of granules, fine pebbles, or sand-sized grains exist in places. Scattered subangular silt or fine sand grains are common within these maroon clay/silt layers which rarely are massive. Clots of poorly sorted siltstone (rarely sandstone) exist in some maroon, clay/silt layers. These clots are generally equidimensional and about 1 mm across (locally up to 7 mm), with diffuse boundaries. The larger ones deform underlying, and are draped by overlying, laminae. They are interpreted as till pellets (Ovenshine, 1970; Clark *et al.*, 1980).

The *fine laminites* are similar to the maroon, finer grained layers of the coarse laminite subfacies (Fig. 3.7*b*; see also Fig. 9 in Rankin, 1993). Laminae are up to 2 mm thick and parallel. Most show no grading, are defined by grain-size alternations, and have sharp top and bottom contacts. However, normal grading is present in some of both the silt and clay laminae. In general, the clay laminae are of equal or greater thickness than the silt laminae, but in places clay

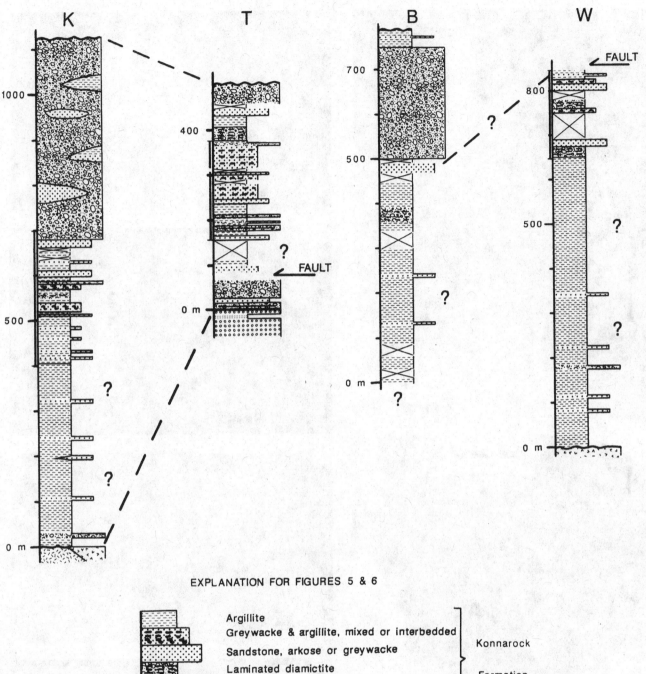

EXPLANATION FOR FIGURES 5 & 6

Argillite
Greywacke & argillite, mixed or interbedded
Sandstone, arkose or greywacke
Laminated diamictite
Bedded diamictite
Massive diamictite
Pebbly sandstone, conglomerate, siltstone
Volcanic rocks
Cranberry Gneiss
No exposure

Konnarock
Formation

Mount Rogers
Formation

Fig. 3.5. Composite, generalised stratigraphic columns through Konnarock Formation on Shady Valley thrust sheet. For location of columns see Fig. 3.4. *Note:* these sections were constructed partly from detailed measurements and descriptions of good exposures (designated by double lines on left side of column), and partly from estimates based on mapping scattered outcrops. ? = uncertainty.

Fig. 3.6. Composite, generalized stratigraphic sections through Konnarock Formation in Mountain City window. For location of columns see Fig. 3.4. *Note:* these sections were constructed partly from detailed measurements and descriptions of good exposures (designated by double line on left side of column), and partly from estimates based on mapping scattered outcrops. ? = uncertainty.

Fig. 3.7. (*a*) Argillite, including both coarse and fine laminite. Outcrop is about
4 km east of Weaver's (Fig. 3.4) on Shady Valley thrust sheet. (Photo taken by
D.W. Rankin.) (*b*) Coarse laminite showing coarser (light) and finer grained
(dark) layers, forming couplets; Argillite facies, Konnarock section. Note
multiple fine laminae within finer, dark layers; this texture is typical of the fine
laminite subfacies. (*c*) Graded sandstone, with rare argillite rip-up clasts
(arrow), interbedded with coarse laminite. (*d*) Massive diamictite, Big Hill
section. (*e*) Internally stratified diamictite; Bedded diamictite subfacies, Trout
Dale section. Beds are right way up and dip ≈ 50° to left. (*f*) Laminated
diamictite; Konnarock section.

laminae thickness varies proportionally with silt laminae thickness. Convolute lamination is present locally. Scattered silt or fine sand grains are common, as they also are in the *massive mudstone*.

Isolated stones, up to 10 cm across, exist in all three argillite subfacies. The size, setting and contact relationships of many of these stones show that they are dropstones.

Argillite is volumetrically the most abundant facies of the KF and dominates the lower part of the formation (Figs. 3.5, 3.6). Thicknesses are hard to estimate. Mapping suggests the facies ranges up to about 650 m thick (Figs 3.5, 3.6), but indistinguishable thrusts may have thickened it. Fine laminites appear to be more common lower, and dropstones more common higher, in the section. Argillite is interbedded with all Konnarock facies; sandstone interbeds are particularly common towards the top of the section (Section K, Fig. 3.5). Gray layers in the coarse laminite subfacies are transitional to thin graded sandstone beds, and in places maroon argillite is intimately mixed with sandstone within a single bed (see below). As laminated argillite becomes more pebbly it grades into the laminated diamictite subfacies (Fig. 3.3a).

Interpretation

Laminites in the KF are remarkably similar in their overall appearance to some Pleistocene varves (e.g., Ashley, 1975; Ashley *et al.*, 1985) and Late Paleozoic varvites (Rocha Campos *et al.*, 1981; Levell *et al.*, 1988). Because many laminae, particularly in the fine laminite subfacies, are not graded and have sharp contacts between the sand/silt and clay layers, they are more similar to classic glaciolacustrine varves than to glaciomarine silt-clay laminites, or cyclopels. The latter are normally graded and have gradational coarse to fine contacts (Mackiewitz *et al.*, 1984; Powell, 1988; Cowan and Powell, 1990).

In the KF *fine laminites*, the silt laminae are interpreted as deposited either by density underflows with a variable sediment content or by suspension settling from overflow/interflows in a thermally stratified lake during the summer. The clay laminae represent suspension settling during the winter and/or after overturning of the lake (Ashley, 1975; Ashley *et al.*, 1985; Sturm, 1979). Silt-clay thickness ratios of the fine laminites are comparable to Group I and II varves of Ashley (1975), suggesting deposition in locations relatively distant from sediment sources. *Massive mudstones* were deposited from suspension either at places above the thermocline or at times when the lake water was not thermally stratified (Sturm and Matter, 1978).

Dropstones, till pellets, and scattered angular sand grains indicate that the Konnarock lake(s) were in contact with the ice front at least some of the time. Ashley *et al.* (1985) suggest that in ice-contact lakes sedimentation is dominated by underflows. Proportional variation in the thicknesses of silt and clay laminae in some KF fine laminites is consistent with a dominance of underflows during certain periods (Ashley *et al.*, 1985).

The coarser, gray, graded, layers within the *coarse laminite* subfacies were deposited by low density turbidity currents and represent episodic, or surging, underflows entering the glacial lake (Lowe, 1982; Ashley *et al.*, 1985). These turbidites were deposited by similar mechanisms but in more distal locations than the graded sandstone beds (see below). Soft-sediment deformation attests to some slope instability within the lake.

The overall upward coarsening evident in some argillite sections (e.g., Section K, Fig. 3.5) suggests increasing proximity to sediment sources as the KF basin filled. Additionally, the remarkably great thickness of argillite implies rapid basin subsidence and/or lacustrine conditions over a long period of time.

Sandstone

Sandstones within KF are subdivided, on the basis of sedimentary structures and petrographic composition, into: (i) massive, (ii) graded, and (ii) sandstone mixed with argillite.

The *massive sandstones* (petrographically fine- to coarse-grained arkosic arenites; Withington, 1986), characteristically form thick, structureless beds about 1 m thick. Bed contacts are sharp and locally erosive. Parallel laminae, defined by detrital magnetite, are visible near the tops of some beds. Rarely these sandstones contain scattered pebbles or argillite rip-up clasts, have pebbly bases, or become more muddy towards the bed top where they are interbedded with or grade up into argillite. Near the top of Section W (Fig. 3.6), several meters of very fine-grained arkosic arenite with abundant current ripple cross-lamination occur.

Graded sandstones (petrographically very fine- to coarse-grained arkosic wackes, Withington, 1986) usually show good graded bedding and sharp basal contacts, as well as some convolute bedding and slump folds, and rare parallel laminae, ripple marks, sole marks and load casts (Fig. 3.7c). Bed thickness is very variable (5 cm to 1 m); the thinner graded sandstone beds are transitional to the coarser, gray layers of the coarse laminite subfacies. Rare scattered pebbles exist; where pebbles are more abundant these sandstones are transitional with the bedded diamictite subfacies. Argillite rip-up clasts (up to 1 by 25 cm in size) are abundant, particularly near bed tops.

In the *mixed sandstone and argillite* subfacies, all different proportions of sandstone (generally arkosic wacke, rarely arkosic arenite) to argillite exist. Typically the ratio is about 1:1 or lower (more argillite), although the subfacies grades into graded sandstone with abundant argillite rip-up clasts. Texture is characteristically chaotic with very irregular contacts between the two lithologies. Beds are 1 cm to about 1 m thick. Centimeter and millimeter-scale laminae are locally visible and in place folded within the argillite patches, which suggests that mixing occurred after deposition of each lithology.

All sandstone subfacies are typically associated and interbedded with the three argillite subfacies and become more abundant up section (Figs. 3.5, 3.6). They are themselves interbedded locally. A massive sandstone is present as a mappable unit near the top of the formation, below the thick massive diamictite (e.g., Sections K, T, B of Figs. 3.5 and 3.6). East of the Konnarock section (Fig. 3.4), this sandstone is lenticular, varying from a few to about 30 m thick over a strike distance of a few hundred metres. Lenses of massive sandstone are also present within the top thick diamictite (Section K, Fig. 3.5).

Interpretation

Massive sandstones of the KF are similar to massive sands, which rarely show faint horizontal lamination and may contain scattered pebbles, described from certain Quaternary successions and interpreted as subaqueous outwash or delta deposits (Rust and Romanelli, 1975; Rust, 1977, 1988; Sturm and Matter, 1978; Diemer, 1988). These massive sands form both sheets and channel fills and probably originated as debris flows or liquefied flows and evolved into density-modified grain flows (Rust, 1977, 1988; Lowe, 1976a). The horizontal laminations present in some KF beds may be consolidation lamination (Lowe, 1975), or may indicate deposition from flows transitional between modified grain flows and high-density turbidity currents, as in the subaqueous toe region of a fan delta described by Nemec et al. (1984). In all these examples, the structureless nature of the sands is evidence for deposition by sediment gravity flows in a subaqueous setting.

Most *graded sandstones* were deposited by low-density turbidity currents and many were modified by liquefaction (Lowe, 1975, 1982). Since many turbidites form isolated beds amidst argillite and show only the graded A-division of the Bouma sequence, the turbidity currents may have decelerated rapidly (Banerjee, 1977). The thickness of each graded bed probably depended upon its proximity to terrigeneous sediment sources, i.e., meltwater influxes into the lake. Sturm and Matter (1978) describe similar deposits in a lake in the Swiss Alps where turbidites, up to 1.5 m thick, thin laterally and distally into coarse layers indistinguishable from the summer half-couplets of the varves.

Mixed sandstone–argillite beds were deposited by liquefied sediment flows and/or modified after deposition by liquefaction and/or fluidization (Lowe, 1975, 1976b). As Lowe (1976b) points out, there is a continuous gradation between true liquefied flows and turbidity currents, and liquefied flows may evolve directly into turbidity currents. Thus similar mechanisms deposited the graded sandstones, the mixed sandstone–argillite beds, and the coarse, gray layers within the coarse laminite subfacies in this glacial lake.

Diamictite

The diamictite facies is subdivided according to bedding character into: (i) massive, (ii) bedded, and (iii) laminated.

Massive diamictite usually shows no sedimentary structures (Fig. 3.7d). Diffuse bedding and discontinuous wispy layers are rarely present. The uppermost thick diamictite unit is generally relatively clast-rich (e.g., clasts form 20 to 50% of rock) with a muddy matrix; locally, in its lowest few meters in Section K (Fig. 3.5), it includes conglomerate with boulders up to 1.2 m across. Clasts are predominantly basement crystalline rocks, but some volcanic clasts are present. The matrix is composed of poorly sorted, angular, sand-sized grains floating in a clay-rich matrix commonly replaced by hematite. In the Konnarock area, the upper thick diamictite unit contains interbeds or lenses of massive sandstone and massive and laminated argillite, which are 1 to over 30 m thick (Section K, Fig. 3.5). Massive diamictites with sparse clasts, more rip-up clasts, and commonly a sandier matrix form discrete beds up to about 3 m thick lower in the KF.

Bedded diamictite is characteristically bedded on the 1 to 20 cm scale, although some beds are up to 1 m thick. The subfacies varies from internally stratified diamictite (Fig. 3.7e) to diamictite interbedded with variable proportions of sandstone and argillite. Clasts typically form about 15% of the rock. Apparent long axes of most clasts lie parallel to bedding; they are rarely vertical. Dropstones which truncate laminae exist in some argillite interbeds. Sedimentary structures within the sandstone and argillite interbeds are similar to those in the sandstone and argillite facies.

Laminated diamictite describes diamictite laminated on a scale of less than 5 cm (Figs. 3.3b, 3.7f). Clasts, generally up to 5 cm across, are abundant and distributed randomly through the rock, giving it an overall appearance of diamictite. Many clasts are clearly dropstones (Fig. 3.3b); some larger ones (e.g., up to 45 cm across) show splash-up structures. Lamination is defined by alternations of very poorly sorted, fine- to coarse-grained muddy sand or gravel (diamictite), and clay. Clay laminae are typically 1 to 2 mm thick, and may be internally laminated. Coarse laminae are more variable in thickness (1–12 mm) and very rarely include very thin clay laminae separating sandy layers. Some coarse laminae show normal grading, but most are not graded. Most contacts between laminae are sharp; some coarse layers grade into overlying clay laminae. Slump folds exist locally. Overall ratios of diamictite (i.e., coarse layers) to argillite vary from 4:1 to 1:4.

Massive diamictite is commonest and thickest, up to 400–500 m (including interbeds of massive sandstone and argillite), at the top of the KF (Figs. 3.5, 3.6). The unit thins to the northeast and is locally absent, but this may be due to the unconformity above the KF. Bedded diamictite is relatively rare and occurs mostly in eastern sections interbedded with sandstone and argillite facies in the lower part of the formation (Section T, Fig. 3.5). Laminated diamictite occurs in all sections and is typically associated and/or gradational with argillite. It locally contains discrete sandstone and diamictite interbeds and lenses 10 to 30 cm thick. A persistent layer of laminated diamictite, about 30 m thick, occurs near the top of the KF in the Trout Dale area (Section T, Fig. 3.5). The diamictite types, particularly the bedded and laminated diamictite, grade into one another and the graded sandstone and argillite subfacies.

Interpretation

Facies context of diamictites within the KF indicates deposition in a subaqueous setting, i.e., they are waterlain tills according to the definition and descriptions of Dreimanis (1979). Actual mechanisms of deposition of *massive diamictite* at the top of the KF are, however, difficult to ascertain. This diamictite is texturally very similar to undermelt diamictite described by Gravenor et al. (1984) which is produced by melting at the base of active floating to partially floating ice, and is deposited through a slurry at the base of the ice, possibly near the grounding line (Gravenor et al., 1984). Alternatively, this diamictite could be deposits of thick debris flows, or amalgamated debris flows. Eyles (1987) describes massive diamictites up to 10 m thick, which are interbedded with lenses of gravel and thin-bedded turbidites, deposited by debris flows in large late Pleistocene glacial lakes. Abundant rip-up clasts in thin beds of massive diamictite lower in the KF indicate that they were deposited by debris flows, but, apart from being interbedded with subaqueous sediments, the diamictite at the top of the KF does not contain features definitive of debris flows.

The large thickness of the massive diamictite in western sections is provocative (Figs. 3.4, 3.5, 3.6). Eyles and Lagoe (1990) interpret thick, massive diamictite in a Cenozoic glaciomarine section as formed by rain-out from floating ice under quiet water conditions. Young and Gostin (1991) interpret Neoproterozoic massive diamictites hundreds of meters thick as deposited by rain-out in a tidewater, and active rift, setting where the sediment supply was high. Similar processes may have deposited thick KF diamictite when glaciers laden with sediment calved at a lake margin. Abundant sediment supply and a rapidly subsiding basin are needed in order for such thick diamictite to accumulate. Both these factors are consistent with, and suggest, active tectonism during KF deposition.

The association of KF *bedded diamictite* with turbidites and varved argillite indicates that this diamictite was deposited by subaqueous debris flows, with ice-rafting contributing dropstones. It is in part similar to well-stratified waterlain till in the Pleistocene Catfish Creek Formation, which has been interpreted as subaquatic flow till (Dreimanis, 1979, 1982; Evenson *et al.*, 1977). Locally, where the bedded diamictite overlies pebbly sandstone and conglomerate of the MRF, debris flow deposits may have been winnowed in very shallow water (Fig. 3.7e; Section T, Fig. 3.5).

Seasonal controls probably affected deposition of the KF *laminated diamictite*. The coarse layers were likely deposited by thin debris flows and turbidity currents. The relatively constant thickness of the clay laminae suggests that they represent suspension settling during winter months. Ice-rafting contributed randomly distributed pebbles and cobbles. This subfacies is similar to 'faintly laminated diamictite interbedded with rhythmites and choked with clasts' described as glaciogenic subaquatic slurry flow deposits by Gravenor *et al.* (1984), and is a little coarser grained than the 'diamictic varves' of Bannerjee (1973). Eyles and Eyles (1983) emphasize the importance of both rain-out and resedimentation processes in interpreting a similar facies in a Pleistocene lacustrine section. The laminated diamictite was probably deposited fairly close to the ice front (Gravenor *et al.*, 1984).

Environmental synthesis

Both glaciation and tectonism most likely influenced sedimentation of the KF. Many good dropstones (Fig. 3.5) and varve-like laminae (see also Figs. 9 and 11 in Rankin, 1993) attest to a glacial influence. Because thick massive diamictite at the top of the formation is not closely interbedded with abundant graded sandstone, I infer that it was deposited by or near grounded or floating ice, rather than in distal or basinal locations by sediment gravity flows (cf. Eyles, 1990). The unusually large thickness of both argillite and massive diamictite, particularly considering their restricted areal extent, suggest rapid basin subsidence. Glaciation in a region of active tectonics could lead to such thick sedimentary accumulations.

The topographic relief which existed during deposition of the KF, and the basin in which the formation accumulated, most likely formed through normal faulting (Rankin, 1975; Thomas, 1991). Glacial erosion could have created or accentuated the relief. Neither synsedimentary faults nor glacial pavements have been recognized. Because the KF coarsens upward and records infilling of the basin, probably during glacial advance, I propose that the basin was formed primarily by normal faulting, i.e., tectonic processes, rather than by glacial erosion (Fig. 3.8).

Basal contact relationships of the KF attest to local tectonism before or during sedimentation. Rankin (1993) suggests that the unconformity at the base of the formation south of Konnarock (east of K on Fig. 3.4) could be produced by block faulting and rotation of blocks near MRF volcanic centres. On the other hand, in the Trout Dale region, the KF rests conformably on pebbly sandstone and conglomerate of the MRF (Figs. 3.4, 3.5; Rankin, 1993). These sandstones and conglomerates likely represent braided fluvial deposition on distal parts of an alluvial fan or a braidplain. Abundant volcanic clasts were locally derived from MRF units. The sandstones and conglomerates are locally interbedded with MRF rhyolite flows and were probably deposited on flanks of the MRF volcanic centres. The KF lake lapped up onto these alluvial fans (?)/volcanic centres (Fig. 3.8a).

Argillite facies, which dominate the lower KF, record sedimentation in a thermally stratified lake(s) which extended at least 35 km. Dropstones indicate that it was an ice-contact lake at least during mid- to late KF deposition (Fig. 3.8b). The existence of a lake, rather than sea water, is based primarily upon the nature of the sand/silt to clay couplets (varves) in the fine laminite subfacies. Fine sediment carried across the lake by overflows and interflows settled seasonally from suspension, producing varves. This process was punctuated by underflows and surge/turbidity currents which deposited thicker, coarser laminae and graded sandstone beds. Lake sedimentation was therefore both seasonally controlled (varvites) and episodic (turbidites). Note that the presence of turbidites need not detract from a varved origin for the fine laminites. Sturm and Matter (1978) show that mechanisms of turbidite deposition and clastic varve formation are integrally related. They emphasize that over- and interflows as well as turbidites, with seasonal thermal stratification, are needed to produce laminated clastic sediments in lakes.

More powerful turbidity currents, density-modified grain flows, and debris flows deposited thicker graded sandstone, massive sandstone, and diamictite beds, respectively, in the lake. Effects of liquefaction and units mixed and resedimented by liquefied flows, attest to rapid sedimentation and sediment instability after deposition. As ice floating on the lake surface melted, dropstones and till pellets were scattered throughout the bedded lake deposits. Rare paleocurrent data from ripple cross-lamination and occasional slump folds show dominantly eastward transport within the lake. However, the dearth of evidence for reworking by traction currents

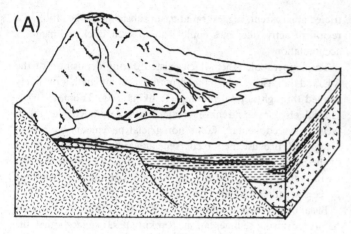

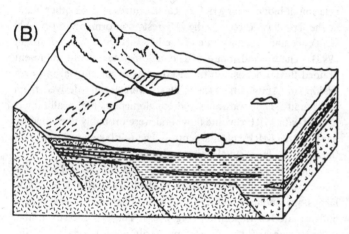

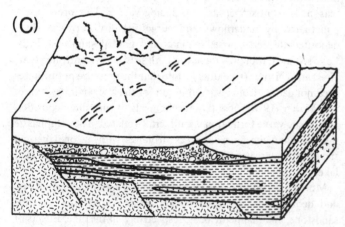

Fig. 3.8. Block diagrams illustrating development of Konnarock Formation basin: (A) Konnarock lake laps up onto alluvial fans and/or volcanic centres of Mount Rogers Formation; (B) Glaciers locally enter lake; (C) Ice covers most of region advancing across lake. For explanation of lithologic (facies) symbols see Fig. 3.5.

suggests quiet bottom conditions interrupted chiefly by sediment gravity flows and ice-rafting.

Abundance of argillite versus sandstone and diamictite depended upon location within the lake and proximity to incoming meltwater sources. The overall sequence, with fine laminites commoner near the base, and sandstone and diamictite interbeds more abundant up section, shows increasing proximity to sediment sources as the KF basin filled, i.e., it is a progradational sequence.

Abundant diamictite at the top of the KF indicates proximity to the ice margin and advance of ice across the region (Fig. 3.8c). Deltaic sands and gravels are not obvious, thus ice cliffs or ice ramps, rather than deltas and subaqueous fans, probably dominated the lake margin (Shaw and Ashley, 1988). Because laminated diamictite is generally associated with argillite, it was likely deposited close to the ice front in places relatively distant from channelized meltwater input. Sediment-laden meltwater, discharged into the lake from subaqueous englacial or subglacial channels, deposited massive sands in channels close to the ice front or as sheets a little farther away. Conglomerate at the base of the topmost diamictite unit could represent deposition near the mouth of a subglacial channel.

Thick, massive diamictite was probably deposited at or very close to the ice grounding line. It could represent sublacustrine moraine deposited below an ice ramp (Holdsworth, 1973; Barnett & Holdsworth, 1974), rain-out from floating ice, and/or thick, subaqueous debris flows. Interbeds and lenses of massive sandstone and laminated argillite within it indicate that subaqueous conditions existed whenever or wherever diamictite was not accumulating. Nevertheless, parts of this diamictite could be lodgement till, deposited when grounded ice advanced over the infilled lake basin.

The KF therefore records glacial advance into and over a lake basin. Glaciers may have advanced from the west, because the top diamictite thins and is locally absent towards the east (Figs. 3.5, 3.6). In places, lacustrine conditions may have resumed later (Section B, Fig. 3.6). However, because an unconformity separates the KF from the overlying Chilhowee Group, an unknown amount of KF sediment could have been removed by erosion.

Alpine glaciers may have deposited the KF, because (1) significant topographic relief existed during KF time, and (2) there is no geographically widespread record of Neoproterozoic glaciation in autochthonous eastern North America, compared, for example, to that of western North America (Hambrey and Harland, 1981). According to Visser's (1989) criteria, the varied stratigraphy and complex vertical and lateral facies relationships in the KF suggest deposition during mountain, as opposed to shelf, glaciation. Clast types do not help make the distinction in the KF because most are of basement rocks and far-travelled ones have not been recognized. Also, the limited areal extent of the KF hampers this distinction. But if uplift was associated with an aborted phase of Iapetan rifting, then ice could have accumulated in mountains created by that uplift.

Speculative continental reconstructions for the latest Neoproterozoic (Bond et al., 1984; Murphy and Nance, 1989, 1991) show the KF in an interior location within a supercontinent. Coeval glaciogenic rocks of Newfoundland and Massachusetts lie within a peripheral orogen, i.e., bordering the supercontinent (Murphy and Nance, 1991). This explains the continental margin and slope setting of the latter diamictites, which are interbedded with, and under- and overlain by, turbidites. In contrast, the KF overlies subaerial volcanic and sedimentary rocks, is overlain by an unconformity, and contains glaciolacustrine facies deposited in a continental rift.

Conclusion

Sedimentary facies and paleotectonic setting of the Neoproterozoic Konnarock Formation both support deposition in a continental setting, most likely a graben within a larger rift system. Most KF rocks accumulated in a thermally stratified, ice-contact lake in a basin that was probably subsiding rapidly. The formation records filling of the lake basin, culminating in ice advance across the area. These rocks demonstrate an interplay of the effects of both glaciation and tectonics. In addition, they show the importance of processes of suspension settling, ice-rafting, sediment gravity flow, and basal ice melting, with significant seasonal control, during glaciolacustrine sedimentation in a rift valley.

Acknowledgements

This research was supported by the Vanderbilt University Research Council. I thank Doug Rankin for introducing me to the area and sharing much experience and many insights about the rocks. I am grateful to Dorothea Withington and Barbara Waugh for assistance in the field, and to participants in the IGCP-260 field trip to the area in July 1989 for helpful discussions on the outcrops. Nick Eyles, Doug Rankin, and Grant Young provided helpful reviews.

References

Aleinikoff, J.N., Zartman, R.E., Rankin, D.W., Lyttle, P.T., Burton, W.C. & McDowell, R.C. (1991). New U-Pb zircon ages for rhyolite of the Catoctin and Mount Rogers formations: more evidence for two pulses of Iapetus rifting in the central and southern Appalachians: *Geological Society of America, Abstracts with Programs*, 23, no. 1, p. 2.

Ashley, G.M. (1975). Rhythmic sedimentation in glacial Lake Hitchcock, Massachusetts-Connecticut. In *Glaciofluvial and Glaciolacustrine Sedimentation*, ed. A.V. Jopling & B.C. McDonald, Society of Economic Paleontologists and Mineralogists Special Publication 23, pp. 304–20.

Ashley, G.M., Shaw, J. & Smith, N.D. (1985). Glacial sedimentary environments. Society of Economic Paleontologists and Mineralogists, Short Course No. 16, 246 pp.

Badger, R.L. & Sinha, A.K. (1988). Age and Sr isotopic signature of the Catoctin volcanic province: implications for subcrustal mantle evolution: *Geology*, 16, 692–5.

Banerjee, I. (1973). Sedimentology of Pleistocene glacial varves in Ontario, Canada: *Geological Survey of Canada Bulletin*, 226, 1–44.

Banerjee, I. (1977). Experimental study on the effects of deceleration on the vertical sequence of sedimentary structures in silty sediments: *Journal of Sedimentary Petrology*, 47, 771–83.

Barnett, D.M. & Holdsworth, G. (1974). Origin, morphology and chronology of sublacustrine moraines, Generator Lake, Baffin Island, Northwest Territories, Canada. *Canadian Journal of Earth Sciences*, 11, 380–408.

Blondeau, K.M. (1970). Sedimentation and stratigraphy of the Mount Rogers Formation, Virginia. M.S. Thesis, Lousiana State University, 117 pp.

Blondeau, K.M. & Lowe, D.R. (1972). Upper Precambrian glacial deposits of the Mount Rogers Formation, Central Appalachians, USA. 24th International Geological Congress, Montreal, Section 1: Precambrian Geology, pp. 325–332.

Bond, G.C., Nickeson, P.A. & Kominz, M.A. (1984). Break up of a supercontinent between 625 Ma and 555 Ma: new evidence and implications for continental histories: *Earth and Planetary Science Letters*, 70, 325–45.

Clark, D.L., Whitman, R.R., Morgan, K.A. & Mackey, S.D. (1980). Stratigraphy and glacial-marine sediments of the Amerasian Basin, Central Arctic Ocean. Boulder, Colorado, *Geological Society of America Special Paper*, 181, 57 pp.

Cowan, E.A. & Powell, R.D. (1990). Suspended sediment transport and deposition of cyclically interlaminated sediment in a temperate glacial fjord, Alaska, USA. In *Glaciomarine Environments: Processes and Sediments*, eds. J.A. Dowdeswell & J.D. Scourse, London, Geological Society Special Publication 53, 75–89.

Dickinson, W.R. & Suczek, C.A. (1979). Plate tectonics and sandstone compositions. *American Association of Petroleum Geologists Bulletin*, 63, 2164–2182.

Diemer, J.A. (1988). Subaqueous outwash deposits in the Ingraham ridge, Chazy, New York. *Canadian Journal of Earth Science*, 25, 1384–96.

Dreimanis, A. (1979). The problems of waterlain tills. In *Moraines and Varves*, ed. Ch. Schluchter, Rotterdam, Balkema, pp. 167–77.

Dreimanis, A. (1982). Two origins of the stratified Catfish Creek Till at Plum Point, Ontario, Canada. *Boreas*, 11, 173–80.

Evenson, E.B., Dreimanis, A. & Newsome, J.W. (1977). Subaquatic flowtill: a new interpretation for the genesis of some laminated till deposits. *Boreas*, 6, 115–33.

Eyles, C.H. & Eyles, N. (1983). Sedimentation in a large lake: a reinterpretation of the late Pleistocene stratigraphy at Scarborough Bluffs, Ontario, Canada. *Geology*, 11, 146–52.

Eyles, C.H. & Lagoe, M.F. (1990). Sedimentation patterns and facies geometries on a temperate glacially influenced continental shelf: the Yakataga Formation, Middleton Island, Alaska. In *Glacimarine Environments: Processes and Sediments*, eds. J.A. Dowdeswell, J.D. Scourse, London, Geological Society, Special Publication 53, 363–86.

Eyles, N. (1987). Late Pleistocene debris flows in large glacial lakes. *Sedimentary Geology*, 53, 33–71.

Eyles, N. & Eyles, C.H. (1989). Glacially influenced deep-marine sedimentation of the Late Precambrian Gaskiers Formation, Newfoundland, Canada. *Sedimentology*, 36, 601–20.

Eyles, N. (1990). Marine debris flows: Late Precambrian 'tillites' of the Avalonian–Cadomian orogenic belt. *Palaeogeography, Palaeoclimatology, Palaeoecology*, 79, 73–98.

Gravenor, C.P., Von Brunn, V., & Dreimanis, A. (1984). Nature and classification of waterlain glaciogenic sediments, exemplified by Pleistocene, Late Paleozoic and Late Precambrian deposits. *Earth Science Reviews*, 20, 105–66.

Hambrey, M.B. & Harland, W.B., eds. (1981). Earth's Pre-Pleistocene Glacial Record: Cambridge, Cambridge University Press, 1004 pp.

Hatcher, R.D., Jr., Thomas, W.A., Geiser, P.A., Snoke, A.W., Mosher, S. & Wiltschko, D.V. (1989). Alleghanian orogen: in *The Appalachian–Ouachita Orogen in the United States*, eds. R.D. Hatcher, Jr., W.A. Thomas, G.W. Viele, Boulder, Colorado, Geological Society of America, F-2, 233–318.

Holdsworth, G. (1973). Ice calving into the proglacial Generator Lake, Baffin Island, N.W.T., Canada. *Journal of Glaciology*, **12** (65), 235–50.

Levell, B.K., Braakman, J.H. & Rutten, K.W. (1988). Oil-bearing sediments of Gondwana glaciation in Oman. *American Association of Petroleum Geologists Bulletin*, **72**, 775–96.

Lowe, D.R. (1975). Water escape structures in coarse-grained sediments. *Sedimentology*, **22**, 157–204.

Lowe, D.R. (1976a). Grain flow and grain flow deposits. *Journal of Sedimentary Petrology*, **46**, 188–99.

Lowe, D.R. (1976b). Subaqueous liquefied and fluidized sediment flows and their deposits. *Sedimentology*, **23**, 285–308.

Lowe, D.R. (1982). Sediment gravity flows. II. Depositional models with special reference to the deposits of high-density turbidity currents. *Journal of Sedimentary Petrology*, **52**, 279–97.

Mackiewicz, N.E., Powell, R.D., Carlson, P.R. & Molnia, B.F. (1984). Interlaminated ice-proximal glaciomarine sediments in Muir Inlet, Alaska. *Marine Geology*, **57**, 113–47.

Miller, J.M.G. (1989). Glacial and glaciolacustrine sedimentation in a rift setting: Upper Proterozoic Mount Rogers Formation, S.W. Virginia, USA. Washington D.C., 28th International Geological Congress Abstracts, 2, 2-436–2-437.

Murphy, J.B. & Nance, R.D. (1989). Model for the evolution of the Avalonian–Cadomian belt. *Geology*, **17**, 735–8.

Murphy, J.B. & Nance, R.D. (1991). Supercontinent model for the contrasting character of Late Proterozoic orogenic belts. *Geology*, **19**, 469–72.

Nemec, W., Steel, R.J., Porebski, S.J. & Spinnangr, A. (1984). Domba Conglomerate, Devonian, Norway: Process and lateral variability in a mass flow-dominated, lacustrine far-delta: In *Sedimentology of Gravels and Conglomerates*, eds. E.H. Koster & R.J. Steel, Canadian Society of Petroleum Geologists Memoir 10, 295–320.

Ovenshine, A.T. (1970). Observations of iceberg rafting in Glacier Bay, Alaska, and the identification of ancient ice-rafted deposits. *Geological Society of America Bulletin*, **81**, 891–4.

Plumb, K.A. (1991). New Precambrian time scale. *Episodes*, **14**, 139–40.

Powell, R.D. (1988). Processes and facies of temperate and subpolar glaciers with tidewater fronts. Short Course Notes. Geological Society of America Centennial Annual Meeting, Denver, Colorado, 114 pp.

Rankin, D.W. (1967). Guide to the Geology of the Mt Rogers Area, Virginia, North Carolina and Tennessee. Carolina Geological Society, Field Trip Guidebook, 48 pp.

Rankin, D.W. (1969). Late Precambrian glaciation in the Blue Ridge Province of the southern Appalachian Mountains (abstract), Boulder, Colorado, Geological Society America Special Paper 121, 246.

Rankin, D.W., Stern, T.W., Reed, J.C., Jr., & Newell, M.F. (1969). Zircon ages of felsic volcanic rocks in the upper Precambrian of the Blue Ridge, Appalachian Mountains. *Science*, **166**, 741–4.

Rankin, D.W. (1970). Stratigraphy and structure of Precambrian rocks in northwestern North Carolina. In *Studies of Appalachian Geology: Central and Southern*, ed. G.W. Fisher, F.J. Pettijohn, J.C. Reed, Jr. & K.N. Weaver, New York, Wiley pp. 227–45.

Rankin, D.W., Espenshade, G.H. & Newman, R.B. (1972). Geologic map of the west half of the Winston–Salem quadrangle, North Carolina, Virginia, and Tennessee. US Geological Survey Miscellaneous Geologic Investigations Map I-709-A.

Rankin, D.W., Espenshade, G.H. & Shaw, K.W. (1973). Stratigraphy and structure of the metamorphic belt in northwestern North Carolina and southwestern Virginia: a study from the Blue Ridge across the Brevard fault zone to the Sauratown Mountains anticlinorium. *American Journal of Science*, Cooper volume, **273-A**, 1–40.

Rankin, D.W. (1975). The continental margin of eastern North America in the Southern Appalachians: the opening and closing of the Proto-Atlantic Ocean. *American Journal of Science*, **275-A**, 298–336.

Rankin, D.W. (1976). Appalachian salients and recesses: Late Precambrian continental breakup and opening of the Iapetus Ocean. *Journal Geophysical Research*, **81**, 5604–19.

Rankin, D.W., Stern, T.W., McLelland, J., Zartman, R.E. & Odom, A.L. (1983). Correlation chart for Precambrian rocks of the eastern United States. United States Geological Survey Professional Paper, 1241-E, 18.

Rankin, D.W., Drake, A.A. Jr., Glover, L., III, Goldsmith, R. Hall, L.M. Murray, D.R. *et al.*, (1989). Pre-orogenic terranes: In *The Appalachian-Ouachita orogen in the United States*: eds. R.D. Hatcher Jr., W.A. Thomas, G.W. Viele, Boulder, Colorado, Geological Society of America, The Geology of N. America, F-2, 7–100.

Rankin, D.W. (1993). The volcanogenic Mount Rogers Formation and the overlying glaciogenic Konnarock Formation – two Late Proterozoic units in southwestern Virginia. *US Geological Survey Bulletin*, 2029, 26 pp.

Rexroad, R.L. (1978). Stratigraphy, sedimentary petrology, and depositional environments of tillite in the Upper Precambrian Mount Rogers Formation, Virginia: M.S. Thesis, Louisiana State University, 164 pp.

Rocha-Campos, A.C., Ernesto, M. & Sundaram, D. (1981). Geological, palynological and paleomagnetic investigations on Late Paleozoic varvites from the Parana Basin, Brazil: Atas c. 3° Simposio Regional de Geologia, Curitiba, Novembro 1981, Sociedade Brasileira de Geologia, Nucleo de Sao Paulo, 2, 162–75

Rust, B.R. & Romanelli, R. (1975). Late Quaternary subaqueous outwash deposits near Ottawa, Canada: In *Glaciofluvial and Glaciolacustrine Sedimentation*, eds. A.V. Jopling, B.C. McDonald, Society of Economic Paleontologists and Mineralogists Special Publication 23, 177–192.

Rust, B.R. (1977). Mass flow deposits in a Quaternary succession near Ottawa, Canada: diagnostic criteria for subaqueous outwash. *Canadian Journal of Earth Sciences*, **14**, 175–84.

Rust, B.R. (1988). Ice-proximal deposits of the Champlain Sea at South Gloucester, near Ottawa, Canada. In *The Late Quaternary Development of the Champlain Sea Basin*, ed. N.R. Godd, Geological Association of Canada Special Paper 35, 37–45.

Schermerhorn, L.J.G. (1974). Late Precambrian mixtites: glacial and/or nonglacial? *American Journal of Science*, 673–824.

Schwab, F.L. (1976). Depositional environments, provenance and tectonic framework: upper part of the late Precambrian Mount Rogers Formation, Blue Ridge Province, southwestern Virginia. *Journal Sedimentary Petrology*, **46**, 3–13.

Schwab, F.L. (1981). Late Precambrian tillites of the Appalachians: In *Earth's Pre-Pleistocene Glacial History*, eds. M.J. Hambrey, W.B. Harland, Cambridge, Cambridge University Press, pp. 751–5.

Schwab, F.L. (1986). Latest Precambrian – earliest Paleozoic sedimentation, Appalachian Blue Ridge and adjacent areas: review and speculation: In *The Lowry Volume: Studies in Appalachian Geology*, eds. R.C. McDowell & L. Glover, III, Virginia Tech., Dept of Geological Sciences Memoir 3, 115–37.

Secor, D.T., Snoke, A.W. & Dallmeyer, R.D. (1986). Character of the Alleghanian orogeny in the southern Appalachians. III. Regional tectonic relations. *Geological Society of America Bulletin*, **97**, 1345–53.

Shaw, J. & Ashley, G.M. (1988). Glacial Facies Models: Continental Terrestrial Environments. *Geological Society of America Short Course*, 121 pp.

Simpson, E.L. & Sundberg, F.A. (1987). Early Cambrian age for synrift deposits of the Chilhowee group of southwestern Virginia. *Geology*, **15**, 123–6.

Simpson, E.L. & Eriksson, K.A. (1989). Sedimentology of the Unicoi Formation in southern and central Virginia: Evidence for late Proterozoic to Early Cambrian rift-to-passive margin transition. *Geological Society of America Bulletin*, **101**, 42–54.

Socci, A.D. & Smith, G.W. (1987). Evolution of the Boston Basin: a sedimentologic prospective. In *Sedimentary Basins and Basin-Forming Mechanisms*, eds. C. Beaumont & A. Tankard, Canadian Society of Petroleum Geologists Memoir 12, 87–99.

Sturm, M. (1979). Origin and composition of clastic varves: In *Moraines and Varves*, ed. Ch. Schluchter, Rotterdam, Balkema, pp. 281–5.

Sturm, M. & Matter, A. (1978). Turbidites and varves in Lake Brienz (Switzerland): deposition of clastic detritus by density currents. In *Modern and Ancient Lake Sediments*. ed. A. Matter, M.E. Tucker, International Association of Sedimentologists Special Publication 2, 147–68.

Thiesmeyer, L.R. (1939). Varved slates in Fauquier County, Virginia. *Virginia Geological Survey Bulletin*, **S1-D**, 105–18.

Thomas, W.A. (1991). The Appalachian–Ouachita rifted margin of southeastern North America. *Geological Society of America Bulletin*, **103**, 415–31.

Visser, J.N.J. (1989). The distinction between ancient mountain and shelf glaciations [abstract]. Washington, D.C., *28th International Geological Congress Abstracts*, 3, 3-303–3-304.

Walker, D. & Driese, S.G. (1991). Constraints on the position of the Precambrian–Cambrian boundary in the southern Appalachians. *American Journal of Science*, **291**, 258–83.

Wehr, F. & Glover, L.I. (1985). Stratigraphy and tectonics of the Virginia–North Carolina Blue Ridge: evolution of a late Proterozoic–early Paleozoic hinge zone. *Geological Society of America Bulletin*, **96**, 285–95.

Wehr, F. (1986). A proglacial origin for the upper Proterozoic Rockfish Conglomerate, central Virginia, USA. *Precambrian Research*, **34**, 157–74.

Williams, H. & Hatcher, R.D., Jr. (1982). Suspect terranes and accretionary history of the Appalachian orogen. *Geology*, **10**, 530–6.

Withington, D.B. (1986). Petrography and provenance of sedimentary rocks from the upper Mount Rogers Formation, Upper Proterozoic, southwestern Virginia. M.S. Thesis, Vanderbilt University, 102 pp.

Young, G.M. (1991). The geologic record of glaciation: relevance to the climatic history of Earth. *Geoscience Canada*, **18**, 100–8.

Young, G.M. & Gostin, V.A. (1991). Late Protrozoic (Sturtian) succession of the North Flinders Basin, South Australia: an example of temperate glaciation in an active rift setting. In *Glacial Marine Sedimentation; Paleoclimatic Significance*, eds. J.B. Anderson & G.M. Ashley, Boulder, Colorado, Geological Society of America Special Paper 261, 207–22.

4 Glaciogenic deposits of the Permo-Carboniferous Dwyka Group in the eastern region of the Karoo Basin, South Africa

VICTOR VON BRUNN

Abstract

Depositional styles of the Late Palaeozoic Dwyka Group in the eastern part of South Africa were governed by the nature of the palaeotopography which, in turn, was largely controlled by the tectonic setting of the region. Much of the Dwyka succession is preserved in an Early Palaeozoic linear downwarp that trended subparallel to the present-day coastline and which originally developed as an incipient crustal rupture preceding the break-up of Gondwana. During the Dwyka glaciation ice flowing south-southwest down the axial gradient of the trough-shaped depression, which continued to subside in Dwyka time, was augmented by glaciers entering the depression from its elevated flanks. Following the Mesozoic fragmentation of Gondwana, only the northwestern flank of the trough remained on the African subcontinent. The sloping region, situated on the southeastern part of the stable Kaapvaal craton, is characterized by an uneven pre-Dwyka surface related to complexly deformed Archaean basement rocks, and a northwest–southeast structural lineament. Limited tectonic activity prevailed during the Dwyka sedimentation. Ice-flow to the southeast was controlled by the palaeoslope while the irregular topography profoundly influenced the nature of sedimentation. Debris-filled glacially incised valleys, excavated along zones of crustal weakness, extended northwards into the periphery of a bordering highland region down which ice flowed and which was later subjected to limited post-glacial isostatic rebound.

The predominant lithotype of the Dwyka Group is a bluish-black diamictite which originated in a marine setting primarily by ice-rafting and settling of suspended fines, on the floor and flanks of the subsiding regional trough. A heterolithic sedimentary assemblage, overlying the diamictite in the zone of uneven palaeorelief on the northwestern flank of the downwarp, testifies to an oscillatory glacial recession associated with fluctuating sea levels. These glaciogenic deposits comprise argillaceous and arenaceous diamictite, conglomerate, sandstone, siltstone and mudrock. Sedimentation towards the close of the glacial episode was dominated by outwash processes and pulses of subaqueous sediment gravity flow that produced extensive arenaceous and rudaceous deposits on the regional palaeoslope. A complex evolutionary spectrum of mass movement included turbidity currents and debris flows. During the final stages of deglaciation, ice receded and disintegrated on the highland to the north where large volumes of meltwater reworked glaciogenic debris under mainly subaerial conditions. Fine outwash particles, transported basinwards, finally mantled glaciomarine deposits of the Dwyka Group to the south.

Introduction

The Permo-Carboniferous Dwyka Group of southern Africa was deposited in the structural Karoo Basin and comprises glaciogenic rocks, up to 800 m thick, that constitute the lowermost stratigraphic unit of the Karoo Sequence (Visser, 1989). This contribution is focused on the Dwyka Group in the eastern region of the Karoo Basin (Fig. 4.1) where Matthews (1970) recognized three broad palaeotopographic zones which he referred to as a pre-Dwyka highland surface, an intermediate zone of irregular palaeo-relief and a pre-Dwyka lowland plain.

This communication adheres to these subdivisions and aims to apply more recent concepts relating to the influence on the character and arrangement of sedimentary facies by palaeotopography, the tectonic setting, and processes associated with changing conditions during the period of glaciation.

Pre-Dwyka highland surface (Northern Region)

Matthews (1970) has shown that a gentle southeastward descent of an ancient upland surface in the north, referred to here as the Northern Region (Fig. 4.1), coincides with the regional Late Palaeozoic ice-flow direction illustrated by Du Toit (1921). This trend is reflected by the arrangement of pre-Dwyka valleys along the incised and indented southeastern periphery of the elevated region where it falls sharply to the south. The Dwyka Group is dispersed over a broad arc in the Transvaal and extends south into northern Natal (Fig. 4.1). Erosional remnants of the Late Palaeozoic glaciogenic beds rest on an undulatory palaeoplain bevelled across Archaean granitoid-greenstone terrane and Early Proterozoic successions. The thickness reaches a maximum of only a few tens of metres (Mellor, 1905b; Matthews, 1970; Von Brunn et al., 1988).

The Dwyka deposits are represented mainly by arenaceous to

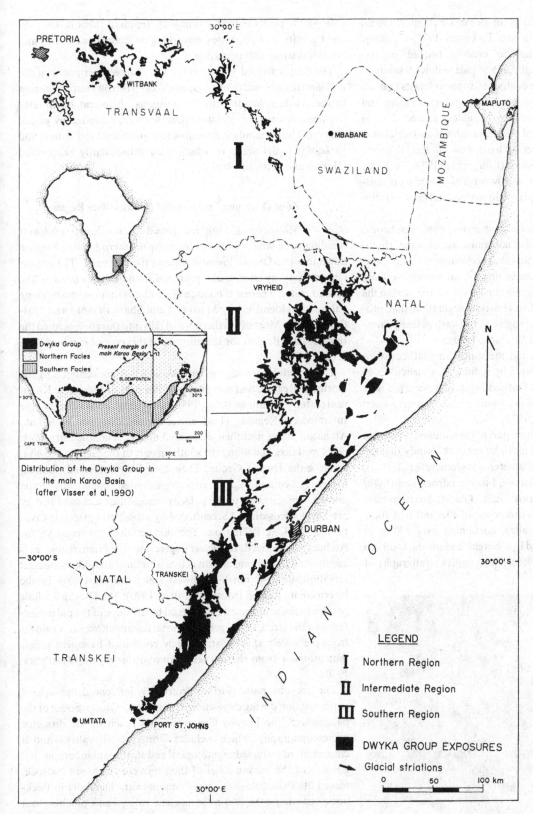

Fig. 4.1. Locality map demonstrating the distribution of Dwyka Group exposures in the eastern part of the Karoo Basin. Subdivision into three palaeotopographic regions is based on Matthews (1970). Inset map shows the extent of Northern (valley/inlet) and Southern (platform) facies of the Dwyka Group of Visser (1986) and Visser et al. (1990) in the main Karoo Basin.

rudaceous rocks described as glacial conglomerate (Mellor, 1905*a*) and diamictite (Le Blanc Smith and Eriksson, 1979; Le Blanc Smith, 1980). This structureless to crudely bedded matrix-supported sedimentary rock is light grey to pale yellow. A sandy to silty matrix, characteristic of the diamictite, supports clasts that are randomly orientated and subrounded to angular. Cobbles and pebbles are most common, but some boulders exceed 2 m in diameter (Mellor, 1905*b*). Several clasts are faceted and striated.

The diamictite is the product of mass flow of glacial debris released from a retreating ice sheet (Cairncross, 1979; Le Blanc Smith, 1980). Local reworking by meltwater is indicated by percussion marks on clasts of associated conglomerates and cross-stratification of interbedded sandstone.

The diamictite is succeeded by conglomerate, pebbly sandstone, massive and cross-stratified sandstone, laminated siltstone, rhythmites and shale with ice-rafted clasts that accumulated in fluvioglacial outwash and glaciolacustrine settings. Coal seams overlying diamictite and associated glaciogenic deposits are attributed to the development of shallow-rooted tundra-type vegetation that produced peat bogs on abandoned proglacial outwash plains (Cairncross, 1979; Le Blanc Smith and Eriksson, 1979).

The nature of these deposits and the south- to southeastward palaeocurrent trends, demonstrated by sedimentary structures of braided fluvial deposits, suggest that sedimentation occurred near the northern margin of the Karoo Basin where relatively steep gradients prevailed (Cairncross, 1979).

Ancient valleys along the southern part of the highland region are partly re-excavated and contain partially reworked sandy diamictite with boulders up to 3 m in diameter, conglomerate (Fig. 4.2), massive to cross-laminated sandstone, black mudrock and rhythmically banded dropstone-bearing shale. Glacial striations are preserved on steep valley flanks composed of Precambrian rock. One southeast-trending palaeovalley, containing over 300 m of glaciogenic rocks, was intersected by a borehole along the southern border of Swaziland (Hunter, 1969). The complex stratigraphy of

Fig. 4.2. Glaciogenic debris exposed in a pre-Dwyka valley. Granitoid boulder protrudes from arenaceous diamictite. Partial reworking is evident from clast concentrations and conglomerate in background.

valley-fill deposits indicates slumping, repeated subaqueous sediment gravity-flow processes, reworking by meltwater currents and glaciolacustrine sedimentation.

The deeply incised valleys along the southern periphery of the highland region, and their extension as deep furrows into the region to the south, coincide with the positions of ancient northwest–southeast zones of crustal weakness that were exploited by glacial erosion. These include Archaean shear and thrust structures, and strike-slip faults some of which were subsequently reactivated (Hunter and Wilson, 1988).

Zone of irregular palaeorelief (Intermediate Region)

Matthews (1970) recognized a northeast–southwest trending zone which descends from the northern highland region down to a pre-Dwyka lowland plain in the southeast. The zone is characterized by an irregular palaeorelief and extends north to the upland region where it is bounded by a discontinuous south-facing palaeoscarp identified by Matthews and Sharrer (1968) and Matthews (1970). Most of northern Natal falls into this zone, termed the Intermediate Region for the purpose of this communication (Fig. 4.1).

The granitoid-greenstone terrane, which is encountered in the Northern Region, and which constitutes part of the stable Kaapvaal craton (cf. Tankard *et al.*, 1982), also forms the basement of the Intermediate Region. These ancient rocks, together with Late Archaean strata and their associated major magmatic intrusions, developed here and along the southern part of the highland region, underlie the Dwyka Group. Their complexly deformed (faulted, sheared, tilted and folded) nature contributed significantly to the uneven configuration of the palaeosurface which was modified by pre-Dwyka erosion and accentuated by subsequent glacial excavation in the Late Palaeozoic. Tectonic activity that began in the Archaean, continued, to a lesser degree, into the Phanerozoic, as is exemplified by faulting and an active structural arch-like basement elevation along a north-northwest–south-southeast axis in the Intermediate Region (Matthews, 1970, 1990). The Dwyka ice-flow pattern to the southeast was governed by the slope of the palaeosurface and the strike of the general crustal lineament which is evident from the widely spaced, structurally controlled basement inliers that protrude from the Dwyka Group in this region (Matthews, 1970).

The irregular palaeosurface profoundly influenced the style of sedimentation during deposition of the Dwyka Group. South of the palaeoscarp, the Dwyka Group mantles much of the dissected palaeotopography, which includes former glacial valleys, and is characterized by its sedimentological and stratigraphic complexity. Because of the uneven relief of the subjacent Archaean bedrock, these Late Paleozoic glaciogenic deposits vary markedly in thickness, attaining over 300 m in ancient valleys and pinching out against some basement highs.

The Dwyka Group thus rests unconformably on Archaean rocks but towards the southern and southeastern limits of the Intermediate Region it disconformably overlies Early Palaeozoic sediment-

Fig. 4.3. Massive argillaceous diamictite with dispersed clasts overlain by Dwyka sandstone with a conglomeratic base. Large granitoid clast supported in diamictite (bottom centre) measures 1.5 m. Height of cliff: 20 m.

ary strata. The regional southeastward ice-flow direction, being a continuation of that documented to the north, is evident from the common occurrence of glacial pavements on resistant basement rock. Soft-sediment glacial grooves have been recorded within the Dwyka succession by Von Brunn (1977) and Von Brunn and Talbot (1986). Black argillaceous shales of the Permian Ecca Group, draped over an undulatory Dwyka surface, follows this glaciogenic succession conformably.

The lower part of the Dwyka Group comprises bluish-black diamictite and subordinate mudrock. This is followed abruptly by a heterolithic sequence of interstratified sedimentary rocks, up to 150 m thick, which include beds of diamictite identical to that at the base of the succession (Fig. 4.3). Stratigraphic sections, illustrated by Von Brunn (1987), display the marked vertical and lateral facies changes that characterize these sedimentary assemblages.

In the Intermediate Region a basic distinction can be made between a diamictite–mudstone facies association and a sandstone–arenaceous diamictite–conglomerate facies association.

Diamictite–mudrock facies association

Homogeneous bluish-black *argillaceous diamictite* constitutes up to 70% of the Dwyka Group of the Intermediate Region. The term argillaceous refers to the extremely fine-grained, dense matrix with microscopic comminuted mineral and lithic particles. This matrix supports randomly dispersed clasts, ranging in size from granules to boulders up to 2.5 m in diameter. The clasts, some of which are faceted and striated, vary in shape from angular to subrounded. The most common clast lithologies are basement granitoid rocks. The argillaceous diamictite is mostly massive but locally displays crude stratification. The northernmost exposures occur along the base of the palaeoscarp along the border of the upland region (Von Brunn, 1987).

Stratified diamictite has affinities with dropstone-bearing fissile and laminated *mudrock* into which it grades. The mudrock is subordinate and occurs mostly in the upper, heterolithic part of the Dwyka succession where some rhythmites are also developed. Parting planes of laminated argillites are commonly studded with clasts, and arthropod trackways are preserved on some surfaces (Anderson, 1981).

Interpretation

The *diamictite* originated primarily by ice-rafting, settling of suspended fines, redeposition by gravity flow of the rained-out sediment (cf. Eyles *et al.*, 1985) and lodgement. The overall textural homogeneity of massive diamictite can be attributed to passive flow into subaqueous topographic depressions and mixing of the accumulating debris.

The degree of grading of *mudrock* in and out of diamictite suggests that the processes of sedimentation, particularly suspension settling, were essentially similar for these lithofacies. Much of the argillite, supporting clasts of various sizes, originated as what Powell (1984) has termed bergstone mud. The presence of arthropod trails indicates a suitable habitat for bottom-dwelling fauna in a low-energy proglacial setting.

Sandstone–arenaceous diamictite–conglomerate facies association

This association, constituting most of the heterolithic sequence, forms a broad lithological spectrum in which boundaries are generally diffuse. The main lithotypes comprise fine-grained sandstone, siltstone, arenaceous diamictite and conglomerate. These deposits are developed high up in the Dwyka succession where they commonly drape basement highs and are dominant in the northernmost part of the outcrop region, south of the palaeoscarp. Sandstone, in association with sandy diamictite, forms discontinuous horizontal ridges fringing valley- and hillsides, cliffs and streamed exposures. In the extreme southern part of the Intermediate Region outcrops of the arenaceous–rudaceous association diminish and are replaced by stratified lithologies, including argillites, that make up the topmost part of the Dwyka Group in the Southern Region.

Fig. 4.4. Laterally extensive overlapping lenses of structureless to horizontally stratified fine-grained glaciogenic sandstone.

The *fine sandstone and associated siltstone* is light grey to pale yellow and is the most widely distributed component of the association. It typically has a lens-, sheet- to broad channel-like geometry and units persist laterally from several tens of metres to a few kilometres (Fig. 4.4). Marked thickness variations, ranging from <0.5 to about 25 m, where overlapping sets with amalgamated, or fused, contacts are developed, characterize laterally extensive bodies. Whereas the sandstone in many exposures appears to be massive, ripple cross-lamination and flat lamination are commonly preserved. Sedimentary structures are often found to be obscured by the homogeneity of the sandstones and siltstones that are generally well sorted. Sand particles, in thin section, are mostly subangular to angular, and closely packed. They consist mainly of quartz and subordinate fresh feldspar. Dispersed pebbles and cobbles are irregularly distributed in the sandstone which commonly merges with arenaceous diamictite. Basal contacts are commonly sharp. Dewatering structures have been observed in some exposures.

Arenaceous diamictite is common in the upper part of the Dwyka Group in the Intermediate Region and is lithologically similar to the sandy diamictite of the Northern Region. The rock is a poorly sorted matrix-supported conglomerate which is distinguished from the argillaceous diamictite by its light colour, typical arenaceous character of the matrix, textural inhomogeneity and higher clast content. It contains clast clusters, lenses of coarse sandstone, and sandy wisps. The arenaceous diamictite grades over short distances (tens of metres) into the associated lithologies, with end-members being clast-supported conglomerate on one extremity of the spectrum, and sandstone with sparsely dispersed clasts on the other. The diamictite has a lenticular geometry and its lateral extent is far more restricted than that of the sandstone. Thickness of units, which are commonly stacked, rarely exceeds 10 m. Clast sizes range from 2 mm (an arbitrary lower limit) to boulders up to 3 m in diameter. Shapes are mostly rounded to subrounded. Clasts are represented by a wide variety of lithotypes of which granitoid basement rocks predominate. The diamictite is massive to poorly bedded. Basal contacts of the arenaceous diamictite are invariably sharp.

Junctions between clast-rich diamictite and conglomerate are generally difficult to define because of the lithological gradation. *Conglomerate* is regarded here as a rudaceous rock with clast packing density in excess of 25% (Visser, 1986) and includes both matrix- and clast-supported varieties. Clast-supported conglomerate accounts for a relatively minor proportion (about 5%) of the arenaceous–rudaceous association. It is most commonly developed as lenses, pockets, irregularly shaped clusters, basal lag, nests and stringers occurring within the sandstone or arenaceous diamictite. The conglomerate is generally poorly sorted. Subrounded cobbles and pebbles are the most common constituents. Deposits are massive to poorly stratified. Over 75% of the clasts are represented by granitoid lithologies.

Interpretation

The sandstone–arenaceous diamictite–conglomerate facies association reflects a complex evolutionary continuum of subaqueous sediment gravity flow processes that include debris flow and turbidity current sedimentation (cf. Middleton and Hampton, 1973; Lowe, 1979; 1982).

The sediment was derived by undermelt (Gravenor et al., 1984) near the grounding line of a marine ice sheet where large quantities of subglacial detritus were deposited during glacial retreat. Morainal mounds, that became increasingly unstable by continued accumulation of debris, were subjected to slumping, remobilization and gravity flow. The marked facies variations, occurring over short distances, that typify these Dwyka deposits, is a feature also noted by McCabe (1986) in Pleistocene sediments left in front of retreating glacier margins.

Subaqueous outwash processes and density currents were involved in the deposition of extensive sheet sands and silts. Turbidity flows, transporting cohesionless particles winnowed from proglacial debris, which was moving downslope under gravitational influence, contributed substantially to the development of these deposits. The mechanism was probably similar to that reported by Wright and Anderson (1982) who described similar processes occurring on the Weddell Sea continental shelf, on which mass flow was induced by a rugged topography. Massive, and commonly overlapping lobes of Dwyka *sandstone* were produced by rapid suspension sedimentation of sand and silt from high-density turbidity currents and clouds of fine fractions supported by fluid turbulence (cf. Lowe, 1982). Horizontal limination and ripple-drift cross-lamination resulted from the deceleration of traction currents during the depositional process. Channel fills of apparently massive sandstone, originating from swift deposition by sediment-laden meltwater currents discharged from the base of grounded ice, are to some extent reminiscent of the Quaternary subaqueous outwash sands described by Rust and Romanelli (1975). Rip-up clasts, occasionally encountered in the sandstone, are indicative of scour.

The *arenaceous diamictite* represents deposits from cohesive debris flows (cf. Lowe, 1979) and processes of resedimentation (Fig. 4.5). During downslope flow the cohesiveness of the sediment–water matrix had sufficient strength to support large clasts (cf. Lowe, 1982), thus accounting for the presence of boulders in the

Fig. 4.5. Debris flow arenaceous diamictite in contact with Archaean diabase (on right of dotted line) of a basement high, over which it is draped.

diamictite. The arenaceous and clast-rich character of this rock is attributed to the removal of very fine sand to clay-sized particles by winnowing (cf. Miller, 1989). The general roundness of many of the clasts reflects abrasion in the tractional zone of the ice sheet as described by Boulton (1978). Stacking of arenaceous diamictite beds suggests pulses of gravity flow events. Locally developed undulatory flow structures in the diamictite and the sharp base of individual units indicate emplacement by debris flow.

Conglomerate is the product of gravelly high-density turbidity currents (cf. Lowe, 1982), subaqueous outwash and reworking of glaciogenic debris. The overall poorly sorted nature of the conglomerate, and local slump structures suggest ice-proximal sedimentation. Lag deposits and pockets of conglomerate in diamictite and associated sandstone are ascribed to prolonged depletion of fines by the winnowing activity of traction currents.

The uneven palaeosurface was a significant factor in promoting and controlling subaqueous sediment gravity flow processes, and contributed to the evolution of the various sedimentary facies that characterize the Dwyka Group of the Intermediate Region.

Pre-Dwyka lowland plain (Southern Region)

To the south and southeast the zone of irregular palaeotopography merges into a pre-Dwyka lowland region which forms part of a broad trough-shaped topographic depression with an axial gradient to the south-southwest (Matthews, 1970). Here the Dwyka Group was laid down on a flat-lying Early Palaeozoic sedimentary succession comprising the Natal Group. The Dwyka deposits overstep these sedimentary beds to rest on the subjacent rocks of the 1000 Ma-old Natal Metamorphic Province (cf. Tankard *et al.*, 1982) which constitute the basement south of the Kaapvaal craton. The cover of Ecca Group shale extends into the Southern Region where it thickens and continues to overlie the Dwyka Group conformably. Post-Dwyka block faulting, seaward tilting, monoclinal flexure and erosion, following the fragmentation of Gondwana, has led to the development of a double Dwyka outcrop belt separated by exposures of metamorphic basement rocks (Fig. 4.1).

The linear downwarp of the pre-Dwyka lowland region, with its axis subparallel to the present-day coastline, was initiated in the Early Palaeozoic and marks the line of an incipient crustal rupture that developed in response to attenuation induced by tensional stress prior to the break-up of this part of Gondwana in the Mesozoic. The subsiding trough behaved as a failed rift in which initially fluvial sediments of the Natal Group accumulated (Hobday and Von Brunn, 1979). Tectonism of this regional depression continued during Dwyka and post-Dwyka Karoo sedimentation (Matthews, 1970; Whateley, 1980) and sustained subsidence contributed to thickening of the Dwyka Group in the south. The south-southwest plunge of the trough is confirmed by the Dwyka ice-flow direction (Fig. 4.1), inferred from widely developed glacial striae, which coincided with the palaeocurrent trend defined by sedimentary structures in the subjacent Natal Group. The southeast-sloping Intermediate Region comprises the northwestern flank of the broad topographic depression. The flank was mirrored by an elevated region that bordered the lowland region to the east of the present Natal coast and which provided an additional provenance for the transport of glacial sediment into the downwarp (Matthews, 1970). Because Mesozoic break-up occurred approximately along the trough axis, only the northwestern flank was preserved on the African subcontinent. A marked difference in the lithostratigraphic character of the Dwyka Group on the elevated flank (the Intermediate Region) and the adjoining pre-Dwyka lowland reflects the influence of basin tectonics on glacial sedimentation in this region during the Late Palaeozoic.

Argillaceous diamictite constitutes an estimated 90% of the Dwyka Group in the Southern Region. The rock is typically homogeneous and massive. Lithologically it is identical to that encountered in the Intermediate Region to the north, except for a higher matrix-to-clast ratio and an abundance of quartzite occurring together with granitoid clasts. Maximum boulder size rarely exceeds 1 m. Crude stratification occurs in several localities (Du Toit, 1946; Thomas *et al.*, 1990). Unlike the region to the north, a heterolithic interstratification with arenaceous and rudaceous sedimentary rocks is not developed. Subordinate shale and sandstone, at different levels within the diamictite, have been reported by Kent

(1938), Gevers (1941), Du Toit (1946) and King and Maud (1964). These horizons are of limited lateral extent, and range from a few centimetres to less than 2 m in thickness.

The basal part of the Dwyka Group, where it rests on the Early Palaeozoic substrate, commonly has a marked sandy character, such that it becomes almost indistinguishable from the subjacent arenite (cf. Mountain and Hodson, 1933). The zone can be several metres thick and grades up into argillaceous diamictite. In other localities the diamictite is separated from the pre-Dwyka substrate by a 3 to 10 m-thick sequence of sandstone, sandy to shaley siltstone containing dropstones, rhythmite and shale (Mountain and Hodson, 1933; King, 1948; King and Maud, 1964; Von Brunn and Marshall, 1989). In the southern outcrop region Thomas *et al.*, (1990) noted that the lowermost 60 m of the 450 m-thick Dwyka Group comprises diamictite units, 2 to > 20 m thick, interstratified with shaley dropstone-bearing diamictite and finer sandstone associated with rhythmites.

The top 20 m of the Dwyka succession becomes stratified, as massive, homogeneous diamictite is followed by silty to sandy diamictite which progressively passes up into fissile shale (Von Brunn and Gravenor, 1983). Boulder- to pebble-sized ice-rafted clasts are widely dispersed in this sequence whereas the overlying Ecca shale is devoid of clasts.

The maximum thickness of the Dwyka Group is 100 to 200 m over the greater part of the Southern Region (Du Toit, 1931; Kent, 1938; King, 1948; Von Brunn and Gravenor, 1983) but increases to between 450 and 600 m in the south (Gevers, 1941; Du Toit, 1946; Thomas, 1988; Thomas *et al.*, 1990).

Glacial pavements are widespread and record regional ice-flow to the south-southwest (Du Toit, 1921). Extensive glacial abrasion of the sandstone floor has produced the sandy character at the base of the Dwyka succession and has contributed to the smoothness of the palaeosurface (King, 1948). Local irregularities of the pre-Dwyka surface, including fault-controlled valleys, troughs, and hollows, are filled with glaciogenic sediment (Mountain and Hodson, 1933; Von Brunn and Marshall, 1989).

Interpretation

As in the Intermediate Region, the argillaceous diamictite originated mainly by iceberg-rafting, hemipelagic suspension settling of fines from turbid plumes, and subaqueous sediment gravity flow. The paucity of large clasts may indicate ice-distal sedimentation (cf. Eyles *et al.*, 1985; Miall, 1985; Visser and Loock, 1987). The overall monotonous appearance of this homogenous diamictite, and its great thickness that increases to the south, suggests a prolonged period of low-energy glaciomarine sedimentation on a subsiding basin floor. The appreciable glaciogenic sedimentation rates that are possible in modern high latitude oceans was demonstrated by Powell (1984) while Eyles *et al.* (1985) drew attention to the blanket-like geometry, significant regional extent and the homogeneous character of marine diamictons. Shales intercalated with thick argillaceous diamictites have also been reported from the Main Karoo Basin by Visser (1983) who, noting that their extent is not basinward, suggested that they may be indicative of intersta-

dials. Interbedded arenites, particularly those occurring towards the base of the Dwyka Group, represent material entrained by ice that planed bedrock obstructions. Some sandstone lenses could have resulted from intermittent reworking during periods of reduced glacial sedimentation.

The upward fining sequence, forming the topmost strata of the Dwyka Group, was ascribed by Von Brunn and Gravenor (1983) to the recession of a marine ice sheet towards the close of the Late Palaeozoic glacial episode.

Sequence of events

During the Permo-Carboniferous glaciation, which lasted some 60 Ma (Visser, 1990), ice masses flowing from spreading centres in the north, south and east, coalesced to form an extensive cover in southwestern Gondwana (Visser, 1989). Marine conditions prevailed during sedimentation of the Dwyka Group (Visser, 1989). The broad events and processes pertaining to the Dwyka glaciation, identified by Visser (1987; 1989; 1991), can, to a large degree, be traced to the eastern region of the Karoo Basin.

During *initial advance* of the ice into the eastern part of the Karoo Basin in the Late Palaeozoic, glaciers scoured the pre-Dwyka bedrock surface. There is no evidence, however, to prove that the glaciated pavements, preserved today on older rocks of Natal and southeastern Transvaal, mark the early stages of glaciation and may equally record subsequent glacial advance. The two major regional sets of striations, one trending northwest–southeast, the other indicating ice flow to the southwest (Fig. 4.1), led Du Toit (1921) to postulate the existence of two separate ice sheets, one originating from a spreading centre off the present-day Natal coast, the other centred on a pre-glacial highland to the northwest. Matthews (1970), however, recognized the significance of the tectonic framework of this region as a controlling factor in the transport and location of glacial debris. He proposed that ice, accumulating on high ground along a northeastern margin of the linear pre-Dwyka downwarp of Natal, prior to break-up of Gondwana, expanded to highlands in the west and southeast while an ice lobe began to migrate down the south-southwest axis of this regional depression. With further expansion the flowing ice was augmented by ice moving down the northwestern (the Intermediate Region) and eastern flanks of the lowland trough.

At the time of *maximum glaciation* ice masses from extensively glaciated elevated regions flowed downslope and overdeepened pre-glacial river valleys and excavated zones of structural weakness in the Archaean basement surface. Subsidence of the Early Palaeozoic trough, furthered by isostatic depression related to the advancing ice, brought on a marine incursion from the south where the linear downwarp opened into the main Karoo depository which formed part of a foreland basin (Viser, 1991). Bedrock erosion was sustained by the ice grounded below sea level. Limited deposition occurred in subglacial depressions.

A progressive rise in sea level, following *climatic amelioration*, led to decoupling of the ice sheet in deepening parts of the basin. A major depositional phase ensued as prolonged debris rain-out from floating ice and resedimentation on the basin floor resulted in the accumulation of thick diamictons, particularly in the Southern Region. These sediments blanketed large areas and infilled topographic depressions. Because of continued slow subsidence of the downwarp, and being removed from erosion processes, diamictons had a high preservation potential.

Where the marine ice sheet was partly grounded on the irregular palaeosurface flanking the regional trough, considerable volumes of debris were rafted to sea by icebergs, calved from tidewater glaciers, and contributed significantly to the accumulation of thick homogeneous diamictons in submarine depressions. Fine fractions settling from overflow plumes, and hemipelagic suspension, contributed to deposition of diamictic mud. Wet-based conditions near grounding lines induced shedding of glaciogenic debris, originating from the base of retreating ice, to accumulate in the proglacial zone. Except possibly for the Intermediate Region, where basement highs could have anchored floating ice, it seems unlikely that ice shelves played any significant role in the origin of the argillaceous diamictite occurring over such a large region. Ice shelves would probably also not have survived the prevailing subpolar conditions proposed by Visser (1989). Considering the constraints relating to lodgement and melt-out depositional processes, it is difficult to assess to what degree subglacial sedimentation of locally grounded ice in the Intermediate Region contributed to accumulation of the diamictons. In very few localities here do diamictites reveal the criteria diagnostic of deposition by lodgement, such as sheared fabric.

Intensified deglaciation towards the close of the Permo-Carboniferous glacial period resulted in development of the multiple sedimentary assemblages noted today in the upper part of the Dwyka succession in the Intermediate Region. An oscillatory retreat of the ice, related to major eustatic fluctuations and adjustments in the glacial budget, produced these complexly interbedded lithofacies. The effects of possible tectonic adjustment on the north-northwestern flank of the subsiding regional trough cannot be ascertained from the Dwyka strata. However, moderate flexuring towards the depression may have influenced downslope sediment transport.

The uneven configuration of the palaeosurface and regional southeastward slope were major factors in promoting episodic subaqueous gravity flows of sediment released from rapidly disintegrating ice. Sand, winnowed from glacial debris, was dispersed by meltwater and turbidity currents to be deposited as extensive sheets and southeastward prograding apron-fan systems. Stacking and subsequent amalgamation of sheet sands, and associated sandy diamictons originating from dilute debris flow slurries, resulted from successive gravity flow surges and lateral shifting of sediment-charged currents. Subaqueous outwash fans developed near grounding lines where meltwater debouched from subglacial tunnels. From the upland region large volumes of debris were channelled southward by ice streams flowing down glacial valleys incised into the periphery of the highland region. Farther south this sediment load was dispersed beyond valley outlets to contribute to the subaqueous sediment fan and lobe complexes of the Intermediate Region.

Periodic readvance of the ice produced subglacial diamictons that were interbedded with mass flow deposits left by glacial retreat. Overriding ice locally gouged soft sediment glacial grooves into previously-deposited proglacial arenaceous beds that were subsequently covered by diamicton (Von Brunn, 1977; Von Brunn and Talbot, 1986). Disintegration of grounded ice, retreat of the glacial margin and subsequent release of debris were accelerated by sea-level rise.

During the *final stages of glaciation* a sharp rise in sea-level led to rapid break-up and eventual collapse of the ice sheet. Moderate isostatic rebound, following reduction of the ice load, affected mainly the elevated Northern Region (cf. Visser, 1987) resulting in extensive denudation of glacial deposits from the highland surface under subaerial conditions. Waterlogged debris, melting out from stagnant ice, flowed into depressions on the landsurface or was reworked by increasing volumes of meltwater. Large quantities of debris also slumped into former glacial valleys and topographic depressions, where redeposition by subaqueous sediment gravity flow and current reworking ensued, and where glaciolacustrine sediment finally accumulated (cf. Le Blanc Smith and Eriksson, 1979).

During this time the extensive subaqueous glaciomarine sedimentary cover south of the elevated region remained virtually undisturbed and was eventually preserved under a blanket of accumulating mud and silt. These deposits, now forming the extensive Ecca Group argillites that overlie the Dwyka succession, originated from the basinward transport of fine particles derived from glaciogenic debris that was being reworked to the north.

In the final phase of the glacial episode small ice caps were confined to highland areas (Visser, 1989, 1991). A tundra-type vegetation flourished and peat bogs developed on proglacial outwash plains below to produce the present-day coal occurrences above the glaciogenic deposits of the Northern Region. Subsequent subaerial processes eroded these glacially derived sedimentary rocks, remnants of which are now found as widely dispersed deposits in the southeastern parts of the Transvaal. They attain thicknesses of up to only a few tens of metres, and constitute mere vestiges of sediment deposited during the Permo-Carboniferous glaciation. In contrast, a substantial thickness of Dwyka diamictite is preserved in the major tectonic downwarp of Natal which was initiated well before glaciation and which remained unaffected by post-glacial isostatic rebound.

Discussion

Within the context of the Karoo Basin the clast-poor predominantly homogeneous, massive diamictite in the southern part of Natal and the Transkei can be regarded as belonging to the Southern ('platform') facies of Visser *et al.* (1990) (see inset of Fig. 4.1). This platform region borders on the elevated glacially incised terrain of a Northern ('valley and inlet') facies (Visser *et al.*, 1990) which comprises heterolithic sequences, similar to those exposed in northern Natal (Intermediate Region).

The stratigraphy of the Intermediate Region is, however, anomalous in that it contains elements of both of the lithologically differing broad regional facies divisions in the Dwyka Group identified by Visser (1986). In northern Natal the lower part of the succession, being dominated by argillaceous diamictite, is an extension of the diamictite of the Southern facies, whereas the overlying strata of this region are heterolithic and distinctive of the Visser *et al.*, (1990) northern facies. The anomaly in northern Natal could reflect the tectonic influence exerted on the flank of the subsiding regional trough during the period of glacial sedimentation.

The region in the eastern part of southern Africa, where the Dwyka Group, discussed in this paper, is exposed, has always been regarded as belonging to the Karoo Basin. This major basin constitutes the type area comprising what Rust (1975) referred to as the 'Karoo tectono-sedimentary terrain' (p. 538) of southern Africa, and which is 'characterised by broad, open basins with intervening swells' (p. 540). Rust draws attention to the contrast between this region and his 'Zambezian terrain' in the eastern region of southern Africa which is 'distinguished by its tensional fault tectonics' (p. 540) and in which graben-type deposits occur. As suggested earlier, deposition of the main body of the Dwyka Group discussed here was controlled by the presence of a subsiding trough which was initiated in the Early Palaeozoic as a failed rift resulting from tensional crustal stresses (Hobday and von Brunn, 1979). Further, the influence of graben tectonics in the northern part of the regional downwarp on the depositional history of the Permian Ecca Group, overlying the Dwyka succession, was demonstrated by Whateley (1980). The question therefore arises whether the Dwyka Group of Natal can indisputably be regarded as being part of the Dwyka succession belonging to the main Karoo Basin and whether it should tentatively be regarded as an accumulation in a depository resembling those of the 'Zambezian tectono-sedimentary terrain' (Rust, 1975) and which merges south into the main Karoo Basin. In this regard cognizance should perhaps also be taken of the fact that the Lesotho Rise, to the west of Natal, was tectonically a mildly positive element over a long period of time (Rust, 1975), and may have contributed to differences in sedimentation between in the main Karoo Basin and the accumulation of glaciogenic deposits to the east during the Dwyka glaciation.

Many unresolved questions concerning the Dwyka Group in Natal remain to be addressed. Amongst these is the apparent lack of evidence for interglacials recognized by Visser (1983, 1987, 1991). The effects of such events may have been masked by continued subsidence of the trough in Natal, while traces of interglacials could have been obliterated in the elevated regions bordering this regional downwarp by reworking and resedimentation of glaciogenic sediments. Further research should also focus on the processes contributing to the origin of the remarkably thick homogeneous diamictite and examine how different types may be related to derivation from separate source areas. Another question revolves around the abrupt contact in the Intermediate Region, between the topmost Dwyka exposures, some still bearing northwest–southeast striking soft-sediment glacial grooves (Tavener-Smith *et al.*, 1988), and the conformably overlying black shale of the Ecca Group which is devoid of ice-rafted material. The phenomenon could testify to rapid eustatic change coupled with tectonism rather than sudden climatic warming during the closing stages of the Dwyka glaciation.

Acknowledgements

Financial assistance, provided by the Foundation for Research and Development (FRD) and the University Research Fund, is gratefully acknowledged. Reviewers of the original manuscript are thanked for their constructive suggestions. Distribution of Dwyka Group exposures in Fig. 4.1 is based on the 1:1000000 Geological Map of South Africa (1984).

References

Anderson, A.M. (1981). The *Umfolozia* arthropod trackways in the Permian Dwyka and Ecca Series of South Africa. *Journal of Paleontology*, **55**, 84–108.

Boulton, G.S. (1978). Boulder shapes and grain-size distributions of debris as indicators of transport paths through a glacier and till genesis. *Sedimentology*, **25**, 773–9.

Cairncross, B. (1979). Depositional framework and control of coal distribution and quality, Van Dyks Drift area, northern Karoo Basin. Ph.D. thesis, University of Natal, Pietermaritzburg, (unpublished), 88 pp.

Du Toit, A.L. (1921). The Carboniferous glaciation of South Africa. *Transactions of the Geological Society of South Africa*, **24**, 188–227.

Du Toit, A.L. (1931). The geology of the country surrounding Nkandhla, Natal. An explanation of Sheet 109, Geological Survey of South Africa. Pretoria, Government Printer, 105 pp.

Du Toit, A.L. (1946). The geology of parts of Pondoland, East Griqualand and Natal. An explanation of sheet 119, Geological Survey of South Africa, Pretoria, Government Printer, 31 pp.

Eyles, C.H. Eyles, N. & Miall, A.D. (1985). Models of glaciomarine sedimentation and their application to the interpretation of ancient glacial sequences. *Palaeogeography, Palaeoclimatology, Palaeoecology*, **51**, 15–84.

Gevers, T.W. (1941). Carbon dioxide springs and exhalations in northern Pondoland and Alfred County, Natal. *Transactions of the Geological Society of South Africa*, **44**, 233–301.

Gravenor, C.P., von Brunn, V. & Dreimanis, A. (1984). Nature and classification of waterlain glaciogenic sediments, exemplified by Pleistocene, Late Paleozoic and Late Precambrian deposits. *Earth-Science Reviews*, **20**, 105–66.

Hobday, D.K. & von Brunn, V. (1979). Fluvial sedimentation and paleogeography of an early Paleozoic failed rift, southeastern margin of Africa. *Palaeogeography, palaeoclimatology, Palaeoecology*, **28**, 169–84.

Hnter, D.R. (1969). An occurrence of the Dwyka Series near Goedgegun, southern Swaziland. *Transactions of the Geological Society of South Africa*, **72**, 31–35.

Hunter, D.R. & Wilson, A.H. (1988). A continuous record of crustal evolution from 3.5 to 2.6 Ga in Swaziland and northern Natal. *South African Journal of Geology*, **91**, 57–74.

Kent, L.E. (1938). The geology of a portion of Victoria County, Natal. *Transactions of the Geological Society of South Africa*, **41**, 1–36.

King, L.C. (1948). *The geology of Pietermaritzburg and environs*. Pretoria, Government Printer, 20 pp.

King, L.C. & Maud, R.R. (1964). *The geology of Durban and environs*. Bulletin of the Geologial Survey of South Africa, **42**, Pretoria, Government Printer, 52 pp.

Le Blanc Smith, G. (1980). Logical-letter coding system for facies nomenclature: Witbank coalfield. *Transactions of the Geological Society of South Africa*, **83**, 301–11.

Le Blanc Smith, G. & Eriksson, K.A. (1979). A fluvioglacial and glaciolacustrine deltaic depositional model for Permo-Carboniferous coals of the northeastern Karoo Basin, South Africa. *Palaeogeography, Palaeoclimatology, Palaeoecology*, **27**, 67–84.

Lowe, D.R. (1979). Subaqueous liquefied and fluidized sediment flows and their deposits. *Sedimentology*, **23**, 285–308.

Lowe, D.R. (1982). Sediment gravity flows. II. Depositional models with special reference to the deposits of high-density turbidity currents. *Journal of Sedimentary Petrology*, **52**, 279–97.

Matthews, P.E. (1970). Paleorelief and the Dwyka glaciation in the eastern region of South Africa. In *Second Gondwana Symposium. Proceedings and Papers*, ed. S.H. Haughton, Pretoria, Council for Scientific and Industrial Research, 491–9.

Matthews, P.E. (1990). A plate tectonic model for the late Archaean Pongola Supergroup in southeastern Africa. In *Crustal Evolution and Orogeny*, ed. S.P.H. Sychanthavong, Calcutta, Oxford & IBH Publishing Co., 41–73.

Matthews, P.E. & Sharrer, R.H. (1968). A graded unconformity at the base of the Early Precambrian Pongola System. *Transactions of the Geological Society of South Africa*, **71**, 257–71.

McCabe, A.M. (1986). Glaciomarine facies deposited by retreating meltwater glaciers: an example from the Late Pleistocene of Northern Ireland. *Journal of Sedimentary Petrology*, **56**, 880–94.

Mellor, E.T. (1905a). The glacial (Dwyka) conglomerate of South Africa. *American Journal of Science*, **20**, 107–18.

Mellor, E.T. (1905b). A contribution to the study of the glacial (Dwyka) conglomerate in the Transvaal. *Quarterly Journal of the Geological Society of London*, **61**, 679–89.

Miall, A.D. (1985). Sedimentation on an early Proterozoic continental margin under glacial influence: the Gowganda Formation (Huronian), Elliot Lake area, Ontario, Canada. *Sedimentology*, **32**, 763–88.

Middleton, G.V. & Hampton, M.A. (1973). Sediment gravity flows: mechanics of flow and deposition. In *Turbidites and Deep Water Sedimentation*, eds. G.V. Middleton & A.H. Bouma, Anaheim, California, Society of Economic Paleontologists and Mineralogists, Pacific Section, Short course, 1–38.

Miller, J.M.G. (1989). Glacial advance and retreat sequences in a Permo-Carboniferous section, central Transantarctic Mountains. *Sedimentology*, **36**, 419–30.

Mountain, E.D. & Hodson, N.G. (1933). Glacial phenomena on Table Mountain, Pietermaritzburg. *Transactions of the Geological Society of South Africa*, **36**, 89–95.

Powell, R.D. (1984). Glacimarine processes and inductive lithofacies modelling of ice shelf and tidewater glacier sediments based on Quaternary examples. *Marine Geology*, **57**, 1–52.

Rust, B.R. & Romanelli, R. (1975). Late Quaternary subaqueous outwash deposits near Ottawa, Canada. In *Glaciofluvial and Glaciolacustrine Sedimentation*, eds. A.V. Jopling & B.C. McDonald, Tulsa, Oklahoma, Society of Economic Paleontologists and Mineralogists Special Publication, **23**, 177–91.

Rust, I.C. (1975). Tectonic and sedimentary framework of Gondwana basins in southern Africa. In *Gondwana Geology*, ed. K.S.W. Campbell, Canberra, Australian National Press, 537–64.

Tankard, A.J., Jackson, M.P.A., Eriksson, K.A., Hobday, D.K., Hunter, D.R. & Minter, W.E.L. (1982). *Crustal Evolution of Southern Africa*. New York, Springer-Verlag, 523 pp.

Tavener-Smith, R., von Brunn, V. & Smith, A.M. (1988). *Selected Exposures in the Dwyka Formation and the Ecca Group of Natal*. Guide book to field excursion B, 22nd Earth Science Congress of the Geological Society of South Africa, Durban, University of Natal, 49 pp.

Thomas, R.J. (1988). The geology of the Port Shepstone area. An explanation of Sheet 3030, Geologial Survey of South Africa. Pretoria. Government Printer, 136 pp.

Thomas, R.J., von Brunn, V. & Marshall, C.G.A. (1990). A tectono-sedimentary model for the Dwyka Group in southern Natal, South Africa. *South African Journal of Geology*, **93**, 809–17.

Visser, J.N.J. (1983). Glacial-marine sedimentation in the Late Paleozoic Karoo Basin, southern Africa. In *Glacial-Marine Sedimentation*, ed. B.F. Molnia, New York, Plenum Press, pp. 667–701.

Visser, J.N.J. (1986). Lateral lithofacies relationships in the glacigene Dwyka Formation in the western and central parts of the Karoo Basin. *Transactions of the Geological Society of South Africa*, **89**, 373–83.

Visser, J.N.J. (1987). The palaeogeography of part of southwestern Gondwana during the Permo-Carboniferous glaciation. *Palaeogeography, Palaeoclimatology, Palaeoecology*, **61**, 205–19.

Visser, J.N.J. (1989). The Permo-Carboniferous Dwyka Formation of southern Africa: deposition by a predominantly subpolar ice sheet. *Palaeogeography, Palaeoclimatology, Palaeoecology*, **70**, 377–91.

Visser, J.N.J. (1990). The age of the late Palaeozoic glacigene deposits in southern Africa. *South African Journal of Geology*, **93**, 366–75.

Visser, J.N.J. (1991). A reconstruction of the late Palaeozoic ice sheet in southwestern Gondwana. Abstracts. Eighth International Symposium on Gondwana, p. 89. Hobart, University of Tasmania.

Visser, J.N.J. & Loock, J.C. (1987). Ice margin influence on glaciomarine sedimentation in the Permo-Carboniferous Dwyka Formation from the southwestern Karoo, South Africa. *Sedimentology*, **34**, 929–41.

Visser, J.N.J., von Brunn, V. & Johnson, M.R. (1990). Dwyka Group. In *Catalogue of South African Lithostratigraphic Units*, ed. M.R. Johnson, South African Committee for Stratigraphy, 2, Pretoria Government Printer, 15–17.

Von Brunn, V. (1977). A furrowed intratillite pavement in the Dwyka Group of northern Natal. *Transactions of the Geological Society of South Africa*, **80**, 125–30.

Von Brunn, V. (1987). A facies analysis of Permo-Carboniferous glacigenic deposits along a paleoscarp in northern Natal, South Africa. In *Gondwana Six: Stratigraphy, Sedimentology, and Paleontology*, Geophysical Monograph 41, ed. G.D. McKenzie, Washington, D.C., American Geophysical Union, 113–22.

Von Brunn, V. & Gravenor, C.P (1983). A model for the late Dwyka glaciomarine sedimentation in the eastern Karoo Basin. *Transactions of the Geological Society of South Africa*, **86**, 199–209.

Von Brunn, V. & Talbot, C.J. (1986). Formation and deformation of subglacial intrusive clastic sheets in the Dwyka Formation, South Africa. *Journal of Sedimentary Petrology*, **56**, 35–44.

Von Brunn, V. & Marshall, C.G.A. (1989). Glaciated surfaces and the base of the Dwyka Formation near Pietermaritzburg, Natal. *South African Journal of Geology*, **92**, 420–426.

Von Brunn, V., Smith, R.G. & Sleigh, D.W.W. (1988). Outliers of the Dwyka Formation in the Piet Retief area, southeastern Transvaal. Extended Abstracts, 22nd Earth Science Congress of the Geological Society of South Africa, Durban, University of Natal, 693–6.

Whateley, M.K.G. (1980). Deltaic and fluvial deposits of the Ecca Group, Nongoma graben, northern Zululand. *Transactions of the Geological Society of South Africa*, **83**, 345–51.

Wright, R. & Anderson, J.B. (1982). The importance of sediment gravity flow to sediment transport and sorting in a glacial marine environment: Eastern Weddell Sea, Antarctica. *Geological Society of America Bulletin*, **93**, 951–63.

5 Itararé Group: Gondwanan Carboniferous-Permian of the Paraná Basin, Brazil

ALMÉRIO BARROS FRANÇA

Abstract

Glacial deposits have recently been attracting the attention of petroleum geologists. Oman has about 3.5 billion bbl of oil in place in Carboniferous Gondwanan glacial sandstones; Bolivia produces about 10 MMbbl (oil and gas) annually, approximately 70% comes from Carboniferous glacial deposits: Australia has 22 gas fields in glacial Early Permian reservoirs.

The Paraná Basin in Brazil has the largest Gondwanan Carboniferous-Permian deposits, covering more than 700 000 km^2. To date the Itararé Group has had several gas finds and some subcommercial gas production. Currently, Petrobrás, the Brazilian oil company, is putting a great deal of effort, mostly seismic surveys and some drilling, in an attempt to find commercial hydrocarbons in the Paraná Basin. Being an important target in this exploratory process the Itararé Group deserves detailed study aiming at sand body geometry (reservoir-rocks), diamictite distribution (seal-rocks) and their trapping mechanisms. The main purpose of this paper is to present an overview of the stratigraphy and hydrocarbon potential of the Itararé Group based upon well logs, core description, and outcrop information.

The Itararé Group is subdivided into three main formations: Lagoa Azul, Campo Mourão, and Taciba. A red bed unit, called the Aquidauana Formation, predominates in the northern realm of the Paraná Basin.

Introduction

The Paraná Basin is located in southern Brazil with extensions into Argentina, Paraguay, and Uruguay. In Brazil alone it covers an area of approximately 1 000 000 km^2 (Fig. 5.1).

The Itararé Group is probably the most continuous and thickest temperate glacial-marine sequence in South American Gondwana. It covers an area larger than 700 000 km^2 and it is thicker than 1300 m in the central part of the Paraná Basin (Fig. 5.2). A minor part of the Itararé Group crops out in the northern and south-southeastern outcrop belt of the basin. The largest volume of rock, however, is underneath thick Jurassic-Cretaceous lava flows that cover almost entirely the Paraná Basin. Several previous studies have been done on the Itararé Group in the outcrop belt, such as White (1908); Oliveira (1927); Gordon Jr. (1947); Bigarella et al. (1967); Schneider et al. (1974); Rocha-Campos et al. (1976); Gravenor and Rocha-Campos (1983); Caetano and Landim (1987).

The present work is based chiefly on well data, some of which were drilled deep in the central part of the Paraná Basin. More than 1000 m of cores were described, and over 100 well logs were analysed.

The Itararé Group is composed mostly of sandstones and diamictites. Sandstones vary from a few metres up to 300 m thick. Soft-deformed sandstones predominate among other types such as structureless massive and graded sandstones. Conglomerates up to 1.5 m thick are commonly interbedded with sandstones. Turbidity currents are the main sedimentary process associated with downslope resedimentation, possibly triggered by high rate of deposition and/or contemporaneous seismic activity. Submarine channel filling and submarine fans may be important basin filling processes during the deposition of sandstones of the Itararé Group.

Petrographically the sandstones of the Itararé Group are quartz-feldspathic-lithic arenites/wackes, according to Dott's 1964 classification. Porosity may be as high as 20% and it is mostly secondary by dissolution of carbonate and sulfate cements, and unstable grains (França and Potter, 1991).

Diamictite is an important lithology in the Itararé Group. Diamictite bodies can be either thin and areally confined or thick (150 m) and widespread, covering hundreds of kilometers. Massive diamictites were deposited by rain-out process. Subsequently, some massive diamictites could be downslope resedimented, creating stratified to crudely stratified diamictites.

A northeastern structural lineament network seems to have had important influence on the deposition of the Itararé Group by controlling thickness and distribution of sand and diamictite of its three main formations.

Tectonic setting

The Paraná Basin is a cratonic basin (Zalán et al., 1990). It is, however, a polyhistory basin and is classified as an Interior Fracture – IF basin (Kingston et al., 1983) in its early stages,

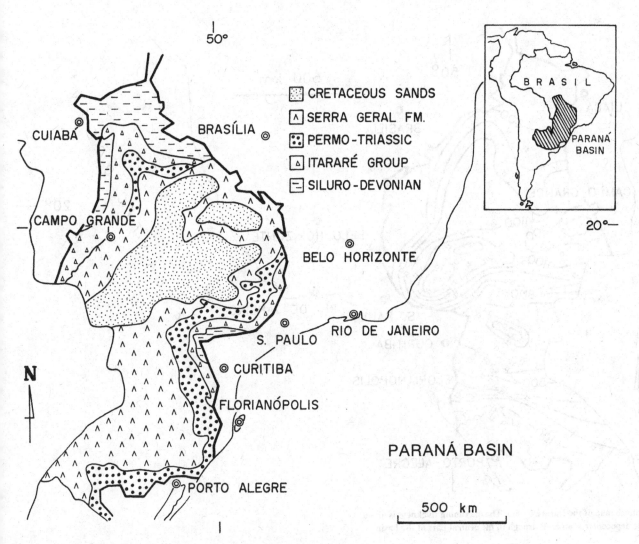

CRETACEOUS SANDS

SERRA GERAL FM.

PERMO-TRIASSIC

ITARARÉ GROUP

SILURO-DEVONIAN

PARANÁ BASIN

500 km

Fig. 5.1. Geological map of the Paraná Basin. Most of the sedimentary sequence – over 5000 m thick – is covered by lava flows of the Serra Geral Formation. The Itararé Group crops out in the eastern and northwestern (red beds) outcrop belts.

probably from Late Cambrian up to the Devonian time. From the Carboniferous until the Cretaceous, the Paraná Basin was an Interior Sag – IS basin (Kingston *et al.*, 1983). Figure 5.3 shows the main structural lineaments present in the Paraná Basin. Two major directions can be observed: northwest–southeast (NW–SE), and northeast–southwest (NE–SW). Most of the lineaments are fault zones identified chiefly in seismic and aeromagnetic surveys. The fault zones may be hundreds of kilometers long and tens of kilometers wide.

Although the two sets of intersecting linear structures seem closely related to each other, it is believed that they are different both in timing and origin. The NW–SE ones are probably Meso-zoic, or at least had main activity during that time. They are intruded by numerous Jurassic-Cretaceous diabase dikes and, according to Zalán *et al.* (1990), structures associated with strike-slip reactivations are commonly associated with the NW–SE linea-ments. The NE–SW lineaments are older than the NW–SE ones:

their onset was probably during the Transbrasiliano tectonic cycle (700–650 MA) and were active during most of the Paleozoic. The northeastern lineaments were important controls on the deposition of the Itararé Group. These lineaments commonly do not contain diabase dikes and are pervasive in the Precambrian basement near Curitiba (Fig. 5.3). Zalán *et al.* (1990) have a more complete and detailed analysis of these lineaments, and of the tectonic framework of the Paraná Basin as well.

Stratigraphy

Figure 5.4 is a generalized stratigraphic column of the Paraná Basin. According to Daemon and Quadros (1970) and Daemon *et al.* (1991) the Itararé Group ranges in age between the Late Carboniferous (Stephanian) and the Early Permian (Artins-kian). Hence the glacial record in the Paraná Basin lasted for about 30 million years, whereas, the entire Late Palaeozoic Ice Age in

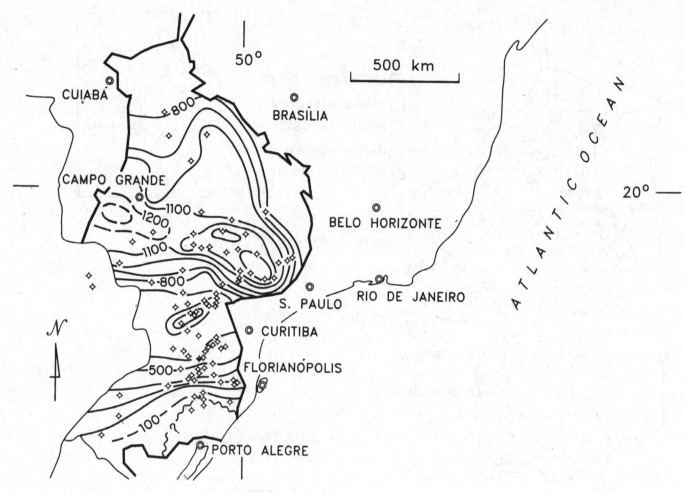

Fig. 5.2. Isopach map of the Itararé Group. The maximum thickness is about 1400 m. The depocenter is an E–W trough in the central part of the basin.

Gondwana lasted for about 55 million years (González, 1990) from the Namurian (C) to Artinskian (P).

The Itararé Group is about 1400 m thick in the central-eastern part of the basin (Fig. 5.2), and it is subdivided into four units: Lagoa Azul, Campo Mourão, and Taciba formations in the central-southern part of the basin (França and Potter, 1991), and the Aquidauana Formation (red beds) in the northern realm of the basin. The Aquidauana Formation is still not well understood and it might even be considered a distinct Carboniferous unit as suggested by its fossil content and correlation with its counterpart: the Escarpment Formation in neighbouring Bolívia. Figure 5.5 shows the three main formations Lagoa Azul, Campo Mourão, and Taciba with their log characteristics and brief comments on lithology and possible depositional environments.

Figures 5.6 and 5.7 show the distribution of the main units in the subsurface.

Lagoa Azul Formation

The Lagoa Azul Formation is composed of a widespread sheet-like deposit in its basal section, called the Cuiabá Paulista Member, and a diamict-silty upper unit called the Tarabai Member

(Figs. 5.5–5.7). The isopach map of the Lagoa Azul Formation (Fig. 5.8) shows its distribution for about 300 000 km² and suggests that at the onset of the deposition of the Itararé Group the sedimentation was probably controlled by two tectonic lineaments: the Jacutinga Fault and a minor control by the Araçatuba Lineament. The northern depositional limit of the Lagoa Azul Formation is uncertain because of the low concentration of wells in that area, and mostly because the correlation with the red beds of the Aquidauana Formation is still doubtful.

Based on cross-sections (Figs. 5.6 and 5.7) it is very unlikely that the Lagoa Azul Formation crops out. However, because the basin margin was controlled by faults, we cannot rule out the possibility of isolated outcrops in uplifted blocks.

The Lagoa Azul Formation commonly overlies the black shales of the Devonian Ponta Grossa Formation, from which it is always separated by an erosional unconformity. The shales of the Ponta Grossa Formation are the source rock for the subcommercial gas and condensate shows found so far in the Itararé Group. Hence, there is a direct relationship between source rocks of the Ponta Grossa Formation and reservoir rocks of the Itararé Group.

No macrofossils have been observed in the Lagoa Azul Formation: its paleontological content, according to Daemon and Quad-

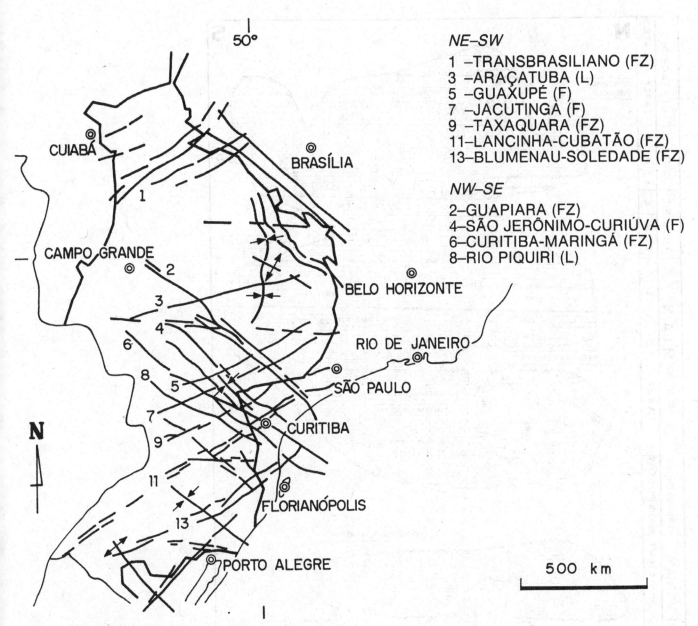

Fig. 5.3. Main structural framework in the Paraná Basin. F, single fault; FZ, fault zone: L, lineament of unknown nature. (After Zalán *et al.*, 1991.)

ros (1970) and Daemon *et al.* (1991), is restricted to spores of the *Potonieisporites* microflora of Stephanian to Sakmarian age, besides reworked fossils of the Ponta Grossa Formation (Devonian).

Turbidites, debris-flow, and some rain-out deposits dominate in the lowermost unit of the Itararé Group (Fig. 5.5; Eyles *et al.* 1993).

Campo Mourão Formation

The Campo Mourão Formation is mostly a sandy unit; locally it can be subdivided into a sandy basal unit and an upper diamict unit (Figs. 5.6 and 5.7) Typical thickness of the Campo Mourão Formation is about 450m and a maximum around 900m.

Figure 5.9 is a sand isolith map of the Campo Mourão Forma-

tion, which covers approximately 640 000 km² and is absent only in the southern Rio Grande do Sul. However, the correlation in the northern realm of the basin is also doubtful because of scattered well spacing control and of an uncertain correlation with the red beds of the Aquidauana Formation.

Two important features should be emphasized on the isolith map: (1) the expansion of the southern border of the basin when compared with the previous Lagoa Azul Formation; and (2) the prevalence of thick sandstone bodies near the main fault zones, suggesting tectonic control on their deposition. During the sedimentation of the Lagoa Azul Formation, the southern basin margin was along the Jacutinga Fault, and moved probably by steps, first to Lancinha–Cubatão Fault Zone and then to the Blumenau–Soledade Fault Zone (Fig. 5.9).

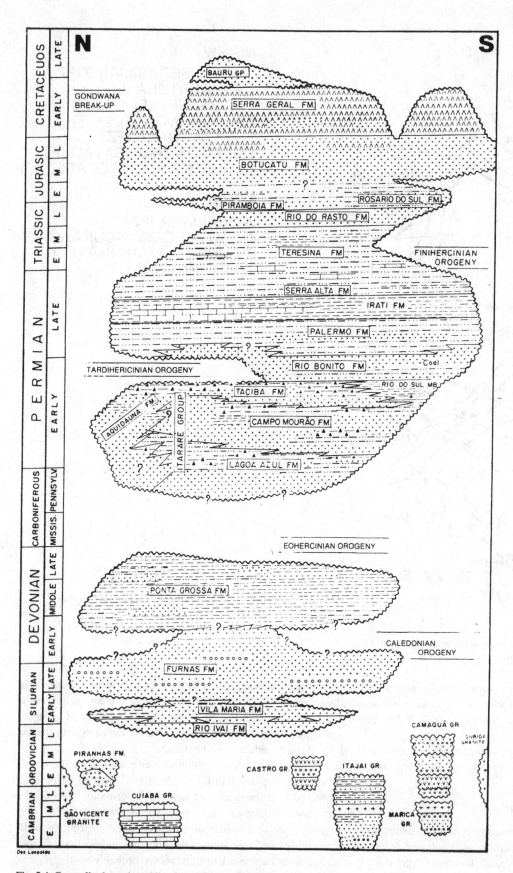

Fig. 5.4. Generalized stratigraphic column for most of the central part of the Paraná Basin.

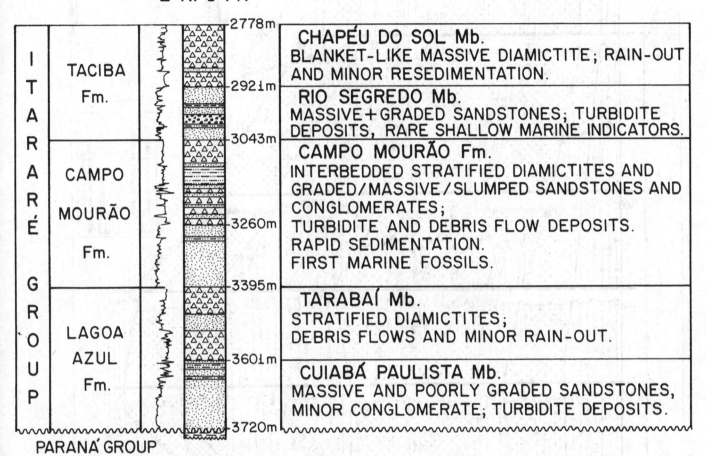

Fig. 5.5. Typical section of the Itararé Group in the central part of the Paraná Basin in 2-RI-1-PR well. In the southern realm the diamictites of the Chapéu do Sol Member interfinger with shales of the Rio do Sul Member. (Modified from França and Potter, 1991.)

In the central part of the basin the Campo Mourão Formation conformably overlies the Lagoa Azul Formation. Because the basin has quickly expanded during the deposition of the Campo Mourão Formation, this unit unconformably overlies the Ponta Grossa Formation, and then the Precambrian basement (Figs. 5.6 and 5.7).

The sandstones of the Campo Mourão Formation had important gas and condensate shows, sourced in the Devonian black shales of the Ponta Grossa Formation. Because the Campo Mourão Formation is locally in direct contact with source rocks it has been considered an important exploratory target in the Paraná Basin.

The principal seal rock in the Campo Mourão Formation is diamictite. Figure 5.10 is a diamictite isolith map and it shows that diamictites are fairly widespread in the basin and that three main source areas for diamictites seems to have existed: two in the western side of the basin (Mato Grosso Lobe and Santa Catarina Lobe, França and Potter, 1991) and one in the eastern side, perhaps linked to the Kaokoveld Lobe in Africa (Crowell and Frakes, 1975). Figure 5.10 suggests also that the diamictite sedimentation was controlled by several fault zones.

The fossil content of the Campo Mourão Formation is predomi-nantly spores and pollens, with minor *Tasmanites* sp., some bivalves and brachiopods (Daemon and Quadros, 1970; Daemon *et al.*, 1991).

Turbidites, debris-flows, and slumped sandstones predominate in the Campo Mourão Formation (Fig. 5.5; Eyles *et al.* 1993).

Taciba Formation

The Taciba Formation is composed, in most of the basin, of a sandy basal unit called the Rio Segredo Member and a diamict unit called the Chapéu do Sol Member. In the southern portions of the basin the Chapéu do Sol Member interfingers with shales of the Rio do Sul Member (Figs 5.6 and 5.7). The Chapéu do Sol Member covers about 700 000 km² and is the most continuous diamictite unit in the Gondwana record (Fig. 5.11). The isolith map shows that (1) tectonic control has decreased towards the uppermost unit of the Itararé Group, and (2) the three main source areas that seem to have existed in the lower Campo Mourão Formation (Fig. 5.10), were still active during the Chapéu do Sol Member deposition.

The Taciba Formation is the main unit of the Itararé Group to

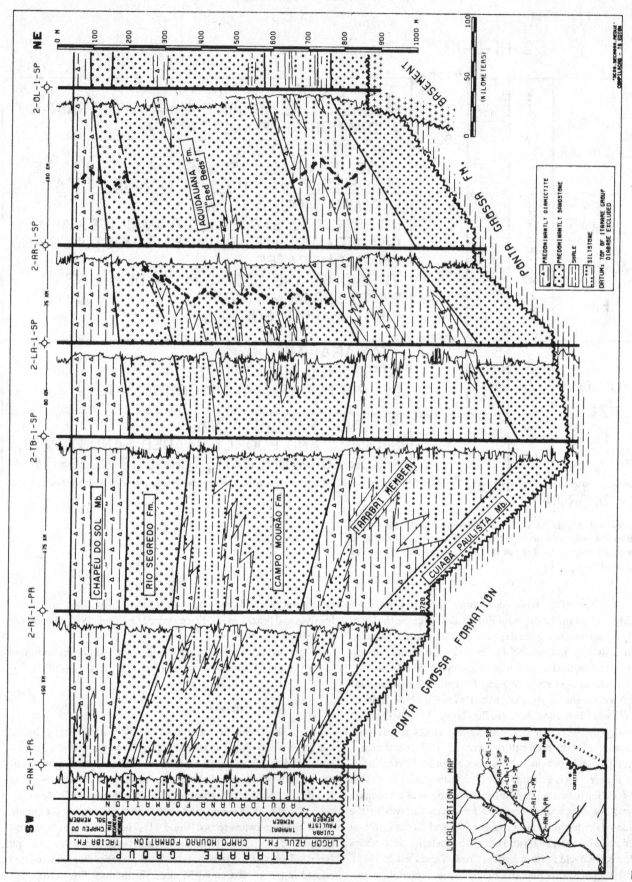

Fig. 5.6. Cross-section AB.

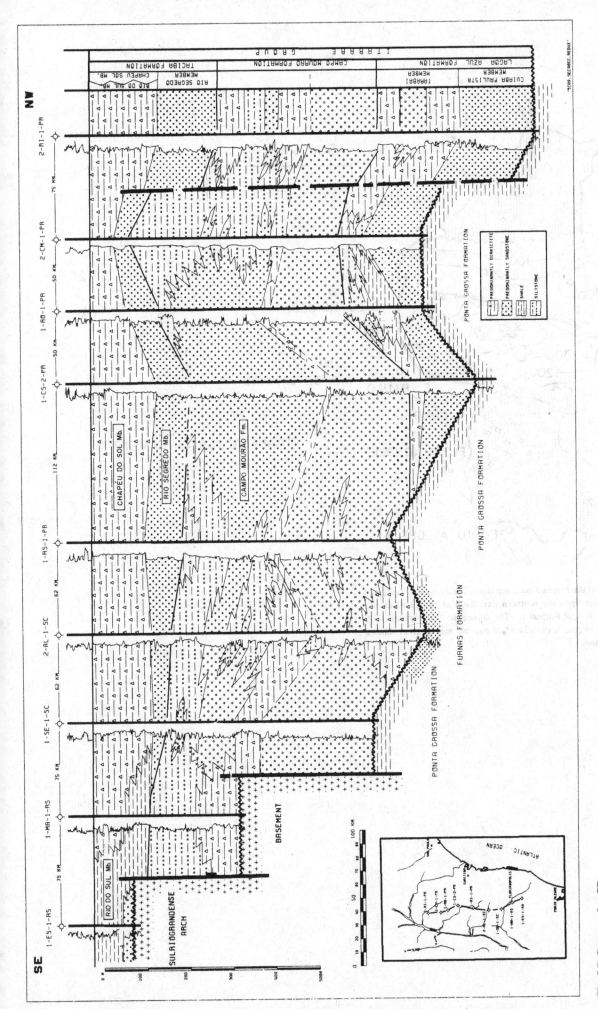

Fig. 5.7. Cross-section CD.

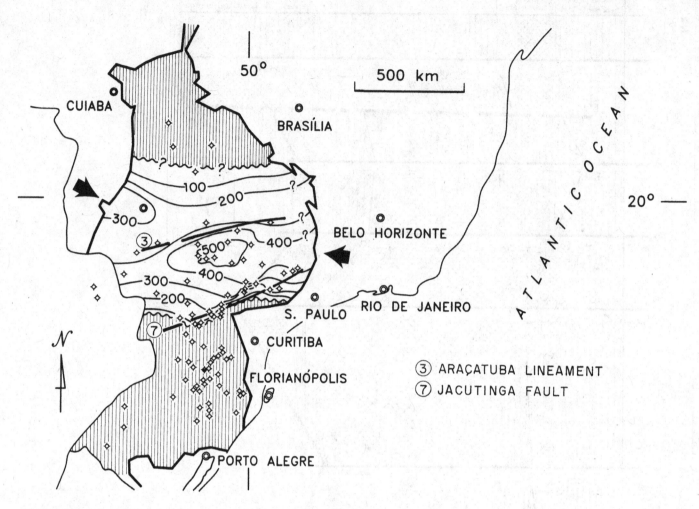

Fig. 5.8. Isopach map of the Lagoa Azul Formation. The Jacutinga Fault was an important constraint upon the southern expansion of the basin during deposition of the Lagoa Azul Formation. The depocenter was controlled by NE–SW structural features. Arrows indicate possible source area; dashed areas show no deposition.

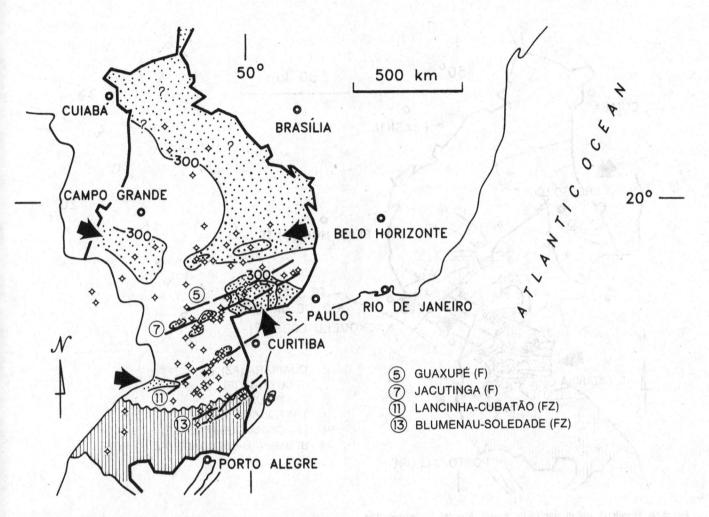

Fig. 5.9. Sand isolith map of the Campo Mourão Formation. Sandstones predominate in the north-northwestern areas of the Paraná Basin, and along fault zones. The basin has expanded, probably by steps, from the Jacutinga Fault to the Blumenau–Soledade Fault Zone.

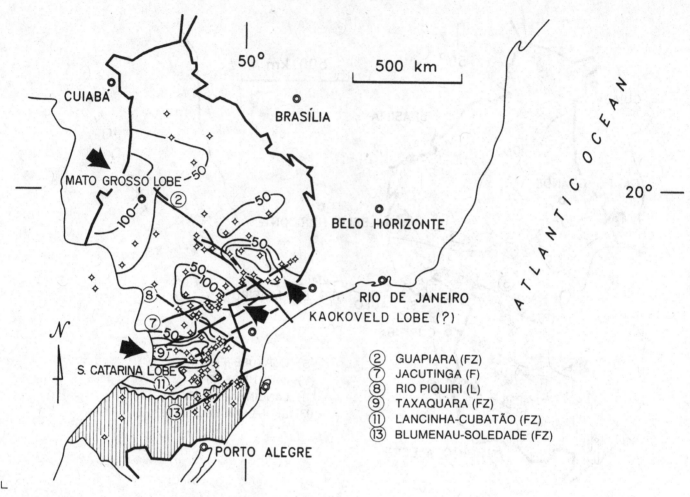

Fig. 5.10. Diamictite isolith map of the Campo Mourão Formation. The diamictites had about the same source area as the sandstones (Fig. 5.9). Tectonic control is evident as suggested by the isoliths.

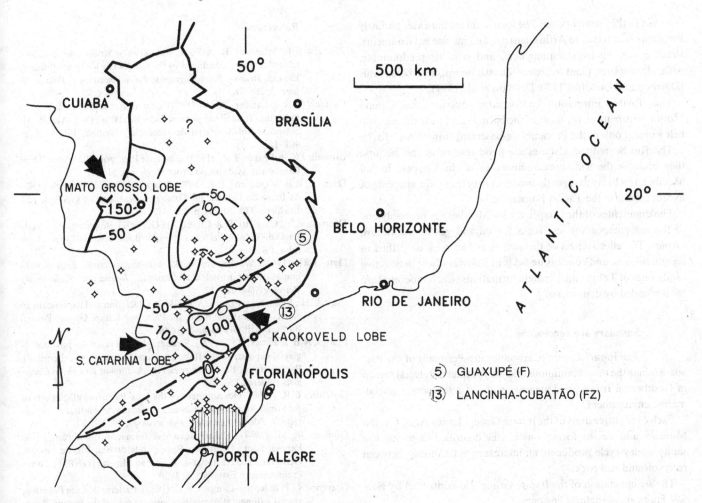

Fig. 5.11. Diamictite isolith map of the Chapéu do Sol Member. The diamictites of this member cover almost the entire basin, it is a blanket-like deposit with less tectonic control than the older diamictites.

5 GUAXUPÉ (F)

13 LANCINHA-CUBATÃO (FZ)

crop out in the Paraná Basin. The fossil content indicates an Early Permian (Sakmarian to Artinskian) age and marine environments. Brachiopods, bivalves, foraminifers, and crinoids predominate, with minor spores, plant remnants, *Tasmanites* sp., and trace fossils (Daemon and Quadros, 1970; Daemon *et al.*, 1991).

The Taciba Formation conformably overlies the Campo Mourão Formation, for most of the basin except near the outcrop belt where it onlaps the Precambrian basement (Figs. 5.6 and 5.7).

The Rio Segredo sandstones are good reservoirs and because they underlie the widespread diamictites of the Chapéu do Sol Member, it is likely that gentle anticlines may trap large amounts of hydrocarbons in the Taciba Formation.

The diamictites of the Chapéu do Sol Member are probably one of the most important seal rocks for oil and gas in the Itararé Group. The effectiveness of diamictites as cap rock is verified in several oil, gas, and condensate fields in Bolívia, where productive sandstones of Tarija and Tupambi formations (Carboniferous) are mostly sealed by diamictites.

Summary and conclusions

The Itararé Group (Carboniferous–Permian) of the Paraná Basin is the most continuous and maybe thickest glacial record in Gondwana. It is a good ancient record of a temperate glacial-marine environment.

Each of the three units of the Itararé Group: Lagoa Azul, Campo Mourão, and Taciba formations can be described in terms of a sandy/shaley cycle producing an interesting relationship between reservoir and seal rocks.

The sedimentation of the Itararé Group was controlled by NE–SW faults and structural lineaments.

Diamictites were desposited mostly by rain-out processes and frequently resedimented downslope, due to rapid sedimentation rate and/or seismic activity.

Sandstone bodies are turbidites, commonly filling submarine channels, ending probably in large submarine fans, not identified yet due mostly to poor well spacing.

The Itararé Group is in direct contact with source rocks of the Ponta Grossa Formation, which is a favorable factor for vertical migration.

The production of gas and condensate in Bolívia, from Carboniferous glacial deposits, sourced in the Devonian black shales, and subcommercial shows of gas and condensate in the Itararé Group, makes this unit an important target for hydrocarbon exploration in the Paraná Basin.

Acknowledgements

I wish to thank Nick Eyles and Carolyn Eyles for the assistance on core description, field work, discussions, and critical comments which have improved this work. Thanks are due to Elmo Fernandes de Oliveira and Edton Araujo Barbosa for drawing the figures. I also would like to thank PETROBRAS for the clearance to publish the present paper.

References

Bigarella, J.J., Salamuni, R. & Fuck, R.A. (1967). Striated surfaces and related features, developed by the Gondwana ice sheets (State of Paraná, Brazil). *Paleogeography, Paleoclimatology, Palaeoecology*, **3**, 265–76.

Caetano, M.R. & Landim, P.M.B. (1987). Os sedimentos glaciais da base do Subgrupo Itararé nas proximidades de Itararé (SP). Atas do III Simpósio Sul-Brasileiro de Geologia, Curitiba, PR, v. 1, p. 401–12.

Crowell, J.C. & Frakes, L.A. (1975). The Late Paleozoic glaciation. Third Gondwana Symposium Proceedings, pp. 313–31.

Daemon, R.F. & Quadros, L.P. (1970). Bioestratigrafia do Neopaleozóico da Bacia do Paraná. In: Congresso Brasileiro de Geologia, 24, Brasilia, 1970. Anaps, p. 359–412.

Daemon, R.F., Casaletti, P. & Ciguel, J.H.G. (1991). Biopaleogeografia da Bacia do Paraná. Petrobrás/Nexpar/Seint, Internal Report. Curitiba, PR, Brazil.

Dott, R.H. (1964). Wacke, graywacke and matrix: what approach to immature sandstone classification?: *Journal of Sedimentary Petrology*, **34**, 625–32.

Eyles, C.H., Eyles, N. & França, A.B. (1993). Glaciation and tectonics in an active intracratonic basin: the Paleozoic Itararé Group, Paraná Basin, Brazil. *Sedimentology*, **40**, 1–25.

França, A.B. & Potter, P.E. (1991). Stratigraphy and reservoir potential of glacial deposits of the Itararé Group (Carboniferous–Permian), Paraná Basin, Brazil: *American Association of Petroleum Geologists Bulletin*, **75**, 62–85.

González, C.R. (1990). Development of the Late Paleozoic glaciations of the South American Gondwana in western Argentina. *Paleogeography, Paleoclimatology, Paleoecology*, **79**, 275–87.

Gordon, Jr. M. (1947). Classificação das formações gondwanicas do Paraná, Santa Catarina e Rio Grande do Sul. Rio de Janeiro, Departamento Nacional da Produção Mineral (DNPM). Notas Preliminares e Estudos, 38, 1–20.

Gravenor, G.P. & Rocha-Campos, A.C. (1983). Patterns of Late Paleozoic glacial sedimentation on the southeast side of the Paraná Basin, Brazil. *Paleogeography, Paleoclimatology, Paleoecology*, **43**, 1–39.

Kingston, D.R., Dishroon, C.P. & Williams, P.A. (1983). Global Basin classification system: *American Association of Petroleum Geologists Bulletin*, **67**, 2175–93.

Oliveira, G.P. (1927). Geologia de recursos minerais do Estado do Paraná. Rio de Janeiro, SGM, monografia, 6.

Rocha-Campos, A.C., Oliveira, M.E.C.B. de, Santos, P.R. dos & Saad, A.R. (1976). Boulder pavements and the sense of movement of Late Paleozoic glaciers in central eastern São Paulo State, Paraná Basin, Brazil. Boletim IG. Instituto de Geociências – Universidade de São Paulo, USP, 7, 149–60.

Schneider, R.L., Muhlmann, H., Thomasi, E., Medeiros, R.A., Daemon, R.F., & Nogueira, A.A. (1974). Revisão estratigráfica da Bacia do Paraná: Anais do XXVIII Congresso Brasileiro de Geologia, Porto Alegre, pp. 41–65.

White, I.C. (1908). Relatório sobre as 'Coal Measures' e rochas associadas do sul do Brasil. Imprensa Nacional – Rio de Janeiro, 727 pp. (in Portuguese and English). Reprint for the VII Gondwana Symposium, São Paulo, 1988, by the DNPM, Brasilia, DF.

Zalán, P.V., Wolff, S., Astolfi, M.A.M., Vieira, I.S., Conceição, J.C.J., Appi, V.T. *et al.* (1990). The Paraná Basin, Brazil. In *Interior Cratonic Basin*, eds. Leighton *et al.*, American Association of Petroleum Geologists Memoir 51, 681–708.

6 The interpretation of massive rain-out and debris-flow diamictites from the glacial marine environment

JOHAN N.J. VISSER

Abstract

Lithofacies and clast fabric analyses of four massive diamictites from the glacigene Dwyka Formation show two (Floriskraal and Douglas) to be primarily of rain-out origin and the other two (Elandsvlei and Kransgat River) to have formed by subaqueous sediment gravity flow. Ancient massive diamictites, which exhibit no bedding or other internal structures, thus may have a multiple origin as they can form directly by dense rain-out or indirectly by resedimentation of glacial material. Rain-out and resedimentation can occur simultaneously, and distinction between the two processes in ancient diamictites is problematical. For the interpretation of ancient massive diamictites a lithofacies analysis must therefore be a first priority, because results of fabric analyses and other criteria are often inconclusive. Diagenesis, compaction and dewatering can obliterate diagnostic features necessary for the successful interpretation of massive diamictites.

Introduction

Glacial sequences from Precambrian to Late Palaeozoic in age consist mostly of homogeneous diamictites lacking sedimentary structures as well as observable grain fabrics. Such sequences attain thicknesses of several hundreds of metres over large areas and their interpretation always poses a problem to glacial geologists.

A massive diamictite is here defined as being a homogeneous rock body without bedding or other internal structures on a macro- and mesoscale (Fig. 6.1). A diamictite sequence may, however, be thickly bedded although bedding planes are commonly absent in diamictites up to several tens of metres thick. According to Pettijohn (1975) truly massive beds are probably very rare and radiography of such seemingly homogeneous beds has in many cases revealed internal bedding structures. Some diamictites are seemingly massive on a macroscale and one has to beware of a false impression of homogeneity. The term 'massive' should therefore only be used in field observations where any noticeable internal arrangement within the rock body is absent.

This paper addresses diamictites which contextual data indicate are subaqueous deposits and which accumulated in a glacially influenced marine setting. In ancient subaqueously deposited (waterlain) diamictites the currently favoured mechanisms are rain-out and debris-flow and a combination thereof. It is, therefore, the objective of this paper to describe criteria whereby these depositional processes can be recognized in ancient glacial sequences.

Although there are numerous publications dealing with the genesis of massive diamictite deposits, only literature pertaining to well-indurated sequences deposited subaqueously will be referred to in the text. It was long believed that all waterlain diamictites or glaciomarine sediments are bedded (Dreimanis, 1979; Domack, 1984). The view that massive diamictites can form by rain-out was, however, expressed recently (Eyles & Eyles, 1983a; Young & Gostin, 1988; Visser, 1989a; Eyles & Lagoe, 1990). It was argued that massive deposits can result from influxes of suspended sediment and ice-rafting with minor lithofacies variability produced by episodic traction current or sediment gravity flow.

In the subaqueous environment massive diamictites can also form by sediment gravity flow. Examples of such glacigene deposits are described by Visser (1983a), Martin et al. (1985), Young &

Fig. 6.1. Outcrop of north-dipping massive diamictite at Floriskraal. Bedding becomes more evident towards the top of the lower unit where a boulder bed (arrow) is present. The interbedded shale (Sh) can be followed for several kilometres along strike.

Gostin (1988) and Collinson *et al.* (1989). The sediment source for the debris flows could be mobile subglacial debris pumped into the water column at a collapsing ice margin (Eyles & McCabe, 1989), rain-out material or subaqueous morainal banks (Visser, 1983a). A variation on a debris-flow origin is the deposition of undermelt diamicton subaqueously from the sole of the glacier, probably near the grounding line, as a slurry by continuous basal melting (Gravenor *et al.*, 1984). The clast fabrics which may be strong to poor and at variance with the ice flow direction, can be ascribed to downslope mass movement.

Massive diamictites from the Dwyka Formation

In an attempt to define satisfactory criteria for the absolute identification of massive diamictites in a glacial marine setting, four selected outcrops (Floriskraal, Elandsvlei, Kransgat River and Douglas; Fig. 6.2) of the Permo-Carboniferous Dwyka Formation were studied in detail. There were several reasons for the choice of the four outcrops studied. The diamictites and associated facies are well exposed, the diamictites are sufficiently clast-rich for fabric analysis, and good stratigraphic control on the studied sections was possible. In two outcrops (Kransgat River and Douglas) sedimentary features associated with the massive diamictites enhanced their interpretation, whereas clast fabrics of a dropstone argillite at Nieuwoudtville were also done to serve as a reference study. Three-dimensional blocks for fabric studies of the smaller clast sizes were cut from diamictite samples collected at Floriskraal and nearby Matjiesfontein. A reference plane (bed surface) for each sample was obtained by marking the regional bedding of the diamictite unit on the outcrop before the sample was taken. Since it is impossible to determine the clast shape (e.g. blade vs. disc) from a vertical plane alone clast orientations in the three-dimensional blocks are described in terms of ab-plane fabrics.

Floriskraal

The outcrops studied form part of a 50 m thick massive diamictite unit in the upper part of the Dwyka Formation which is about 650 m thick on the farm Floriskraal, about 8 km south of the town of Laingsburg (Fig. 6.2). The massive diamictite studied lies between stratified diamictite and mudrock at the base and an interglacial mudrock unit at the top (Fig. 6.3). Lithological contacts are sharp and undeformed, and individual facies can be traced laterally for more than 50 km. Ill-defined bedding planes may be present on a macroscale. The diamictite consists of angular to subangular clasts, up to 35 cm in diameter, in an arenaceous to argillaceous matrix. In thin section the diamictite matrix shows a porphyric fabric (Van der Meer *et al.*, 1983), textural variation and an apparent grain long axis orientation more or less parallel to the regional bedding (Fig. 6.4).

The polished diamictite samples (one from Floriskraal and two from the same stratigraphic horizon at nearby Matjiesfontein) show good to poor grain fabrics (Figs. 6.5–6.7). The Florisbad sample has a faintly deformed bed surface with a large percentage of clasts lying with their ab-planes almost perpendicular to the apparent bedding as seen in a vertical section (Fig. 6.5). Sample 1 from Matjiesfontein shows a good clast ab-plane fabric at an angle of between 25° and 35° with the regional bedding (Fig. 6.6). Some of the small clasts and grains lie along well-defined lines. In Matjiesfontein sample two grains in the inferred bed surface have a crude long-axis alignment with ab-planes parallel as well as almost perpendicular to the bedding (Fig. 6.7).

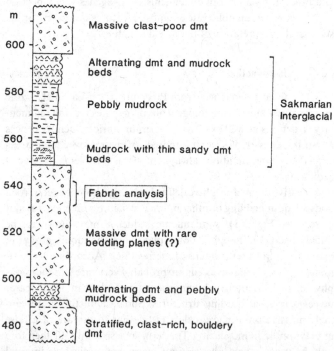

Fig. 6.3. Part of the glacial sequence at Floriskraal, 8 km to the south of Laingsburg. Massive diamictite between 498 and 548 m studied in detail. dmt, diamictite.

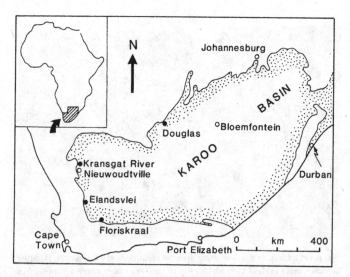

Fig. 6.2. Map showing the distribution of the Dwyka Formation in the main Karoo Basin and the location of the outcrops discussed in the text.

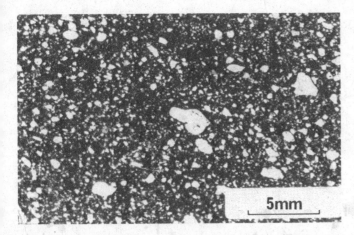

Fig. 6.4. Photomicrograph of the arenaceous matrix of the massive diamictite from Floriskraal. The matrix in the lower right hand section of the photograph has a porphyric fabric with the elongated sand-size grains showing a crude orientation. Deformation of the matrix texture with a corresponding increase in the clay content is evident in the upper left hand section. Disruption of the texture is probably due to soft-sediment flow or dewatering.

A fabric analysis of prolate clasts in the diamictite shows a bimodal clast long-axis distribution (Fig. 6.8) as well as a multimodal distribution for the long-axis dip angles (Fig. 6.9). The fabric diagram shows widely dispersed clast orientations with 53% of the clasts having long-axis dip angles of more than 20°.

Interpretation

The relationship of the massive diamictite with the over- and underlying mudrocks, the sharp undeformed contacts, possible bedding planes, and wide distribution of the facies indicate subaqueous deposition. The widely dispersed clast orientations showing occasional steep dips are consistent with those reported by Domack & Lawson (1985) and Dowdeswell & Sharp (1986) for either rain-out or sediment gravity flow deposits. Part of the probability distribution curve of the clast long-axis dip angles (Fig. 6.9) falls within the envelope for ice-rafted diamicton as illustrated by Domack & Lawson (1985). The crude grain alignment and textural variation as seen in thin section could best be explained by rain-out processes (Fig. 6.4). The deformed laminae and the apparent subvertical to vertical fabrics of clasts in the polished blocks (Figs 6.5–6.7) can be attributed to soft-sediment deformation.

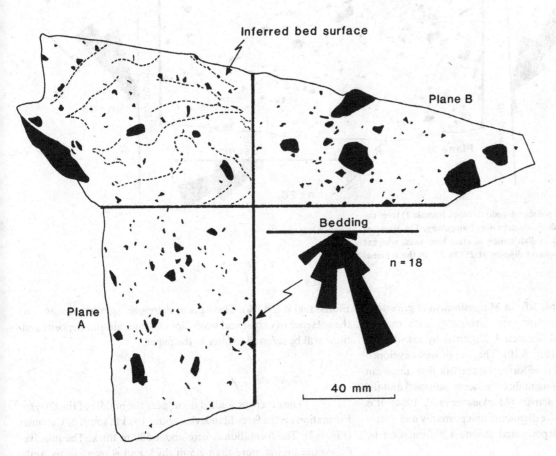

Fig. 6.5. Unfolded diagram of a polished diamictite block from Floriskraal. The inferred bed surface shows faint laminae which are deformed. In a vertical section a poorly defined ab-plane fabric is almost perpendicular to the inferred bedding.

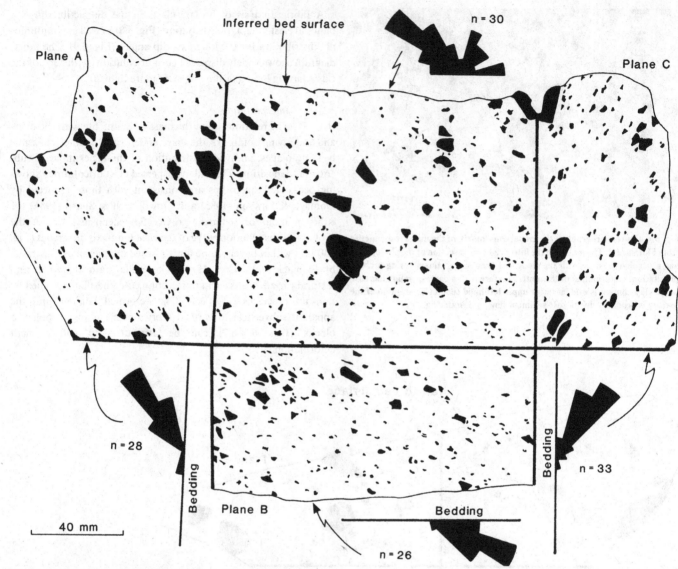

Fig. 6.6. Unfolded diagram of a polished diamictite block (sample 1) from the same stratigraphic horizon at Matjiesfontein near Laingsburg. The apparent bed surface shows a near random distribution of clast long axes, whereas vertical sections show clast ab-planes dipping at 25° to 35° to the regional bedding.

The grain fabrics in sample 1 from Matjiesfontein (e.g. vertical section B in Fig. 6.6) compare very favourably with that of laminated to thinly bedded diamictite deposited by low-density rain-out at Nieuwoudtville (Fig. 6.10). The lack of well-developed laminations in the samples from Floriskraal and Matjiesfontein can be attributed to a higher sediment flux (i.e. dense rain-out) and the absence of episodic current activity (Mackiewicz et al., 1984). It is thus suggested that the massive diamictite unit primarily had a rain-out origin with some syndepositional sediment deformation by gravity flow.

The fabric analysis of a dropstone argillite from Nieuwoudtville, which was studied as a reference sample for rain-out processes, differs strongly from that of Floriskraal largely because of 92% of the clasts having long-axis dip angles of less than 20° (Fig. 6.11; at

Floriskraal it is 47%). There must, therefore, be other reasons for the presence of occasional steep dips in some rain-out deposits and these will be referred to later in the paper.

Elandsvlei

The outcrops studied occur near the middle of the Dwyka Formation on the farm Elandsvlei, about 100 km south of Calvinia (Fig. 6.2). The formation is here about 580 m thick. The massive diamictite unit is more than 2.5 m thick and is overlain by well-stratified diamictite having bed thicknesses of between 10 and 50 cm (Fig. 6.12). Large isolated clasts of more than 1 m in diameter truncate bedding in the stratified diamictite. Towards the top of the massive diamictite unit erosional bed surfaces which cut up to 1 m

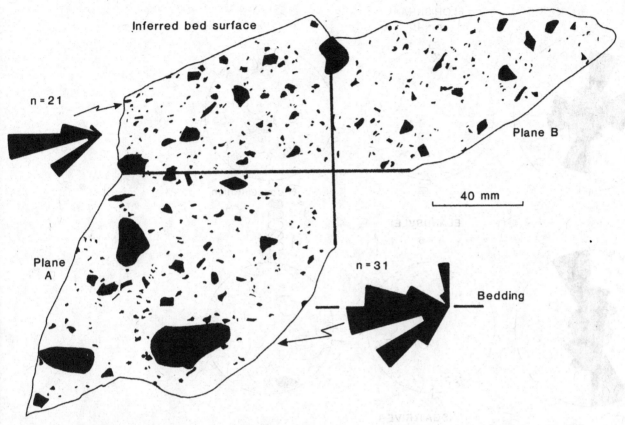

Fig. 6.7. Unfolded diagram of a polished diamictite block (sample 2) from the same stratigraphic horizon at Matjiesfontein near Laingsburg. The apparent bed surface shows a preferred clast long-axis orientation. In vertical section clasts lie with their ab-planes from parallel to vertical with respect to the bedding.

into the underlying massive diamictite, can be seen (Fig. 6.12). The diamictite consists of subangular to rounded clasts (maximum diameter 30 cm) in a structureless argillaceous matrix. In thin section the diamictite matrix shows an unstructured porphyric fabric (Van der Meer *et al.*, 1983). Within the massive unit horizontal, tubular, as well as cross-cutting dyke-like, carbonate-cemented diamictite bodies are present.

A fabric analysis of prolate clasts in the diamictite shows a preferred clast long-axis distribution (Fig. 6.8), but the long-axis dip angles have a wide distribution with 31% of the clasts having dip angles of more than 20° (Fig. 6.9).

Interpretation

The relationship of the massive diamictite with the stratified diamictite and the erosive bed surfaces within the diamictite suggest a possible subaqueous origin. The stratified diamictites show flow banding (Fig. 6.13) and contain large ice-rafted stones, indicating their formation in a subaqueous ice-marginal setting.

Comparison of the fabric diagrams with published data (Domack & Lawson, 1985; Dowdeswell & Sharp, 1986) suggests a possible debris-flow or rain-out origin. The probability distribution plot shows no correspondence with published data (Fig. 6.9). The fabric analysis indicates transport of the material from the south,

whereas ice flow for the Elandsvlei glacials was towards the west (soft sediment and striated boulder pavements) (Fig. 6.8).

The dispersion of clast axes and the possible perpendicular transport direction to the inferred ice flow, may indicate a subaqueous debris-flow deposit (Lawson, 1979). The carbonate-cemented, tubular diamictite bodies, parallel to the flow direction (Fig. 6.8), can probably be attributed to flow shear during emplacement of the viscous sediment. The carbonate-cemented dyke-like bodies in the diamictite may represent water-escape structures (Collinson *et al.*, 1989). The massive diamictite is thus interpreted as possible resedimented rain-out material on account of the ice-rafted material in the overlying stratified facies. The textural differences between the massive and stratified diamictite facies can be attributed to diverse pore water contents and possible flow distances during deposition (Visser, 1983*a*).

Kransgat River

The outcrops studied occur near the base of the Dwyka Formation on the farm Koringhuis, about halfway between Nieuw-oudtville and Loeriesfontein (Fig. 6.2). The formation is about 250 m thick. The massive diamictite has a thickness of about 7 m and is overlain by a succession of black shale, siltstone and fine-grained

88   J.N.J. Visser

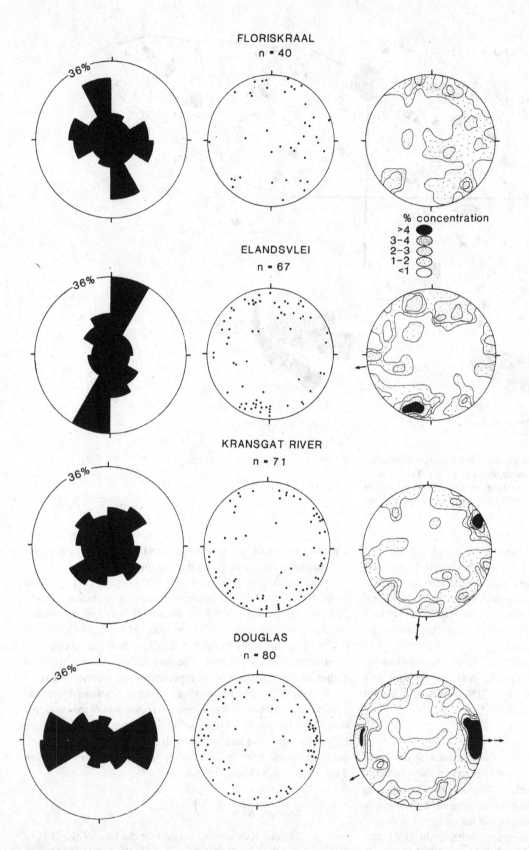

Fig. 6.8. Fabric diagrams of massive diamictites from Floriskraal, Elandsvlei,
Kransgat River and Douglas. Ice-flow direction where determined from
striated pavements, indicated by arrows. Orientation of elongated carbonate-
cemented diamictite bodies at Elandsvlei (solid lines), depositional slope at
Kransgat River (double arrow) and bedrock slope at Douglas (double arrow)
are also shown.

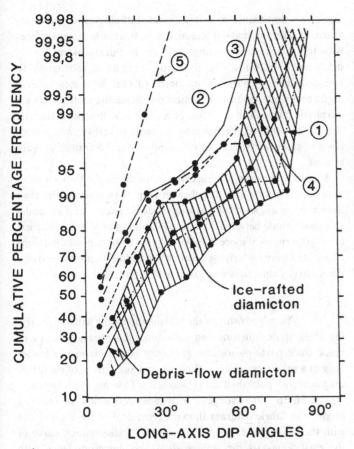

Fig. 6.10. Laminated to thinly bedded diamictite typical of low-density sediment rain-out from meltwater plumes and icebergs. Dwyka Formation near Nieuwoudtville.

Fig. 6.9. Probability distribution plots of clast long-axis dip angles for Dwyka diamictites, and debris-flow (Dowdeswell & Sharp, 1986) and ice-rafted (after Domack & Lawson, 1985) diamictons. 1, Floriskraal; 2, Elandsvlei; 3, Kransgat River; 4, Douglas; 5, Nieuwoudtville.

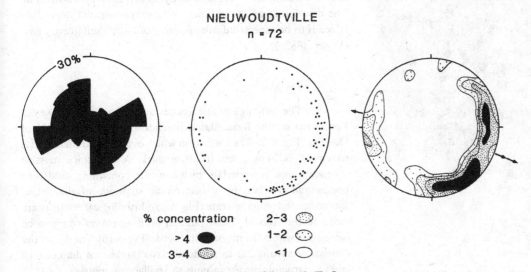

Fig. 6.11. Fabric diagrams of dropstone argillite from near Nieuwoudtville. Ice flow direction determined from striated pavements, is indicated by an arrow and bedrock slope by a double arrow.

Fig. 6.12. Well-stratified diamictite deposited by debris flows overlying massive diamictite at Elandsvlei. The thick massive unit contains erosional bed surfaces (arrow) towards its top.

Fig. 6.13. Polished sample of stratified diamictite from Elandsvlei showing flow banding. Deposit formed by resedimentation of possible rain-out sediment.

sandstone containing ice-rafted stones. Lithological contacts are sharp. Within the massive diamictite undulated bedding surfaces with thin black shale partings (<5 cm in thickness) are locally developed (Fig. 6.14). The diamictite consists of subangular to rounded clasts (maximum diameter 40 cm) in a structureless argillaceous matrix. In thin section the diamictite matrix shows a porphyric fabric (Van der Meer *et al.*, 1983). Within the diamictite vertical to inclined, carbonate-cemented, dyke-like, arenaceous bodies, some of which show compactional deformation, are present.

A fabric analysis of prolate clasts in the diamictite shows a weak bimodal clast long-axis distribution (Fig. 6.8), whereas the clast long-axis dip angles show a strong mode between 1° and 10° and a secondary mode between 40° and 60° resulting in 22% of the clasts having dip angles of more than 20° (Fig. 6.9). The depositional slope as deduced from underlying bedrock and bedded deposits overlying the massive diamictites was towards the south (Fig. 6.8).

Interpretation

The relationship of the massive diamictite with the overlying black shale, siltstone and sandstone, the presence locally of black shale partings and the geometry of the diamictite bodies suggest a subaqueous debris-flow origin. Comparison of the fabric diagrams with published data (Domack & Lawson, 1985; Dowdeswell & Sharp, 1986) suggests a possible debris-flow or rain-out origin. The fabric diagram shows a great deal of correspondence with the Elandsvlei fabrics. The probability distribution curve of the clast long-axis dip angles shows no correspondence with published data (Fig. 6.9).

The approximate upslope imbrication of the long axes (Fig. 6.8) lends further support for a debris-flow origin of the massive diamictite (Lawson, 1979). The carbonate-cemented, dyke-like sandstones represent dewatering structures (Collinson *et al.*, 1989). Flow of water through the structures may have flushed out some of the clays leaving more arenaceous dykes before compaction set in. The dewatering of the sediment also supports rapid deposition which is to be expected during freezing of a sediment gravity flow (Visser, 1983a).

Douglas

The outcrops studied occur at the base of the Dwyka Formation on the farm Blaauwboschdrift, about 4 km west of Douglas (Fig. 6.2). The formation which overlies a striated glacial pavement, is from a few to 40 m thick. A 3 m thick massive diamictite unit is underlain by diamictite containing sandstone bodies and overlain by a heterolithic sequence of diamictite, sandstone and conglomerate (Fig. 6.15). Lithological contacts are sharp to transitional. Faint parallel bedding planes can also be recognized within the massive diamictite (Fig. 6.16). The diamictite consists of subangular to rounded clasts (maximum diameter 15 cm) in a structureless arenaceous to argillaceous matrix. A fabric analysis of prolate clasts in the diamictite shows a preferred long-axis orientation (Fig. 6.8) and the clast long-axis dip

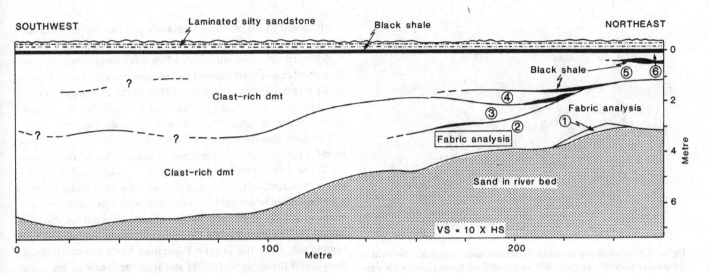

Fig. 6.14. Section of the basal diamictite along the Kransgat River between Nieuwoudtville and Loeriesfontein. The numbers referred to different diamictite beds. Thicknesses of interbedded shales are exaggerated. dmt, diamictite.

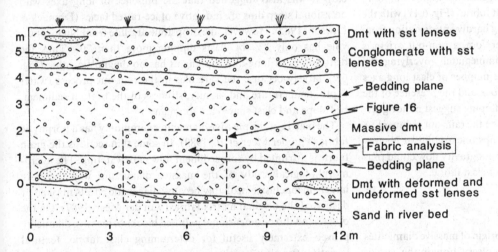

Fig. 6.15. Basal section of the Dwyka Formation about 4 km to the west of Douglas. Massive diamictite in the middle of the section was studied in detail. sst, sandstone; dmt, diamictite.

Fig. 6.16. Faint bedding planes (indicated by arrows) within the massive diamictite at Douglas.

angles form a strong mode between 1° and 10° with only 20% of the axes having a dip of more than 20° (Fig. 6.9). The probability distribution plot of the clast long-axis dip angles shows no correspondence with published data (Fig. 6.9). The clast fabrics weakly suggest ice flow from the east which is approximately the direction of flow determined from the underlying striated pavement. The bedrock surface slopes towards the east (Fig. 6.8).

Interpretation

The presence of parallel bedding planes, lag conglomerates and small sandstone lenses which represent starved ripples (Fig. 6.17) within the diamictite indicates subaqueous deposition. The dispersion of clast long axes weakly supports a rain-out origin for the diamictite (Domack & Lawson, 1985; Dowdeswell & Sharp, 1986), but the clast long-axis orientations almost parallel to the ice flow direction, cast doubt on such an interpretation. However,

Fig. 6.17. Starved ripples (at hammer) forming small sandstone lenses within the massive diamictite at Douglas. Note how difficult it is to laterally trace the bedding plane on which the ripples had developed.

studies of dropstone argillites by Domack & Lawson (1985) and Visser (1989b) show weak to strong clast fabrics (Fig. 6.11) with the principal mode in some samples showing hardly any deviation from the established ice flow direction. Therefore, a rain-out origin is inferred for the massive diamictite immediately overlying the striated pavement. The fact that a large number of clast long axes are parallel to the dip slope of the bedrock and the presence of the deformation of sandstone bodies by slumping suggest minor gravity flow, induced by the bedrock slope, on the rain-out sediment. A similar observation was made in the dropstone argillite at Nieuwoudtville (Fig. 6.11) and downdip sediment gravity flow could thus have caused modification of the rain-out clast fabric.

Discussion

This study illustrates that the origin of massive diamictites can be established, although with some reservation, in the case of rain-out and debris-flow deposits. Such a situation is not uncommon because wherever rain-out occurred the possibility of resedimentation by sediment gravity flow is always present (Eyles & Eyles, 1983b), and the two processes often occur simultaneously (Visser, 1983b). Where stratification is absent the recognition of subaqueous debris flow deposits in the glacial environment remains a problem (Visser, 1983b). Under such circumstances the geometry of the diamictite body may contain clues to the origin of the deposit. Rain-out deposits have a much more even geometry, often draping the substrate, whereas debris flow diamictites are mostly lenticular and uneven in thickness (Fig. 6.14). Flow-banding, where present, is very diagnostic of sediment-gravity flow deposits (Visser, 1983b; Fig. 6.13). Lack of it may be attributed to a high flow velocity, high porewater pressures within the sediment and a long transportation path. High porewater pressures are also commonly present during dense rain-out which may explain partly the overlap in characteristics between sediment gravity flow and rain-out deposits in the glaciomarine environment.

Rain-out deposits are commonly laminated (Fig. 6.10) which can

be attributed to the presence of episodic current activity by underflows debouching from the ice front. However, where the ice has a high basal debris concentration and there is a low meltwater input, ice-rafting can be predominant and massive diamictites can form in very proximal settings (Domack, 1984). In the absence of internal laminations undeformed grain or clast layers within the massive diamictite are an indicator of rain-out processes (Fig. 6.6).

The study also illustrates that fabric analysis on its own cannot be used for the ultimate interpretation of massive diamictites because the fabric data are often inconclusive. It can be argued that more elaborate statistical treatment of the present data (e.g. eigenvalues) may have given better answers, but where such data are drawn from bimodal or multimodal populations of axial orientations, misleading results can be obtained (Mark, 1973). Only one of the four sample sets from the Dwyka Formation has a unimodal distribution and therefore no further statistical treatment of the orientation data was undertaken. Both rain-out and sediment-gravity flow processes resulted in a wide dispersion of clast long axes (Fig. 6.8). It was also suggested that the presence of long axes with occasional steep dips are indicative of ice-rafted facies (Domack & Lawson, 1985). However, the dropstone argillite from Nieuwoudtville has only 8% of clasts dipping more than 20°. Thus the distribution of long-axis dip angles has no diagnostic value (Fig. 6.9), largely because the variables in the rain-out process (water depth, temperature and salinity, stiffness of the bottom sediment, bed slope and clast shape) are too many.

The matrix of the massive diamictites commonly has a porphyric fabric (Van der Meer et al., 1983). In some thin sections the grain fabric is disrupted by soft-sediment flow or dewatering. The micromorphological study of the massive diamictites was therefore largely inconclusive probably because diagenetic effects masked or destroyed the original matrix fabric (Van der Meer, 1987). On the other hand three-dimensional polished blocks of massive diamictite proved extremely useful for determining clast fabric, textural variation and clast layering.

Ancient massive diamictites have a multiple origin. They can form directly by dense rain-out or indirectly by resedimentation of high-density rain-out, subglacial diamictons, or distal bergstone mud deposits (Fig. 6.18). No supporting evidence for the formation of massive diamictites by resedimentation of lodgement and melt-out diamictons has so far been recorded in the study area. In ancient diamictites factors like diagenesis, compaction and dewatering can obliterate diagnostic features or modify grain fabric.

The application of fabric analysis together with a study of the bedding characteristics can be useful tools in the recognition of rain-out, debris flow or a combination of these mechanisms in the formation of ancient glacial marine diamictites. However, there is no single criterion diagnostic in the interpretation of ancient massive diamictites. The study further emphasizes the importance of a lithofacies analysis in the field as a first priority in the interpretation and, if combined with systematic laboratory and clast fabric analyses as part of a multicriteria approach, it will achieve a better understanding of the genesis of massive diamictites from the glacial environment (Dowdeswell et al., 1985).

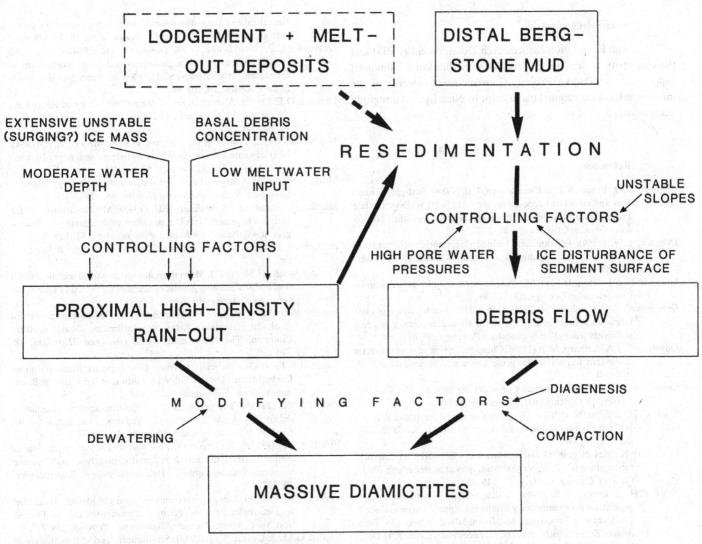

Fig. 6.18. Flow diagram of the origin of ancient massive diamictites in the glacial marine environment. The contribution of lodgement and melt-out diamictons to the formation of thick, massive, resedimented diamictites has not yet been proved in the study area.

Acknowledgements

The Foundation for Research Development (FRD) and the University of the Orange Free State are thanked for financial support of the Dwyka Project. Constructive comment on the improvement of the original manuscript by Nick Eyles is gratefully acknowledged.

References

Collinson, J.D., Bevins, R.E. & Clemmensen, L.B. (1989). Post-glacial mass flow and associated deposits preserved in palaeovalleys: the late Precambrian Moraeneso Formation, North Greenland. *Meddelelser Grønland, Geoscience*, **21**, 1–27.

Domack, E.W. (1984). Rhythmically bedded glaciomarine sediments on Whidbey Island, Washington. *Journal of Sedimentary Petrology*, **54**, 589–602.

Domack, E. & Lawson, D.E. (1985). Pebble fabric in an ice-rafted diamicton. *Journal of Geology*, **93**, 577–91.

Dowdeswell, J.A., Hambrey, M.J. & Wu, R. (1985). A comparison of clast fabric and shape in late Precambrian and modern glacigenic sediments. *Journal of Sedimentary Petrology*, **55**, 691–704.

Dowdeswell, J.A. & Sharp, M.J. (1986). Characterization of pebble fabrics in modern terrestrial glacigenic sediments. *Sedimentology*, **33**, 699–710.

Dreimanis, A. (1979). The problems of waterlain tills. In *Moraines and Varves*, ed. C. Schlüchter, Rotterdam, Balkema, pp. 167–77.

Eyles, C. H. & Eyles, N. (1983a). Glaciomarine model for upper Precambrian diamictites of the Port Askaig Formation, Scotland. *Geology*, **11**, 692–4.

Eyles, C.H. & Eyles, N. (1983b). Sedimentation in a large lake: a reinterpretation of the late Pleistocene stratigraphy at Scarborough Bluffs, Ontario, Canada. *Geology*, **11**, 146–52.

Eyles, C.H. & Lagoe, M.B. (1990). Sedimentation patterns and facies geometries on a temperate glacially influenced continental shelf: the Yakataga Formation, Middleton Island, Alaska. In *Glacimarine Environments: Processes and Sediments*, eds. J.A. Dowdeswell, J.D. Scourse, Geological Society of London Special Publication, **53**, 363–86.

Eyles, N. & McCabe, A.M. (1989). Glaciomarine facies within subglacial tunnel valleys: the sedimentary record of glacio-isostatic downwarping in the Irish Sea Basin. *Sedimentology*, **36**, 431–48.

Gravenor, C.P., Von Brunn, V. & Dreimanis, A. (1984). Nature and classification of waterlain glaciogenic sediments, exemplified by Pleistocene, late Paleozoic and late Precambrian deposits. *Earth-Science Review*, **20**, 105–66.

Lawson, D.E. (1979). A comparison of the pebble fabrics in ice and the deposits of the Matanuska Glacier, Alaska. *Journal of Geology*, **87**, 629–45.

Mackiewicz, N.E., Powell, R.D., Carlson, P.R. & Molnia, B.F. (1984). Interlaminated ice-proximal glacio-marine sediments in Muir Inlet, Alaska. *Marine Geology*, **57**, 113–47.

Mark, D.M. (1973). Analysis of axial orientation data, including till fabrics. *Geological Society of America Bulletin*, **84**, 1369–74.

Martin, H., Porada, H. & Walliser, O.H. (1985). Mixtite deposits of the Damara sequence, Namibia, problems of interpretation. *Palaeogeography, Palaeoclimatology, Palaeoecology*, **51**, 159–96.

Pettijohn, F.J. (1975). *Sedimentary rocks*. New York, Harper & Row, 628 pp.

Van der Meer, J.J.M. (1987). Micromorphology of glacial sediments as a tool in distinguishing genetic varieties of till. Special Paper of the Geological Survey of Finland, 3, 77–89.

Van der Meer, J.J.M., Rappol, M. & Semeijn, J.N. (1983). Micromorphological and preliminary X-ray observations on a basal till from Lunteren, The Netherlands. *Acta Geologica Hispanica*, **18**, 199–205.

Visser, J.N.J. (1983a). Submarine debris flow deposits from the upper Carboniferous Dwyka Tillite Formation in the Kalahari Basin, South Africa. *Sedimentology*, **30**, 511–23.

Visser, J.N.J. (1983b). The problems of recognizing ancient subaqueous debris-flow deposits in glacial sequences. *Transactions of the Geological Society of South Africa*, **86**, 127–35.

Visser, J.N.J. (1989a). The Permo-Carboniferous Dwyka Formation of southern Africa: deposition by a predominantly subpolar marine ice sheet. *Palaeogeography, Palaeoclimatology, Palaeoecology*, **70**, 377–91.

Visser, J.N.J. (1989b). Stone orientations in basal glaciogenic diamictite: four examples from the Permo-Carboniferous Dwyka Formation, South Africa. *Journal of Sedimentary Petrology*, **59**, 935–43.

Young, G.M. & Gostin, V.A. (1988). Stratigraphy and sedimentology of Sturtian glacigenic deposits in the western part of the North Flinders Basin, South Australia. *Precambrian Research*, **39**, 151–70.

7   Neoproterozoic tillite and tilloid in the Aksu area, Tarim Basin, Xinjiang Uygur Autonomous Region, Northwest China

LU SONGNIAN and GAO ZHENJIA

Abstract

Precambrian sequences of the Aksu region, which is situated near the northwestern margin of the Tarim Block, are composed of the Mesoproterozoic Aksu Group and Neoproterozoic Qiaoenbulak and Wushinanshan Groups. Whole-rock Rb-Sr isochron ages of 962 Ma and 944 Ma indicate that the protolith of the Aksu Group is older than about 960 Ma. The Qiaoenbulak Group comprises a large thickness of clastic rocks, representing turbidite deposits on a submarine fan. The Wushinanshan Group is made up of glacial deposits of the Umainak Formation, red sandstone of the Sugiatbulak Formation and littoral carbonate rocks of the Qigebulak Formation.

Two types of diamictite exist in the Qiaoenbulak and Wushinanshan Groups. The lower diamictite (Qiaoenbulak), previously described as a tillite, consists mainly of debris flow deposits. The upper diamictite (Wushinanshan) is a continental glacial deposit. Although they seem similar in appearance, appreciable differences in clast composition, roundness and sorting, as well as in facies association and contact relationships, allow the two to be distinguished.

Introduction

Precambrian sequences including the Mesoproterozoic Aksu, Neoproterozoic Qiaoenbulak and Wushinanshan Groups are exposed in the Aksu region along the northwestern margin of the Tarim Block (Fig. 7.1). Precambrian tillites were first recognized there by Zhang Zhenhua and Gao Zhenjia et al. of Geological Brigade No. 13 of the Ministry of Geology (Zhang et al., 1957). Later, several investigations concentrating on the Precambrian sequences were performed by Gao et al. (1982, 1983, 1985), Gao and Qian (1985) and Wang and Gao (1986). These researchers described two layers of diamictite and proposed a glacial origin for both. In 1986 the authors of this paper had an opportunity to investigate the Precambrian sequences in the Aksu region. They noticed many features characteristic of debris flows in diamictite in the Qiaoenbulak Group and of glacial deposits in diamictite in the Umainak Formation of the Wushinanshan Group. In this paper, we clarify the stratigraphic positions of the two diamictites, describe their principal sedimentary characteristics, and interpret their environment of deposition.

The Precambrian stratigraphic sequences

Previously the Precambrian sequence near Aksu was usually divided into two parts, namely lower metamorphic rocks and upper sedimentary rocks. The lower part, the Aksu Group, was assigned to the Lower Proterozoic (The Compiling Group on Stratigraphy of Xinjiang Uygur Autonomous Region, 1984); and the upper part was included in the Sinian System. Gao Zhenjia et al. (1982, 1985) subdivided the upper sedimentary rocks into the Qiaoenbulak, Umainak, Sugaitbulak and Qigebulak formations, in ascending order, as shown in Table 7.1.

In this investigation, a significant angular unconformity between the Qiaoenbulak Formation and the overlying formation was found. In addition, a basic dike intruded into the Qiaoenbulak Formation is unconformably truncated by the overlying formations. Moreover, there is an abrupt sedimentary facies change between the Qiaoenbulak and the overlying formations. The Qiaoenbulak Formation is composed of a large thickness of turbidites, whereas the Umainak Formation contains glacial deposits at the base. A well-preserved glacial pavement was found on the surface of the sandstone of the Qiaoenbulak Formation. In this paper, therefore, we divide the Precambrian sequences into three parts: the Aksu, Qiaoenbulak and Wushinanshan Groups. The Wushinanshan Group is subdivided into the Umainak, Sugaitbulak and Qigebulak formations (Table 7.1).

The Aksu Group is composed of complexly deformed metamorphic rocks including blue schist, (blue) actinolite schist, epidote-actinolite schist, albite-quartz schist and minor liptite. We have obtained two whole-rock Rb-Sr ages of 962 ± 12 Ma (where $Sr_i = 0.7079$) and 944 ± 12 Ma (where $Sr_i = 0.7043$) on albite-quartz schist and alternating blue schist and epidote-actinolite schist, respectively. We interpret these to indicate the time of the last metamorphic episode effecting the Aksu Group (Lu and Gao, 1990).

The Qiaoenbulak Group comprises a series of flysch deposits, up

Table 7.1. *Stratigraphic subdivisions of Precambrian rocks of the Aksu Region*

Gao *et al.* (1982, 1985)		Lu & Gao (1990)			
Upper Proterozoic (Sinian System)	Qigebulak Fm.	Neoproterozoic	Sinian	Wushinanshan Group	Qigebulak Fm.
	Sugaitbulak Fm.				Sugaitbulak Fm.
	Umainak Fm.				Umainak Fm.
	Qiaoenbulak Fm.		Qingbaikou System	Qiaoenbulak Group	
Lower Proterozoic Pt₁	Aksu Group	Middle Proterozoic	Aksu Group		

(b)

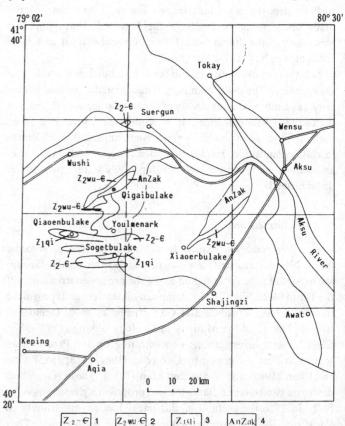

(a)

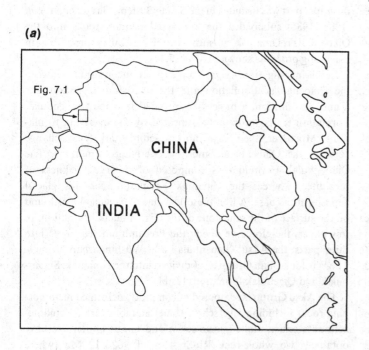

Fig. 7.1. Sketch geological outcrop map of the Aksu-Wushi area of Xinjiang. 1, Terminal Sinian-Cambrian; 2, Wushinanshan Group; 3, Qiaoenbulak Group; 4, Aksu Group.

Fig. 7.2. Bidirectional cross-bedding of the Sugaitbulak Formation.

to 2000 m thick, consisting of grayish green arkose, feldspathic lithic sandstone and siltstone. The rocks have low textural maturity, are poorly-sorted, and have a grain size distribution similar to that of abyssal turbidites. Bouma sequences are common. A diamictite layer occurs in the middle part of the group. The Qiaoenbulak Group is inferred to be early Neoproterozoic in age.

The Wushinanshan Group, including the Umainak, Sugaitbulak, and Qigebulak formations, varies from 310 to 1193 m in thickness. This group, composed of terrestrial or shallow water clastic and carbonate rocks is distinctly different from the Qiaoenbulak Group. The Umainak Formation is composed of glacial deposits unconformably overlying the Qiaoenbulak Group. Disconformably or unconformably overlying the Umainak Formation, the Sugaitbulak Formation consists of purplish-red to red sandstone with excellent bidirectional cross-bedding (Fig. 7.2) suggesting deposition in an intertidal zone. The Qigebulak Formation consists of shallow water carbonates. The overlying Yuertusi Formation is lowest Cambrian in age (Meishueun Stage) because of the occurrence of *Anabarites trisalcartus*. Because of its stratigraphic position below Cambrian rocks, and because it included glacial deposits, the Wushinanshan Group is assigned to the Sinian system.

Sedimentary features of diamictites in the Qiaoenbulak Group

The Qiaoenbulak Formation was previously subdivided into four members, in ascending order: the 'Sandstone Member', the 'Tillite Member', the 'Calcic Sandstone Member', and the 'Sand-Conglomerate Member' (Wang and Gao, 1986). The Sandstone Member is 1715 m thick. Bouma sequences within it are commonly 50–100 cm thick. Sandstone with graded bedding (A division) is overlain by siltstone showing even bedding (B division). Locally micro cross-lamination is visible (C division), but C and E divisions are generally not present. In the Sand-Conglomerate Member single Bouma sequences are up to 60–120 cm thick and graded beds are composed of fine conglomerate and coarse-,

medium-, and fine-grained sandstone. The Tillite Member is variable in thickness ranging up to 97 m. It comprises massive diamictite with slumping structures. It was inferred to be an oceanic deposit formed by ice-rafting. Based on features described below, we consider this tilloid to have been deposited by debris flows.

The tilloid is composed of a lower conglomerate and an upper diamictite. The conglomerate varies from 15 m to 20 m thick and is interbedded with 3 to 5 layers of sandstone, each sandstone layer less than 1 m thick. The conglomerate matrix is composed of lithic, feldspar and quartz grains. The interbedded sandstones have the same composition. Clasts are mostly about 8 to 12 cm in diameter, decreasing in size upwards and show the following features: high degrees of rounding (Fig. 7.3); good sorting; a preferred orientation that generally parallels bedding; and a varied composition which includes mainly medium to basic metamorphic volcanic rocks and a lesser amount of granite. These features indicate that the conglomerate clasts were transported a long distance and by water, not by ice (Brenchley and Williams, 1985).

The upper diamictite comprises a mixture of pebbles, sand and mud about 70 m thick. Clasts are unevenly distributed and show

Fig. 7.3. Well-rounded clasts from massive diamictite of the Qiaoenbulak Group.

random orientation. Some clasts show fissures caused by compression, even though only striations are visible on the clast surface. The diamictite, in appearance, is similar to tillite. However, careful field observation reveals that the clasts, although unevenly distributed, resemble those in the underlying conglomerate. For instance, they show excellent roundness and diameters generally about 10 cm. Although randomly oriented, the diamictite clasts are well-sorted and identical to clasts of the underlying conglomerate in composition.

Another important feature is that sandstone blocks within the diamictite show complex sedimentary structures. Some sandstone blocks preserve regular bedding. Others are folded or show convolute bedding, which reflects soft-deformation.

No evidence for a sedimentary hiatus was recognized between the tilloid and overlying or underlying rocks, so the tilloid should form in an environment similar to adjacent rocks. We infer that all were deposited by sediment gravity flow. The features of the tilloid itself indicate that it is composed of sediment reworked from underlying rocks. Soft sediment deformation in the sandstone block suggests that the tilloid was formed by slumping of underlying rocks before consolidation (Lopez-Gamundi, 1987).

Features of the diamictite of the Umainak Formation (Wushinanshan Group)

The Umainak Formation is variable in thickness with a maximum thickness of over 70 m. Gao et al. (1985) described glacial features within the formation.

Glacial pavements are one of the most important forms of evidence of glaciation. The glacial pavement below the Umainak Formation in the Aksu area is developed on several different layers of feldspathic sandstone or stratified siltstone at the top of the Qiaoenbulak Group. We dug out a glacial pavement at three sites each about 1 m² near Umainak Spring. The surface of the pavement, because of glacial abrasion, is smooth and slippery. Grooves and striations exist on the pavement (Fig. 7.4). According to the

Fig. 7.4. Pavement with striations on sandstone of the Qiaoenbulak Group.

orientation of the striations the ice moved towards a direction between S30°W and S40°W.

The Umainak Formation mainly includes two facies: massive diamictite and thin-bedded rhythmite. Massive diamictite dominates. Clasts within it are generally less than 10 cm, and rarely up to 50 cm, in diameter. Clasts 2.5 to 6 m across have been found in the basal part of the formation in western Umainak. The clasts are diverse in composition and randomly distributed irrespective of size. Flatiron-shaped clasts commonly with striated and polished surfaces as well as shallow holes formed by compression are visible, particularly in the lower and upper parts of the formation.

The rhythmite is 2 to 4 m thick and forms an interlayer between the lower and upper massive diamictites. It comprises thin-bedded siltstone, with individual beds millimeters to centimeters thick. Even bedding and graded bedding are common. Clasts, which are more than 10 cm in diameter, are preserved in the rhythmite and look like dropstones.

Massive diamictite may occur in following environments: (1) the basal part of terrestrial glaciers; (2) the lower portion of a glacial tongue extending into a lake or ocean; and (3) a submarine fan and margins of the abyssal basin due to rapid ablation of ice bergs. In the Umainak Formation, the glacial pavement, the distinct thickness changes along strike, and the high proportion of massive diamictite (up to 85% of the formations thickness) are evidence that the diamictite is a subglacial deposit. We infer that the massive diamictite represents sediments derived directly from the lower part of a glacier.

The rhythmite facies reflects deposition in a basin in front of the glacier. It is quite thin and therefore unlikely to be a glacial-marine deposit. The rhythmite probably represents glaciolacustrine deposits with dropstones, formed in a proglacial lake (Visser & Loock, 1987). Thus the Umainak Formation, as a whole, is a product of terrestrial glaciation (Von der Borch et al., 1988; Vorren et al., 1989).

Comparison between diamictites of the Qioenbulak Group and the Wushinanshar Group

Both the debris flow developed in the upper part of the Qiaoenbulak Group and the glacial deposits in the Umainak Formation of the Wushinanshan Group are composed mainly of diamictite which is massive and shows random clast orientations. The two diamictites formed, however, by different geological processes. The former was deposited by high density sediment gravity flow and the latter by glacial activity. A comparison of the main features of the diamictites is given in Table 7.2.

We emphasize that debris flows are a type of sedimentary gravity flow. Once conditions are favorable, for example, when a slope is too steep to hold the sediment that has settled on it, a sediment gravity flow will form. Sometimes the flow will spread over a large area. This is primarily controlled by the size of the basin. Therefore, debris flows, as a special facies, can be used for correlation within a basin, but they are not isochronous outside the basin. Glacial sediments contrast with sediment gravity flow deposits in that they

Table 7.2. *Composition of sedimentary features of the tilloid of the Qiaoenbulak Group and the glacial deposits of the Wushinanshan Group*

	Tilloid (clastic flow deposit) (Qiaoenbulak Group)	Glacial Deposit (Wushinanshan Group)
Clast roundness	Excellent	Not good, mainly angular or subangular
Clast sorting	Good, most about 10 cm in diameter	Poor, varies from large to small
Clast composition	Diverse	Very diverse
Clast orientation	Random	Random
Sedimentary structures	Primary stratification still preserved, some soft sediment deformation, convolute bedding and slump features	Massive, resulting from mixing of gravel, sand and mud
Basal contact	Continuous with sediments in section	Glacial pavement seen on basement
Facies association	Associated with submarine fan gravity flow deposits and valley conglomerate facies	Associated with fine rhythmites with dropstones
Proportion of total formation thickness	Massive diamictite amounts to less than 5%	Massive diamictite totals up to 85%
Geological agent of transportation	Sediment gravity flow	Glacier
Sedimentary environment	Near feeding course on sea fan	Various parts of terrestrial glacier

record a climatic event in Earth history. Glacial sediments reflect a cold climate. Neoproterozoic glacial deposits have been found on all continents except Antarctica. We believe that glacial sediments and other glacial features can be used as approximate marker horizons in stratigraphic correlation between different tectonic units (Lu *et al.*, 1982, 1983). Therefore, gravity flow deposits are distinct from glacial deposits in geological significance. However, sediment gravity flows are very common in many glacial sedimentary environments. So sediment gravity flow deposits related to glaciation must be carefully distinguished from non-glacigenic gravity flow deposits. The two sediment types are not difficult to recognize. But this distinction must be made extremely carefully because of the different geological significance of the two deposits. Massive diamictite is commonly formed by both sediment gravity flow and glaciation. Thus its genesis easily becomes a focus of heated discussion. The authors have emphasized repeatedly that in order to determine the genesis of diamictite at least three factors in particular must be considered. First, the regional geological background should be studied and data on the regional distribution and variations in the massive diamictite, as well as the associated strata, should be collected. Second, sedimentary facies association and stratigraphic contact relationships should be noted. Third, the composition, source, sorting, roundness, shape and arrangement of clasts, as well as included blocks or intercalated beds within the diamictite, should be carefully examined (Lu and Gao, 1984; Roser and Korsch, 1986; Young and Gostin, 1988).

Acknowledgements

The authors acknowledge financial support from Project No. 305 of Xinjiang and cooperation from Peng Changwen, Qin Qhenyong, Xiao Ping *et al.* We had helpful discussions with Professor J.C. Crowell and express our thanks to all, including every unit or researcher, who have helped us during the work.

References

Brenchley, P.J. & Williams, B.P.J. (1985). *Sedimentology: Recent Developments and Applied Aspects.* Geological Society of London, Special Publication, 18, 342 pp.

Gao Zhenjia, Wu Shaozuo *et al.* (1982). Research on stratigraphy of the Sinian–Cambrian Sequences, Aksu-Keping Region, Xinjiang. *Kaxue Tong Bo*, **27**, 524–7.

Gao Zhenjia, Wang Wu-Yan & Peng Chang-Wen (1983). Sinian System on Aksu-Wushi Region, Xinjiang. Xinjiang Publishing House (in Chinese), 184 pp.

Gao Zhenjia *et al.* (1985). Sinian System on Xinjiang. Xinjiang Publishing House (in Chinese), 173 pp.

Gao Zhenjia & Qian Jianxin (1985). Sinian glacial deposits in Xinjiang, Northwest China. *Precambrian Research*, **29**, 143–7.

Lopez-Gamundi, O.R. (1987). Depositional models for the glaciomarine sequences of Andean Late Paleozoic basins of Argentina. *Sedimentary Geology*, **52**, 109–26.

Lu Songnian & Ding Baolan (1982). Features of the grain size-frequency distribution for matrix of till and tillite in some regions. Bulletin of the Tianjin Institute of Geology and Mineral Resources, Chinese Academy of Geological Sciences, 5, 67–73 (in Chinese).

Lu Songnian, Ma Guogan *et al.* (1983). Primary Research on Glacigenous Rocks of Late Precambrian in China. *Precambrian Geology*, **1**, 1–74 The Collected works of Late Precambrian Glacigenous Rocks in China (in Chinese).

Lu Songnian & Gao Zhenjia (1984). Proterozoic diamictites in China. Scientific papers on Geology for international exchange. Prepared for the 27th International Geological Congress (1), Geological Publishing House, Beijing, 205–14 (in Chinese).

Lu Songnian & Gao Zhenjia (Research party on Precambrian and its ore-bearing properties in northern Xinjiang) (1990). Subdivision of

Precambrian in northern Xinjiang. *Geological Sciences of Xinjiang*, **1**, 88–100 (in Chinese).

Roser, B.P. & Korsch, R.J. (1986). Determination of tectonic setting of sandstone–mudstone suites using SiO_2 content and K_2O/Na_2O ratio. *Journal of Geology*, **94**, 635–50.

Visser, J.N.J. & Loock, J.C. (1987). Ice margin influence on glaciomarine sedimentation in the Permo-Carboniferous Dwyka Formation from the southwestern Karoo, South Africa. *Sedimentology*, **34**, 929–41.

Von der Borch, C.C., Christie-Blick, N. & Grady, A.E. (1988). Depositional sequence analysis applied to Upper Proterozoic Wilpena Group, Adelaide Geosyncline, South Australia. *Australian Journal of Earth Sciences*, **35**, 59–71.

Vorren, T.O., Lebesbye, E., Andreassen, K., Larsen, K.B. (1989). Glacigenic sediments on a passive continental margin as exemplified by the Barents Sea. *Marine Geology*, **85**, 251–72.

Wang Wunyan & Gao Zhenjia (1986). Sinian system on northwest margin of Tarim Basin. *Precambrian Geology of China*, **3**, 195–202 (in Chinese).

The Compiling Group on Stratigraphy of Xinjiang Uygur Autonomous Region (1984). *Xinjiang Regional Stratigraphy Scale*. Xinjian Publishing House (in Chinese).

Young, G.M. & Gostin, V.A. (1988). Stratigraphy and sedimentology of Sturtian glacigenic deposits in the western part of the North Flinders Basin, South Australia. *Precambrian Research*, **39**, 151–70.

Zhang Zhenhua, Gao Zhenjia *et al.* (1957). Regional geological investigation. Report of Keping-Akegu Area, scale 1:200 000. Geological Brigade No 13, Ministry of Geology.

8 Lithology, sedimentology and genesis of the Zhengmuguan Formation of Ningxia, China

ZHENG ZHAOCHANG, LI YUZHEN, LU SONGNIAN and LI HUAIKUN

Abstract

In the central segment of the Helan Mountain of Ningxia three Proterozoic sedimentary units are underlain by the Huangqikou Granite (K/Ar ages: 1440–1839 Ma). The Proterozoic sequences include, in ascending order, the Huangqikou Group composed of shallow marine clastic deposits, the Wangquankou Group consisting of carbonates and yielding a K/Ar age of 1289 Ma on glauconite from its base. The youngest formation is the Zhengmuguan Formation, which rests unconformably on the underlying beds. Still higher, the Cambrian Suyukou Formation, which contains trilobites, disconformably overlies the terminal Proterozoic to Early Cambrian Zhengmuguan Formation.

The Zhengmuguan Formation can be subdivided into two members. The lower member is composed of diamictite containing carbonate clasts, and the upper one is made up of stratified fine clastic deposits and argillites. The diamictite is mainly subglacial in origin. Intercalations of laminated rocks with dropstones within the diamictite member suggest ice sheet fluctuation and repeated climatic changes. The upper fine clastic and argillaceous rocks with abundant trace fossils were deposited in a relatively deep and stable periglacial basin.

REE data from shales of the upper Zhengmuguan Formation illustrate that they have a high REE abundance (208.7 ppm), a high $(La/Yb)_N$ ratio and an obvious negative Eu anomaly. Their REE patterns are quite similar to those of post-Archean shales in western Europe, America and Australia.

Introduction

The Zhengmuguan Formation crops out in the middle segment of the Helan Mountain within a limited area of about 300 km² (Fig. 8.1). This sequence is named after the type section located in Zhengmuguan (Compiling Group of Stratigraphical Timescale of Ningxia Hui Autonomous Region, 1980). It is overlain by the Cambrian Suyukou Formation containing tribolite fossils, such as *Msuaspis* and *Ningxiaspis* and rests unconformably on stromatolite-bearing dolomite of the Wangquankou Group. The Zhengmu-guan Formation has obvious stratigraphic subdivisions, clear top and bottom boundaries, and specific lithologies. Its lower part is dominated by diamictite that passes gradually upward into sandstones and siltstones of the upper member. It is highly analogous in lithology, stratigraphy, and sedimentary features to the Luoquan Formation in Henan and Shaanxi provinces (Wang *et al.*, 1981; Mou, 1981; Li *et al.*, 1983; Guan *et al.*, 1983). The Zhengmuguan Formation is the depositional record of Luoquan Glacial Epoch at the western margin of the North China Block (Lu *et al.*, 1983).

Tectonic and stratigraphic setting of the Zhengmuguan Formation

Helan Mountain is situated in the western part of the North China Block; it is adjacent to the Alashan Block to the northwest and Qilian Block to the southwest. Proterozoic sequences preserved in Helan Mountain include, from bottom to top, the Middle Proterozoic Huangqikou and Wangquankou Groups, and the Zhengmuguan Formation.

The Huangqikou Group is unconformably underlain by the Archean Helan Group or gneissic biotite-plagioclase granite, and overlain by the Wangquankou Group with paraunconformity. The Huangqikou Group is 383 m thick and consists of quartz sandstone, siltstone, shale, and minor stromatolite-bearing dolostone. K/Ar ages for biotite ranging from 1440 to 1839 Ma have been obtained from the underlying granite (Zhao *et al.*, 1980). The granite may be the result of a thermotectonic event of Early to Middle Proterozoic age (*c.* 1800 Ma). The Wangquankou Group, up to 1000 m thick, is mainly composed of dolostones, dolostones with bedded chert, and limestones containing abundant stromatolites. Glauconite from a basal sandstone gives a K/Ar age of 1289 Ma.

The Zhengmuguan Formation occurs only in the middle part of the Helan Mountain area. It can be subdivided into lower and upper members. The lower member comprises diamictite, 7 to 144 m thick, and the upper member contains fine-grained clastic rocks, 8 to 161 m thick. There is no sedimentary record of the Lowermost Cambrian (Meishucunian and Qiongzhusian Stages). Between the

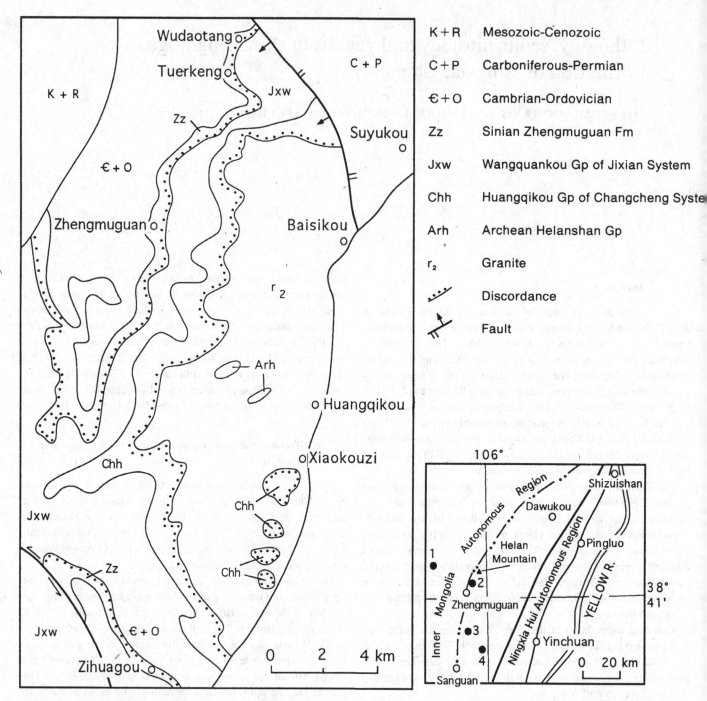

Fig. 8.1. Sketch map of geology and geography of the study area. Places marked by numbers are: 1, Alashanzuoqi; 2, Wudaotang; 3, Zihuagou; 4, Jingdiquan.

Zhengmuguan Formation and the Suyukou Formation, which consists of phosphorous-bearing strata of the Middle-Lower Cambrian (Canglangpuan Stage), a paraunconformity is preserved.

The sedimentary facies and their associations in the Zhengmuguan Formation

The Zhengmuguan Formation is approximately 300 m thick. The lower diamictite (lower member) is a series of rock associations related genetically to glaciation with abrupt thickness variations. For instance, in a lateral distance of 25–30 km, its thickness may range from several meters to 140 m. The upper rocks (upper member), 10 m to 160 m thick, are mainly a suite of dark grey and greyish green siltstones and shales, intercalated with medium to fine-grained sandstones. At the base thin beds of carbonaceous shales can be found.

According to field investigations, at least five main lithofacies have been identified, ie. massive diamictite, poorly stratified diamictite, bedded diamictite, dropstone-bearing laminated rock, and thin-bedded to banded-laminated siltstone-mudstone facies. Conglomerate and sandstone facies are also present but are less important.

Massive diamictite facies

This forms the dominant lithofacies in the lower member of the Zhengmuguan Formation, which is well exposed in five measured sections. It is closely associated with poorly stratified diamictite facies and usually underlies the bedded diamictite facies. Its thickness varies greatly, ranging from several meters to more than 100 m.

Massive diamictites are structureless deposits with a mainly dolomitic matrix containing scattered stones. The stones are of highly variable composition. They show poor sorting, random arrangement, and have variable shapes. The massive diamictite facies is obviously the result of rapid accumulation. Stones are dominantly derived from the underlying basement, especially from various kinds of carbonate rocks and cherts of the Wangquankou Group. Besides abundant carbonate stones, fragments of clastic rocks, granitoids, diorite, and gabbro have also been found. Stones in a massive diamictite show a wide range of shapes, including tabular, triangular, pentagonal, flatiron, saddle-like and irregular. Stone diameters range from several centimetres to more than 109 cm. For example, the biggest stone (dolostone) in the base of the Zihuagou Section is 100×80 cm, and in the Suyukou Section 96×44 cm (white quartzite) (Fig. 8.2). The diamictite matrix is poorly sorted. Under the microscope, silt-sand size material consists of angular dolostone, sandstone lithoclasts and quartz, chert and feldspar grains. The matrix consists of clay and dolomitic materials; it has a high CaO (6.30–17.64%) and MgO (4.38–14.26%) content (Table 8.1).

Poorly stratified diamictite facies

This facies is identical with the massive diamictite facies in composition. The main differences between them are that this facies

Fig. 8.2. A large boulder (96×64 cm) found in the Suyukou section of the Zhengmuguan Formation. It is composed of white quartzite of the Huangqikou Group.

is characterized by weak stratification and stones with some preferred orientation. The poorly stratified diamictite facies usually overlies the massive diamictite facies and underlies the bedded diamictite facies in the sequence. Boundaries between them are commonly gradational.

Bedded diamictite facies

This facies shows clear medium to thick stratification. The clasts show more obvious preferred orientation than clasts in the poorly stratified facies. Some giant boulders were noted. For example, two dolomite boulders ($62 \times 34 \times 32$ cm and $72 \times 48 \times 27$ cm) and one sandstone boulder ($145 \times 110 \times 75$ cm) were found in the Tuerkeng Section.

Dropstone-bearing laminated lithofacies

The rock is similar in appearance to grey to light-grey laminated dolostone but comprises various kinds of clasts in a dolomitic and argillaceous matrix. The lamination is relatively well developed and individual laminae are from less than one to several millimetres in thickness. Under the microscope, the rock has a clastic texture. The mineral composition of sand-size clasts is quite complex and includes dolostone, mudrock, sandstone, siltstone, quartz, chert and plagioclase, with angular to subangular shapes. The lithofacies is characterized by stones of different composition in laminated beds. The stone's long axes lie parallel, oblique, or perpendicular to lamination. Some of these stones show features typical of dropstones. Stones in the laminated lithofacies are less abundant and smaller than stones in the three diamictite facies described above.

This facies is from several to tens of centimetres thick. It occurs above the bedded diamictite facies and forms the top unit of individual sedimentary cycles, which commonly begin with a lower massive diamictite, overlain by the poorly stratified facies and the bedded diamictite, and capped by the laminated facies with dropstones. Massive diamictite beds of another cycle, with a sharp contact at their base, overlie the laminated facies.

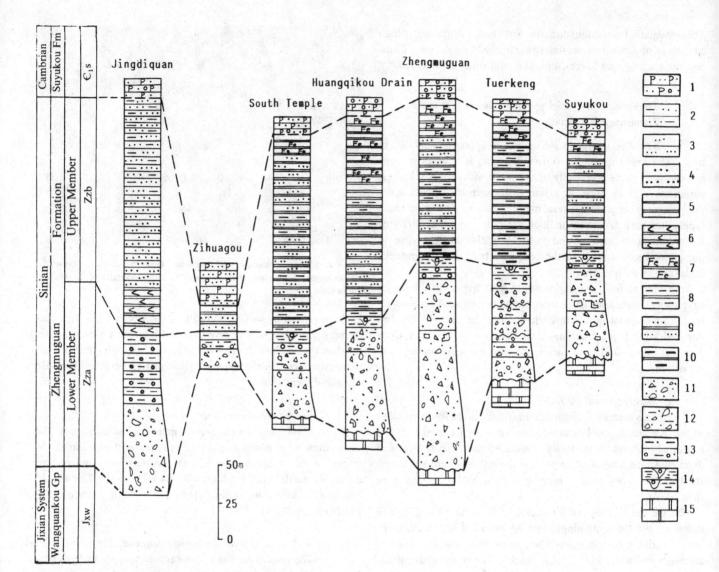

Fig. 8.3. Correlation of lithofacies associations of the Zhengmuguan Formation among sections. 1, phosphatic conglomerate; 2, muddy siltstone; 3, medium to thin-bedded siltstone; 4, medium to fine-grained sandstone; 5, slate; 6, siliceous slate; 7, iron-band-bearing slate; 8, muddy slate; 9, silty slate; 10, carbonaceous slate; 11, massive diamictite; 12, poorly stratified diamictite; 13, bedded diamictite; 14, dropstone-bearing laminated facies; 15, medium- to thick-bedded dolomite.

Thin-bedded to banded-laminated siltstone-mudstone facies

The upper part of the Zhengmuguan Formation is composed of this lithofacies. It includes mainly siltstone and silty shale (most are now slate owing to low-grade metamorphism). Generally speaking, units with a coarser grain-size have thicker laminations. Horizontal bedding is the dominant structure in this lithofacies. In the northern part of our research area, pyrite layers and lenses are common in siltstone-mudstones. Various kinds of trace fossils are present. These fossils are believed to have been formed by organisms living in a relatively deep and stable basin (Yang and Zheng, 1985).

The five major lithofacies listed above make up various facies

associations at different sites in the Helan Mountain area. Some representative sections described below demonstrate the characteristic facies associations (Fig. 8.3). Localities of most sections are shown in Fig. 8.1.

The Wudaotang Section in Helan County

The facies association of the Zhengmuguan Formation is quite simple in the Wudaotang section located north of Zhengmuguan (Fig. 8.1). Its lower part consists of massive diamictite facies, up to 66 m thick. A sedimentary break between this lithofacies and the underlying thick-bedded dolostone of Wudaotang Group is easily identified. Five meters of bedded diamictite overlie the massive diamictite with a sharp contact. The bedded diamictite is characterized by normal grading; in general the lower part of the bed contains coarser and more abundant stones. The upward transition from massive to bedded diamictite reflects glacial regression and a change from subglacial to proglacial sedimentary environments. The bedded diamictite passes gradually upward into the siltstone-mudstone facies that is 92 m thick. The latter is characterized by even laminations with a high carbon content and widespread microdisseminated or thin-bedded pyrite. Some trace fossils reflecting relatively deep-water environments such as *Neonerietes uniserialis, Taenioichnus zhengmuguanensis* occur in the middle part of the siltstone-mudstone facies. Therefore, the whole sequence from subglacial to relatively deep euxinic environments reflects a glacial retreat and/or rising sea-level.

The Tuerkeng Section

Facies associations in the Tuerkeng Section (Fig. 8.3), which is about 2 km south of Wudaotang (Fig. 8.1), are more complex. There are three cyclic facies associations in this section. The lower association consists of massive diamictite, poorly stratified diamictite and a 'varved' fine clastic facies. Contacts between these facies are gradational. A few stones of both acid and basic magmatic origin, which must have been derived from distant sources, were noted in the massive diamictite. The 'varved' fine clastic facies consists of laminated mud-rich shales of variable thickness and thin-bedded fine granule conglomerate beds which display characteristics of sediment gravity flows. The middle association comprises massive and poorly stratified diamictites with some intercalations of thin-bedded fine conglomerates. The upper association is made up of massive diamictite, poorly stratified diamictite and dropstone-bearing laminated facies. Boulders with diameters greater than 1 m exist in all diamictite facies. Dropstones are common in the fine clastic facies. Deep water fossils are abundantly preserved in the upper member, which is similar to that of the Wudaotang section, except that pyrite is less common.

The Zhengmuguan Section

In the lower part of the Zhengmuguan Section (Fig. 8.3), which is located near the middle of our research area (Fig. 8.1), a complete sedimentary cycle comprising massive diamictite, poorly stratified diamictite, bedded diamictite and fine-grained laminated facies, has been recognized. Owing to its thickness and resistance to weathering, the massive diamictite forms a sharp precipice more than 100 m high (Fig. 8.4). As in the Wudaotang Section, the upper siltstone facies of this section contains carbonaceous material and pyrite. Plentiful trace fossils have been collected from the top of the section.

The Jingdiquan Section

The Jingdiquan Section (Fig. 8.3) up to 268 m thick, is located in the southern part of the research area (Fig. 8.1). Its lower member consists of massive and bedded diamictite facies. The

Fig. 8.4. A sharp precipice, more than 100 m high, formed by the massive diamictite of the Zhengmuguan Formation. Zhengmuguan section.

massive diamictite can be subdivided into lower grey and upper light brown parts with an abrupt contact between them. This is interpreted to mean that there were two glacier advances and the earlier glacier deposits were subjected to erosion by the later glaciation. The clast content is higher but the clasts are commonly smaller in the bedded diamictite than in the massive diamictite. Some bigger stones, however, are preserved in the mid to upper part and at the top of the bedded diamictite, which shows evidence of resedimentation (Fig. 8.5). Pyrite occurs in siltstone of the upper member but no trace fossils have been discovered. Sandstone is more common in the upper member here than it is in sections in the northern part of the studied area (Fig. 8.3).

From the above brief description of lithofacies associations of different sections it is evident that the massive diamictite is the most important lithofacies occurring in every sedimentary cycle of all the studied sections. A complete cyclothem consists of, from base to top, massive diamictite, poorly stratified diamictite, bedded diamictite, and the dropstone-bearing laminated fine clastic rock. Each

Table 8.1. *Chemical composition of the matrix of massive diamictites of the Sinian Zhengmuguan Formation in Helan Mountain*

Sample Number	Location	Lithology	Major elements (Chemical composition, wt. %)											Minor elements (Chemical compositions, wt. %)			
			CaO	MgO	SiO2	Fe$_2$O$_3$	FeO	Al$_2$O$_3$	M nO	TiO$_2$	Na$_2$O	K$_2$O	P$_2$O$_5$	Cr$_2$0$_3$	SrO	BaO	V$_2$0$_5$
Ydh-1	Tuerkeng	Matrix	17.64	12.86	33.36	1.69	1.28	2.01	0.02	0.00	0.37	1.46	0.009	?	?	?	?
Zhen-1	Wudaotang	of	10.03	7.32	50.00	?	1.65	8.43	0.00	0.23	0.05	3.53	0.00	0.31	0.00	0.31	0.00
Zhen-2	Wudaotang	Massive	6.30	4.83	65.74	?	1.97	6.55	0.00	0.11	0.01	2.69	0.00	0.30	0.00	0.00	0.02
Zhen-3	Wudaotang	Diamictite	11.14	14.26	33.41	?		2.097	14.26	0.02	0.78	0.61	3.32	0.00	0.08	0.00	0.17	0.14

Table 8.2. *REE data of the fine clastic rocks of the Zhengmuguan Formation*

No.	Lithology	La	Ce	Pr	Nd	Sm	Eu	Gd	Tb	Dy	Ho	Er	Tm	Yb	Lu	Y	ΣREE	ΣLREE	ΣHREE	ΣLREE/ΣHREE	Eu/Eu*	(La/Yb)$_N$
1	Silty mudstone	149.20	88.56	72.17	54.94	31.77	11.11	16.99	14.89	9.85	11.11	11.27	15.15	17.31	18.75	15.15	183.2	167.0	16.2	10.309	0.456	8.6
2	Carbonaceous mudstone	171.11	104.55	84.35	63.48	33.33	4.17	18.92	14.89	10.77	9.72	8.45	12.12	14.42	15.63	16.44	208.7	193.2	15.5	12.465	0.159	11.9

Fig. 8.5. A large boulder (72 × 48 × 27 cm), made of dolomite of the Wang-quankou Group, occurs at the top of a diamictite bed, which is overlain by a 30 cm thick-bedded diamictite bed. Jindiguan section.

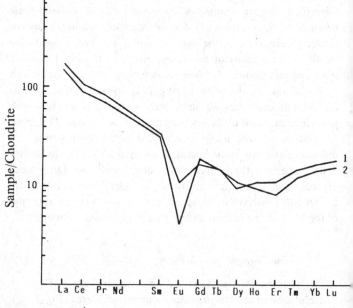

Fig. 8.6. REE patterns of the fine clastic rocks of the Zhengmuguan Formation (data are normalized to chondrite). 1, silty mudstone; 2, carbonaceous mudstone.

cycle may reflect one episode of glacial advance and retreat. Because of different positions relative to glacier bodies and different degrees of erosion, the described sections consist of one or several cycles. The correlation of lithofacies associations among sections is shown in Fig. 8.3.

REE geochemical characteristics of the Zhengmuguan Formation

REE analyses of 2 samples (silty mudstone and carbonaceous mudstone) from the upper member of the Zhengmuguan Formation are listed in Table 8.2. REE abundance of these two samples is high, up to 183.2 and 208.7 ppm respectively. The LREE values of 167 and 193.2 ppm, the HREE values of 16.2 and 15.5 ppm, and the LREE/EHREE ratios of 10.309 and 12.465 indicate

that these rocks are rich in LREE. The (La/Yb)$_N$ ratios are also high, 8.6 and 11.9. REE abundance patterns are roughly identical (Fig. 8.6). The curves are slightly depressed with a steep LREE part and relatively gentle HREE part. Obvious negative Eu anomalies are present in both, but that of the carbonaceous mudstone is particularly pronounced. Eu/Eu* ratios are 0.456 and 0.159. After being normalized to PAAS, these REE patterns are almost the same as those of post Archean shales of Western Europe, America and Australia (Fig. 8.7) (Nance and Taylor, 1976).

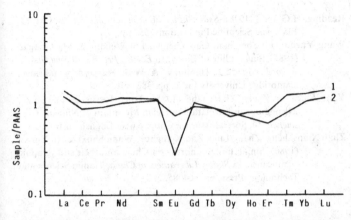

Fig. 8.7. REE patterns of the fine clastic rocks of the Zhengmuguan Formation (normalized to PAAS). 1, silty mudstone; 2, carbonaceous mudstone.

Sedimentary facies and environmental analysis

Four of the five sedimentary facies mentioned above are genetically related to glaciation (Lu and Gao, 1991). The massive diamictite facies is usually interpreted as subglacial deposits. However, it can also be formed under glacier tongues that extend out into a basin, i.e. a so-called subaqueous tillite which is massive (Hambrey and Harland, 1981). Therefore, both the regional facies changes and the facies sequences of sections must be considered to reach a good facies interpretation.

In the research area, the lower member of the Zhengmuguan Formation consists of a complete cycle of sedimentation from a lower massive diamictite through poorly stratified and bedded diamictite to an upper laminated dolomitic facies with dropstones. In some sections such as the Binggou Section in Huangiqikou (Fig. 8.1), the facies sequence is quite simple; its lower member consists of only one cyclothem. However, in other complex sections, such as the Tuerkeng Section (Fig. 8.3), the lower member is formed of several cyclothems. The massive diamictite varies regionally in both thickness and facies sequence among sections. The massive diamictite displays a random arrangement of stones, which are poorly sorted and some at least are far-travelled. Because of the well-developed erosional profile at the top of the underlying basement, the massive diamictite of the Zhengmuguan Formation is believed to be the product of subglacial deposition (Reading, 1978). During the period of glacial retreat, stronger meltwater currents produced stratification of the diamictite. Thus the poorly stratified diamictite can form near glaciers and bedded diamictite farther away. Bedded diamictite showing features of outwash deposits, such as distinct bedding and preferred orientation of stones, may be subaqueous, deposited in front of glacial tongues on a basin margin. However, the arrangement of stones, which are commonly concentrated in the middle to upper part and even at the top of the diamictite beds, suggests the activity of debris flow deposits. Most of the bedded diamictites described in this paper are attributed to sediment gravity flows.

The widely distributed dropstone-bearing laminated lithofacies is mainly preserved at the top of the lower member. It not only

shows that the glaciation came to an end, but also indicates a larger basin at that time. The occurrence of dropstones in laminated silty-sandy rocks indicates that small icebergs carrying stones existed then.

The fine-grained clastic rocks, including mudstone, siltstone, and sandstone in the upper member of the Zhengmuguan Formation, are coarser grained in the south and finer grained in the north. In addition, the distribution of pyrite indicates relatively deep water and a reducing environment in the north. Trace fossils include burrows, food-seeking traces, and fecal pellets, which all indicate relatively deep water and a low energy environment. All the evidence suggests that the water in the north was deeper than that in the south. The fine-grained clastic rocks appear to be postglacial deposits.

Conclusions

The Zhengmuguan Formation in the Helan Mountain of Ningxia overlies the middle Proterozoic Wangquankou Group and underlies the Cambrian tribolite-bearing Suyukou Formation. It can be stratigraphically correlated with the Luoquan Formation both in Henan and Shaanxi provinces.

The Zhengmuguan Formation consists of two parts. The lower is dominated by diamictite associations, whereas the upper is mainly composed of thin-bedded siltstones with plentiful trace fossils. The former consists mainly of glaciogenic rock associations formed in a cold climate; the latter is fine clastic rocks deposited in relatively deep water during the post-glacial transgression.

The lower member of the Zhengmuguan Formation can be divided into four major lithofacies, massive diamictite, poorly stratified diamictite, bedded diamictite, and dropstone-bearing laminated lithofacies. These four lithofacies usually form a regular facies sequence that reflects a process of glacial retreat.

Like the Luoquan Formation on the southern margin of the North China Block, the diamictite member of the Zhengmuguan Formation is one of the widespread deposits of the Luoquan Glacial Epoch of terminal Sinian to lowermost Cambrian age.

References

Compiling Group of Stratigraphical Timescale of Ningxia Hui Autonomous Region (1980). Regional stratigraphical timescale of North-West China, the part of Ningxia Hui Autonomous Region. Geological Publishing House, Beijing (in Chinese).

Guan Baode, Gent Wuchen, Rong Zhiquan & Du Huiying (1983). On the age of Luoquan Formation in Henan Province, *Precambrian Geology*, 1, 183–206. Geological Publishing House, Beijing (in Chinese with English abstract).

Hambrey, M.J. & Harland, W.B., eds (1981). *Earth's Pre-Pleistocene Glacial Record*. Cambridge, Cambridge University Press, 1004 pp.

Li Qinzhong, Wang Yingzhang & Jia Jinchang (1983). The age and sedimentary facies of Luoquan Formation in southern margin of North China Platform (the part in Shaanxi Province), *Precambrian Geology*, 1, 163–81. Geological Publishing House, Beijing (in Chinese with English abstract).

Lu Songnian, Ma Guogan, Gao Zhenjia & Lin Weixing (1983). Primary research on glacigenous rocks of Late Precambrian in China, *Precambrian Geology*, **1**, 1–65, Geological Publishing House, Beijing (in Chinese with English abstract).

Lu Songnian & Gao Zhenjia (1991). A comparative study on the characteristics of the debris flow and glacial deposits of Late Proterozoic in the Aksu area of Xinjiang. Bulletin 562 Geo. Brigade, CAGS, No. 9, 1–16.

Mou Yongji. (1981). Luoquan tillite of the Sinian system in China. In *Earth's Pre-Pleistocene Glacial Record*, eds. Hambrey, M.J. and W.B. Harland, Cambridge, Cambridge University Press, pp. 402–13.

Nance, W.B. & Taylor, S.R. (1976). Rare earth element patterns and crustal evolution-1, Australia Post-Archaean sedimentary rocks. *Geochimica Cosmochimica Acta*, **40**.

Reading, H.G. (ed.) (1978). *Sedimentary Environment and Facies*. Oxford, Blackwell Scientific Publications, 557 pp.

Wang Yuelun, Lu Songnian, Gao Zhenjia, Lin Weixing & Ma Guogan (1981). Sinian tillite of China. In *Earth's Pre-Pleistocene Glacial Record*, eds. M.J. Hambrey, & W.B. Harland, Cambridge, Cambridge University Press, pp. 386–401.

Yang Shipu & Zheng Zhaochang (1985). Trace fossils of the Sinian Zhengmuguan Formation in Helan Mountain of Ningxia, *Geoscience*, **10**, Special Issue (in Chinese with English abstract).

Zhao Xiangsheng, Zhang Luyi, Zou Xianghua, Wang Shuxi & Hu Yunxu (1980). Sinian tillite in northwest China and their stratigraphic significance. In *Sinian Suberathem in China*, Tianjin Science and Technology Press, pp. 164–85.

9 Architectural styles of glacially influenced marine deposits on tectonically active and passive continental margins

MICHAEL R. GIPP

Abstract

The ancient glacial record is dominated by glacial marine sediments which have accumulated on continental shelves and slopes. Stratigraphic models showing the gross architecture of the principal glacial geological deposits are developed from onshore and offshore studies of Late Cenozoic deposits in the Gulf of Alaska and across the Scotian Shelf and Slope. These should be applicable to other areas, and should be of use in establishing the tectonic setting of glacially influenced marine deposits in the ancient record. The processes of deposition (including ice-rafting, suspension rain-out, and turbidity currents) are the same on both tectonically active and passive margins, yet the large-scale depositional architecture of glacial marine deposits on both types of margin differs because sediment preservation is strongly influenced by the impact of tectonics on local relative sea-level changes. Ice sheets can destroy any glacial marine deposits on the continental shelf, provided they can advance across it. Eustatic sea-level falls during world-wide glaciations, so that ice sheets may advance across passive margin shelves, whereas rapid subsidence at active margins may restrict glacial advance into the offshore, thereby preserving shelf deposits. As a result, over a long sequence of glacial and interglacial events, slope deposits will be selectively preserved on passive margins, whereas both shelf and slope deposits will be preserved on active margins. Furthermore, as the shelf builds up and out in the active margin, the importance of shelf deposits may be expected to increase upsection.

Introduction

Although subglacial terrestrial deposits are locally important, the thickest accumulations of glaciogenic sediments in the rock record are represented by glacially influenced marine deposits, such as those of the Early Proterozoic Gowganda Formation of Ontario, Canada (Miall, 1985), the late Proterozoic (Sturtian) glacial marine sediments in the North Flinders Basin of South Australia (Young and Gostin, 1991), and the Late Cenozoic Yakataga Formation, Alaska (Eyles *et al.*, 1991). Typical deposits consist of poorly sorted, rain-out diamicts, debris flows and turbi-dites. Deposition and facies successions are controlled in complex fashion by tectonic setting, sea-level fluctuations, and climate change (Boulton, 1990).

This paper presents models of the large-scale architecture of glaciated shelf and slope systems in contrasting tectonic settings. The models have been developed from two Late Cenozoic glaciated shelf and slope systems: the active margin of the Gulf of Alaska and the tectonically passive Scotian Shelf and Slope of eastern Canada.

In the Gulf of Alaska, a long and complete record of Cenozoic climatic deterioration, glacial marine deposition and active tectonism is clearly exposed in outcrops of the 5 km thick Yakataga Formation, which can be correlated with offshore boreholes and seismic data (Eyles *et al.*, 1991). On the continental margin of Atlantic Canada, decades of hydrocarbon exploration and the Frontier Geoscience Program have resulted in a substantial data base, consisting of thousands of kilometres of seismic data and many boreholes and cores. These two areas can be argued to be representative of much of the ancient glacial record, and thus provide data which aid in the interpretation of other glaciated basins.

The Gulf of Alaska

The southern continental margin of Alaska (Fig. 9.1) is composed of numerous allochthonous terranes which have been assembled into their present-day positions during the Mesozoic and Cenozoic by the northward motion of the Pacific Plate (Plafker, 1987). The Yakutat terrane is currently colliding with and accreting onto the North American Plate in the Gulf of Alaska area (Bruns, 1983). The Yakutat Block is undergoing oblique subduction along the Fairweather–Queen Charlotte transform fault (Bruns, 1983), resulting in the rapid uplift of coastal mountains, and the initiation of Late Cenozoic glaciation (Eyles *et al.*, 1991). Glaciation resulted in the deposition of the 5 km thick Late Miocene to Recent Yakataga Formation in a complex forearc basin. Dating control for the Yakataga Formation is provided by planktic foraminifera in Cape Yakataga and from Deep Sea Drilling Project (DSDP) boreholes in the Gulf of Alaska (Lagoe *et al.*, 1993). After glaciation in the latest Miocene, there followed a warm interval which lasted

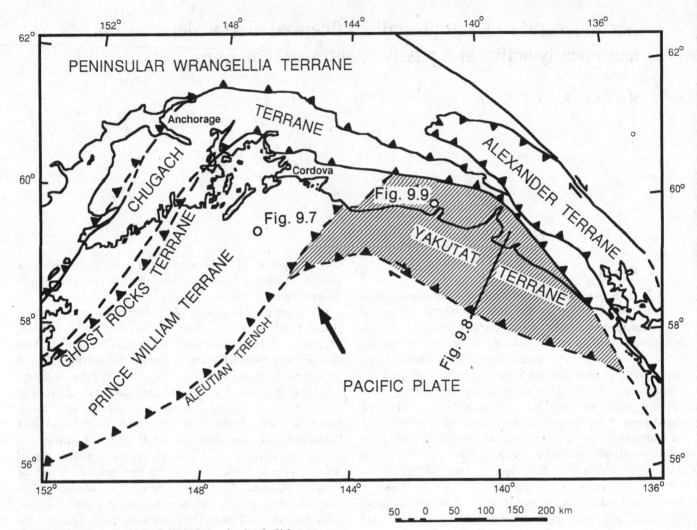

Fig. 9.1. Tectonic setting of the Gulf of Alaska, showing the Yakutat terrane being docked against the Chugach and Prince William terranes by the Pacific Plate, after Plafker (1987). Also shown is the location of Fig. 9.7, on Middleton Island, the location of the cross-section in Fig. 9.8, and the site of the outcrop depicted in Fig. 9.9.

through the mid-Pliocene, until glaciation resumed in the Late Pliocene, presumably related to increased northern hemisphere glaciation. Exposures of the Yakataga Formation between Cape Yakataga and Icy Bay, which lie upon the Yakutat Block, have undergone about 25% crustal shortening, and so are intensely deformed (Plafker, 1987). On Middleton Island, which lies on the Prince William Terrane, deformation has been comparatively mild.

Field work was carried out in the Cape Yakataga to Icy Bay section in the summer of 1989, consisting principally of attempting to produce a photomosaic of the Yakataga Formation outcrops in the Robinson Mountains. Fieldwork on Middleton Island was carried out in the summer of 1991.

The Scotian margin

Rifting along the Scotian continental margin (Fig. 9.2) began in the Late Triassic to Early Jurassic. Seafloor spreading has

occurred since the Jurassic, and the margin has been passive, but subsiding, throughout the Cenozoic (Welsink et al., 1989). Fluvial systems incised channels on the Scotian Shelf and Slope, and provided sediments throughout the Tertiary, particularly during times of lowered sea-level, until the first glacial influences in the Pleistocene (Piper et al., 1987). The Tertiary fluvial channels have been overdeepened by Pleistocene glaciation, and glacial sediments have draped the inner shelf bedrock highs, and partially filled the deeper mid-shelf troughs (King and Fader, 1986). The outer Scotian Shelf was a major depocentre during the Wisconsinan, and Quaternary sediments overlie gently dipping Tertiary strata (Boyd et al., 1988).

Data has been collected in several phases. Airgun seismic data was collected on the Scotian Slope in the spring and summer of 1990.

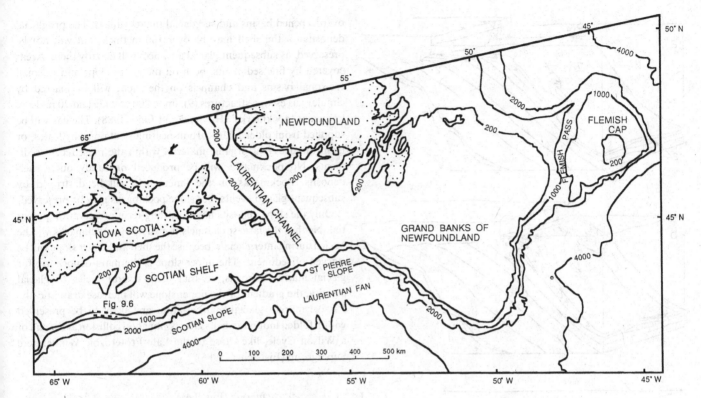

Fig. 9.2. The southeastern continental margin of Canada. The Scotian Shelf and Slope are at left. The location of the seismic lines in Fig. 9.6 is depicted.

Models

On the basis of multichannel seismic data, and exposures of sediment along coastal mountains, models of glacially influenced continental margins are produced: one for passive margins, based on observations made on the eastern Canadian margin, and another for active margins, based on the observations in the Gulf of Alaska.

Passive margin (Fig. 9.3)

Ice advance (Fig. 9.3a) occurs during eustatic sea-level fall. A thin sheet of glacial ice advances across a passive shelf, crossing seaward-dipping beds representing early (possibly non-glacial) shelf progradation. A thin layer of pre-existing proglacial sediments (1), deposited during a previous glacial retreat or ahead of the advancing ice, acts as a deforming bed to aid ice advance. The advancing ice planes off the most recent sediments, locally eroding into bedrock, although in many places moves along a layer of deforming sediments. Pre-existing structural weaknesses may be overdeepened. Sediments bulldozed by the ice will gradually be moved to the outer shelf and, if the ice reaches the shelf edge, will be pushed over the shelf break and onto the upper slope. Much of this sediment may move downslope in the form of debris flows (2), but there will be an increase in slope as some of the sediment remains

piled up. Other sediment will be provided by subglacial meltwater (3) and icebergs (4).

At lowest sea-level, the ice reaches the shelf edge (Fig. 9.3b). Ahead of major exit points for meltwater streams, ephemeral channels will be cut into the upper slope by turbidity currents (6). Pre-existing channels will focus these turbidity currents, becoming more deeply incised into the slope. The channel systems consist of numerous small, deep cut-and-fill channels as small as a few metres deep and 10 m wide (6), which coalesce downslope into major channels several kilometres wide and up to 500 m deep (7). These large channels start to become broader and shallower in water depths greater than about 1 km, until at depths of ~3 km, they can hardly be distinguished at all, for two possible reasons: (1) the channels are filled in by the debris flows and turbidites; or (2) the flows disperse as they flow downslope, so their erosive power decreases, and the channels widen downslope (8). The lower slope will be characterized by numerous lenticular debris flow deposits downslope of these channels – in even deeper waters, the sediments will be mainly turbidites.

Stacked debris flows (2), known as 'till tongues', form along the glacier margin (King & Fader, 1986; Mosher *et al.*, 1989). These sediments grade downslope into finer debris flows (5), probably containing less gravel, and turbidites, draped by hemipelagic sediments. The upper slope deposits will consist of debris flows, composed of eroded bedrock, proglacial sediments, and ice-rafted

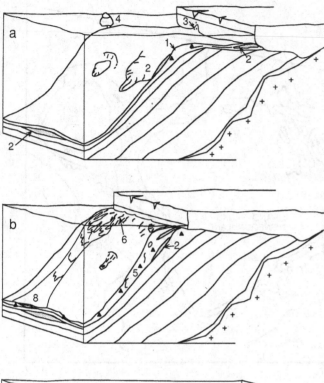

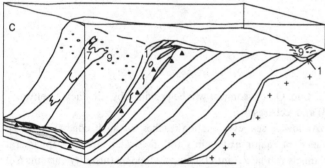

Fig. 9.3. (*a*)–(*c*) **Model of shelf progradation on a passive margin. Numbers on the diagram label processes referred to in the text, as follows: (1) proglacial muds, (2) massive, lenticular debris flows, (3) subglacial meltwater stream outlet, (4) icebergs and ice-rafted debris, (5) small debris flows (of uncertain geometry), (6) cut-and-fill channels, (7) large submarine channels, (8) submarine channels on the middle slope, losing channel geometry, (9) gas-escape structures (pockmarks).**

debris, with cut-and-fill channels, filled with gravels and diamicts, overlain by finer silts and muds. Sediments on the upper and mid-slope can be remobilized downslope independently of the channel systems and the glacial meltwater by storms or tectonic activity.

As the ice retreats (Fig. 9.3*c*), the entire sequence will be blanketed by a seaward-thinning prism of proglacial muds (1). Small feeder channels are likely to be completely filled during this phase, but the largest slope channels (7) will only be partially filled. Sediments in the large channels will be susceptible to reworking at this phase due to downvalley surges caused during storms (Shepard, 1979). Heavily bioturbated proglacial sediments will cover the inner shelf and partially infill shelf depressions, which will include

overdeepened basins and subglacial tunnel valleys. The proglacial deposits on the shelf may be over 100 m thick, but will not be preserved, as subsequent glacial advances will destroy them. Areas covered by fine sediments, both on the upper slope and in small erosional basins and channels on the shelf, will be marred by circular gas escape structures (9), up to 400 m wide and 20 m deep, known as 'pockmarks' (Hovland and Judd, 1988). The gas will be supplied from older hydrocarbon-bearing strata, gas hydrates, or from the decay of organic material within the most recent sediments. No pockmarks will be produced on sandy outer shelf deposits. Pockmarks on the inner shelf will be destroyed by subsequent glacial events, but slope pockmarks may be preserved.

Only the slope deposits have any long-term preservation potential. Neither the largest channels nor the slope pockmarks will be buried during interglacials, because the shelf and upper slope will be starved of sediment. The upper slope is the main depocentre for glacial marine deposits, so that after several cycles of glacial activity, the gradient of the upper slope will increase dramatically. Passive margin glacial marine strata are likely to be preserved within folded mountain belts after continental collision at the end of a Wilson Cycle, like those of the Late Proterozoic Windermere Supergroup (Eisbacher, 1985).

Active margin (Fig. 9.4)

Large volumes of sediment are continually supplied because of rising mountain ranges, which also are suitable sites for alpine-type glaciers. The absence of a broad continental shelf prevents the ice from advancing far offshore (Fig. 9.4*a*). Eustatic sea-level changes do not significantly alter the ability of the glacier to advance, because initially, the forearc basin is several kilometres deep. Sediment will be piled up in front of the glacier terminus, which will be resedimented downslope in the form of debris flows (2) and turbidites. Submarine channels (7) will be incised on the upper slope by the coalescing of feeder channels (6) which drain coarse-grained turbidity currents from the ice margin. These channels will likely coalesce downslope into large channels, which will broaden downslope. These channels will not necessarily follow regional slope, but will drain towards the centre of the forearc basin, the location of which is tectonically dependent (Dobson *et al.*, 1991).

Sedimentary processes are the same as on the passive margin. Ice-rafting (4) supplies coarse material into the basin, as dropstones in otherwise fine-grained sediments. Repeated tectonic shock causes sediments deposited on the upper slope to be remobilized downslope as debris flows and turbidity currents, where they fill up the basin (5). Thus, the basin fills up with a horizontally stacked succession of remobilized sediments with no apparent progradation of the slope (Fig. 9.4*b*). Small feeder channels continue to feed submarine channels (7) with sediment-laden water from beneath the glacier. The submarine channels will eventually cross the forearc area and carry sediments towards the trench, thus showing possible changes in regional orientation at different stratigraphic intervals. During deglaciation and interglacial events, there is enough sedi-

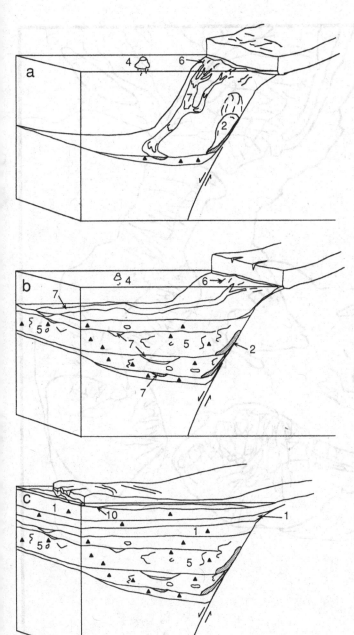

Fig. 9.4. (a)–(c) Model of shelf progradation into a subsiding forearc basin. Depiction of major glacial marine depositional processes, numbered as in Fig. 9.3, except for (10) ice grounding on the outer shelf.

outer shelf (10), abrading lag gravels to form striated boulder pavements (Eyles, 1988a).

Pockmarks will probably not be developed, even in the presence of hydrocarbon seepages from bedrock, because sedimentation rates remain high during interglacials, and pockmark accumulation is only possible when sediment rates are low.

Thus, the earliest sediments in the sequence will be dominated by debris flows, but sediments which rained out of suspension or were melted out of floating ice subsequently increase in stratigraphic importance. Debris flows and turbidites will still be important before the forearc basin is completely filled. Once the shelf extends all the way across the previous forearc basin, resedimentation is only of local importance. The continental slope will be on the seaward side of the forearc basin. As long as the margin keeps subsiding, grounded ice sheets will be unable to advance across the new shelf, thus preserving the shelf deposits. As long as sediment is supplied to the basin faster than it can subside, the basin will gradually fill. If subsidence is faster than sediment supply, then the case in Fig. 9.4b represents the end result: the slope does not advance significantly, and extensive shelf deposits are not developed.

The preservation potential of these sediments is also likely to be quite high. They are deposited in the forearc basin landward of the subduction margin. If they are on a stable substratum, they should not be crushed beyond recognition into the accretionary wedge (e.g. Eyles, 1990).

Similarities

Glacial marine sedimentary processes are identical on both types of margins. The resulting geometry differs because of the tectonic motion on the active margin, and the related increase in sediment availability. Although glaciation causes an increase in sedimentation rate, this increase is much more important on a passive margin. On an active margin, mountain building would have already increased sedimentation rates.

An important result of the increased sedimentation is a steepening of the upper slope. As sediment piles on the outer shelf/upper slope, a downslope-thinning wedge develops. Downslope remobilization is usually modelled as a diffusive process, meaning that the rate of sediment transport is a function of the slope (e.g. Syvitski, 1989). Low sediment rates result in low slopes, because equilibrium can be maintained by slow removal of sediments downslope. As sedimentation rate increases, the slope must increase in order to remove sediments at about the rate of supply. At even higher rates of supply, channels must be developed.

Around the Atlantic Ocean, submarine channels are closely correlated with areas of high Neogene sediment accumulation (Emery and Uchupi, 1984). Not all such areas occur on glaciated margins – they occur near the mouths of large river systems which have received glacial runoff, such as the Hudson River (Fig. 9.5).

The formation of these channels on glaciated margins is largely a function of subglacial melt patterns. The distribution of channels on the Scotian Slope suggests that these channels grow over a

ment coming off rising coastal mountain chains that the ephemeral feeder channels and even the largest submarine channels may be buried.

As the forearc basin fills up (Fig. 9.4c), a broad shelf may be built. This shelf will differ in geometry from the broad shelf on the passive margin, however, as it will largely be aggradational. Proglacial shelf sediments will rain out of suspension (1), but during the next glacial advance, the shelf will have subsided sufficiently to prevent the ice from advancing across the shelf, thus preserving them. Ice may cover the shelf in the form of a floating ice shelf, grounded on the

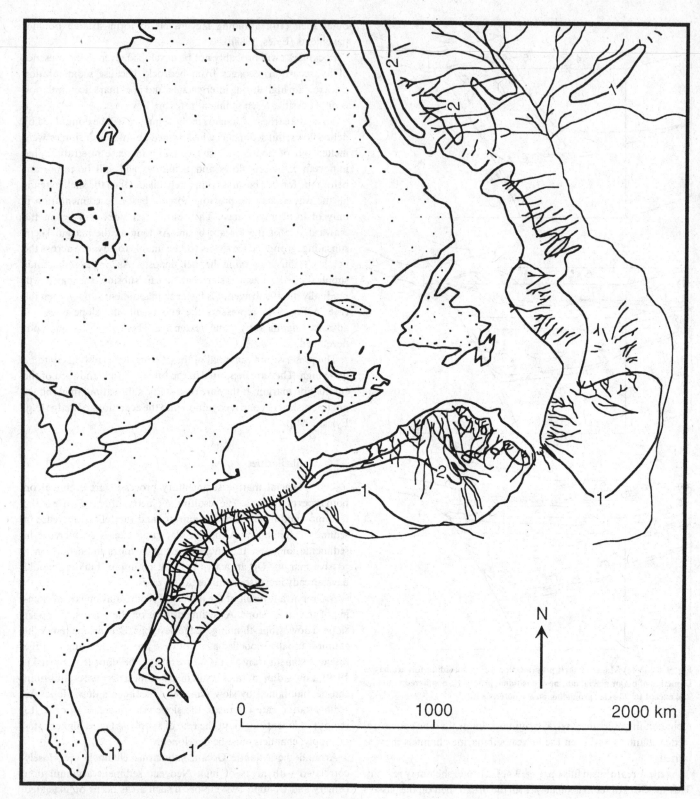

Fig. 9.5. Figure showing connection between high Neogene sedimentation and submarine channel formation. 1, 2, 3, Neogene sediment thickness (km). (Modified after Emery and Uchupi, 1984.)

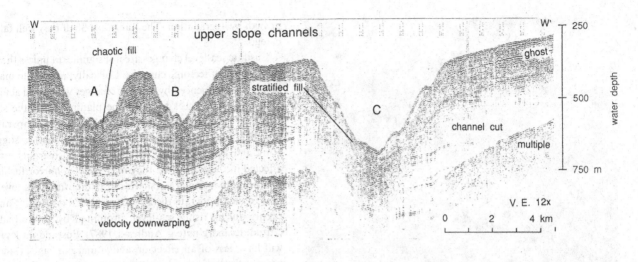

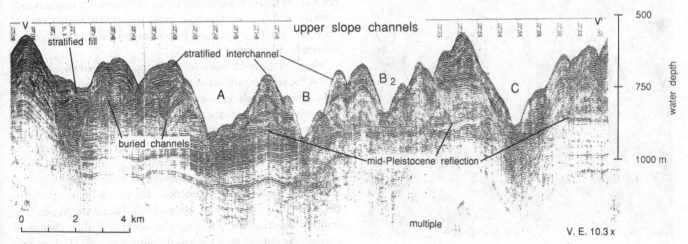

Fig. 9.6. Seismic lines from east coast offshore showing submarine channels (see Fig. 9.2 for location). Shallow water line (W–W') shows that on the uppermost slope, only large channels are developed, without any apparent feeder channels. In deeper water (V–V'), channels of all sizes are developed, and buried channels are apparent. Labelled channels have been correlated from one line to the other. Note that channels, A, B, B2, and C are all incised to different depths.

period of time, possibly over several glacial periods. Both large (> 2 km across and > 350 m deep) and small channels (< 1 km across, < 200 m deep) are observed at the present-day seafloor (Fig. 9.6; see Fig. 9.2 for location). Only small channels are seen in the subsurface, implying that, on the Scotian Shelf, once channels reach a certain size, they cannot be completely buried. The importance of small cut-and-fill channels, which are too small to be observed on seismic records, is based on observations of a submarine channel system observed on Middleton Island (Eyles, 1987). Small 'feeder channels' (Fig. 9.7) are ephemeral features, possibly lasting only for the duration of the turbidity current. These feeder channels coalesce into larger channels which are cut during times of lowered sea-level. When sea-level rises again, and the glacier retreats, the channels will be blanketed by proglacial sediments. Small submarine channels may be completely filled in, whereas the largest channels are not. During times of high sea-level, storm waves may cause downvalley

surges in them (Shepard, 1979), remobilizing some of the proglacial infill downslope. During subsequent glaciations, surviving channels become the foci of newly developed drainage systems. New feeder channels develop to feed the old submarine channels, which will be cut both deeper and farther upslope. After a few glacial sequences, large channels will, through headward erosion, reach the uppermost slope, where they will be the only channels which can be resolved on airgun records (Fig. 9.6).

Resedimentation is an important process on both types of margins. On passive margins, it is likely caused by instability of the sediments and cyclical loading during storms, whereas on active margins, it is principally caused by seismic activity. Earthquake generated slides cannot be ruled out on otherwise passive margins – the Grand Banks 1929 earthquake results in massive turbidity currents, and moved 175 km^3 of sediment downslope (Piper et al., 1988).

Fig. 9.7. Side wall of a cut-and-fill channel (the hammer lies on coarse channel-fill) in Yakataga Formation sediments on Middleton Island. The top of the photograph is upsection. These channels show a scale of development invisible to seismic investigation.

Distinguishing features in the models

The principal control on the growth of the continental margin is tectonic setting. Shelf growth on active margins is principally by aggradation, which is possibly due to subsidence, leading to a deepening of water on the shelf, restricting the ice's advance across the shelf. Figure 9.8 shows a line drawing of an interpretation of multichannel seismic data shot in the Gulf of Alaska (see Fig. 9.1 for location). A forearc basin lying atop the Yakutat Block has been undergoing infill throughout the Tertiary, but only since the onset of deposition of Yakataga Formation has the shelf break advanced significantly seaward: approximately 100 km since 6 Ma. Internal reflections are tilted slightly upwards towards the sea, reflecting the tilting undergone by the Yakutat Block (Plafker, 1987). Shelf growth on the passive margin is by progradation only, but since subsidence is the key, even 'passive margins' which are undergoing post-rift subsidence may show some

growth by aggradation. The Labrador Shelf may well fall into this category.

Eustatic sea-level changes are not significant unless they occur in the absence of tectonic changes. Logically, temperate margins will be strongly influenced by eustatic changes which will act in concert with glacial advances to make the planing off of the shelf more effective. This is because an ice advance across a temperate margin is unlikely during a global retreat (although this is suggested by King and Fader (1986)). Polar continental margins may not be affected in the same way: they may react in mirror fashion to the temperate margins of ice sheets, so that eustatic sea level changes may in fact moderate the impact of glacial erosion, if an advance occurs during rising sea-level, prompted by the retreat of mid-latitude ice margins (e.g. Andrews, 1987). Eustatic sea-level changes will have less of an effect on active margins, as subsidence may prevent the ice from advancing to the shelf edge. Ice may advance onto the shelf at times of glacio-eustatic lows (Eyles, 1988b), but will only be able to erode any significant amount of sediment if subsidence is very slow.

Although the mechanism for the formation of submarine channels is the same regardless of tectonic setting, the configuration of submarine channels may be tectonically controlled. Submarine channels mapped by sidescan sonar along the Aleutian arc, which drain towards forearc basins, do not always follow regional slope (Dobson et al., 1991). This is in line with observations by Eisbacher et al. (1974) that the drainage patterns in a compressive tectonic regime were prone to change direction from downslope to along-slope, as compression proceeded. Simple drainage patterns, oriented directly towards the depositional basins are proposed to only exist during the first stages of uplift, but then change to produce curved and even U-shaped drainage systems (Eisbacher et al., 1974). Similar observations appear to be the case for the cross-shelf submarine valleys in the Gulf of Alaska, some of which are partially controlled by faults (Carlson et al., 1982).

There is some evidence for a change in direction of the large submarine valleys near the 'type section' of the Yakataga Formation (Armentrout, 1983; Eyles et al., 1991). Photographic analysis of the section exposed along the Robinson Mountains between Cape Yakataga and Icy Bay reveal numerous large channels, originally interpreted as fjords by Armentrout (1983), but reinterpreted as submarine channels by Eyles et al. (1991). On the basis of the aspect of the channels as they intersect the plane of the outcrop, some of the channels appear to be oriented differently from channels lower in the section (Fig. 9.9).

Faulting and folding of sediments will be much more important on the active margins than the passive ones. In the Icy Bay area, most of the deformation has occurred at the leading edge of the Yakutat block.

The differences in the geometry of the development of the two types of glaciated margins should be reflected as differences in their expected facies successions. On a passive margin (Fig. 9.10), the sequence that develops should coarsen upwards from turbidites to debris flows, to a mature submarine channel system, and be capped by gravel cut-and-fill channels. This anticipated sequence reflects

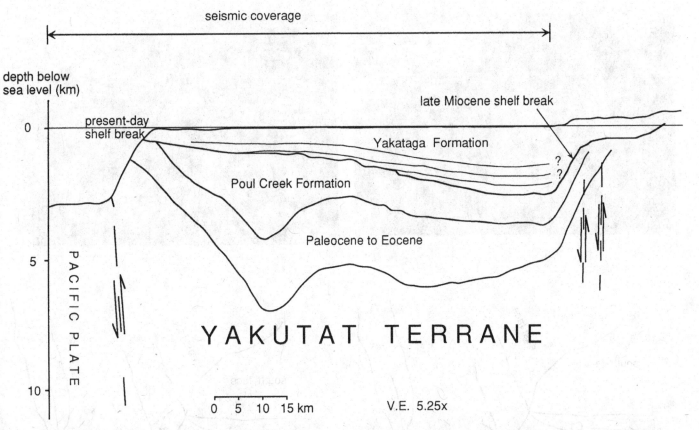

Fig. 9.8. Interpreted cross-section of Yakutat block based on multichannel seismic data and boreholes, showing how the Yakataga Formation has filled in a subsiding forearc basin. Seismic profiles show no sign of slope progradation. The shelf break migrated seaward only slightly (~5 km) from the Paleocene to the Late Miocene, but advanced catastrophically ~100 km by the infilling of the forearc basin. (Modified after Ehm, 1983.)

the progradation of the continental margin expected in such an environment. Paleocurrent directions should not deviate significantly from the average downslope direction. Depending on sea-level change, the final proglacial deposits may be preserved in the form of coarse sands on the outer margin and muds in depressions on the central shelf.

The active margin develops in a manner similar to the passive margin during the initial stages, the main difference being changes in orientation of successive submarine channels (Fig. 9.10). If the basin fills by aggradation, so that the slope advances catastrophically across the forearc basin the cut-and-fill channels may not be developed in the interior of the basin. The active margin sequence should be capped by a thick shelf sequence of repeated diamicts punctuated by coquinas (interglacials) and striated boulder pavements. Shelf diamicts are formed by the rainout of suspended muds, which will be heavily bioturbated, with coarse debris supplied by icebergs.

Other examples

The Pacific margin of the Antarctic peninsula was active prior to a ridge–trench collision starting at about 50 Ma (Larter and Barker, 1991). Multichannel seismic lines on the Pacific margin of

Antarctica suggest that there have been several successive intervals of shelf progradation followed by peneplanation. There have been four major depositional sequences since the major erosional truncation caused by tectonic uplift which was caused by the ridge crest–trench collision. The last two sequences are proposed to have been produced by ice sheets grounded out to the shelf edge (Larter and Barker, 1991). For all sequences, the shelf break is sharp. Although the margin is presently passive, because it is still steadily subsiding, preservation of glacial sequences is good. Larter and Barker (1991) point out that the sequences are produced by ice advances, and that major ice advances may destroy any record of pre-existing glaciers out on the shelf.

The Labrador Shelf and Slope appears to show the general structure of the passive margin, with growth by progradation. Airgun data on the shelf, however, appears to show local aggradation where recent glacial advances have not completely removed pre-existing glacial sediments (Josenhans et al., 1986). No such sediment has been observed on the Scotian Shelf, except possibly at the very edge. Rifting in the Labrador Sea is more recent than in the mid-Atlantic (Emery and Uchupi, 1984), consequently, the Labrador margin is probably subsiding more rapidly than the Scotian margin.

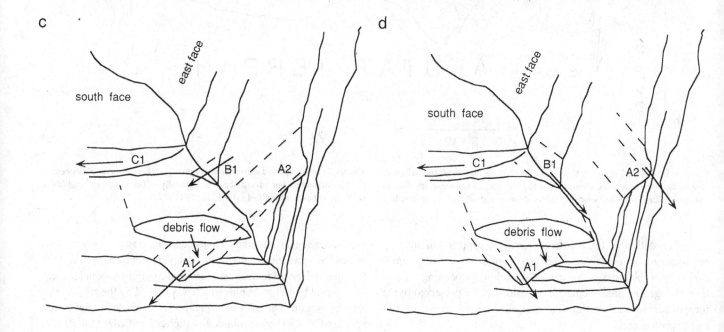

Fig. 9.9. Submarine channel outcrops of Yakataga Formation, near Icy Bay.
Adjacent faces of the same peak are depicted in (a) and (b). Channels are
labelled A1, A2, B1, and C1, and are filled with diamicts and turbidites. A
thick, channelized debris flow is also labelled. Line drawings (c) and (d)
represent possible interpretations of the orientations of channels and downslope movement of debris flows. (c) shows the most likely interpretation: that A1
and A2 are expressions of the same channel (because they are both incised to
the same stratigraphic horizon, whereas incision-depth of different channels is
different in Fig. 9.7), oriented NE–SW, and B1 is likely parallel to it. C1 is
oriented approximately E–W. The debris flow must be oriented towards the
south. If it had been oriented NE–SW, it should be expressed on the eastern
face. (d) Alternatively, if A1 and A2 are not expressions of the same feature,
then they must be oriented approximately NW–SE. B1 and the debris flow may
be approximately parallel to them, but C1 would still have to be approximately
perpendicular to them. In either case, not all of the channels can have the same
orientation.

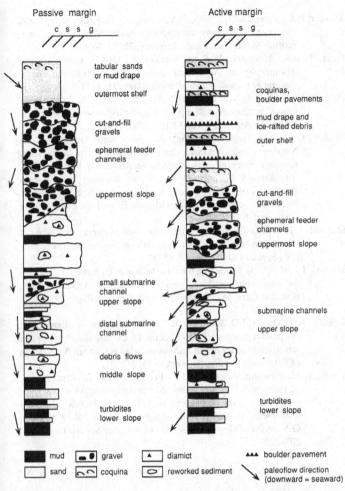

Fig. 9.10. Schematic contrast of the facies successions possible for both passive and active glaciated margins. The initial development of both margins is probably similar. Major differences might be noted in the variability of paleocurrent directions shown by large submarine channels, and the development of a thick sequence of shelf deposits on the active margin.

Summary

The impact of tectonic setting on the architecture of glacial marine deposits is felt through its effects on preservation. Continental shelf deposits are much less likely to be preserved on tectonically passive margins, as ice sheets are able to advance across to the shelf break. Thus, passive margins will be characterized by prograding continental slope deposits, although it is possible that a veneer of shelf deposits, representing the last glaciation will be preserved. On tectonically active margins, rapid subsidence prevents ice sheets from advancing to the shelf edge, allowing repeated sequences of continental shelf sediments to be preserved as the shelf builds by both aggradation and progradation. Both types of margins are likely to have well-developed systems of submarine channels. However, the orientations of the channel systems on the active margin will vary complexly through time, whereas channel systems on passive margins are likely to remain constant through time.

Acknowledgements

Funding for this project was provided by NSERC grants to Dr. N. Eyles, and by the University of Toronto (University of Toronto Open Fellowship). Field expenses to Alaska were met by NSERC. Shipboard expenses were met by the Geological Survey of Canada using funds from the Panel on Energy Research and Development. Thanks are due to Don and Lahoma Leishman for their hospitality in Alaska, and the officers and crew of the CSS Dawson and CSS Hudson for their assistance at sea. This paper has benefitted from discussions in the lab and in the field with N. Eyles, D.J.W. Piper, and C.H. Eyles.

References

Andrews, J.T. (1987). The Late Wisconsin glaciation and deglaciation of the Laurentide Ice Sheet. In *North America and Adjacent Oceans During the Last Deglaciation*, eds. W.F. Ruddiman, H.E. Wright, Jr., Geological Society of America, The Geology of North America, vol. K-3, 13–37.

Armentrout, J.M. (1983). Glacial lithofacies of the Neogene Yakataga Formation, Robinson Mountains, southern Alaska Coast Range, Alaska. In *Glacial Marine Sedimentation*, ed. B.F. Molnia, New York, Plenum Press, pp. 629–66.

Boulton, G.S. (1990). Sedimentary and sea level changes during glacial cycles and their control on glacimarine facies architecture. In *Glacimarine Environments: Processes and Sediments*, eds. J.A. Dowdeswell, J.D. Scourse, Geological Society of London Special Publication 53, 15–52.

Boyd, R., Scott, D.B. & Douma, M. (1988). Glacial tunnel valleys and Quaternary history of the outer Scotian Shelf. *Nature*, 333, 61–64.

Bruns, T.R. (1983). Model for the origin of the Yakutat block, an accreting terrane in the northern Gulf of Alaska. *Geology*, 11, 718–21.

Carlson, P.R., Bruns, T.R., Molnia, B.F. & Schwab, W.C. (1982). Submarine valleys in the northeastern Gulf of Alaska: characteristics and probable origin. *Marine Geology*, 47, 217–42.

Dobson, M.R., Scholl, D.W. & Stevenson, A.J. (1991). Interplay between arc tectonics and sea-level changes as revealed by sedimentation patterns in the Aleutians. In *Sedimentation, Tectonics and Eustasy: Sea-level Changes at Active Margins*, ed. D.I.M. MacDonald, International Association of Sedimentologists Special Publication 12, 151–163.

Ehm, A. (1983). Oil and gas basins map of Alaska. Division of Geological and Geophysical Surveys Special Report 32, Alaska Department of Natural Resources.

Eisbacker, G.H. (1985). Late Proterozoic rifting, glacial sedimentation, and sedimentary cycles in the light of Windermere deposition, western Canada. *Palaeogeography, Palaeoclimatology, Palaeoecology*, 51, 231–54.

Eisbacher, G.H., Carrigy, M.A. & Campbell, R.B. (1974). Paleodrainage pattern and late orogenic basins of the Canadian Cordillera. In *Tectonics and Sedimentation*, ed. W.R. Dickinson, Society of Economic Paleontologists and Mineralogists Special Publication 22, 143–66.

Emery, K.O. & Uchupi, E. (1984). *The Geology of the Atlantic Ocean*. New York, Springer-Verlag, 1050 pp.

Eyles, C.H. (1987). Glacially influenced submarine-channel sedimentation in the Yakataga Formation, Middleton Island, Alaska. *Journal of Sedimentary Petrology*, 57, 1004–1017.

Eyles, C.H. (1988a). A model for striated boulder pavement formation on glaciated, shallow-marine shelves: an example from the Yaka-taga Formation, Alaska. *Journal of Sedimentary Petrology*, **58**, 62–71.

Eyles, C.H. (1988*b*). Glacially and tidally influenced shallow marine sedimentation of the Late Precambrian Port Askaig Formation, Scotland. *Palaeogeography, Palaeoclimatology, Palaeoecology*, **68**, 1–25.

Eyles, C.H., Eyles, N. & Lagoe, M.B. (1991). The Yakataga Formation: a six-million-year record of temperate glacial marine sedimentation in the Gulf of Alaska. In *Glacial Marine Sedimentation: Paleoclimatic Significance*, eds. J.B. Anderson, G.M. Ashley, Geological Society of America Special Paper 261, 159–80.

Eyles, N. (1990). Marine debris flows: Late Precambrian 'tillites' of the Avalonian–Cadomian orogenic belt. *Palaeogeography, Palaeoclimatology, Palaeoecology*, **79**, 73–98.

Hovland, M. & Judd, A.G. (1988). *Seabed pockmarks and seepages*, London, Graham and Trotman, 293 pp.

Josenhans, H.W., Zevenhuizen, J., & Klassen, R.A. (1986). The Quaternary geology of the Labrador Shelf. *Canadian Journal of Earth Sciences*, **23**, 1190–213.

King, L.H. & Fader, G.B. (1986). *Wisconsin Glaciation of the Atlantic Continental Shelf of Southeast Canada*. Geological Survey of Canada Bulletin 363, 72 pp.

Lagoe, M.B., Eyles, C.H. & Eyles, N. (1993). Neogene tidewater glaciation in the far North Pacific Ocean: chronostratigraphic calibration, depositional record and regional climatic framework. *Geological Society of America Bulletin* (in press).

Larter, R.D. & Barker, P.F. (1991). Neogene interaction of tectonic and glacial processes at the Pacific margin of the Antarctic Peninsula. In *Sedimentation, Tectonics, and Eustasy: Sea-level Changes at Active Margins*, International Association of Sedimentologists Special Publication 12, 165–86.

Miall, A.D. (1985). Sedimentation on an early Proterozoic continental margin under glacial influence: the Gowganda Formation (Huronian), Elliot Lake area, Ontario, Canada. *Sedimentology*, **32**, 763–88.

Mosher, D.C., Piper, D.J.W., Vilks, G.V., Aksu, A.E. & Fader, G.B. (1989). Evidence for Wisconsinan glaciations in the Verill Canyon area, Scotian Slope. *Quaternary Research*, **31**, 27–40.

Piper, D.J.W., Normark, W.R. & Sparkes, R. (1987). Late Cenozoic stratigraphy of the Central Scotian Slope, eastern Canada. *Bulletin of Canadian Petroleum Geology*, **35**, 1–11.

Piper, D.J.W., Shor, A.N. & Hughes Clarke, J.E. (1988). The 1929 'Grand Banks' earthquake, slump, and turbidity current. Geological Society of America Special Paper 229, 77–92.

Plafker, G. (1987). Regional geology and petroleum potential of the Northern Gulf of Alaska continental margin. In *Geology and Resource Potential of the Continental Margin of Western North America and Adjacent Ocean Basins: Beaufort Sea to Baja, California*, eds. D.W. Scholl, A. Grantz, J.G. Vedder, Circum-Pacific Council for Energy and Mineral Resources, Earth Science Series, vol. 6, 229–68.

Shepard, F.P. (1979). Currents in submarine canyons and other types of seavalleys. Society of Economic Paleontologists and Mineralogists Special Publication 27, 85–94.

Syvitski, J.P.M. (1989). Modelling the fill of sedimentary basins. In *Statistical Applications in the Earth Sciences*, ed. F.P. Agterberg, G.F. Bonham-Carter, Geological Survey of Canada, Paper 89–9, 505–15.

Welsink, H.J., Dwyer, J.D. & Knight, R.J. (1989). Tectono-stratigraphy of the passive margin off Nova Scotia. In *Extensional tectonics and stratigraphy of the North Atlantic margins*, American Association of Petroleum Geologists Memoir 46, 215–31.

Young, G.M. & Gostin, V.A. (1991). Late Proterozoic (Sturtian) succession of the North Flinders Basin, South Australia; An example of temperate glaciation in an active rift setting. In *Glacial marine sedimentation: paleoclimatic significance*, eds. J.B. Anderson, G.M. Ashley, Geological Society of America Special Paper 261, 207–222.

10 Marine to non-marine sequence architecture of an intracratonic glacially related basin. Late Proterozoic of the West African platform in western Mali

JEAN-NOËL PROUST and MAX DEYNOUX

Abstract

Elementary stratigraphic events, or parasequences, are generally understood as progradational events bounded by flooding surfaces. Major sediment accumulation and preservation occur during baselevel fall. Baselevel rise is recorded in the marine environment by a veneer of sediments that is generally neglected. Field evidence from the Late Proterozoic glacially related deposits of the West African craton suggests that this concept must be revised to be applied to a continental–marine transitional zone.

Strata deposited during periods of both baselevel fall and baselevel rise, and their volumetric proportions, vary as a function of their paleogeographic position relative to the shoreline. Accordingly, an elementary building block, or depositional genetic unit is defined (Fig. 10.5). It is composed of three kinds of architectural elements bounded by four kinds of erosional or non-depositional regionally correlative surfaces. Each architectural element (Ae) corresponds to a typical association of facies, which allow the distinction of a 'progradational wedge' (Ae1) made up of lower to upper shoreface wave- to storm-dominated shales and sandstones, a 'continental wedge' (Ae2) mostly composed of non-marine rocks comprising ephemeral fluvial deposits, eolian sand sheet and dune deposits, and lagoonal to backshore sandy or carbonaceous deposits, and a 'transgressive wedge' (Ae3) made up of high-energy, upper shoreface, clean, well-sorted sandstones. The distinction of the bounding surfaces is based on the geometric relationships and the nature of the facies tracts they bound. The non-depositional hiatal surface, or maximum flooding surface (MFS), forms the upper and lower limits of each genetic unit; it marks the maximum marine extension inland, and forms basinward a downlap surface for the 'progradational wedge' (Ae1) of the overlying unit. The emersive erosional bounding surface (ES) is located at the base of the 'continental wedge' (Ae2) and down cuts into the 'progradational wedge' (Ae1). The ravinement erosional bounding surface (RS) forms the lower boundary of the 'transgressive wedge' and truncates landward the Ae1 and Ae2 architectural elements. The ravinement surface (RS), the maximum flooding surface (MFS), and the emersive surface (ES) merge together inland to form the intra-eolian super bounding surfaces (ISS). Representative genetic units are described and a very large and continous span of depositional environments is illustrated ranging from eolian dunes, lacustrine deposits, and fluvial systems, to wave-dominated barrier islands, nearshore bars, outer-shelf storm deposits and sandridges.

On account of the differential development of these architectural elements, the three-fold depositional genetic unit evolves through space and time leading to the definition of different orders of stacked sequences. Each of these orders exhibits a specific volumetric partition of sediments that is interpreted in terms of sediment fluxes linked with baselevel fluctuations through time. The respective impact of superimposed short-term climatically controlled and long-term tectonically driven baselevel fluctuation cycles is determined, and their role in net sediment accumulation and preservation on the platform is discussed.

Introduction

The concepts of sequence stratigraphy have been scarcely applied to pre-Pleistocene glacially related sedimentary basins (e.g., Christie-Blick et al., 1988; Lindsay, 1989). This paper gives an example of stratigraphic analysis in terms of chronostratigraphic and dynamic genetic units that may be used to trace in the rock record the sedimentological response to glacial fluctuations.

The study concerns the Late Proterozoic West African epicratonic Taoudeni Basin, which was active during the Varangian glacial period (c. 630 Ma). Its particular geodynamic and paleogeographic settings allow the following specific considerations:

1. The tectonically passive epicratonic setting limits the number of factors that may control the sedimentation and leads to a more accurate, undiluted, climatic signal. In such a setting and during glacial periods, the number of allogenic factors can be reduced to the global impact of glacioeustatism and the local effect of glacial isostasy. Other parameters represent only a background noise and can be neglected when considering a short time span.
2. The marginal setting of the studied area with respect to the inferred paleocenter of the West African Late Proterozoic ice sheet, results in a thick pile of glacial or glacially

related deposits much more representative of the glacial period than the generally thin veneer of subglacial continental lag deposits.

3. The deposits consist of shallow marine to non-marine transitional sediments. In such a zone changes in sedimentary flux generate easily recognizable facies changes.

Following considerations concerning the baselevel concept and a brief positioning of the studied sequences within the structural and stratigraphic framework of the West African platform, a depositional genetic unit will be briefly defined and then justified by field descriptions. In the discussion, which forms the second part of this paper, sequential characteristics of the genetic unit will be emphasized and its differential evolution through space and time will be presented in terms of baselevel changes related to glacioeustatism and tectonism.

Baselevel concept

The baselevel concept was introduced by Powell (1875) as an imaginary surface that determines the depth of erosion. Rice (1897) enlarged the concept to include sediment deposition, so that baselevel was interpreted as an equilibrium surface that intersects the physical surface of the lithosphere (i.e., the topography). Neither deposition, nor erosion take place along this surface, which marks the limit between deposition (below) and erosion (above). Later, Barrell (1917) introduced the modern idea of sediment accommodation space by considering baselevel as a dynamic surface that controls the rate of sedimentation by the rate of space creation. Subsequently, baselevel was considered as a plane or superimposed horizontal or nearly horizontal planes (Hayes, 1899; Dunbar and Rodgers, 1957) despite the difficulties in application. Incoherencies in the model were eliminated in the treatment by Wheeler in a series of papers (Wheeler, 1957, 1958, 1959, 1964, 1966).

Wheeler argued that baselevel is a worldwide constantly undulating or vibrating abstract surface toward which the lithosphere surface is moving. The lithosphere tends to reach the baselevel when deposition occurs. The baselevel goes below the lithosphere surface where erosion occurs. The baselevel intersects the lithosphere surface in areas of sediments bypass (Fig. 10.1). In a closed stratigraphic system (energy, mass, and space conserved), baselevel is a potentiometric surface along which the sediment flux is maintained constant (equilibrium between the rates of sediment supply and sediment removal). When applying this concept to the stratigraphic record, baselevel appears as a descriptive tool, which allows one to consider observed synchronous erosional or depositional areas in marine or non-marine contexts independently from factors controlling the sedimentation (eustasy, tectonics, climate).

In the continental–marine transitional zone, which is the focus of this paper, the baselevel concept is a practical way to integrate marine and non-marine rocks in the same dynamic sense of accumulation/preservation. This concept also avoids the paradox of using sea-level changes to interpret fluvial or eolian depositional processes.

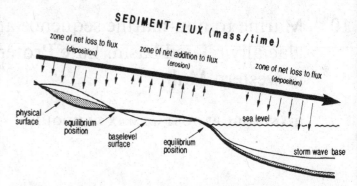

Fig. 10.1. Baselevel surface viewed as an equilibrium surface bounding depositional and erosional areas, and along which the net sediment flux is maintained constant. The loss from the flux in depositional areas is balanced by a net addition to the flux in erosional areas. The baselevel surface intersects the topographic surface in areas of sediment bypass (hiatus). (From T. Cross, work in progress.)

In this paper, major interpretations issuing from sequence stratigraphic analysis will be based on the notion of baselevel formalized by Wheeler (1964) and revised recently by T. Cross (work in progress).

Geological setting

West Africa is made up of a large, 1600 Ma old, cratonic platform covered by a thin skin (3000 m thick on average) of Upper Proterozoic and Paleozoic sedimentary rocks marginally involved in the Panafrican and Caledono-Hercynian fold belts. The Late Proterozoic glacial and related deposits are part of this sedimentary cover, which forms the Taoudeni Basin (Fig. 10.2a).

Paleogeographic reconstructions of the Late Proterozoic glacial period (Deynoux, 1980, 1985) show a primary ice sheet center located towards the North of the Reguibat Shield with possible secondary ice sheet centers on the Leo Shield and on the Kedougou and Kayes exhumed basements to the west and northwest of the studied area (Fig. 10.2a,b). The Late Proterozoic glacial period is recorded in the northern part of the Taoudeni Basin by a thin irregular (0 to 50 m thick) veneer of terrestrial tillites with subordinate proglacial outwash deposits preserved in limited shallow depressions. Towards the south of the platform, the glacial drift thickens (150 to 500 m) in a large diversified accumulation zone such as the shallow intracratonic basin in western Mali filled by sediments of the glacial Bakoye Group (Simon, 1979; Simon et al., 1979; Marchand et al., 1987) (Fig. 10.2b,c).

The Bakoye Group is a complex alternation of shallow marine and continental deposits subdivided into six lithostratigraphic units (Ba1 to Ba6 of Fig. 10.2c). Marine deposits are made up of shales and siltstones including banded hematite-chert intercalations (Ba2), shallow marine sandstones, glaciomarine diamictites, and shales (Ba4), shales and glaciomarine diamictites (Ba5) capped by a thin barite-bearing dolomitic horizon (Ba6). Non-marine rocks (Ba1 and Ba3) consist of a complex interfingering of eolian, fluvial sandstones, and terrestrial tillites. Further details concerning the

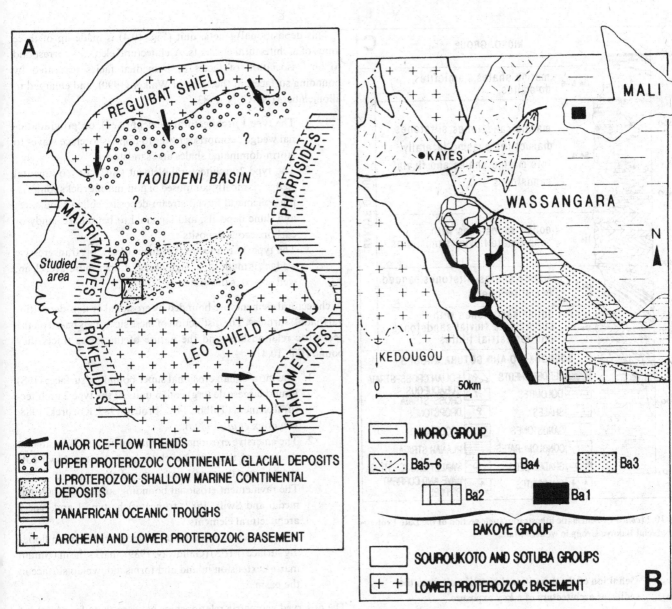

Fig. 10.2. Late Proterozoic glacial deposits on the southwestern part of the West African Craton. A. Paleogeographic map of the West African Platform in Upper Proterozoic times, showing major ice-flow pathways and the location of the studied area in western Mali. B. Geological map of the studied area in western Mali.

stratigraphy and sedimentology of the Bakoye Group can be found in Simon (1979), Rossi (1980), Rossi et al. (1984), Deynoux (1985), Marchand et al. (1987), Proust (1987, 1990), Deynoux et al. (1989a, b, 1990, 1991), Proust et al. (1990a, b).

The Bakoye Group is separated from the underlying Upper Proterozoic sedimentary rocks by a craton-wide erosional and slightly angular unconformity underlain by patches of terrestrial tillites. The Bakoye Group may also rest directly on the Archean and Lower Proterozoic basement. It is unconformably overlain by shales and bedded cherts of the Nioro Group, which represent the post-glacial marine transgression on the whole platform.

The study discussed in this paper deals with the relationships between the continental (fluvial and eolian) deposits of the Ba3 Formation and the shallow marine deposits of the Ba4 Formation in the Wassangara area. During the glacial period this depositional area was located some tens to a hundred kilometres eastward or southeastward from at least periodically glaciated uplands that are now partially represented by the Kedougou and Kayes exhumed basements (Fig. 10.2b).

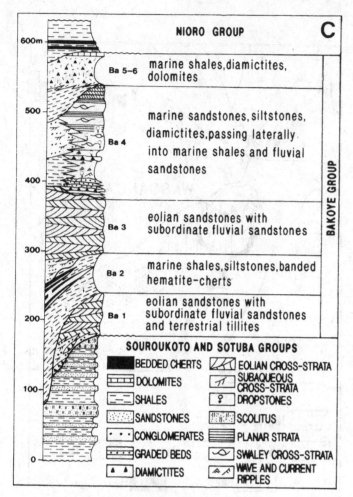

Fig. 10.2 (*cont.*) C. Schematic lithostratigraphic section of the Late Proterozoic glacial Bakoye Group in western Mali.

Definition of the depositional genetic unit from the sediment architecture of a key exposure.

In the Wassangara area, the 150 m thick Ba3 and Ba4 Formations are part of the sedimentary pile that fills a 30 to 50 km wide shallow intracratonic basin. They are characterized by a complex interfingering of facies and structures ranging from off-shore deposits to intracontinental eolian erg deposits (Fig. 10.5). The glacial environment is variously represented, either by specific features (tillites, glaciomarine diamictites, dropstones, periglacial structures) or by complex and rapid facies changes linked to sea-level fluctuations and large sediment input.

The contact between marine and non-marine rocks is particularly well exposed in a 150 m high and 4 km long section, close to the Doungué Village which is located 11 km south-southeast of the Wassangara Village. This exposure (Fig. 10.3) was selected as a type locality in order to define an elementary building block, or deposit-ional genetic unit, which was then used as a reference frame to draw up correlation along transects in the Wassangara Basin.

This depositional genetic unit (Fig. 10.4) is made up of three kinds of architectural elements. Architectural elements correspond to an 'association of facies or individual facies separated by bounding surfaces' in the sense of Walker (1990), and enlarged to allomembers (NACSN, 1983).

1. The type 1 architectural element (Ae1), or 'progradatio-nal wedge', comprises lower to upper shoreface wave- to storm-dominated shales and sandstones.
2. The type 2 architectural element (Ae2), or 'continental wedge', is mostly composed of non-marine rocks compris-ing ephemeral fluvial stream deposits, eolian sand sheet and dune deposits, and lagoonal to backshore sandy or carbonaceous deposits.
3. The type 3 architectural element (Ae3), or 'transgressive wedge', is made up of high-energy, upper shoreface, clean, well-sorted, sandstones.

Architectural elements are bounded by erosional or non-depositio-nal (hiatal) surfaces. The distinction of these surfaces is based on the geometric relationships and the nature of the facies tracts they bound (Fig. 10.4).

1. The intra-eolian erosional super bounding surfaces (ISS) bound largescale (erg) eolian units of the type 2 architec-tural element (Talbot, 1985; Proust, 1987; Kocurek, 1988; Deynoux *et al.*, 1989a).
2. The emersive erosional bounding surface (ES) located at the base of the type 2 architectural element down cuts sharply into the type 1 architectural element.
3. The ravinement erosional bounding surface (RS) (Num-medal and Swift, 1987) truncates the type 1 and type 2 architectural elements.
4. The non-depositional hiatal surface, or maximum flood-ing surface (MFS) (Galloway, 1989), marks the maximum marine extension inland and forms a downlap surface in the basin.

The observed geometric relationships between these four kinds of surfaces (Figs 10.3, 10.4) show that the ravinement surface (RS), the maximum flooding surface (MFS) and the emersive surface (ES) merge together towards the most inland part of the section to form the intra-eolian super bounding surfaces (ISS). Landward and basinward, RS and MFS also merge together when Ae3 wedges out. The MFS/ISS surfaces are the most extensive surfaces that can be traced confidently in the whole Wassangara Basin. They were chosen as the lower and upper boundaries of the depositional genetic units.

Ten vertically stacked depositional genetic units (U1 to U10) have been distinguished in the Doungué section (Fig. 10.3) and traced throughout the rest of the basin (Fig. 10.5). In the whole succession the landward, seaward, and vertical displacements of the depocentre result in changes in the nature and thickness of each individual architectural element, and cause variations in the geo-metry of the genetic unit they form (Figs 10.3, 10.5, 10.6, 10.25).

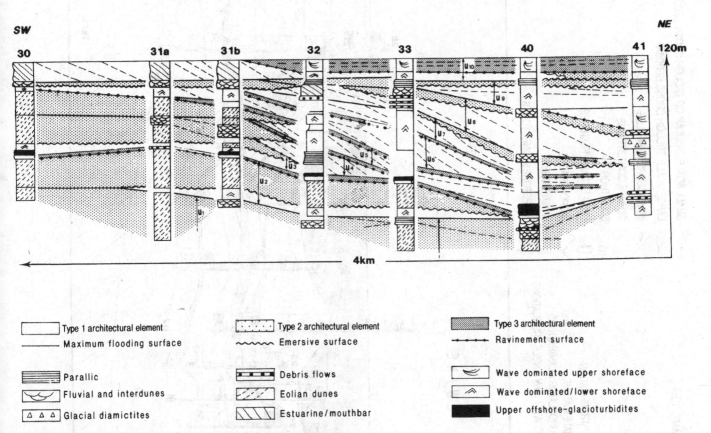

Fig. 10.3. Schematic representation of a shore-normal section showing the interfingering of fluvioeolian (Ba3 Formation) and marine (Ba4 Formation) deposits of the glacial Bakoye Group in the Wassangara Basin. Doungué Village exposure.

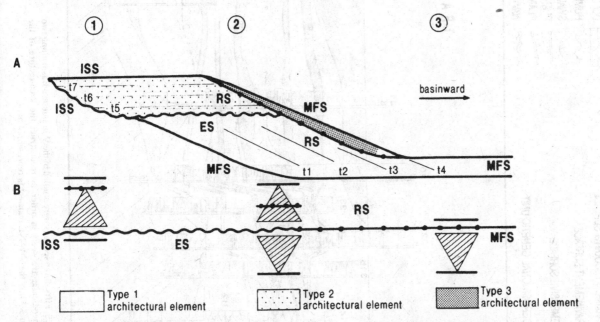

Fig. 10.4. Schematic representation of depositional genetic units in the Wassangara Basin. A, distribution of architectural elements, their bounding surfaces, and B, nature of the deepening up or shallowing up sequences they form, from the more landward (1) to the more seaward (3) locations. ISS, intra-eolian superbounding surface; ES, emersive surface of erosion; RS, erosional ravinement surface; MFS, non-depositional, hiatal, maximum flooding surface, tx = time lines.

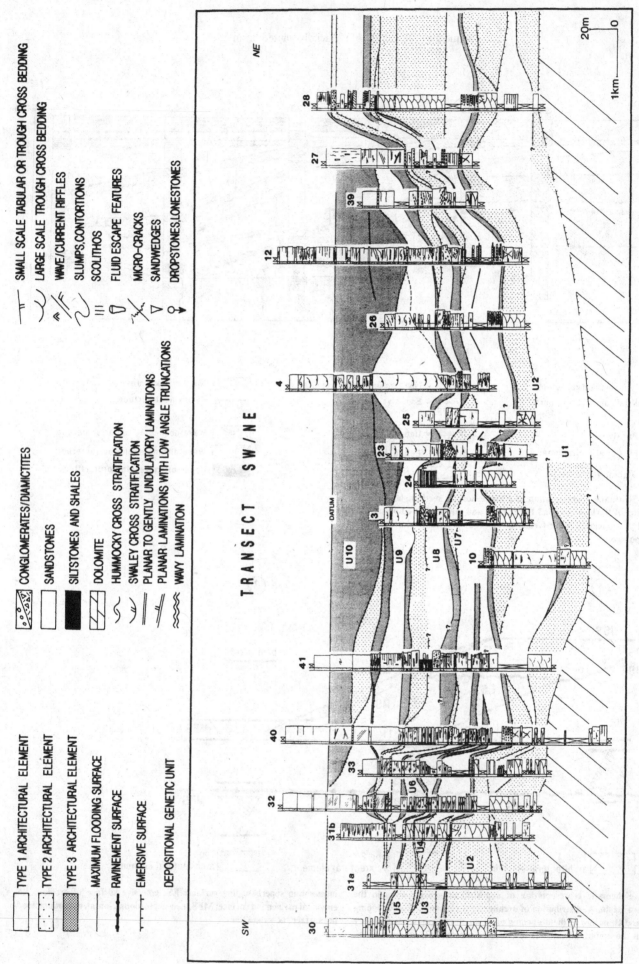

Fig. 10.5. Fifty-three sections were studied in the Wassangara Basin along two subperpendicular transects in order to reconstruct the architecture of the sediments. Only the SW–NE transect is represented here.

TRANSECT SW/NE

Legend:

TYPE 1 ARCHITECTURAL ELEMENT
TYPE 2 ARCHITECTURAL ELEMENT
TYPE 3 ARCHITECTURAL ELEMENT
MAXIMUM FLOODING SURFACE
RAVINEMENT SURFACE
EMERSIVE SURFACE
DEPOSITIONAL GENETIC UNIT

CONGLOMERATES/DIAMICTITES
SANDSTONES
SILTSTONES AND SHALES
DOLOMITE
HUMMOCKY CROSS STRATIFICATION
SWALEY CROSS STRATIFICATION
PLANAR TO GENTLY UNDULATORY LAMINATIONS
PLANAR LAMINATIONS WITH LOW ANGLE TRUNCATIONS
WAVY LAMINATION

SMALL SCALE TABULAR OR TROUGH CROSS BEDDING
LARGE SCALE TROUGH CROSS BEDDING
WAVE/CURRENT RIFFLES
SLUMPS, CONTORTIONS
SCOLITHOS
FLUID ESCAPE FEATURES
MICRO-CRACKS
SANDWEDGES
DROPSTONES, LONESTONES

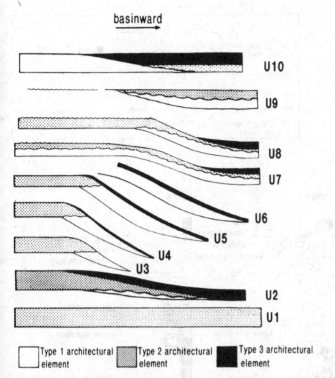

basinward →

U10

U9

U8

U7

U6

U5

U4

U3

U2

U1

☐ Type 1 architectural element   ▨ Type 2 architectural element   ■ Type 3 architectural element

Fig. 10.6. Simplified scheme of the Doungué Village exposure (Fig. 10.3) showing the stacking pattern of depositional genetic units. U1, U2, U9, landward-stepping units; U3, U4, U5, U6, U8, vertically stacked units; U7, U10, seaward-stepping units.

(1) Landward-stepping events (Units 1, 2, and 9) are elongated and asymmetrical. The landward displacement of the depocenter results in a well-developed type 2 architectural element (continental wedge). These can be traced over several kilometres into the whole of the Wassangara Basin.

(2) Vertically stacked events are either shifted landward (Units 3, 4, 5, 6) or seaward (Unit 8). The units are sigmoidal and short, with a well-developed type 1 architectural element (progradational wedge). They extend a few kilometres seaward and wedge out landward.

(3) Seaward-stepping events (Units 7 and 10) are asymmetrical. They are elongated over tens of kilometres and thus can be traced over the whole Wassangara Basin. The progradational wedges (Ae1) are thin and deeply truncated by the emersive bounding surface (ES).

Facies distribution in architectural elements

Facies descriptions of each architectural element are illustrated from examples taken in the Wassangara Basin, in selected landward-stepping (Unit 2), vertically stacked (Unit 8), and seaward-stepping (Unit 7) genetic units.

Landward-stepping unit (Fig. 10.7a)

Type 1 architectural element

The distal part of the type 1 architectural element of Unit 2 (30 m thick), is made up of a thickening upward succession of coarse- to medium-grained sandstone beds. Each bed, 0.3 to 2 m thick, displays a fining upward grain size distribution and an erosional base. Erosional features evolve upward in the succession from large (60 cm), deep (30 cm) and flat-based gutter casts, to shallow rounded base gutter casts (50 cm wide, 15 cm deep) and finally to slightly undulatory erosional surfaces. The gutter casts are asymmetrically filled with coarse material distributed along the toe of low-angle laminations. Internal structures within the beds are represented by large, low-relief hummocky cross-stratification in the basal part of the succession, passing upward into simple undulatory laminations that flatten progressively up section.

The fining upward hummocky cross-stratified sandstone beds, bounded by gutter casts, are indicative of a high-energy, mixed oscillatory and unidirectional overcharged flow, which could be related to episodic storm deposition. According to Guillocheau (1990), the shape and nature of the gutter casts, which unevenly truncate the underlying laminations, the lack of compensation of the pre-existing topography by the overlying laminations, the coarse nature of the material, and the absence of clay to silt size sediment strongly suggest deposition in an offshore/shoreface transitional zone. A gradual shallowing upward trend is also inferred from the general coarsening upward of the sediments, from the progressive flattening of laminations, and from the evolution of deep flat-based gutter casts to shallow rounded-base gutter casts and to simple undulatory erosional bases.

The HCS sandstone bed succession grades upward and laterally into amalgamated swaley cross-bedded sandstones (SCS), which in a landward direction form the bulk of a prograding wedge. Along strike, the SCS sandstones exhibit laterally and vertically stacked sigmoidal clinoforms (Fig. 10.8). Each of these, less than two meters thick and several tens of meters in lateral extent, rests with a slightly erosional contact on the former one. Out-of-phase SCS form the basal parts of the clinoforms, while in-phase SCS form their middle and proximal parts. The clinoforms and their basal erosional surfaces merge upward, giving way to low angle slightly undulatory, flat laminations, and then to very shallow erosional troughs (10 m wide, 1 m deep).

The SCS sandstones, which form the prograding clinoforms, were deposited between lower wave base and foreshore (Duke, 1985; Leckie and Walker, 1982; McCrory and Walker, 1986). The upper flat laminations with low-angle truncations and slightly undulatory laminations of high orbital velocity are respectively characteristic of the beach face or the uppermost part of the shoreface (e.g., Clifton et al., 1971; Davidson-Arnott and Greenwood, 1976; Roep et al., 1979; Harms et al., 1982; Short, 1984; 1986). Such an upper shoreface facies association probably represents a wave-dominated barrier island (Schwartz, 1973; Swift, 1975; Wanless, 1976; Field and Duane, 1976; Halsey, 1979; Rahmani, 1983).

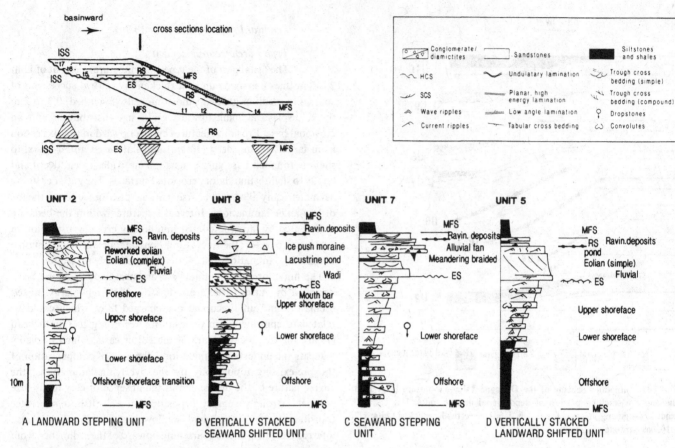

Fig. 10.7. Cross-sections of the four end members of the Ba3/Ba4 stacking cycles (genetic units) of the Bakoye Group in the Wassangara Basin. All sections are drawn at the depositional break. In such a location, the sections are the most complete and exhibit the vertical superimposition of a negative shallowing up and a positive deepening up sequences. ISS, intra-eolian superbounding surface; ES, emersive surface of erosion; RS, erosional ravine-ment surface; MFS, non-depositional, hiatal, maximum flooding surface; tx, time lines.

Fig. 10.8. Laterally and vertically stacked sigmoidal clinoforms of a downlapping type 1 architectural element (progradational wedge). Person for scale.

On the whole, the type 1 architectural element of Unit 2 corresponds to a progradational wedge. It exhibits a gradual shallowing upward trend from storm-dominated offshore/shoreface sequences to upper shoreface prograding barrier island deposits. The latter are capped with foreshore deposits sharply truncated by an emersive erosional surface (ES). No significant volume of marine sediments seems to have been removed below this erosional surface.

Type 2 architectural element

The type 2 architectural element onlaps the underlying progradational wedge. The erosional surface (ES) is overlain by two superimposed crudely fining upward sequences composed of a few meters of medium scale, medium- to coarse-grained trough cross-bedded sandstone covered by a decimetre-thick structureless or rippled siltstone. These lenticular, 'patchy', sheet-like sequences wedge out landward, passing into a thin veneer of evenly dispersed pebble lag. The pebble lag underlies a flat erosional super surface to 6 m thick, a few tens of meters wide) trough cross-bedded sets. In the most inland part, individual sets consist mainly of grainflow cross-strata (Hunter, 1977) extending nearly to the base of each set with only minor wind-rippled bottomsets. The thickness of individual grainflow cross-strata range from 2 to 6 cm. For crescentic transverse dunes, such characteristics suggest dune heights of 15 to 45 m (G. Kocurek, unpublished data). Basinward, wind-rippled laminations, which form the bulk of the cross-strata, are indicative of low-relief dome-shaped dunes without slipfaces.

In the uppermost part of the type 2 architectural element, massive sandstone occurs as channel fills. The channels, up to 10 m thick and several hundreds of meters long in the paleodip direction, downcut sharply into the underlying well-stratified eolian sandstone. These structures, already described in Deynoux *et al.* (1989*a*), contain cobbles to boulders of well-preserved eolian cross-strata up to 10 m across. The massive sandstone and the adjacent undisturbed eolian strata are capped by a horizon showing patches of granitic and metamorphic pebbles and cobbles passing laterally into a few meters of coarse to conglomeratic sandstone. Internal structures within this horizon, consists of shallow trough cross-bedded sets with low-angle laminae and basal lags comprising outsized clasts. These cross-bedded sandstones are very commonly oversteepened along the dune paleotopography and reworked downward into the massive sandstone facies.

The massive sandstone facies that truncates the eolian sets is interpreted to be a mass-flow remobilization of eolian sand early during the marine transgression represented by the overlying type 3 architectural element. This interpretation is based on the massive nature of the deposits, the preservation of slumped eolian blocks, and the sharp contact with eolian dunes that forms steep overhanging walls. Such structures require both wet (cohesion of eolian blocks and walls) and water-saturated (fluidization of massive sands) conditions. They could be ascribed to water transfer in dunes by a rising water table. This is also supported by the occurrence of water escape cylindrical structures in the surrounding dunes (Deynoux *et al.*, 1990). The massive sandstone grades upward into cross-bedded sandstone, representing ephemeral fluvial deposits. These two facies are in close association, but they cannot account for the emplacement of heavy basement rocks boulders. However, such large exotic clasts are fairly common in periglacial shallow water environments where they are rafted by seasonal shore ice (Dionne, 1985). Such an interpretation is supported by the general glacial context of the Bakoye Group and by the landward occurrence of periglacial cracks and sandwedges just below the ISS surface. The upper bounding surface of the eolian deposits is planar landward (ISS) but grades basinward into a soft, wavy topography that represents the paleodune field morphology (Eschner and Kocurek, 1988; Chan and Kocurek, 1988) modified along the ravinement surface (RS) by the overlying transgressive deposits.

Type 3 architectural element

The type 3 architectural element rests on the type 2 along a sharp, erosional surface. It is made up of a single, coarse- to medium-grained, well-sorted, 1 m thick, sandstone bed. Internal

Fig. 10.9. Eolian cross-bedded sandstone of type 2 architectural elements (continental wedge), SW of section 30, close to the Doungué Village.

(Bagnold surface) that, over distances of several tens to a hundred kilometers, separates inland large eolian dunes units or ergs (Deynoux *et al.*, 1989*a*).

These two sequences are interpreted as fluvial channel deposits. The relative scarcity of deep channeling features underlain by pebbly material, the poor development of fine-grained and/or abandonment sediments, and the small thickness of the deposits suggest a very shallow braided stream environment.

The overlying facies is made up in the best exposures of 30 m thick medium-grained eolian cross-stratified sandstone (Fig. 10.9). The sandstone consists of two superimposed cosets, several meters to some tens of meters thick, separated by a sharp erosional surface capped by a patchy lag of coarse, massive to laminated sand including millimetric granules of quartz or sandstone. Each coset corresponds to the migration of an individual eolian draa bounded by a first order erosional surface (Kocurek, 1988) overlain by a veneer of coarse material resulting from eolian reworking of initial waterlaid deposits (required to transport granule-size material in interdune). Within each draa, eolian cross-strata consist of large (5

Fig. 10.10. Medium- to coarse-grained flat laminated sandstone of a type 3 architectural element (transgressive wedge) overlying eolian deposits of a type 2 architectural element (continental wedge) along a sharp erosional surface (pencil). The eolian sandstone comprises vertical sand-cracks and sand-wedges of periglacial origin. Pencil for scale.

Fig. 10.11. Alternation of massive, medium- to coarse-grained, sandstone beds and rippled, argillaceous, fine-grained, sandstone interbeds. Massive beds are structureless or show planar, undulatory, or hummocky laminations. Interbeds exhibit combined flow ripples. This alternation (1), with occasional dropstones (circled), passes upward into a more uniform wave to current rippled sandstone (2) abruptly truncated by a fluvial conglomeratic horizon (3). (1) and (2) form part of a shallowing up type 1 architectural element (progradational wedge), (3) corresponds to the lower part of a type 2 architectural element (continental wedge). Hammer for scale (small arrow).

structures consist of either slightly undulatory laminations, or horizontal, parallel laminations (Fig. 10.10). This sandstone bed is typically lenticular and is confined to paleodepressions at the top of the smooth paleodune field topography of the underlying type 2 architectural element.

This bed is interpreted to be a high energy, upper shoreface deposit squeezed between eolian or fluvial deposits below and lower shoreface 'glacioturbidites' of the overlying type 1 architectural element (Unit 3). This type 3 architectural element, or transgressive wedge, was formed by the permanent shoreface erosion–deposition processes linked with wave action during marine transgression and subsequently preserved in paleodepressions atop non-marine rocks (Abbott, 1985). It could be considered as the trace of the shoreface shift.

Vertical stacking unit (Fig. 10.7b,d)

Two kinds of vertically stacked units have been distinguished in the Wassangara Basin: landward shifted (Fig. 10.7d) and seaward shifted units (Fig. 10.7b). The landward shifted units display similarities with the landward-stepping units (see Proust, 1990 for details), and only the seaward shifted Unit 8, which shows a well-diversified continental wedge (Ae2), will be described here.

Type 1 architectural element

The type 1 architectural element of Unit 8 is 25 m thick and 10 km in lateral extent. Basal shales and siltstones grade upward into an alternation of massive, medium- to coarse-grained, sandstone beds, and rippled, argillaceous, fine-grained, sandstone interbeds (Fig. 10.11). The massive beds, which contain a few skolithous burrows, have sharp, flat, or undulatory bases. These beds are structureless or show planar, undulatory, or hummocky

Fig. 10.12. Three-dimensional view of wave rippled beds of the upper part of a type 1 architectural element (progradational wedge). Pencil for scale.

laminations. The rippled interbeds exhibit combined flow ripple structures. Up-section, the sand to clay ratio increases, the HCS beds give way to well-defined wave-rippled beds (Fig. 10.12) that pass progressively into bidirectional current ripples (herringbone structures) and finally to trochoidal ripples (Harms *et al*, 1982) interbedded with flat-laminated horizons.

This coarsening upward succession corresponds to a gradual shallowing upward trend. The offshore thinly bedded shale and sandstone alternations grade into a sandier lower shoreface emplaced under the mixed influence of permanent fair weather energy (rippled interbeds) and intermittent storm conditions (massive HCS beds). These beds are overlain by coarser-grained wave-current rippled upper shoreface deposits. The uppermost part of this progradational succession suggests a somewhat lower flow regime (embayment?) responsible for the deposition of fine-grained flat-laminated sandstones and lapping trochoïdal ripples.

In some sections, the uppermost part of the succession comprises erosional, elongated, 5 to 6 meters thick, medium- to coarse-grained sandstone bodies of several kilometers in lateral extent. They are composed of simple, prograding, sigmoidal sets with low-angle internal laminations dipping basinward like the basal boundaries of the sets. In places these sandstone bodies are partially or entirely slumped, and downlap onto earlier deposits. In such occurrences the sandstones are either structureless or display remnants of pre-existing laminations, deformed or brecciated by fluidization processes during the downdip remobilization. Numerous debris flows also occur throughout the succession. They are generally 5 to 10 meters thick and of tens to hundreds of meters in lateral extent. Most of the outsized clasts are derived from the neighbouring crystalline basement. Three kinds of debris flows occur from base to top of the section: argillaceous matrix-supported debris flows with centimeter-scale pebbles, sandy matrix-supported debris flows with decimeter- to meter-scale cobbles to boulders, and clast-supported debris flows with large cobbles to boulders.

The association of elongated sandstone bodies, interpreted as mouth bar deposits, and coarse-grained sediment gravity flows into a large, shallow marine, prograding system, is generally encountered in submarine fan delta environments (Holmes, 1965; Wescott and Ethridge, 1980; Allen, 1986; Massari and Colella, 1988; McPherson et al., 1987; Nemec and Steel, 1988). The progradational wedge of Unit 8, however, does not correspond to the classical scheme of tectonically (fault) controlled fan deltas that provide the basin with an important supply of coarse, immature, material, in a very short time. However, the general glacial context, which implies a high sediment supply, may account for the good analogy with humid type of fan deltas.

Type 2 architectural element

The type 1 architectural element is sharply truncated by an emersive erosional surface (ES) underlined locally by periglacial sandwedges (Fig. 10.13). This surface is overlain by a complex non-marine facies assemblage (10 to 20 m thick, tens of kilometers in lateral extent) of the type 2 architectural element.

In an oversimplified vertical succession this architectural element starts with a conglomeratic horizon (Fig. 10.11, 10.13) made up of well-rounded, flat-shaped pebbles and boulders of quartz, granite, carbonate and jasper. Clasts are rarely imbricated but generally parallel to crude horizontal or cross-strata. Remnants of eolian faceting (dreikanters) were observed. This facies is overlain by decimeter to meter thick, thinly bedded, creamy, honeycomb or laminated dolomite (Fig. 10.14) that passes vertically into structureless, decimeter-scale, alternations of fine-grained sandstones and green siltstones.

The laminated dolomites are related to algal mats. These algal

Fig. 10.13. Periglacial conglomeratic wedge (outlined) filled by fluvial material (3) of the basal part of a type 2 architectural element and protruding into the upper part of a type 1 architectural element (1) (2). Same location and similar horizons (1), (2), (3), as for Fig. 10.11. Hammer for scale.

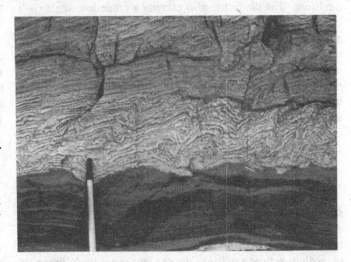

Fig. 10.14. Laminated dolomite of a type 2 architectural element (continental wedge). These locally deformed or broken laminations are related to algal mats in a lacustrine or coastal embayment.

Fig. 10.15. Tillite (B) and conglomeratic outwash sandstone (C) overriding fluvial sandstones (A). A, B and C form a type 2 architectural element (continental wedge) overlain by flat laminated sandstone (D) of a type 3 architectural element (transgressive wedge). Person at the lower right corner for scale.

Fig. 10.16. Cut-and-fill structures in a type 3 architectural element (transgressive wedge). These asymmetrically filled structures truncate planar or slightly undulatory laminations and pass progressively upward into swaley cross-stratifications. Pencil (arrow) for scale.

Fig. 10.17. Swaley cross-stratification of a type 3 architectural element. Ruler (1.2 m) for scale.

laminations, which exhibit dome-shaped structures, are locally broken by desiccation cracks. Gases trapped during water-level fluctuation in a beach face environment may account for the honeycomb structures. This facies association represents fluctuations between fluvial (conglomeratic horizon) and lacustrine or coastal embayment (carbonates and structureless sandstone–siltstone alternations) environments.

At some places in the basin (Fig. 10.15), this succession is truncated and reworked by a 3–5 m thick diamictite along an erosional surface with step fractures (Biju Duval *et al.*, 1974). This diamictite consists of a microconglomeratic sand/clay matrix, with centimeter- to meter-scale clasts of basement rocks, sandstones, or carbonates. Some of these clasts are striated, and large angular pieces were obviously picked off from the immediate underlying sandstone. The diamictite also exhibits a crude low angle (20°) bedding bounded by nearly horizontal shear planes. A boulder pavement and lodgment structures are present in the lower part of the diamictite, which is interpreted as a basal tillite or an ice-pushed moraine. The clast fabric, lodgment imbrications, and the orientation of shear planes and step fractures in the underlying sandstone beds, are all indicative of a south-southeast displacement of the material. The uppermost part of the diamictite becomes stratified and passes into a 1 m thick conglomeratic calcareous outwash sandstone (Fig. 10.15).

Type 3 architectural element

Type 2 architectural element is truncated by a ravinement surface (RS) onlapped by a 1 to 8 m thick type 3 architectural element which extends along tens of kilometers and wedges out basinward. This type 3 architectural element is made up of a coarse- to medium-grained sandstone divided into meter thick, fining up beds that are separated by well sorted, lenticular, decimeter thick, pebbly conglomerates. In the lower part of Ae3, internal structures

consist of planar to three-dimensional undulatory laminations in sets bounded by low-angle planar erosional surfaces and truncated by sparse cut and fill structures (Fig. 10.16). Upward, these structures pass into amalgamated swaley cross-stratifications (Fig. 10.17).

Such a succession is interpreted as a beach face coarse-grained low-angle laminated sandstone (including pebble lags) that grades upward into an upper shoreface swaley cross-bedded sandstone.

Seaward-stepping unit (Fig. 7c)

Type 1 architectural element

Basinward, the type 1 architectural element of Unit 7 consists of an alternation of 0.1 to 1 m thick fining upward sequences of coarse- to fine-grained calcareous sandstone. Above an erosional base underlain by steep-sided scours, a complete

Fig. 10.18. Three sequences (A,B,C) of combined wave-current ripples passing upward into convolute bedding, in clinoforms of a type 1 architectural element (progradational wedge). Note the sharp truncation of the convolutes by the overriding sequence and their deformation in the dip direction. Pencil for scale.

sequence displays from base to top: (1) a structureless or crudely bedded (normally or inversely graded) coarse to conglomeratic horizon, (2) a medium-grained horizon with parallel, flat, or slightly undulating laminations and rare inverse graded microconglomeratic streaks, (3) a thin fine-grained rippled horizon overlain by (4) lenticular brown dolomite and/or shale. Dropstones and sparse diamictite intercalations, with intra- or extra-basinal clasts up to 80 cm across, occur in the succession.

Such sequences, already discussed and termed 'glacioturbidites' in Deynoux *et al.* (1991, their figures 5,6,7), present strong analogies with the Bouma sequences and could be related to high-density gravity flows (Lowe, 1982). However, some features could also be related to storm-graded layers, especially the wide range of orientations of gutter casts and the presence up sequence of polygonal ripples, which may represent small-scale HCS (Brenchley *et al.*, 1986; Guillocheau and Hoffert, 1988). Basinward, the genetic units 3 to 6 are essentially made up of such glacioturbiditic sequences and become indistinguishable from each other and from the overlying Unit 7 (Fig. 10.5).

The 'glacioturbidites' grade landward into a decimeter thick alternation of shales and fine- to medium-grained wave-rippled sandstones, which form the toe of large sigmoidal bedforms about 1 m thick and tens of meters across. Along strike, these laterally stacked sigmoidal bedforms are part of an overall larger (up to 10 m thick, several kilometers long) clinoform that wedges out basinward into the gravity-induced sequences. A vertical section into the middle well-developed part of one of these sigmoidal bedform shows the progressive superposition of type A, B and S combined wave/current ripples (Jopling and Walker, 1968; Boersma, 1970) which pass upward into convolute bedding (Fig. 10.18). However, a strong differentiation occurs along each sigmoidal element. Current ripples form the bulk of the upper proximal part of the sigmoids and pass basinward into combined wave/current ripples and then into well-defined wave ripples. The convolute bedding is usually sharply

truncated by the next sigmoid and typically overturned in the dip direction (Fig. 10.18). As well, sedimentary structures became obscured by slumping towards the toe of the sigmoids.

The sigmoidal bedforms were emplaced under the mixed influence of unidirectional and oscillatory flows, associated with a very rapid sediment supply. Within the ripple-convolute bedsets, the strong unidirectional component responsible in each bedset for the truncation and deposition of the basal current ripples passes upward into a dominant oscillatory flow along with an increase of sedimentation rate. This reflects a very rapid, quasi-instantaneous, deposition with liquefaction effects responsible for the formation of the convolutes (Ten Haaf, 1956; Dott and Howard, 1962; Dzulinsky and Smith, 1963; Cheel and Rust, 1986) and oversteepening of some wave ripple crests.

The overall succession of this progradational architectural element reflects a gradual shallowing up trend from offshore to lower shoreface environment ('glacioturbidites') to lower/middle shoreface environment (well-developed symmetrical wave ripples passing upward into asymmetrical combined wave–current ripples). In the most proximal part of the basin, at the marine to non-marine transitional zone, the middle to upper shoreface facies are missing. These missing facies (about 10 m of sediments) have been removed by the erosional bounding surface (ES and/or RS) of the Ae2 architectural element.

Type 2 architectural element

This architectural element, 0 to 20 m thick and a few kilometers across, occurs sporadically above an erosional surface, which can be traced over the whole Wassangara Basin. The type 2 architectural element cuts down into the type 1 architectural element, and basinward cuts progressively into the underlying units. It consists mostly of heterogranular medium- to coarse-grained sandstone with interbedded polygenic breccias in its upper part.

The basal part of the section is made up of decimeter thick tabular planar cross-sets with few exotic angular pebbles or boulders. This facies is overlain by laterally and vertically stacked large scale (2.5 m thick, more than 10 m across) shallow trough sets alternating with small decimeter-scale trough sets. Within the large sets, cross-laminae are tabular to asymptotic, and internal reactivation surfaces occur. At places, current ripples, climbing obliquely downstream, are superimposed on foresets. Cosets are locally topped by a centimeter thick siltstone layer, comprising a few isolated centimeter-scale exotic pebbles.

A braided, low-sinuosity stream, with superimposed transverse bedforms is proposed for the tabular sets at the base of the section. Meandering streams could have formed the overlying sequence composed of channel floor dunes (small trough sets) and laterally accreted bedforms (large shallow trough sets) overlain by channel abandonment deposits (silty layers).

The upper part of the section is bounded at its base by a sharp irregular erosional surface with deep and narrow erosional scours, 5 to 10 m wide to 25 m deep. It is composed of four lithofacies, which grade laterally into each other: (1) clast-supported breccias, (2)

Fig. 10.19. Clast-supported breccia of a type 2 architectural element (continental wedge) showing clasts of all sizes and orientations, some of them broken, and fluidized, squeezed, deformed, shearing laminations.

matrix-supported breccias, (3) massive sandstones with minor gravels, and (4) laminated sandstones. Clasts of various sizes (from 1 cm to 1.6 m across) include granite, gneiss, dolerite, dolostone, and sandstone. The sandstone clasts are in high position and include large blocks of eolian material.

The clast-supported breccias are typically stratified with, vertically and laterally fining and thinning individual beds, 1 to 3 meters thick. Beds are separated from each other by siltstone layers, a few millimeters thick, which can be followed over great distances. Clasts are variously oriented. Imbricated clasts, clasts in a vertical position, or broken and splashed out in several pieces, are common. Fluidized, squeezed, and deformed shearing laminations occur rarely (Fig. 10.19).

The matrix-supported breccias are always structureless and massive (Fig. 10.20). The massive sandstone lithofacies is a well-sorted medium- to coarse-grained structureless sandstone with some millimetric lonestones. Polygenic boulders and pebbles (1–40 cm across) are scattered over the base of the beds. Some of them are vertical or imbricated.

According to the literature (e.g. Larsen and Steel, 1978; Lowe, 1979; Nemec *et al.*, 1980; Postma, 1986), matrix-supported breccia results from a subaerial debris flow, and clast-supported breccia from a subaquatic (lacustrine owing to the absence of active hydrodynamic reworking) debris flow passing distally to the massive sandstone lithofacies.

The laminated sandstone lithofacies is made up of well sorted medium- to coarse-grained thinly bedded sandstone. Individual beds, a few centimeters to a few decimeters thick, overlie an erosional base and generally fine upward. The beds exhibit flat parallel laminations with irregular inverse- or normal-graded coarse-grained streaks, overlain by climbing current ripples laminations (Fig. 10.21). The latter grade laterally into thin planar tabular cross-strata, which are sometimes folded and recumbent into the paleoflow direction. Such thin beds, which individually show an

Fig. 10.20. Structureless and massive matrix-supported breccia of a type 2 architectural element (Ae2). Ruler (2 m) for scale.

Fig. 10.21. Centimeter- to decimeter-scale alternation of flat laminated and rippled sandstone of an ephemeral sheet flood deposit in a type 2 architectural element (continental wedge). Ruler (30 cm) for scale.

Fig. 10.22. Large undulatory laminations (pseudo-HCS) in medium- to coarse-grained sandstone of a type 3 architectural element (transgressive wedge). These structures are rather like large type B climbing ripples formed in an upper shoreface environment characterized by a high sediment input. Hammer (circled) for scale.

upward-decreasing flow strength and fluvial characters, appear to correspond to ephemeral sheet floods or levee deposits.

The above briefly described association of braided streams, sheet floods or levees, subaerial to subaquatic debris flows, and sand flows, is compatible with a subaerial to lacustrine alluvial fan environment as reported in the literature (e.g. Larsen and Steel, 1978; Nemec et al., 1980; Postma and Roep, 1985).

Type 3 architectural element

This architectural element is composed of a lenticular, medium- to coarse-grained, sandstone body a few meters thick, and over 15 kilometers in lateral extent. Rare, outsized, angular, basement clasts deform the pre-existing laminations. Sedimentary structures include flat laminations or three-dimensional undulatory laminations of decimeter- to meter-scale wavelength. Individual laminated strata are a few millimeters to 4 centimeters thick and normally graded. Their thickness grows upward as a positive function of the wavelength. Short wavelength undulatory laminations are usually sharply truncated by swaley or cut-and-fill structures. In some places, the undulatory laminations are very large, with an amplitude of 2 m and wavelength of 10 to 15 m. They resemble the laminations of very large HCS (Fig. 10.22). However, no undulatory erosional surfaces are visible, and thus these structures are rather like type B climbing ripples according to the terminology of Jopling and Walker (1968).

Affinities of the 3D undulatory laminations with both HCS and B-type climbing ripples require specific hydrodynamic conditions such as the combination of strong aggradation, very slow unidirectional migration (Allen and Underhill, 1989), and an oscillatory component (Duke et al., 1991). These structures appear in close association with upper flow regime flat laminations, SCS, and cut-and-fill structures. They were probably formed in an upper shore-

face environment characterized by a high sediment input. No vertical trends were noticed in this transgressive architectural element.

Discussion

Bimodal facies architecture and time/space distribution of sediments and bounding surfaces within genetic units

In a vertical section, which comprises the three architectural elements (Ae1, Ae2, Ae3), a genetic unit is formed by the superposition of a negative and a positive sequence (location 2 in Fig. 10.4). Such a bimodal package is bounded below and above by maximum flooding surfaces (MFS), which correspond in the field to a sharp, non-erosional contact between high-energy, clean, well-sorted, shallow marine sandstones of the transgressive wedge (Ae3) and downlapping deeper marine, low-energy, shaly siltstones and sandstones of the progradational wedge (Ae1). The negative and the positive sequences are bounded by an emersive surface (ES) marked by a sharp erosional contact between marine deposits of the progradational wedge (Ae1) and generally non-marine deposits of the continental wedge (Ae2). The positive sequence is truncated by an erosional ravinement surface (RS) between fully continental to paralic deposits (Ae2) and shallow marine deposits (Ae3) emplaced by wave action during transgression in a shoreface environment.

The bimodal package grades laterally into a simple positive or negative sequence when architectural elements wedge out in a landward or basinward direction (locations 1 and 3 in Fig. 10.4). In the same way, the emersive surface of erosion (ES) passes laterally to an erosional intra-eolian supersurface (ISS) in a landward direction, and to an erosional ravinement surface (RS) and to a non-depositional maximum flooding surface (MFS) in a basinward direction. These four kinds of surfaces represent in the field the multiple expressions of a simple physical surface.

The transposition of this sequential organization in a time/space diagram, using hypothetical time increments, allows the comparison between periods of deposition and periods of slight to non-deposition (hiatus) or erosion, according to specific locations in the basin and to baselevel changes (Fig. 10.23).

The negative sequence, which corresponds to the type 1 architectural element (Fig. 10.4), is typically made up of a coarsening up thickening up, shallowing up, progradational, wave- or storm-dominated, linkage of facies. These facies evolve gradually from offshore marine above the maximum flooding surface (MFS) to shallowest upper shoreface, or even foreshore to beachface environments (e.g., Unit 2, Fig. 10.7a), sharply truncated by the emersive surface of erosion (ES). Sigmoïd clinoforms in a downlapping pattern (Fig. 10.8) are generally overemphasized, and remobilization of material (debris flows, slumpings, etc.) along the progradational fronts is common.

Such a succession obviously records a period of baselevel fall. Progradational deposits are fed by a penecontemporaneous landward erosion, which forms the ISS/ES surface in inland areas (Fig. 10.24a). A bypassing zone (hiatus) between erosional and depositional areas migrates basinward. This bypass zone corresponds to

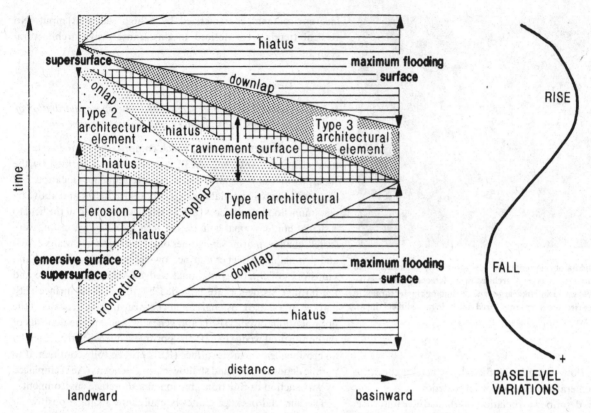

Fig. 10.23. Time–space distribution of sediments in a standard depositional genetic unit during a single baselevel fluctuation cycle.

the basinward migration of the intersection point between the baselevel line and the topographic surface (Fig. 10.1, 10.24a).

The positive sequence, which corresponds to types 2 and 3 architectural elements (Fig 10.4), is strongly retrogradational, so that marine influence increases gradually upsection. In landward-stepping units (U1, U2, U9), the succession evolves from proximal braided stream deposits to progressively more complex eolian dune deposits, reworked eolian sandstones, and transgressive upper shoreface deposits. The source of the sands (braided fluvial) is progressively buried by the resultant eolian deposits which are themselves the source of their capping shoreface deposits. The same retrogradational pattern can be drawn in vertically stacked units either shifted landward (U3, U4, U5, U6: braided streams, simple eolian dune, upper shoreface ravinement deposits), or shifted seaward (U8: fluvial, lacustrine, or embayment deposits, and beachface to upper shoreface sandstone). In seaward-stepping Unit 7 the succession is formed by two unconformably stacked retrogradational events (braided to meandering fluvial deposits and subaerial to subaquatic alluvial fan to upper shoreface deposits) in an overall 'transgressive' sequence. In the seaward-stepping Unit 10, Ae2 and Ae3 form a typical valley fill assemblage above a strongly erosional emersive surface (ES).

The retrogradational character of the Ae2–Ae3 positive sequences, which appears early during continental deposition, and the sharp erosional contact between marine and non-marine

deposits, suggest a period of baselevel rise as depicted on Fig. 10.24b, using baselevel concept. The distal part of the basin is underfed and experiences non-deposition or slight deposition (condensed section) above the incipient migrating landward MFS surface. The deposition of type 2 and 3 architectural elements, as well as the formation of the ravinement surface (RS), occur contemporaneously and postdate each other during their landward migration. A bypass zone at the intersection point of the baselevel line and topographic surface also occurs between the continental deposits and the transgressive ravinement surface (RS) (Fig. 10.24b).

Figure 10.23 suggests that intra-eolian supersurfaces (ISS), which in the landward direction bound two superimposed units, represent the amalgamation of hiatus and erosion occurring during either baselevel rise or baselevel fall. The baselevel rise gap wedges out into a landward direction as the baselevel fall gap widen. The baselevel rise bounding surface represents the shortest time gap in littoral areas. It could be considered as a valuable time line which may be confidently used for correlations into such an environment.

The deposition and preservation of the sedimentary record in each unit is related to a single baselevel fall and rise cycle, and thus these units can be considered as significant elementary events controlled by allogenic processes. The shift between the fall and rise periods is figured in the field by the complex physical surface of erosion: ISS, ES and RS *proparte*. This unconformity represents the

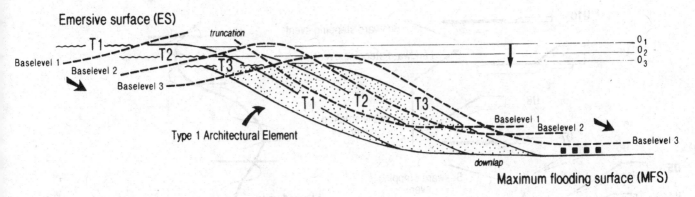

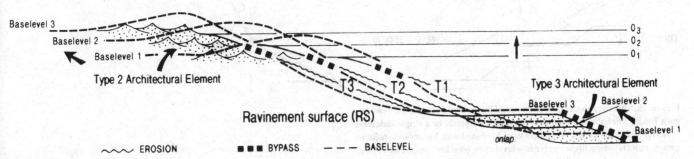

Fig. 10.24. Relationships between erosion, bypass and deposition of the three kinds of architectural elements (Ae1, Ae2, Ae3) in a continental to marine transitional zone during (A) baselevel fall (Ae1 deposition) and (B) baselevel rise (Ae2, Ae3 deposition).

main surface of sediment bypass and marks a drastic change in the location of the main sediment storage shifted from marine (in Ae1) to predominantly non-marine environments (in Ae2 and Ae3).

Stacking cycle

The vertical, landward, or basinward stacking pattern of genetic units in the Wassangara Basin records a baselevel fluctuation of higher magnitude than that which corresponds to the genetic unit cycle described above. This is also reflected by variations in the sediment distribution and shape of genetic units (Fig. 10.25).

The landward-stepping units (U1, U2, U9) are thick (30–50 m), elongated (up to 50 km) and made up of asymmetric clean and well-sorted sediment packages. Most of the sediment volume is stored within the progradational (Ae1) and continental (Ae2) wedges. The progradational (Ae1) and transgressive (Ae3) wedges are poorly diversified. Ae1 is strictly wave-dominated with SCS prograding clinoforms (barrier islands) passing basinward into HCS storm deposits. Ae3 exhibits high-energy planar-laminated reworked sandstone. A retrogradational succession from braided fluvial stream deposits to well-developed eolian dune deposits forms the type 2 architectural element. Reworking of former deposits below the ES surface is reduced to a few meters. As well, most of the modified paleodune field topography was protected from the subsequent wave erosion (RS) during transgression.

The vertically stacked units (U3, U4, U5, U6, U8) are short, sigmoid-shaped and symmetric (equal volume of sediment preserved in each architectural element). Most of the sediments are poorly sorted and enriched in fine-grained material. Sandstones and shales of the progradational wedge (Ae1) are wave-dominated and are inferred to represent a submarine fan delta emplaced under the influence of glacially related high sediment supply ('glacioturbidites'). The shoreface gradient is high, and a large volume of sediment is removed below the RS surface (at least 10 meters) by wave action during transgression. The result is the deposition of a relatively thick and coarse-grained transgressive wedge (Ae3). In the landward-shifted vertically stacked units (U3, U4, U5, U6), the continental wedges (Ae2) are made up of poorly developed dune fields (small simple dunes) with subordinate braided stream deposits. In contrast to landward-stepping units, eolian topography is not preserved in Ae2, and dunes are sharply truncated during the marine transgression along a flat surface (RS). In the

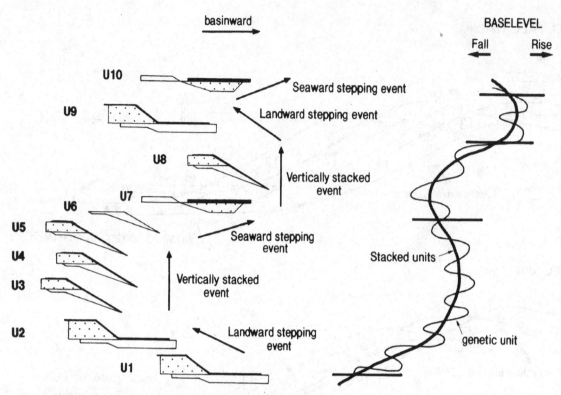

Fig. 10.25. **Superimposed orders of baselevel fluctuation cycles in the Wassan-gara Basin. Each depositional genetic unit corresponds to a single baselevel cycle (genetic unit cycle). The successive landward and basinward shifts of genetic units reveals a higher order baselevel cycle (stacked genetic unit cycle) superimposed on an overall rising trend.**

vertically stacked Unit 8, which is shifted seaward, the type 2 architectural element fluctuates between fluvial, lacustrine, coastal embayment, and glacial (tillite) deposits.

The seaward-stepping units (U7, U10) are elongated and asymmetric. The progradational wedges (Ae1) are thin and deeply truncated by an erosional surface (ES) located at the base of thick continental wedges (Ae2), that are shifted basinward. The facies in Ae2 are well diversified and fluvial-dominated. The ES surfaces of these units mark in the whole basin a sharp change in depositional environment, and determine a major continental gradient with much evidence of erosion, transportation, and reworking. The ES surface of Unit 7 corresponds in the whole succession to the following changes: (1) from eolian-influenced to fluvial-dominated sequences, and (2) from distal glacial influences (diamictites, drop-stones, periglacial wedges, etc.) to proximal glacial conditions (ice-push moraines, overcharged alluvial debris flows, etc.). In Unit 10, the ES surface marks a change from glacially controlled sedimentation to non-glacial conditions. However, the ES surface in Unit 10 is of greater lateral extent than the ES surface in Unit 7.

Accordingly, the following observations can be made for the stacked genetic unit baselevel cycle in the Wassangara Basin (Fig. 10.25). The initial baselevel rise period is characterized by: (1) a progressive landward-stepping of Units 1 and 2 onto the basal Ba2/Ba3 unconformity, (2) a thickening upward trend of these units, (3)

an increase in shale content and water depth that reach a maximum at the base of the vertically stacked units, and (4) a very high preservation potential of sediments, especially those from eolian dune fields. Subsequent baselevel fall is recorded by: (1) short, vertically stacked Units 2 to 6, (2) a crude thinning upward trend of these units, (3) an increase in the unsorted character of the sediment, (4) a fair potential of preservation of sediments, but only accumulations of small simple dunes. During this period, fluvial and glacial influences increase progressively. The peak of baselevel fall is recorded by the seaward-stepping Unit 7, which is characterized by: (1) the most important erosion related to baselevel fall, (2) very coarse and poorly sorted sediments, (3) stream rejuvenation, and (4) onset of proximal glacial conditions. The overlying Units 8 and 9 record an incipient second baselevel rise period with the superposition of a vertically stacked seaward shifted Unit 8 and a landward-stepping Unit 9. These units are sharply truncated by the uppermost basin-wide baselevel fall unconformity at the base of the U10 valley fill.

Comparison with genetic units described in the literature

Numerous 'elementary stratigraphic events' have been described in the literature (see reviews by Busch and West, 1987 and Cross, 1988). The most widely recognized are parasequences (Van

Wagoner, 1985; Van Wagoner et al., 1988), PACs (Goodwin and Anderson, 1980a,b, 1985; Anderson et al., 1984; Goodwin et al., 1986), cyclothems (Wanless and Weller, 1932), depositional events (Frazier, 1974), genetic increment of strata, and genetic sequences (Bush, 1959, 1971, 1974), mesothems (Ramsbottom, 1979), fourth order cycles (Ryer, 1984), or more widely the progradational events (Cross, 1988). Most of these 'elementary stratigraphic events' are asymmetrical, progradational, shallowing upward, time signifi-cant, and allocyclic. They are bounded in the marine environment by the transgressive or flooding surfaces of Van Wagoner (1985) or by the climate change surfaces in non-marine environment (Busch and West, 1987). They form negative sequences similar to the marine distal part of our genetic units.

As for our genetic units, the vertical stacking of progradational 'elementary stratigraphic events' forms higher order cycles, which are correlative at a regional scale. These cycles correspond to: the genetic sequences of strata (Buch, 1971, 1974), the depositional episodes (Frazier, 1974), PACs sequences (Goodwin and Ander-son, 1980a,b, 1985; Anderson et al., 1984; Goodwin et al., 1986), echelon and stacked cycles (Ryer, 1984), and the most widely recognized parasequence sets and sequences (Van Wagoner, 1985; Van Wagoner et al., 1988; Vail et al., 1977; Posammentier and Vail, 1988). Following the latter terminology, the stacking cycles of genetic units in the Wassangara Basin may represent two (third order?) depositional sequences (Fig. 10.25). The first one is com-posed of a transgressive system tract (landward-stepping units 1 and 2) and a highstand system tract (vertically stacked landward shifted Units 3 to 6). The second one is composed of a lowstand system tract (seaward-stepping Unit 7 proparte and vertically stacked Unit 8) and a transgressive system tract with the landward-stepping Unit 9. A third depositional sequence starts within the seaward-stepping Unit 10.

Estimated baselevel cycles duration

The lack of chronostratigraphic controls in the entirely unfossiliferous Proterozoic deposits of the Wassangara Basin do not allow the direct estimation of the duration of baselevel cycles. However, their duration can be tentatively estimated by compari-son with similar sequences emplaced during Phanerozoic times.

As already stated above, the geometry depicted by the stacked genetic units (Fig. 10.25) produces the general patterns of three regionally correlative third order transgressive/regressive sequences. The proposed duration for such third order cycles is 0.5 to 5 Ma (Vail et al, 1991) or 3 to 5 Ma (Cross, 1988), whereas 0.1 to 0.5 Ma is suggested for elementary buildings blocks (T. Cross personal communication, 1990). In the synglacial context of the Permo-Carboniferous Appalachian deposits, Busch and West (1987) proposed to relate their elementary transgressive/regressive units to the Vail et al. (1977) fifth order cycles (0.3 to 0.5 Ma), and their stacking cycles to the fourth order cycle (0.6 to 3.6 Ma). The revised values for the duration of the fourth and fifth order cycles match now with respectively 0.1 to 0.5 Ma and 0.01 to 0.1 Ma. The Carboniferous mesothems and cyclothems in Europe (Ramsbot-tom, 1979) are also consistent with a duration of about 0.1 Ma (elementary events), and $n*0.1$ Ma (transgressive/regressive stack-ing cycles). The elementary event durations fit fairly well with the standing or residence times of ergs during the last glaciation (Kocurek et al., 1991; Ahmed Benan, 1991) and in the pre-Pleistocene rock record (G. Kocurek, personal communication, 1989) where they have a periodic occurrence of $n*0.01$ Ma.

If these durations are extrapolated to the Late Proterozoic synglacial epicratonic sedimentary record, the following values could be proposed: 0.1 Ma for the genetic unit and $n*0.1$ Ma for a transgressive/regressive stacked unit. Three transgressive/regres-sive stacked units (the third one represented only by Unit 10 in this study) built up the whole Ba3–Ba4 sequence whose duration may correspond to about 1 to a few Ma. A time span of less than 10 Ma might then include the entire Bakoye Group, which represents the overall synglacial deposition on the craton itself, and which is made up of three stacked sequences (Proust, 1990).

Inferred factors controlling sedimentation

The net accumulation of sediments at the surface of the Earth results from the interaction of sediment input, substrate subsidence or uplift (subsidence sensu lato), and oscillations of the water/air interface (eustatism sensu lato). The cumulative effect of these three parameters is responsible for the observed baselevel fluctuations. Apart from basin-scale mechanisms like subsidence, the global phenomena responsible for baselevel fluctuations could be referred as 'external causes' (orbital forcing) or 'internal causes' (plate tectonic).

The climatic hypothesis

The orbital forcing signature is a composite signal com-posed of a 0.021 Ma cycle of precession, a 0.04 Ma cycle of obliquity, and a 0.1 Ma cycle of eccentricity (Milankovitch, 1941; Berger, 1984). The main consequences of these variations are either geodynamic or climatic (Morner, 1976, 1979, 1980, 1984a,b). The geodynamic consequences include changes in the shape of the geoid and in the magnetic field. These changes induce local, low ampli-tude (tens of metres), rapid (10 m/1000 y) sea-level variations (Pitman, 1978). The climatic effect caused by the variations of the incidence angle between the sun and the surface of the Earth is considered to be one of the controlling factors of glacial/interglacial cycles. These variations cause global (eustatic), rapid (10 to 100 m/ 1000 y), high amplitude (150–250 m) fluctuations of sea-level (Pitman, 1978). Among all these variations, only the 0.1 Ma cycles are global and constitute a typical signal in a climatically and glacioeustatically controlled sedimentary record (Fig. 10.26) (Morner, 1984a; Posamentier et al., 1988).

The estimated 0.1 Ma duration for the deposition of a single genetic unit is located in the waveband of eccentricity. Except for Units 10 and 11, several indications of the direct glacial input were recorded in each unit and at least two of them (Units 7 and 8) appear directly controlled by glacial advance and retreat. Thus, a global glacioeustatic sea-level change, induced by an elementary glacial

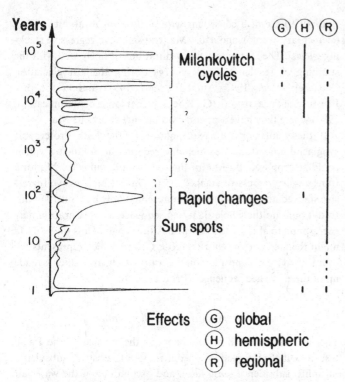

Fig. 10.26. Different orders of climatic fluctuations during the Quaternary and their inferred origin (semi-quantitative amplitude). Only the 0.1 Ma cycle represents global effects (modified from Morner, 1984b, in Guillocheau, 1990).

advance and retreat cycle, is a good candidate for the origin of genetic units in the Wassangara Basin.

The comparison of the evolution of glacial indicators and baselevel fluctuations at the scale of the transgressive/regressive stacked unit cycles (Fig. 10.25) gives evidence of higher order climatic variations. During the initial baselevel rise period (land-ward-stepping Units 1 and 2, and vertically stacked landward shifted Units 3 to 6) the glacial features are represented by ice-wedge casts in non-marine environments, and by dropstones, glacioturbi-dites, sparse marine diamictites, and debris flows in the marine realm. These features are related respectively to periglacial climate inland and distal glaciomarine conditions basinward. During base-level fall (seaward-stepping Unit 7 and vertically stacked seaward shifted Unit 8), ice-push moraines or tillites, high-density debris flows with numerous basement rocks, outsized clasts, and glacial marine diamictites, are all indicative of a proximal glacial environ-ment. The overlying baselevel cycle (landward-stepping Unit 9, and seaward-stepping Unit 10) does not present any evidence of gla-cially related phenomena and indicates 'normal' non-glacial con-ditions. In the stacking unit cycle, baselevel fall fits fairly well with the 'glacial maximum'. This kind of association was described by Frazier (1974) in the Quaternary deposits of the Gulf of Mexico (1.6 to 2 Ma duration). Each glacial stage corresponds to a seaward-stepping event sometimes overlain by a vertically stacked seaward-shifted depositional episode very similar to the stacking pattern of Units 7 and 8. The interglacial stages are characterized by land-ward-stepping and landward shifted vertically stacked motifs like the Unit 1 to Unit 6 evolution. In the same way, each regressive

maximum corresponds to a main glaciation (Riss, Würm, etc.) in the Mediterranean Quaternary stratigraphic record.

Although not proven conclusively, the glacioeustatic component may interplay with other mechanisms at a larger scale than those inferred from Quaternary studies. Chumakov (1985) depicted in the geological record glacial eras (200 to 300 Ma), glacial periods (some tens of Ma, Bakoye Group?), and glacial epochs (0.1 Ma to a few Ma). The latter include, for example, the six globally transgressive polycyclic (stacked genetic units?) Quaternary glaciations (genetic units?) of the northern hemisphere. These 'glacial epochs' could then be compared to our Ba3–Ba4 succession.

The tectonic hypothesis

Plate motions at the surface of the Earth change the size and shape of the oceanic basins, and cause deformations at their rims and in intraplate domains. The volumetric changes through time in the worldwide ocean (at a constant volume of water) generate low frequency (tens of Ma) eustatic variations of sea-level (tectonoeustatism), with an amplitude of 50 to 300 m and a rhythm of 0.5 to 2 m/Ma (Pitman, 1978, Vail and Eisner, 1989). These effects are generally slightly out of phase with the original causes. Brittle and ductile deformations of the crust are respectively at the origin of a rapid (at the geological scale) reactivation of previous anisotropies of the crust (ancient epeirogenies?) (Cloetingh, 1986, 1988a,b; Lambeck et al., 1987) and a slow, continuous deformation of the crust ('in plane stress tectonic' and 'plateau uplift'; Karner, 1986). In the last case, and for a 100 km wide, 4 km deep basin, and a compressional strain of $6.25 \ 10^{12}$ N/m the amplitude of the subsidence/uplift is about 100 m with a velocity of 1 cm/1000 y and a duration of 1 to 10 Ma (Karner, 1986).

Some features in the Bakoye Group deposits argue for either a large waveband tectonoeustatic control (oceanic volumetric varia-tions) or short waveband intraplate stresses. These features include:

1. The craton-wide angular unconformities at the base of the Bakoye Group and below Ba5–6 formations (Fig. 10.2c), which are related respectively to the Panafrican I and II orogenies (Villeneuve, 1988; 1989; Deynoux et al., 1991) in the nearby fold belts (Fig. 10.2a).

2. The banded iron deposits (hematitites) in the Ba2 Forma-tion. According to the fairly common occurrence of ferruginous sediments in Late Proterozoic glacially related deposits in the world, these deposits could be considered as synrift to early drift hydrothermal brine precipitations after remobilization by cold waters (Young, 1988).

3. The close relationship, which may exist between Vail's (1977) systems tracts deposition and tectonics in an extensional setting (Calvet et al., 1990) (Fig. 10.27), such as the extensional period inferred for post-Ba2 deposits (see above). This hypothesis might explain the unconfor-mities between Unit 1 and the Ba2 Formation and between Units 9 and 10. But it does not account for the strongly glacially controlled unconformity below the sea-ward-stepping Unit 7 and Unit 8.

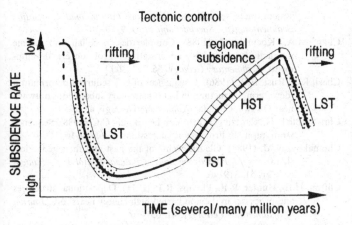

Tectonic control

LST Lowstand system tract
HST Highstand system tract
TST Transgressive system tract

Fig. 10.27. Interpretation in terms of tectonics of depositional sequences (third order cycles) of Posamentier and Vail (1988). Modified from Calvet *et al.*, (1990). Following these authors 'superimposed on these third order (10^6 y) relative changes of sea level, which give rise to depositional sequences, there may well be time-scale, fourth and fifth order (10^5 and 10^4 y) sea-level changes, perhaps resulting from the effects of Milankovitch rhythms, which give rise to parasequences and parasequence sets'.

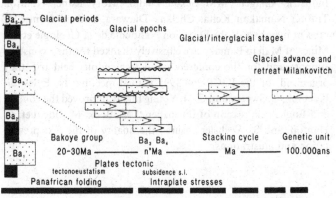

Fig. 10.28. Schematic representation of tectonic versus climatic control of sedimentation as inferred for the Late Proterozoic glacial Bakoye Group. The climatic signal is recorded in high-frequency events, whereas the tectonic control occurs preferentially in the large scale sequences of the low-frequency sedimentary time band.

These arguments suggest that deposition of the sediments of the intracratonic Bakoye Group was controlled either by tectonoeustasy (points 1 and 2) or by intraplate stress tectonics related to plate motion in the peripheral parts of the craton (point 3 above). However, these phenomena are generally slow (0.5 to 2 m/1000 y for tectonoeustasy and 1 cm/1000 y for intraplate stress), compared to the short-term climatically controlled phenomena envisaged for the genetic unit cycles and regressive/transgressive stacking genetic unit cycles (10 to 100 m/1000 y), and might be neglected.

Conclusion

This study is one of the first attempts to obtain, in non-dated Late Proterozoic glacially related deposits, a stratigraphic analysis in terms of genetic, chronostratigraphically significant units. These units, which comprise proximal inland to distal marine facies associations, and their different scale stacking patterns, show that most of the preserved intracontinental ergs and fluvial deposits, as well as most of the subglacial to glacial outwash sediments, were deposited during periods of early baselevel rise. Most of the sediment fluxes, that are oriented basinward in non-marine environments during baselevel fall, pass into marine deposits that are buried by early baselevel rise sediments.

The following four hierarchies of baselevel cycles are proposed for the Late Proterozoic synglacial deposits in Western Mali (Fig. 10.28):

1. A baselevel cycle of 0.1 Ma in duration (Milankovitch cycles waveband) related to elementary glacial advance and retreat rhythms (genetic unit cycle)
2. A baselevel cycle of less than 1 Ma, linked with glacial to interglacial 'stages' (stacked genetic unit cycle), that cannot be strictly related to any specific climatic or tectonic control
3. A baselevel cycle of few Ma in duration (Ba3–Ba4 formations cycle), which may be related to glacial epochs, but which could be also controlled by high frequency intra-plates stresses and deformations
4. A general lowering of baselevel on the order of 10 Ma during the Late Proterozoic glacial sedimentation (Bakoye Group cycle) on the craton. This cycle was also controlled by tectonic events in the marginal Panafrican fold belt.

With such an approach it may be possible to obtain chronostratigraphic data from unfossiliferous, undated sediments (such as most of the Proterozoic glacial deposits), and to reconstruct ancient sediment fluxes from analysis of the stratigraphic record. Moreover, it may also provide a means of investigating the relative importance of climate and tectonics in the control of sediment distribution and storage at the surface of the Earth.

Acknowledgements

This paper represents part of the results of several years of investigation in western Mali and an overview of a doctoral thesis by Jean Noêl Proust. Thus, it is also the result of long-term financial support from the Centre National de la Recherche Scientifique via various programs (ASP and RCP 'Afrique', DBT program contribution No 504), and from contracts with private companies such as the Compagnie Française des Petroles TOTAL, and Klöckner Industrie-Anlagen. MM Sékou Diallo, Mory Kané, Hilarion Traoré, Namakan Keita, Cheikna Diawara, and all former or present members of the Direction Nationale de la Géologie et des Mines of Mali in Bamako are effusively thanked for their constant support and for the conference and tremendous field trip they organized for the IGCP-260 1991 annual meeting. N. Eyles, G. Kocurek, C. Swezey and G.M. Young highly improved the content and English expression of this paper, J.M. Bock, M. Bocquet, E. Hanton, and S. Wechsler ensured the quality of the text presentation and illustrations.

References

Abbott, W.O. (1985). The recognition and mapping of a basal transgressive sand from outcrop, subsurface and seismic data. In *Seismic stratigraphy II, an integrated approach*, ed. R. Berg, D.G. Woolverton, American Association of Petroleum Geologists Memoir, 39, 157–67.

Ahmed Benan, C.A. (1991). Processus sédimentologiques en milieu désertique: exemple de l'Erg Akchar en Mauritanie occidentale. Mémoire Diplome Etudes Approfondies, Strasbourg, Université Louis Pasteur, 33 pp.

Allen, G.P. (1986). Environnements sédimentaires de la côte aquitaine: mécanismes, faciès et séquences de dépôt. Livret Guide Excursion, Association Française de Sedimentologie et Compagnie Française des Pétroles, 56 pp.

Allen, P.A., Underhill, J.R. (1989). Swaley cross-stratification produced by unidirectional flows, Bencliff Grit (Upper Jurassic), Dorset, UK. *Journal of Geological Science*, 146, 241–52.

Anderson, E.J., Goodwin, P.W., Sobieki, T.H. (1984). Episodic accumulation and the origin of formation boundaries in the Helderberg Group of New York State. *Geology*, 12, 120–3.

Barrell, J. (1917). Rhythms and the measurement of geologic time. *Geological Society of America Bulletin*, 28, 745–904.

Biju-Duval, B., Deynoux, M., Rognon, P. (1974). Essai d'interprétation des 'fractures en gradins' observées dans les formations glaciaires précambriennes et ordoviciennes du Sahara. *Rev. Géogr. Phys. Géol. Dyn.*, (Paris), 16, 503–12.

Berger, A. (1984). Accuracy and frequency stability of the earth's orbital elements during the Quaternary. In: *Milankovitch and Climate, part I*, ed., A.L. Berger *et al.* Dordrecht: Reidel, pp. 3–39.

Boersma, J.R. (1970). Distinguishing features of wave-ripple cross-stratification and morphology. Ph.D. Thesis, University of Utrecht, 65 pp.

Brenchley, P.J., Romano, M., Gutierrez-Marco, J.C. (1986). Proximal and distal hummocky cross-stratified facies on a wide Ordovician shelf in Iberia. In: *Shelf Sands and Sandstones*, eds., R.J. Knight, J.R. McLean, Canadian Society of Petroleum Geologists Memoir, 11, 241–55.

Busch, D.A. (1959). Prospecting for stratigraphic traps. *American Association of Petroleum Geologists Bulletin*, 43, 829–1154.

Busch, D.A. (1971). Genetic units in delta prospecting. *American Association of Petroleum Geologists Bulletin.*, 55, 1137–54.

Busch, D.A. (1974). Stratigraphic traps in sandstones, exploration techniques. *American Association of Petroleum Geologists Memoir*, 21, 174 pp.

Busch, R.M., West, R.R. (1987). Hierarchal genetic stratigraphy: a framework for paleoceanography. *Paleoceanography*, 2, 141–64.

Calvet, F., Tucker, M.E., Henton, J.M. (1990). Middle Triassic carbonate ramps systems in the Catalan Basin, northeast Spain: facies,

system tracts, sequences and controls. *International Association Sedimentologists, Special Publication*, 9, 79–107.

Chan, M.A., Kocurek, G. (1988). Complexities in eolian and marine interactions: processes and eustatic controls on erg developments. *Sedimentary Geology*, 56, 283–300.

Cheel, R.J., Rust, B.R. (1986). A sequences of soft sediment deformation (dewatering) structures in Late Quaternary subaqueous outwash near Ottawa, Canada. *Sedimentary Geology*, 47, 77–93.

Christie-Blick, N., Grotzinger, J.P., Von Der Borch, C.C. (1988). Sequence stratigraphy in Proterozoic successions. *Geology*, 16, 100–4.

Chumakov, N.M. (1985). Glacial events of the past and their geological significance. *Palaeogeography, Palaeoclimatology, Palaeoecology*, 51, 319–46.

Clifton, H.E., Hunter, R.E., Phillips, R.L. (1971). Depositional structures and processes in the non barred high-energy nearshore. *Journal of Sedimentary Petrology*, 41, 651–70.

Cloetingh, S. (1986). Intraplate stresses: a new tectonic mechanism for fluctuations of relative sea-level. *Geology*, 14, 617–20.

Cloetingh, S. (1988a). Intraplate stresses: a new element in basin analysis. In: *New perspectives in basin analysis*, eds. K.L. Kleinspehn, C. Paola. Berlin: Springer-Verlag, pp. 205–30.

Cloetingh, S. (1988b). Intraplate stresses: a tectonic cause for third-order cycles in apparent sea level? In: *Sea-level change, an integrated approach*, eds. C. Wilgus *et al.*, *Society of Economic Paleontologists and Mineralogists Special Publication*, 42, 19–29.

Cross, T.A. (1988). Controls on coal distribution in transgressive–regressive cycles, Upper Cretaceous, Western Interior, USA, In: '*Sea-level change, an integrated approach*', eds. C. Wilgus *et al.* *Society of Economic Paleontologists and Mineralogists Special Publication*, 42, 371–80.

Davidson-Arnott, R.G.O., Greenwood, B. (1976). Facies relationships on a barred coast, Kouchibouguac Bay, New Brunswick, Canada. In *Beach and nearshore sedimentation*, eds. R.A. Davis, R.L. Ethington, *Society of Economic Paleontologists and Mineralogists Special Publication*, 24, 149–68.

Deynoux, M. (1980). Les formations glaciaires du Précambrien terminal et de la fin de l'Ordovicien en Afrique de l'Ouest. Deux exemples de glaciation d'inlandsis sur une plate-forme stable. Travaux des Laboratoires des Sciences de la Terre, St-Jérôme, Marseille, (B), 17, 554 pp.

Deynoux, M. (1985). Terrestrial or waterlain glacial diamictites? Three case studies from the late Precambrian and Late Ordovician glacial drift in West Africa. *Palaeogeography, Palaeoclimatology, Palaeoecology*, 51, 97–141.

Deynoux, M., Kocurek, G., Proust, J.N. (1989a). Late Proterozoic periglacial eolian deposits in the Taoudeni Basin in Western Mali, West Africa. *Sedimentology*, 36, 531–49.

Deynoux, M., Marchand, J., Proust, J.N. (1989b). Notice explicative et synthèse stratigraphique et sédimentologique des cartes géologiques au 1/200.000e. Kossanto, Kayes, Kankossa, Mali Occidental. Direction Nationale de la Géologie et des Mines de Bamako, Projet Mali Ouest, 71 pp.

Deynoux, M., Proust, J.N., Durand, J., Merino, E. (1990). Water transfer structures in the Late Protozoic periglacial eolian deposits in Western Mali (West Africa). *Sedimentary Geology*, 66, 227–42.

Deynoux, M., Proust, J.N., Simon, B. (1991). Late Proterozoic glacially controlled shelf sequences in Western Mali (West Africa). *Journal of African Earth Sciences*, 12, 181–98.

Dionne, J.C. (1985). Formes, figures et facies sédimentaires glaciels des estrans vaseux des régions froides. *Palaeogeography, Palaeoclimatology, Palaeoecology*, 51, 415–51.

Dott, R.H.Jr., Howard, J.K. (1962). Convolute lamination in non-graded sequences. *Journal of Geology*, 70, 114–21.

Duke, W.L. (1985). Hummocky cross-stratification, tropical hurricanes and intense winter storms. *Sedimentology*, 32, 167–94.

Duke, W.L., Arnott, R.W.C., Cheel, R.J. (1991). Shelf sandstones and hummocky cross-stratification: new insights on a stormy debate. *Geology*, **19**, 625–8.

Dunbar, C.O., Rodgers, J. (1957). *Principles of stratigraphy*. New York: Wiley, 365p.

Dzulinsky, S., Smith, A.J. (1963). Convolute lamination, its origin, preservation and directional significance. *Journal of Sedimentary Petrology*, **33**, 616–27.

Eschner, T.B., Kocurek, G. (1988). Origins of relief along contacts between eolian sandstones and overlying marine strata. *American Association of Petroleum Geologists Bulletin*, **72**, 932–43.

Field, M.E., Duane, D.B. (1976). Post-Pleistocene history of United States inner continental shelf: significance to origin of barrier islands. *Geological Society of America Bulletin*, **87**, 692–702.

Frazier, D.E. (1974). Depositional-episodes: their relationship to the Quaternary stratigraphy framework in the northwestern portion of the Gulf Basin. *Bureau of Economic Geology, Austin, Geology Circular*, 74–81, 28 pp.

Galloway, W.E. (1989). Genetic stratigraphic sequences in basin analysis I: architecture and genesis of flooding-surface bounded depositional units. *American Association of Petroleum Geologists Bulletin*, **73**, 125–42.

Goodwin, P.W., Anderson, E.J. (1980a). Punctuated aggradational cycles: a general hypothesis of stratigraphic accumulation. *Geological Society of America, Abstract with Program*, **12**, 436.

Goodwin, P.W., Anderson, E.J. (1980b). Application of the PAC hypothesis to limestones of the Helderberg Group. *Society Economic Paleontogists and Mineralogists, East Section Field Conference Guidebook*, pp. 1–32.

Goodwin, P.W., Anderson, E.J. (1985). Punctuated aggradational cycles: a general hypothesis of episodic stratigraphic accumulation. *Journal of Geology*, **93**, 515–31.

Goodwin, P.W., Anderson, E.J., Goodman, W.M., Saraka, L.J. (1986). Punctuated aggradational cycles: implications for stratigraphic analysis. *Paleoceanography*, **1**, 417–29.

Guillocheau, F. (1990). Stratigraphie séquentielle des bassins de plate-forme: l'exemple dévonien armoricain. *Thèse Univ. Louis Pasteur, Strasbourg*, 257 pp.

Guillocheau, F., Hoffert, M. (1988). Zonation des dépôts de tempêtes en milieu de plate-forme: le modèle des plate-formes nord-gondwanienne et armoricaine à l'Ordovicien et au Dévonien. *C.R. Académie des Sciences, Paris*, **307**, 1909–16.

Halsey, S.D. (1979). Nexus: new model of barrier island development. In *Barrier islands, from the Gulf of St Laurence to the Gulf of Mexico*, ed., S.P. Leatherman New York: Academic Press, pp. 185–210.

Harms, J.C., Southard, J.B., Walker, R.G. (1982). Structures and sequences in clastic rocks. *Society of Economic Paleontologists and Mineralogists, Short Course*, **9**, 249 pp.

Hayes, C.W. (1899). Physiology of the Chatanooga District in Tennessee, Georgia and Alabama. *U.S. Geological Survey Annual Report., Part 2*, 58 p.

Holmes, A. (1965). *Principes of physical geology*, 2nd edition, London: Thomas Nelson, 1288 pp.

Hunter, R.E. (1977). Basic types of stratification in small eolian dunes. *Sedimentology*, **24**, 361–87.

Jopling, A.V., Walker, R.G. (1968). Morphology and origin of ripple-drift cross-lamination, with examples from the Pleistocene of Massachusetts. *Journal of Sedimentary Petrology*, **38**, 971–84.

Karner, G.D. (1986). Effects of lithospheric in-plane stress on sedimentary basin stratigraphy. *Tectonics*, **5**, 573–88.

Kocurek, G. (1988). First order and superbounding surfaces in eolian sequences – bounding surfaces revisited. *Sedimentary Geology*, **56**, 193–206.

Kocurek, G., Havholm, K.G., Deynoux, M., Blakey, R.C. (1991). Amalgamated accumulations resulting from climatic and eustatic changes, Akchar Erg, Mauritania. *Sedimentology*, **38**, 751–72.

Lambeck, K., Cloetingh, S., McQueen, H. (1987). Intraplate stresses and apparent changes in sea-level: the basins of Northwestern Europe. In *Sedimentary basins and basin-forming mechanisms*, eds., C. Beaumont, A.J. Tankard, Canadian Society of Petroleum Geologists Memoir, 12, 259–268.

Larsen, V., Steel, R.J. (1978). The sedimentary history of a debris flow dominated, Devonian alluvial-fan: a study of textural inversion. *Sedimentology*, **25**, 37–59.

Leckie, D.A., Walker, R.G. (1982). Storm and tide-dominated shorelines in Cretaceous Moosebar. Lower Gates interval. Outcrop equivalents of deep basin Gas Trap in Western Canada. *American Association of Petroleum Geologists Bulletin*, **66**, 138–57.

Lindsay, J.F. (1989). Depositional controls on glacial facies associations in a basinal setting, Late Proterozoic, Amadeus Basin, Central Australia. *Palaeogeography Palaeoclimatology Palaeoecology*, **73**, 205–32.

Lowe, D.R. (1979). Sediment gravity flows; their classification and some problems of application to natural flows. In 'Geology of continental slopes', eds., L.J. Doyle, O.H. Pilkey Jr, Society of Economic Paleontologists and Mineralogists Special Publication. 27, 75–82.

Lowe, D.R. (1982). Sediment gravity flows: II. Depositional models with special reference to the deposits of high density turbidity currents. *Journal of Sedimentary Petrology*, **52**, 279–97.

Marchand, J., Bassot, J.P., Simon, B. (1987). Notice explicative de la carte photogéologique au 1/200.000è Kossanto, République du Mali. Direction Nationale de la Géologie et des Mines, Bamako, 28 pp.

Massari, F., Colella, A. (1988). Evolution and types of fan deltas systems in some major tectonic settings. In *Fan deltas, sedimentology and tectonic setting*, eds., W. Nemec, R.J. Steel, London: Blackie, pp. 103–22.

McCrory, V.L., Walker, R.G. (1986). A storm and tidally influenced prograding shoreline upper Cretaceous Milk River Formation of southern Alberta, Canada. *Sedimentology*, **33**, 47–60.

McPherson, J.G., Shanmugan, G., Moiola, R.J. (1987). Fan deltas and braid deltas: conceptual problems. In 'Fan deltas, sedimentology and tectonic settings', eds., W. Nemec, R.J. Steel, London: Blackie, p. 14–22.

Milankovitch, M. (1941). *Kanon der Erdbestrahlung und seine Anwendung auf das Eiszeitenproblem*. Acad. Roy. Serbe, 133, 633 pp.

Morner, N.A. (1976). Eustasy and geoid changes. *Journal of Geology*, **84**, 123–51.

Morner, N.A. (1979). Eustasy and geoid changes as a function of core/mantle changes. In *Earth geology, isostasy and eustasy*, ed., N.A. Morner, New York: Wiley, pp. 535–53.

Morner, N.A. (1980). Relative sea-level, tectono-eustasy, geoidal-eustasy and geodynamics during the Cretaceous. *Cretaceous Research*, **1**, 329–40.

Morner, N.A. (1984a). Eustasy, geoid changes, and multiple geophysical interaction. In *Catastrophes and earth history*, eds., W.A. Berggren, J.A. van Couvering, Princeton University Press, 395–415.

Morner, N.A. (1984b). Climatic changes on a yearly to millenial basis – an introduction. In *Climatic changes on a yearly to millennial basis*, eds., N.A. Morner, W. Karlen, Dordrecht: Reidel, pp. 1–13.

Nemec, W., Porebski, S.J., Steel, R.J. (1980). Texture and structure of resedimented conglomerates: example from Ksiaz Formation (Famenian-Tournaisian) Southwestern Poland. *Sedimentology*, **27**, 519–38.

Nemec, W., Steel, R.J. (1988). What is a fan delta and how do we recognize it? In *Fan deltas: sedimentology and tectonic settings*, eds. W. Nemec, R.J. Steel, London: Blackie pp. 3–13.

North American Commission on Stratigraphic Nomenclature (1983). American stratigraphic code. *American Association Petroleum Geologists Bulletin*, **47**, 841–75.

Nummedal, D., Swift, D.J.P. (1987). Transgressive stratigraphy at sequence-bounding unconformities: some principles derived from Holocene and Cretaceous examples. In *Sea-level fluctuation and coastal evolution*, eds., D. Nummedal, O.H. Pilkey, J.D. Howard, *Society of Economic Paleontogists and Mineralogists., Special Publication*. 41, 241–60.

Pitman, W.C. (1978). Relationship between eustasy and stratigraphic sequences of passive margins. *Geological Society of America Bulletin*, **89**, 1389–403.

Posamentier, H.W., Vail, P.R. (1988). Eustatic controls on clastic deposition, II; Sequence and systems tract models. In *Sea-level changes, an integrated approach*, eds., C.K. 'Wilgus *et al.*, Society of Economic Paleontogists and Mineralogists Special Publication, 42, 125–54.

Posamentier, H.W., Jervey, M.T., Vail, P.R. (1988). Eustatic controls on clastic deposition, I: Conceptual framework. In *Sea-level changes: an integrated approach*, eds., C.K. Wilgus *et al.* Society of Economic Paleontogists and Mineralogists Special Publications, 42, 109–24.

Postma, G. (1986). Classification for sediment gravity flow deposits, based on flow conditions during sedimentation. *Geology*, **14**, 291–4.

Postma, G., Roep, T.B. (1985). Resedimented conglomerates in the bottomsets of Gilbert gravel fan delta. *Journal of Sedimentary Petrology*, **55**, 874–85.

Powell, J.W. (1875). Exploration of the Colorado River of the West and its tributaries. Washington D.C.: Smithsonian Institution, 291 pp.

Proust, J.N. (1987). La sédimentation éolienne: synthèse bibliographique et application à un modèle naturel. Les dépôts éoliens périglaciaires du Protérozoïque terminal au Mali occidental (Afrique de l'Ouest). Mémoire Diplôme Etudes Approfondies, University of Poitiers, 101 pp.

Proust, J.N. (1990). Expression sédimentologique et modélisation des fluctuations glaciaires. Exemple des dépôts du Protérozoïque terminal au Mali occidental (Afrique de l'Ouest). Thèse University Louis Pasteur, Strasbourg, 165 pp.

Proust, J.N., Deynoux, M., Guillocheau, F. (1990a). Anatomie fonctionnelle d'une fermeture de bassin sédimentaire: Protérozoïque terminal, groupe glaciaire du Bakoye, Afrique de l'Ouest. *C. R. Académie des Sciences, Paris*, **310**, 255–61.

Proust, J.N., Deynoux, M., Guillocheau, F. (1990b). Effets conjugués de l'eustatisme et de l'isostasie sur les plates-formes stables en période glaciaire. Exemple des dépôts glaciaires du Protérozoïque supérieur de l'Afrique de l'Ouest au Mali occidental. Bulletin de la Societé de Geologique de France, VI, **4**, 673–81.

Rahmani, R.A. (1983). Facies relationships and paleoenvironments of a Late Cretaceous tide-dominated delta. Drumheller, Alberta. Canadian Society of Petroleum Geologists, Field Trip Guidebook, Mesozoic of North America, 63 pp.

Ramsbottom, W.H.C. (1979). Rates of transgression and regression in the Carboniferous of NW Europe. *Quarterly Journal of Geology*, 136, 147–153.

Rice, W.N. (1897). *Revised Text-book of geology* (by J.D. Dana). New York: American Book Co., 482 pp.

Roep, T.B., Beets, D.J., Dronkert, H., Pagnier, H. (1979). A prograding coastal sequence of wave-built structures of Messinian age, Sorbas, Almeria, Spain. *Sedimentary Geology*, **22**, 135–63.

Rossi, P. (1980). Lithostratigraphie et cartographie des formations sédimüentaires du pourtour du massif du Kaarta, Mali occidental. Précambrien terminal, Paléozoïque inférieur du Sud-Ouest du bassin de Taoudeni. Thèse Univ. d'Aix-Marseille, 274 pp.

Rossi, P., Deynoux, M., Simon, B. (1984). Les formations glaciaires du Précambrien terminal et leur contexte stratigraphique (formations pré et post-glaciaires et dolérites du massif du Kaarta) dans le bassin de Taoudeni au Mali occidental (Afrique de l'Ouest). Sciences Géologiques, Strasbourg, **37**, 91–106.

Ryer, T.A. (1984). Transgressive–regressive cycles and the occurrence of coal in some Upper Cretaceous strata of Utah, USA. In *Sedimentology of coal and coal-bearing sequences*, eds., R.A. Rahmani, R.M. Flores, Oxford: Blackwell, pp. 217–27.

Schwartz, M.L. (1973). *Strondsburg P.A.* New York: Dowden, Hutchinson and Ross, 451 pp.

Short, A.D. (1984). Beach and nearshore facies. Southeast Australia. *Marine Geology*, **60**, 261–82.

Short, A.D. (1986). Sandy shore facies in Southeast Australia. A general sequence of sediments, bedforms and structures. In *Recent sediments in eastern Australia. Marine through terrestrial*, eds., E. Frankel *et al.* Geological Society of Australia, **2**, 87–102.

Simon, B. (1979). Essai de synthèse sur les formations sédimentaires de la partie occidentale du Mali. Rapp. Ined., Laboratoire de Géologie Dynamique, Université d'Aix-Marseille, France, 133 pp.

Simon, B., Deynoux, M., Keita, N., Marchand, J., Rossi, P., Trompette, R. (1979). Le Précambrien supérieur et la base du Paléozoïque de la partie sud-ouest du bassin de Taoudeni. Essai de synthèse. 10e Colloque de Géologie Africaine, Montpellier, France, 25–27 avril, Résumés, pp. 108–9.

Swift, D.J.P. (1975). Barrier island genesis: evidence from the Middle Atlantic shelf of North America. *Sedimentary Geology*, **14**, 1–43.

Talbot, M.R. (1985). Major bounding surfaces in aeolian sandstones- a climatic model. *Sedimentology*, **32**, 257–65.

Ten Haaf, E. (1956). The significance of convolute lamination. *Geol. en Mijnbounw*, **18**, 188–94.

Vail, P.R., Mitchum, R.H.Jr., Todd, R.G., Widmier, J.M., Thompson, S., Sangree, J.B., Bubb, J.N., Hatlelid, W.G. (1977). Seismic stratigraphy and global changes of a sea level. In *Seismic stratigraphy: applications to hydrocarbon exploration*, ed., C.E. Payton, American Association Petroleum Geologists Memoir., 26, 49–212.

Vail, P.R., Eisner, P.N. (1989). Stratigraphic signatures separating tectonic, eustatic and sedimentologic effects on sedimentary sections. 2è Congrè. Français de Sedimentologie, 23–24 Nov., Lyon, Résumés, pp. 62–64.

Vail, P.R., Audemard, F., Bowman, S.A., Eisner, P.N., Perez-Cruz, G. (1991). The stratigraphic signatures of tectonics, eustasy and sedimentation. An overview. In *Cycles and evens in stratigraphy*, eds G. Einsele, W. Ricken, A. Seilacher, Berlin: Springer-Verlag, 617–59.

Van Wagoner, J.C. (1985). Reservoir facies distribution as controlled by sea-level change. Society of Economic Paleontologists and Mineralogists, mid-year meeting, Golden County, Abstract pp. 91–92.

Van Wagoner, J.C., Posamentatier, H.W., Mitchum, R.M., Vail, P.R., Sarg, J.F., Loutit, T.S., Hardenbol, J. (1988). An overview of the fundamentals of sequence stratigraphy and key definitions. In *Sea level changes: an interpreted approach*, eds., C.T. Wilgus *et al.* Society Economic Paleontology Minerologists Special Publication, 42, 39–45.

Villeneuve, M. (1988). Evolution géologique comparée du bassin de Taoudeni et de la chaîne des Mauritanides en Afrique de l'Ouest. *C. R. Académie des Sciences, Paris*, **307**, 663–8.

Villeneuve, M. (1989). The geology of the Madina-Kouta Basin (Guinea-Senegal) and its significance for the geodynamic evolution of the western part of the West African Craton during the Upper Proterozoic period. *Precambian Research*, **44**, 305–22.

Walker, R.G. (1990). Facies modeling and sequence stratigraphy. *Journal of Sedimentary Petrology*, **60**, 777–86.

Wanless, H.R. (1976). Intracoastal sedimentation. In *Marine sediment transport and environmental management*, eds., D.J. Stanley, D.J.P. Swift, New York: Wiley, 221–40.

Wanless, H.R., Weller, J.M. (1932). Correlation and extent of Pennsylvanian cyclothems. *Geological Society of America Bulletin*, **43**, 1003–16.

Wescott, W.A., Ethridge, F.G. (1980). Fan delta sedimentology and tectonic setting. Yallahs fan delta, Southeast Jamaica. American Association of Petroleum Geologists Bulletin, **64**, 374–399.

Wheeler, H.E. (1957). Baselevel control patterns in cyclothemic sedimentation. *American Association of Petroleum Geologists Bulletin*, **41**, 1985–2011.

Wheeler, H.E. (1958). Time-stratigraphy. *American Association of Petroleum Geologists Bulletin.*, **42**, 1057–63.

Wheeler, H.E. (1959). Stratigraphic units in space and time. *American Journal of Science*, **257**, 692–706.

Wheeler, H.E. (1964). Baselevel, lithosphere surface and time stratigraphy. Geological Society of America Bulletin, **75**, 599–610.

Wheeler, H.E. (1966). Baselevel transit cycles. In *Symposium on cyclic sedimentation*, ed. D.F. Merriam, Kansas Geol. Surv. Bull., **169**, 623–630.

Young, G.M. (1988). Proterozoic plate tectonics, glaciation and iron formations. *Sedimentary Geology*, **58**, 127–44.

11 The enigmatic Late Proterozoic glacial climate: an Australian perspective

GEORGES E. WILLIAMS

Abstract

Late Proterozoic glaciation between 800 and 600 Ma represents one of the most puzzling climatic events in Earth history. Over the past three decades, increasing evidence has emerged in several continents for the occurrence of Late Proterozoic ice-sheets near sea-level in low palaeolatitudes. Although some early palaeomagnetic data are equivocal or contentious, palaeomagnetic studies since 1980 consistently have indicated glacial deposition in low palaeolatitudes (0–12°). Positive fold tests on soft-sediment slump folds in fine-grained sandstone from the Marinoan glacial succession in South Australia confirm the primary nature of the stable remanence and low palaeomagnetic inclination ($< 10°$) of these rocks. Coeval glaciation in high palaeolatitudes has not been demonstrated; indeed, the North China block occupied high palaeolatitudes (57–62°) in Late Proterozoic time but affords no evidence of glaciation. Structures interpreted as periglacial sand-wedges of seasonal contraction–expansion origin, clearly displayed with other periglacial features in a fossil permafrost horizon in South Australia, imply mean annual air temperatures as low as -12 to -20 °C or lower in coastal terrain near sea-level and strongly seasonal climates (seasonal temperature *range* as great as ~ 40 °C or more). Time-series analysis of tidal–climatic data from coeval rocks in South Australia provides independent evidence of a powerful annual signal. Furthermore, grounded ice-sheets and glacial pavements formed near sea-level in Australia and other continents during the Late Proterozoic. Palaeomagnetic and palaeoclimatic data, provided most clearly by recent studies of Late Proterozoic rocks in Australia, thus present the enigma of *frigid, strongly seasonal climates, with permafrost and grounded ice-sheets near sea-level, in low palaeolatitudes*.

The evidence for strongly seasonal climates in low palaeolatitudes and the existence of at least one unglaciated region in high palaeolatitudes during the Late Proterozoic together militate against the idea of global (pole-to-pole) refrigeration and glaciation. Possible explanations of glaciation and strongly seasonal periglacial climates in preferred *low* palaeolatitudes include: (1) the geocentric axial dipole model for the Earth's magnetic field is invalid for that time interval; and (2) a reversed climatic zonation and marked seasonality prevailed because of a large obliquity of the ecliptic ($> 54°$) in Late Proterozoic time. Global palaeomagnetic data and the wide distribution of Late Proterozoic glaciogenic rocks argue against a non-axial, non-dipole geomagnetic field during the Proterozoic, whereas Late Proterozoic palaeotidal data from South Australia are consistent with both a low palaeolatitude and a substantial obliquity of the ecliptic.

Detailed studies of Late Proterozoic strata world-wide are required to further test these ideas and provide a sharper picture of the Late Proterozoic global environment.

Introduction

The paradox of widespread glaciation in apparent low palaeolatitudes during the Late Proterozoic has been the subject of much debate among geologists and palaeomagnetists over the past three decades. During those years, increasing evidence for an enigmatic Late Proterozoic glacial environment – frigid, strongly seasonal climates near sea-level apparently in preferred low to equatorial palaeolatitudes – has emerged as new palaeoclimatic and geophysical data have been obtained. The Adelaide Geosyncline–Stuart Shelf region in South Australia (Fig. 11.1), which contains some of the best preserved and most accessible Late Proterozoic rocks in the world, recently has provided particularly valuable palaeoclimatic, palaeomagnetic and palaeogeophysical information. Here I review recent data, with emphasis on the Australian observations, and discuss four hypotheses – global glaciation, equatorial ice-ring system, non-axial geomagnetic field, and large obliquity of the ecliptic – that have been advanced to explain the enigmatic Late Proterozoic glacial climate. The hypothesis of a former large obliquity, which has numerous implications, may best account for the complex character of the Late Proterozoic environment. Further detailed studies of Late Proterozoic rocks, as exemplified by recent work in South Australia, are needed to further test these hypotheses and provide a better understanding of the global environment during the Late Proterozoic.

Late Proterozoic glacial and periglacial climate

Late Proterozoic glaciation, which affected all continents with the possible exception of Antarctica between about 800 and

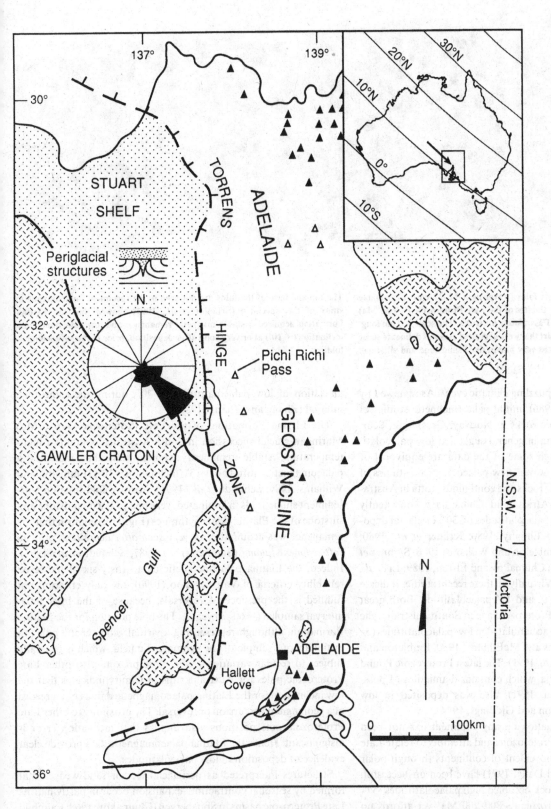

Fig. 11.1. Locality map of the Adelaide Geosyncline region, South Australia, showing the distribution of glacial and periglacial facies of the Late Proterozoic (~650 Ma) Marinoan Glaciation (distribution of glacial facies in the Geosyncline adapted from Coats, 1981). Solid triangles represent tillitic facies; open triangles, dropstone facies; stipple, stable shelves; cross-stitch, cratonic regions. The palaeowind rose diagram (30° class interval, radius of circle = 10 observations) represents 27 directions of maximum dip of foresets in very large-scale cross-bed sets, or mean directions of foreset dip for several cross-bed sets at the same locality, in the periglacial-aeolian Whyalla Sandstone on the southern Stuart Shelf (Fig. 11.5). The inset shows palaeolatitudes for Australia during Late Proterozoic glaciation (adapted from McWilliams and McElhinny, 1980); the arrow indicates the dominant palaeowind direction for the Whyalla Sandstone.

Fig. 11.2. Cuspate soft-sediment folds (wavelength 14–22 cm) of probable gravity-slump origin in tidal rhythmites of the Late Proterozoic (~650 Ma) Elatina Formation, Pichi Richi Pass, South Australia. The scale is 3 cm long. Dark mud drapes about 1 cm apart that were deposited at neaps delineate some 28 fortnightly cycles of laminated very fine-grained sandstone and siltstone. The top and base of the folds are not seen in the specimen. Palaeomagnetic study of this specimen (Schmidt et al., 1991) showed that the Elatina Formation acquired a stable magnetic remanence and low palaeomagnetic inclination (< 10°) at or very soon after deposition, prior to the soft-sediment folding.

600 Ma, represents a most puzzling climatic event. As reviewed by Embleton and Williams (1986), initial palaeomagnetic studies of Late Proterozoic glaciogenic rocks in Norway, Greenland, Scotland and north-west Canada in general suggested low palaeolatitudes of glaciation, although some of the data are equivocal or contentious. Since 1980, however, new palaeomagnetic studies of Late Proterozoic glaciogenic rocks and contiguous strata in Australia,[1] South Africa, West Africa, and China have consistently indicated low to equatorial palaeolatitudes ($\lesssim 30°$) of glacial deposition (McWilliams and McElhinny, 1980; Kröner et al., 1980; Zhang and Zhang, 1985; Embleton and Williams, 1986; Sumner et al., 1987; Perrin et al., 1988; Chumakov and Elston, 1989; Li et al., 1991; Schmidt et al., 1991). Virtually all these recent studies indicate glacial deposition between 0° and 12° palaeolatitude. Both great glacial successions of Late Proterozoic age in South Australia, the Sturtian and the Marinoan, accumulated in low palaeolatitudes (5–15°, mean $\approx 12°$; McWilliams and McElhinny, 1980; Embleton and Williams, 1986; Schmidt et al., 1991). The latest Proterozoic Pound Subgroup in South Australia, which contains diamictites of possible glacial origin (DiBona, 1991), also was deposited in low palaeolatitudes (9°; Embleton and Giddings, 1974).

High-palaeolatitude glaciation on any continent has not been demonstrated for the Late Proterozoic, and attempts to relate Late Proterozoic glaciation to movement of continents through polar regions (e.g., Crawford and Daily, 1971) have been unsuccessful. Indeed, the North China block occupied high palaeolatitudes (57–62°) during the Late Proterozoic (~800–700 Ma) yet affords no evidence of glaciation, whereas the Yangzi block experienced glaciation in low palaeolatitudes (0–20°) during the same time interval (Zhang and Zhang, 1985).

The Elatina Formation, of the Late Proterozoic (~650 Ma) Marinoan glacial succession in South Australia, contains a high-temperature, stable remanent magnetisation that indicates a palaeomagnetic latitude of ~5° (inclination < 10°) (Embleton and Williams, 1986; Schmidt et al., 1991). Positive fold tests on soft-sediment slump folds in laminated, very fine-grained sandstone and siltstone of the Elatina tidal rhythmites (Fig. 11.2) confirm that the remanence was acquired *at or very soon after deposition, prior to soft-sediment folding* (Sumner et al., 1987; Schmidt et al., 1991). Indeed, the Elatina palaeomagnetic data satisfy six of the seven reliability criteria of Van der Voo (1990); the only criterion not fulfilled is the presence of reversals, because of the brief time-interval sampled (~60–70 years). The pole position for the Elatina Formation, although representing a virtual geomagnetic pole (an instantaneous sample of the geomagnetic field, which is generally subject to secular variation), is in close proximity to other Late Proterozoic poles for Australia; this proximity indicates that the low inclination for the Elatina palaeomagnetic data is not a record of a geomagnetic excursion or reversal. The proximity to other Late Proterozoic pole positions also implies that inclination error is insignificant. Hence the Elatina palaeomagnetic data provide clear evidence of deposition in low palaeolatitudes.

Structures interpreted as periglacial ice- or sand-wedges that formed by seasonal contraction–expansion occur in partly marine Late Proterozoic basins in Spitsbergen (Chumakov, 1968; Fairchild and Hambrey, 1984), Scotland (Spencer, 1971, 1985), northern and southern Norway (Edwards, 1975; Nystuen, 1976), Mauritania (Deynoux, 1982), East Greenland (Spencer, 1985) and South Australia (Williams and Tonkin, 1985; Williams, 1986). The origin of sand-wedges in the Port Askaig Formation in Scotland is

---

[1] Late Proterozoic glaciation in Australia occurred between 0° and 30° palaeolatitude (McWilliams and McElhinny, 1980), *not* the 45–60° palaeolatitude incorrectly shown by Chumakov and Elston (1989).

contentious. Spencer (1985) concluded that the presence of vertical bedding and vertically aligned pebbles in deep sandstone wedges, which in one case penetrates a granite conglomerate, indicate that such structures are periglacial sand-wedges. Eyles and Clark (1985), however, argued that all the wedge structures formed by soft-sediment deformation, probably triggered by seismic shocks.

Excellently preserved periglacial structures (Fig. 11.3) on the Stuart Shelf bordering the Adelaide Geosyncline (see Fig. 11.1 for locality) formed in a permafrost horizon *which developed on mid-Proterozoic quartzite bedrock near sea-level* during the Marinoan Glaciation; the Stuart Shelf subsequently was submerged by the Marinoan post-glacial marine transgression (Preiss, 1987). This horizon contains primary sand-wedges (in at least four generations) which outline polygons up to ~10–20 m across, as well as anticlines and tepees, earth mounds, periglacial involutions, frost-heaved blocks, and frost-shattered bedrock breccia (Williams and Tonkin, 1985; Williams, 1986). The sand-wedges characteristically display near-vertical lamination that parallels the wedge margins, and upturning of adjacent material (Fig. 11.3). Identical sand-wedge structures are forming in present arid polar regions such as the Arctic and the dry valleys of Antarctica (e.g., Péwé, 1959; Washburn, 1980; Karte, 1983); the permafrost contracts and cracks during severe winters and the cracks fill with windblown sand, and lateral pressure during summer expansion causes upturning of permafrost adjacent to wedges.

Periglacial features can provide quantitative information on palaeoclimate, such as indications of mean annual air temperature, seasonal temperature range, and mean annual precipitation (Washburn, 1980; Karte, 1983), as well as oscillations of climate on a 10^3-year time-scale. Indeed, periglacial structures probably are the most reliable of palaeoclimatic indicators, because they formed through processes of *physical* weathering and their interpretation thus avoids uncertainties such as the former nature of the atmosphere and biosphere. Ice-wedges and sand-wedges in particular indicate a strongly seasonal palaeoclimate, because only the seasonal temperature cycle, together with rapid drops of temperature, can produce the required contraction cracking. The occurrence of Late Proterozoic *primary sand-wedges*, best exemplified by those on the Stuart Shelf in South Australia (Fig. 11.3), yields important palaeoclimatic information (see Karte, 1983; Williams; 1986; Williams and Tonkin, 1985) for low-lying coastal terrain in low palaeolatitudes:

1. Mean annual air temperatures near sea level were as low as − 12 °C to − 20 °C or lower.
2. Strongly seasonal palaeoclimates prevailed, with mean monthly temperature *ranges* as great as ~40 °C or more (mean monthly temperatures ranging from < − 35 °C in midwinter to < + 4 °C in midsummer).
3. Palaeoclimates were arid (< 100 mm mean annual precipitation) and windy, causing winter contraction-cracks to fill with windblown sand.
4. Oscillations of climate occurred on a 10^3-year time-scale. A mean annual rate of horizontal growth of ~1 mm per year determined for certain modern sand-wedges and ice-wedges (Black, 1973; Washburn, 1980) suggests that the

sand-wedges on the Stuart Shelf in South Australia took up to ~4 ka to form (Williams and Tonkin, 1985). The development of at least four generations of sand-wedges on the Stuart Shelf, alternating with thawing of the permafrost as indicated by distortion of early-formed wedges and deposition of aeolian sand sheets (Fig. 11.3a), therefore suggests climatic cycles of several thousand years duration involving fluctuations as large as ~20 °C or more in mean annual air temperature (from as low as − 20 °C to > 0 °C) in low palaeolatitudes. Such temperature fluctuations greatly exceed the changes of only 1–2 °C in mean surface temperature experienced in low latitudes between the last glacial maximum at 18 ka and the present day (Crowley and North, 1991).

Late Proterozoic glaciogenic sediments provide support for the palaeoclimate indicated by the periglacial structures:

1. A powerful annual signal is revealed by fast Fourier transform of tidal-climatic data obtained from the Elatina rhythmites in the Marinoan glacial succession, South Australia (see next section, and Williams, 1989a,b, 1990, 1991). The presence of varve-like laminites with outsize stones in Late Proterozoic glaciogenic deposits (e.g. Spencer, 1971) also may indicate a strongly seasonal glacial climate. However, many Late Proterozoic glaciogenic sequences are regarded as marine (e.g., Eyles and Clark, 1985; Preiss, 1987) and hence well-developed varves may not normally be expected. Moreover, it can be difficult to demonstrate whether varve-like laminae in ancient sediments represent annual signals or some other frequency, or are random.
2. Arid, windy, Late Proterozoic glacial and periglacial climates are confirmed by the occurrence of extensive periglacial aeolian sandstones in South Australia (Williams and Tonkin, 1985; Preiss, 1987) and northwest Africa (Deynoux *et al.*, 1989), and glacial loessites in North Norway and Svalbard (Edwards, 1979).
3. Relatively abrupt changes of climate are suggested by the stratigraphic proximity of cold- and warm-climate indicators in several Late Proterozoic sequences. The occurrence of micritic, apparently primary dolostones capping Late Proterozoic glaciogenic sequences in Australian basins and immediately underlying such sequences in Spitsbergen is consistent with relatively rapid changes of mean temperature (Williams, 1979; Fairchild and Hambrey, 1984). However, *redeposited* carbonates *interbedded* with Late Proterozoic tillites in Spitsbergen may not be of immediate palaeoclimatic significance (Fairchild *et al.*, 1989). Spencer (1971, 1985) argued for 17 main advances and meltings of the Dalradian ice sheet in Scotland, although an alternative interpretation of the Scottish rocks discussed above (Eyles and Clark, 1985) does not accept the evidence for such numerous glacial and interglacial cycles.

Fig. 11.3. Vertical exposures of Late Proterozoic (~650 Ma) periglacial sand-wedges developed in brecciated bedrock of mid-Proterozoic (~1400 Ma) quartzite and the overlying Late Proterozoic periglacial-aeolian Whyalla Sandstone, Cattle Grid copper mine, Stuart Shelf, South Australia (see Fig. 11.1 for locality). The wedge structures indicate a very cold, strongly seasonal, arid and windy climate. (a) Large sand-wedge 3 m deep (2) truncated by overlying flat-bedded Whyalla Sandstone. A near-vertical diverging lamination that parallels the wedge margins is discernible within the wedge. Distorted sand-wedges of an earlier generation (1) occur within the bedrock breccia. A third-generation wedge (3) occurs within the Whyalla Sandstone and the uppermost part of the large wedge. Bedding in breccia and sandstone is turned upward adjacent to the wedges.

Glacial pavements have been reported below Late Proterozoic tillites in numerous localities, including Western Australia, northwest Africa, southern Africa, Scandinavia, Spitsbergen, Normandy, South America and China (see summaries in Williams, 1975; Edwards, 1978, p. 433; Hambrey and Harland, 1981; also Guan Baode et al., 1986). The cast of a striated surface occurs in South Australia (Preiss, 1987). Excellently preserved Late Proterozoic glacial pavements in the Kimberley region, Western Australia (Dow and Gemuts, 1969; Plumb, 1981), display a variety of features including polished and striated surfaces (Fig. 11.4), chattermarks, crescentic gouges, roches moutonnées and whaleback forms. The pavements in Western Australia indicate the occurrence of grounded ice-sheets close to the edge of marine basins, with transgression accompanying glacial retreat (Plumb, 1981).

Palaeowind data for the Late Proterozoic (Marinoan) periglacial-aeolian Whyalla Sandstone on the Stuart Shelf, South Australia, are summarised in Fig. 11.1. The Whyalla Sandstone is a medium- to very coarse-grained quartzose sandstone as much as ~160 m in thickness and covering ~20 000 km² in outcrop and subcrop (Preiss, 1987). Outcrops on the southern Stuart Shelf display very large-scale cross-bed sets up to ~7 m thick (Fig. 11.5) as well as inversely graded subcritically climbing translatent strata and other sedimentary structures indicating aeolian deposition. The rose diagram in Fig. 11.1 represents 27 directions of maximum dip of foresets in very large-scale cross-bed sets, or mean directions of foreset dip for several cross-bed sets at the same locality. The interpretation of such data in terms of palaeowind directions requires an understanding of the palaeodune forms (Kocurek, 1991). The structure of cross-beds in the Whyalla Sandstone and the consistency of foreset dip directions suggest mainly two- and three-dimensional transverse dune forms. Some forests have steep dips (~30°) and local grainflow deposits. Trough-forms seen in horizontal and vertical sections grew by the migration of curved brinklines in the same general direction as that of maximum dip of adjacent foresets. Furthermore, most foreset dip directions are in the same quadrant over an area of 75 × 60 km. The rose diagram therefore should give an indication of the regional palaeowind direction. Since a relatively low-lying Archaean to mid-Proterozoic craton lay

Fig. 11.3 (*cont.*) (*b*) **Inclined contact between a sand-wedge and bedrock breccia, showing upturning of relict bedding within the breccia. Scale 15 cm long.** (*c*) **Lower portion of a sand-wedge showing relict bedding in adjacent bedrock breccia (by hammer) overturned upward and away from the wedge.** (*d*) **Steeply dipping laminae near the centre of a sand-wedge, truncated by flat-** bedded Whyalla Sandstone. Scale 15 cm long. (*e*) **Third-generation sand-wedge 1.7 m deep developed mainly in flat-bedded Whyalla Sandstone; fourth-generation sand-wedges occur at higher stratigraphic levels within the Whyalla Sandstone. (From Williams and Tonkin (1985); reproduced by permission of Blackwell Scientific Publications.)**

northwest of the Stuart Shelf, the palaeowinds may be regarded as zonal winds related to an atmospheric cell rather than local katabatic winds. Intriguingly, dominant palaeowinds on the Stuart Shelf were *westerlies* relative to contemporary low (~12° N) palaeolatitudes (inset, Fig. 11.1), whereas expected zonal winds in such palaeolatitudes would be easterlies to northeasterlies.

As shown in Table 11.1, the Late Proterozoic palaeoclimate near sea-level in low palaeolatitudes as indicated by periglacial sand-wedge structures in South Australia contrasts markedly, in regard to mean temperatures and seasonal temperature range, with the modern climate in low latitudes and with modelled low-latitude climates for Late Proterozoic and Palaeozoic arrangements of continents (assuming an obliquity the same as that of today). In present low latitudes (0 ± 10°), the mean annual air temperature near sea level is 26.3 °C and the mean monthly temperature range is only 1.8 °C (Oort, 1983). Mean air temperatures near sea level of $\lesssim -20$ °C and seasonal temperature ranges of $\gtrsim 40$ °C, which occurred in low palaeolatitudes during the Late Proterozoic, are found today only in high latitudes. Climate modelling for a

presumed Late Proterozoic arrangement of continents at moderate to low latitudes gives a mean temperature in tropical regions (<8° latitude) of +23.5 °C (Worsley and Kidder, 1991). This mean value is only a few degrees lower than mean surface temperatures in present low latitudes, but is up to 40 °C or more *higher* than mean temperatures experienced near sea level in low palaeolatitudes during Late Proterozoic glaciation (see Table 11.1). The model of Worsley and Kidder (1991) permits glaciation at $\gtrsim 46°$ latitude, whereas available palaeomagnetic data suggest that Late Protero-zoic glaciation occurred mainly at $\lesssim 12°$ palaeolatitude. Moreover, climate modelling for Gondwana during Carboniferous glaciation indicates a mean surface tropical temperature only ~4 °C lower than that of today caused by a presumed lower solar luminosity (Crowley *et al.*, 1991). Modelling also shows that Pangaea had an annual temperature range of only $\lesssim 5$ °C in low to equatorial latitudes during the Permian, whereas annual temperature ranges >40 °C were confined to middle and high southern latitudes (Crowley *et al.*, 1989). Clearly, climates modelled for Late Protero-zoic and Palaeozoic arrangements of continents, based on an

Fig. 11.4. Polished and striated glacial pavement showing two sets of striae, exhumed from beneath the Moonlight Valley Tillite near the junction of the Ord and Negri rivers, Kimberley region, Western Australia. The pavement is formed on mid-Proterozoic sandstone; hammer 30 cm long. The Moonlight Valley Tillite has been broadly correlated with the Marinoan glacial succession in South Australia (Coats and Preiss, 1980).

Fig. 11.5. Periglacial–aeolian Whyalla Sandstone deposited during the Late Proterozoic (~650 Ma) Marinoan Glaciation, South Australia. Two superimposed cross-bed sets each ~7 m thick, which indicate a palaeowind direction toward 140–150°, overlie a basal facies of flat-bedded aeolian sandstone (behind vehicle). The sequence is capped by poorly bedded aeolian sands of Quaternary age. Cattle Grid copper mine, Stuart Shelf.

Table 11.1. *The Late Proterozoic low-palaeolatitude climate near sea level indicated by periglacial sand-wedge structures in South Australia, compared with low-latitude climates modelled for Late Proterozoic and Palaeozoic arrangements of continents (assuming an obliquity of the ecliptic the same as that of today) and with the modern surface climate in low latitudes*

Age, and basis for palaeoclimatic estimates	Mean surface temperature in low latitudes ($\lesssim 8$–$10°$ lat.) (°C)	Seasonal temperature range in low latitudes ($\lesssim 8$–$10°$ lat.) (°C)	Reference
Late Proterozoic (periglacial structures)	$\lesssim -12$ to $\lesssim -20$	$\gtrsim 40$	Karte (1983), Williams (1986)
Late Proterozoic (modelled)	$+23.5$		Worsley & Kidder (1991)
Carboniferous Gondwana (modelled)	$+22$		Crowley *et al.* (1991)
Late Permian Pangaea (modelled)	$+25$ to $+30$	$\lesssim 5$	Crowley *et al.* (1989)
Modern	$+26.3$	1.8	Oort (1983)

obliquity the same as that of today, do not duplicate the enigmatic Late Proterozoic glacial climate.

In summary, available palaeomagnetic and palaeoclimatic data for Late Proterozoic glaciogenic rocks thus present the enigma of *frigid, strongly seasonal climates, with permafrost and grounded ice-sheets near sea level, apparently in preferred low to equatorial palaeolatitudes.* Strong seasonality is an important feature of the Late Proterozoic glacial climate in South Australia, and has global implications. The inferred palaeoclimate is all the more enigmatic because the faster rotation of the Earth in the past would have reduced poleward transport of heat, causing warmer low latitudes and colder high latitudes (see Hunt, 1979). The Late Proterozoic glacial climate is one of the major paradoxes in contemporary Earth science, challenging conventional views on the nature of the geomagnetic field, climatic zonation, and the Earth's planetary dynamics in Late Proterozoic time.

Possible explanations of low-palaeolatitude glaciation

Four hypotheses have been advanced to explain the low to equatorial palaeolatitudes of glaciation and *in situ* frigid climate near sea-level in Late Proterozoic time:

1. Glaciation extended over all latitudes during global refrigeration and very severe ice ages (Harland 1964*a,b*).
2. The Late Proterozoic Earth possessed an equatorial ice-ring system that shielded low latitudes from solar radiation and caused low-latitude glaciation (Sheldon, 1984).
3. The Earth's magnetic field did not conform to a geocentric axial dipole during the Late Proterozoic (Embleton and Williams, 1986).
4. A reversed climatic zonation and marked global seasonality prevailed because of a large obliquity of the ecliptic ($\epsilon > 54°$) (Williams, 1975; Embleton and Williams, 1986).

The presentation and testing of these hypotheses has helped stimulate research on Late Proterozoic glaciogenic sequences, including potentially important lines of study that might otherwise have been overlooked – certainly this is the case in Australia. The four hypotheses are critically reviewed here.

Global refrigeration

Since low palaeolatitudes experienced mean annual air temperatures as low as -20 °C or lower in coastal terrain near sea level, as much as 45 °C or more lower than the mean annual air temperature near sea level in present low latitudes, an average global surface temperature less than -1.9 °C (the freezing point of seawater; Frakes, 1979) and global freezing during the Late Proterozoic are implied by the hypothesis of glaciation over all latitudes. However, the North China block occupied high palaeolatitudes during the Late Proterozoic yet apparently was not glaciated, while the Yangzi block experienced glaciation in low palaeolatitudes (Zhang and Zhang, 1985). The Chinese data appear inconsistent with the concept of global glaciation; observations for other high-palaeolatitude terrains clearly are required.

As noted by Crowell (1983, p. 254), significantly greater iciness of the Earth accompanying world-wide glaciation would have caused 'drastically lowered' sea level. However, detailed palaeogeographic reconstructions of Late Proterozoic South Australia (Preiss, 1987) present no evidence of drastic lowering of sea-level during the Sturtian and Marinoan glaciations.

Climate modelling identifies additional difficulties with the concept of global refrigeration. Importantly, Sellers (1990) found *there is little seasonal variation with global refrigeration* because the very low temperatures inhibit precipitation and sublimation. The very large seasonal temperature-range (as much as ~ 40 °C or more) in low palaeolatitudes indicated by Late Proterozoic periglacial sand-wedge structures, best exemplified in South Australia, therefore argues strongly against global refrigeration.

Furthermore, Sellers (1990) also found that even with 30% reduction in solar luminosity, more than half of the present Earth's land area between ±20° latitude would remain snow-free; this finding presumably depends on factors such as distribution of continents and atmospheric composition. Indeed, a frozen-over Earth may be very difficult to unfreeze; according to North *et al.* (1981), solar radiation is so efficiently reflected by ice and snow that solar luminosity must increase by ~35% above its present value to remove the ice and snow covering a frozen-over Earth. Such an unlikely rapid increase in solar luminosity would exceed the total increase in luminosity during the Sun's 4700-Ma history (see Gough, 1981; Endal and Schatten, 1982).

The survival of long-evolved organisms, including metazoans (Runnegar, 1991) and other shallow-water biota, through Late Proterozoic time may present another objection to the concept of global refrigeration. All known organisms require liquid water during at least some stage of their life cycles (Kasting, 1989), a requirement that would not be met in paralic settings and all other surface waters during the implied global freezing.

For numerous reasons, therefore, the hypothesis of global refrigeration and potential glaciation at all latitudes during the Late Proterozoic is difficult to support on available evidence. Use of the term 'Cryogenian', implying 'global glaciation', for the interval 850–650 Ma (Harland *et al.*, 1990, p. 17) appears premature.

Equatorial ice-ring system

The hypothesis of an equatorial ice-ring system also encounters major difficulties. Ice rings almost certainly could not exist so close to the Sun, even for a younger Sun of slightly lower luminosity. Furthermore, mean annual insolation at the top of the atmosphere of Saturn, a planet with a well-developed equatorial ring system and an obliquity (26.7°) similar to that of the Earth, is maximum at the equator and minimum at the poles (Fig. 11.6; Brinkman and McGregor, 1979). Saturn's equator-to-pole insolation gradient is modified by the ring shadows for latitudes less than ~45°, but is not inverted. Hence, *high latitudes* of an Earth with an equatorial ring system, not low latitudes, would be glaciated preferentially.

Geomagnetic field non-axial

The validity of the geocentric axial dipole model for the Earth's magnetic field in the geological past can be tested by comparing palaeomagnetic data with independent indicators of past latitude. The distribution of palaeoclimate indicators such as glacial deposits, evaporites and coral reefs with respect to palaeolatitudes during the Phanerozoic (McElhinny, 1973; Frakes, 1979; Merrill and McElhinny, 1983) and the frequency distribution of palaeomagnetic inclination angles (Evans, 1976) together strongly support the geocentric axial dipole model for the entire Phanerozoic, and hence the model remains fundamental to palaeomagnetic interpretation. Embleton and Williams (1986) pointed out that any deviation from the geocentric axial dipole model during the Late Proterozoic could not have been short-lived, because of the grouping of Australian poles for the quasi-static interval from Late Proterozoic (~800 Ma) to Middle Cambrian times (see also

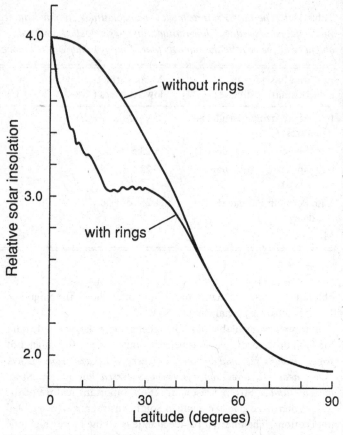

Fig. 11.6. Latitudinal variation of relative mean annual insolation at the top of the atmosphere of Saturn (adapted from Brinkman and McGregor, 1979). The equator-to-pole insolation gradient is modified by the shadows of Saturn's rings for latitudes less than ~45°, but is not inverted.

Schmidt *et al.*, 1991). Invalidation of the geocentric axial dipole model for such an extended interval would raise strong doubts concerning other Precambrian palaeomagnetic data unsupported by independent evidence of past latitude.

The Earth, Jupiter and Saturn exhibit only small dipole tilts with respect to spin axes (11.4°, 9.6° and 0°, respectively; Rädler and Ness, 1990) as well as relatively small quadrupole contributions (<10% of the dipole; Ness *et al.*, 1989). The Voyager 2 spacecraft discovered that the magnetic fields of both Uranus and Neptune are, by contrast, tilted at surprisingly large angles (58.6° and 46.9°, respectively) to the spin axes, and quadrupole moments are comparable to dipole (Connerney *et al.*, 1987, 1991; Ness *et al.*, 1989). The magnetic fields of *two* planets being of such configuration virtually rules out the possibility that they are undergoing transient excursions or reversals, and indicates that non-axial dipole-quadrupole planetary magnetic fields are indeed possible. Rädler and Ness (1990) concluded, however, that simply reorienting an Earthlike or Jupiterlike magnetic field cannot produce the Uranian magnetic field. The magnetic fields of Uranus and Neptune may result from the unique interior composition and state of those planets (Connerney *et al.*, 1987, 1991; Ness *et al.*, 1989).

Idnurm and Giddings (1988) tentatively suggested that the near-

Table 11.2. *Relation between the obliquity of the ecliptic and global climate (modified from Williams, 1975)*

Obliquity (°)	Seasonality	Annual isolation, either pole: equator[a]	Climatic zonation	Preferred latitudes of glaciation (°)
23.5	Moderate	0.4247	'Normal', strong	High ($\gtrsim 50$)
54	Strong	1.0	nil	
90	Very strong	1.5708	Reversed, moderate	low to equatorial ($\lesssim 30$)

Note:
[a] Values from Croll (1875) and Milankovitch (1930).

linear plot of accumulated polar wander angle versus time for 0–3500 Ma, and the low scatter of data points, may indicate that the geocentric axial dipole model is a reasonably good approximation for that long time interval.[2] Furthermore, the tendency for local palaeomagnetic intensity to increase with distance from the palaeomagnetic equator suggests that the palaeomagnetic field may be approximated by a geocentric dipole model as far back as 2500 Ma (Schwartz and Symons, 1969). The distribution of palaeomagnetic inclination angles for the Precambrian also accords with a geocentric dipole model for the interval 600–3000 Ma (Piper and Grant, 1989). Embleton and Williams (1986) noted, moreover, that the hypothesis of a *non-axial* dipole field does not predict the widespread occurrence of Late Proterozoic glaciogenic rocks.

Hence, global palaeomagnetic data and the wide distribution of Late Proterozoic glaciogenic rocks together argue against a non-axial, non-dipole geomagnetic field during the Proterozoic.

Obliquity of the ecliptic > 54°

Fourthly, the paradox of glaciation apparently in preferred low to equatorial palaeolatitudes during the Late Proterozoic may indicate a large obliquity of the ecliptic ($\epsilon > 54°$), assuming the Late Proterozoic geomagnetic field was a geocentric axial dipole (Williams, 1975).

The obliquity of the ecliptic – the Earth's axial tilt of 23.5° or the angle between the equatorial plane and the plane of the Earth's orbit (the ecliptic plane) – exerts a fundamental influence on terrestrial climate. The obliquity controls the seasonal cycle and strongly influences tidal rhythms. In addition, fluctuations of obliquity between 21.5° and 24.5° over a period of 41 ka constitute an important orbital element which, together with precession and variations in eccentricity, drives medium-term (~10–100 ka) climatic cycles (Berger *et al.*, 1984).

As discussed by Williams (1975), the global climate for a large obliquity would be very different from that of today (see Table 11.2):

1. The amplitude of the global seasonal cycle would be much greater than that experienced now. With an obliquity of 60°, for example,

the tropics would be at 60° latitude and the polar circles at 30° latitude. Areas between 30° and 60° latitude would lie within both the tropics and the polar circles! All areas poleward of 30° latitude would endure greatly contrasting seasons, with dark winters of deep cold and torrid summers under a continually circling Sun. Seasonal temperature contrasts would be most extreme in high latitudes, and large seasonal oscillations of temperature would extend into low latitudes.

Furthermore, a monotonic temperature gradient directed from the summer to the winter pole should exist for a large obliquity (Hunt, 1982). Substantial atmospheric circulation across the equator therefore would occur around solstices, when very cold air from the anticyclonic province in the winter hemisphere would flow toward the deep thermal depression in the summer hemisphere. At equinoxes the global climate would display a 'normal' day–night cycle. Hence, low and equatorial latitudes would experience low temperatures and very cold winds around solstices, alternating with more benign equinoctial conditions.

Strong seasonality in low palaeolatitudes is indeed a feature of the Late Proterozoic glacial climate, as demonstrated in South Australia. In addition, the occurrence of extensive periglacial aeolian sandstones and loessites of Late Proterozoic age is consistent with the prediction of frigid solstitial winds.

2. The ratio of radiation received annually at either pole to that received at the equator would be larger than at present (Fig. 11.7). For $\epsilon = 54°$, all latitudes would receive equal solar radiation annually, and the climatic zones would disappear. For $\epsilon > 54°$, low to equatorial latitudes ($\lesssim 30°$) would receive less radiation annually than high latitudes. Fig. 11.7*b* shows that minimum insolation occurs at ±30° latitude for $\epsilon = 60°$, and at the equator for $\epsilon > 60°$. Williams (1975) obtained the critical obliquity of 54°, at which value the climatic zonation reverses, from Milankovitch (1930). This critical value of 54° has been confirmed by more recent studies (Ward, 1974; Toon *et al.*, 1980).

If an Earth with $\epsilon > 54°$ were to enter an ice age through some independent cause such as decline in atmospheric CO_2 partial pressures or moderate decrease in solar luminosity – a large obliquity *per se* is not a cause of glaciation – *low to equatorial latitudes* ($\lesssim 30°$) would be glaciated preferentially. In high latitudes the cold, arid winter atmosphere would allow only limited snowfall which would melt entirely during the very hot summer. Permanent

[2] This interpretation makes the arguable assumption, however, that Precambrian polar wander paths as presently known are substantially complete (P.W. Schmidt, personal communication, 1991).

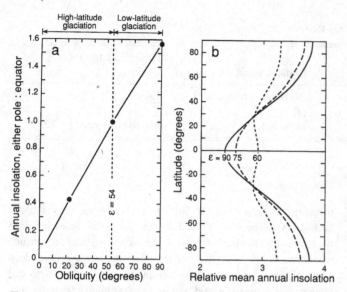

Fig. 11.7. (a) Relation between the obliquity of the ecliptic ε and the ratio of annual insolation at either pole to that at the equator (solid line); the dashed line at ε = 54° separates the fields of potential low-latitude and high-latitude glaciation. Adapted from Williams (1975). (b) Latitudinal variation of relative mean annual insolation of a planet for various values of the obliquity ε, in degrees; solid line for ε = 90°, long-dashed line for ε = 75°, short-dashed line for ε = 60°. Adapted from Van Hemelrijck (1982). The plots in (a) and (b) together illustrate that for ε > 54°, glaciation would occur preferentially in low to equatorial latitudes (≲ 30°).

ice could, however, form in low to equatorial latitudes; this zone, as well as receiving minimum solar radiation annually, would experience the additional cooling effect of frigid solstitial winds and would not be subjected to extreme summer temperatures. The increased albedo due to a snow cover in low to equatorial latitudes would further reduce effective insolation and allow the accumulation of permanent ice. Glaciation for a large obliquity therefore can account for the apparent preferred low palaeolatitude of Late Proterozoic ice sheets and cold climate as suggested by available palaeomagnetic data.

The principle of potential glaciation in low latitudes for a large obliquity is applicable to other planets. Ward (1974) observed that if Mars had an obliquity similar to that of Uranus ($\sim 98°$), the Martian deposits of permanent CO_2 ice would be at the equator and not at the poles. Furthermore, Jakosky and Carr (1985) suggested that when the obliquity of Mars periodically was as high as 45° early in that planet's history, ice could have accumulated in low latitudes by sublimation of ice from the polar caps and transport of the water vapour equatorward; they concluded that polar ice may have formed only for minimum obliquities.

As noted by Williams (1975), the area of a low-latitude zone symmetrical about the equator is $4\pi R^2 \sin\lambda$, and the combined area of two equal polar caps is $4\pi R^2 (1 - \sin\lambda)$, where R is the Earth's radius and λ the limiting latitude. Hence, 50% of the Earth's surface area occurs between latitudes 30°N and 30°S, whereas a *total* of only 23% of the Earth's surface area occurs poleward of latitudes 50°N and 50°S. The area of potential glaciation therefore is much larger for ε > 54° than for ε ≈ 23°. If polar wander caused continents to

move through moderate to low latitudes of potential glaciation for ε > 54°, a misleading impression of *global* glaciation could be gained from the stratigraphic record.

3. The directions of zonal surface winds for ε > 54° would be opposite to those in comparable latitudes today, because the circulation in 'Hadley cells' would be in the opposite direction (Hunt, 1982). Important zonal winds for a large obliquity would be low-latitude *westerlies* and mid-latitude *easterlies*. The directions of zonal palaeowinds indicated by aeolian deposits for ε > 54° therefore should be at wide angles or opposite to expected zonal palaeowind directions based on palaeolatitudes. The dominant palaeowind direction for the Late Proterozoic periglacial-aeolian Whyalla Sandstone on the Stuart Shelf, South Australia (Fig. 11.1), is consistent with this prediction.

4. Climatic zonation would be weaker than at present. The stability of latitude-dependent climates therefore would be lessened, and any medium-term fluctuations in insolation arising from orbital variations, for example, might cause large or abrupt changes of climate over wide areas. Significant changes of mean temperature on a 10^3-year time-scale in low palaeolatitudes during the Late Proterozoic are indeed indicated by the presence of at least four generations of periglacial sand-wedges in South Australia (Williams and Tonkin, 1985), and relatively rapid temperature changes are suggested by the stratigraphic proximity of cold- and warm-climate indicators in Australia and Spitsbergen (Williams, 1979; Fairchild and Hambrey, 1984).

The hypothesis of a large obliquity (ε > 54°), together with an assumed geocentric axial dipolar magnetic field, therefore may provide an explanation of the complex character of the Late Proterozoic glacial environment as revealed by available evidence, namely: (1) the apparent preference for glaciation in low to equatorial palaeolatitudes; (2) *in situ* frigid climate, permafrost, and grounded ice-sheets near sea-level in low palaeolatitudes; (3) a large amplitude of the seasonal cycle in low palaeolatitudes; (4) extensive periglacial aeolian deposits which, in South Australia, indicate a palaeowind direction opposite to the predicted zonal wind direction; and (5) large, perhaps relatively abrupt, changes of mean temperature on a 10^3-year time-scale in low palaeolatitudes. The other hypotheses discussed cannot provide such a broad explanation of the apparent distribution and complex nature of the Late Proterozoic glacial climate.

Other evidence for Late Proterozoic obliquity

Palaeotidal data

Discrimination between the hypotheses of a non-axial geomagnetic field and a large obliquity requires *palaeogeophysical* data that are independent of palaeomagnetism. Such data are provided by tidal rhythmites of Late Proterozoic age (~ 650 Ma) in the Adelaide Geosyncline, South Australia (Fig. 11.8; Williams, 1988, 1989a–c, 1990, 1991). The ~ 60-year long, unsurpassed palaeotidal record of mixed/synodic type provided by the Elatina

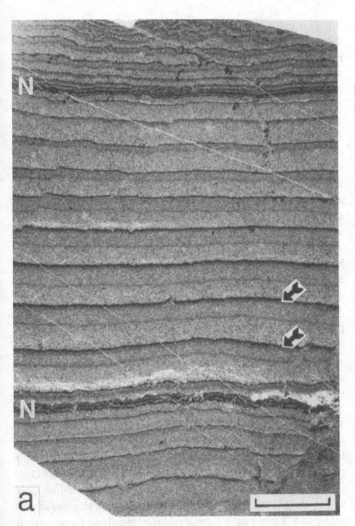

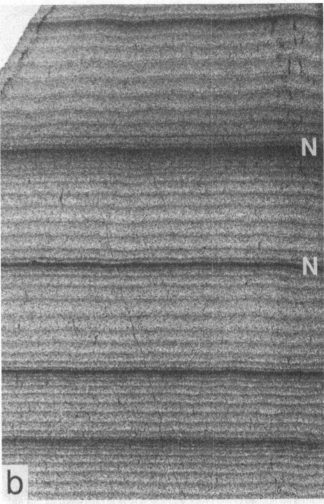

Fig. 11.8. Late Proterozoic (~ 650 Ma) tidal rhythmites, South Australia. Muddy material appears darker than sandy to silty layers. Scale bar 1 cm for both photographs. (*a*) Reynella Siltstone Member from Hallett Cove, showing one thick, fortnightly lamina-cycle that contains 14 diurnal laminae of fine-grained sandstone each topped with a mudstone layer (arrows). Most diurnal laminae comprise two semidiurnal sublaminae of unequal thickness, which reflect the diurnal inequality of the tides. Thinner, more muddy laminae were deposited at neaps (N). (*b*) Elatina Formation from Pichi Richi Pass. Four fortnightly lamina-cycles, each comprising 12 ± 2 graded (upward-fining) diurnal laminae of very fine-grained sandstone and siltstone, are delineated by thin mud-drapes deposited at neaps (N). The lamina-cycles typically are abbreviated by the absence of clastic laminae near neaps.

Formation at Pichi Richi Pass and palaeotidal data also of mixed/synodic type from the Reynella Siltstone Member of the same formation at Hallett Cove near Adelaide, to which a high level of confidence can be attached (Deubner, 1990; Wiliams, 1990, 1991), contain discrete features that together are consistent with a substantial obliquity of the ecliptic in Late Proterozoic time as well as a low latitude of deposition. The features are (1) a clear diurnal inequality of the tides, (2) strong semi-annual and annual signals, and (3) marked abbreviation of fortnightly tidal cycles at equinoctial neaps.

1. The diurnal inequality is best displayed by the Reynella rhythmites (Fig. 11.8a) and is recorded also by the Elatina rhythmites (Williams, 1991). Its presence indicates that in Late Proterozoic time the Earth's equatorial plane was inclined to the Moon's orbital plane. Since the lunar orbital plane is inclined at only 5.15°

to the ecliptic plane, and the inclination probably was little different during the Late Proterozoic (Goldreich, 1966; Mignard, 1982), such a distinct diurnal inequality implies at least a moderate obliquity. The tidal forcing for the diurnal inequality vanishes at the equator and the poles (de Boer *et al.*, 1989), but the arrangement of continents allows the inequality to occur over all latitudes; hence this feature is not latitude-dependent.

2. The semi-annual tidal period of solar declination (the angular distance of the Sun north or south of the Earth's equatorial plane) is clearly shown in a fast Fourier transform (FFT) spectrum of Elatina fortnightly palaeotidal data (Fig. 11.9a). The very strong annual signal in the same spectrum is attributable to a large annual oscillation of sea level. Today, the dominant seasonal oscillation of sea level in Australasia, and the solar annual tidal constituent (S_a) due to the eccentricity of the Earth's orbit, both attain their

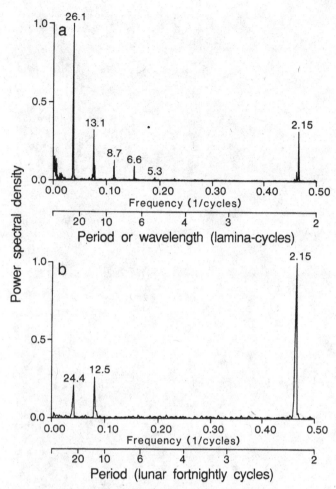

Fig. 11.9. Fast Fourier transform smoothed spectra, with power spectral densities normalised to unity for the strongest peak in each spectrum and with linear frequency scales. (*a*) Spectrum for the Elatina tidal-rhythmite sequence of 1580 fortnightly lamina-cycle thickness measurements (~60-year record). The very strong peak at 26.1 lamina-cycles represents an annual signal, with harmonics at 13.1 (semi-annual), 8.7, 6.6 and 5.3 lamina-cycles. The peak near 2 lamina-cycles (the Nyquist frequency) reflects the monthly inequality of alternate thick and thin fortnightly lamina-cycles. (*b*) Spectrum for the maximum heights of 495 spring tides between January 1, 1966, and December 31, 1985, for Townsville, Queensland. The periods of 24.4 and 12.5 fortnightly cycles represent annual and semi-annual signals. The peak near 2 fortnightly cycles (the Nyquist frequency) reflects the monthly inequality of alternate high and low spring tides. (Tidal data for Townsville supplied by the National Tidal Facility, Flinders University; copyright reserved.)

maximum development in low latitudes (Pariwono *et al.*, 1986; and unpublished data from the National Tidal Facility, Flinders University). Hence the very strong annual signature in the Elatina palaeotidal data accords with the indicated low palaeolatitude for the Elatina Formation.

An idea of the relative power of the semi-annual and annual signals in the Elatina palaeotidal data may be gained by comparing FFT spectra for the Elatina data and for modern tidal data from Townsville, Queensland (Fig. 11.9*a,b*). Comparing these spectra appears justified because each tidal record is of mixed/synodic type and is from low latitudes (Townsville is at 19°16′ S). The spectra in

Fig. 11.9 are for similar, long sequences (20–60 years) of fortnightly data, and show annual, semi-annual and monthly peaks. Normalising the power spectral densities against the monthly peaks in respective spectra indicates that the annual and semi-annual signals in the Elatina spectrum (Fig. 11.9*a*) have ~15 times and ~4 times more power, relative to the related monthly peak, than do the annual and semi-annual signals in the Townsville spectrum (Fig. 11.9*b*). A further difference between the two spectra is the presence, only in the Elatina spectrum, of a sequence of higher harmonics of the annual and semi-annual signals (peaks at 8.7, 6.6 and 5.3 fortnightly cycles in Fig. 11.9*a*). These higher harmonics are attributable to beating among the annual and semi-annual signals and their combination tones. The Late Proterozoic annual and semi-annual oscillations of sea-level in the Adelaide Geosyncline apparently had sufficient power to generate a sequence of higher harmonics in sea-level height and/or tidal range, which was recorded by the Elatina rhythmites. Although differences in geographic settings may well account for some of the distinctions between these ancient and modern tidal records, the important point is that annual and semi-annual signals are very strongly developed in the Late Proterozoic palaeotidal data.

3. The common abbreviation (absence of several diurnal laminae) near the neap part of fortnightly lamina-cycles in the Elatina rhythmites (Fig. 11.8*b*), which persists throughout the ~60-year interval studied, is attributed to small neap-tidal ranges (Williams, 1988, 1989*a–c*, 1990, 1991). The number of diurnal laminae deposited per lamina-cycle, and by inference the range of palaeo-neap tides, were strongly modulated by the semi-annual tidal period (Williams, 1989*a*, 1991). As shown by Williams (1991), abbreviation of fortnightly lamina-cycles was very marked at and near equinoxes (Fig. 11.10), when deposition of diurnal laminae ceased for up to 6–8 lunar days. Comparable fortnightly and semi-annual abbreviation of neap-spring cycles has not been demonstrated for Phanerozoic tidal rhythmites. The inferred Elatina palaeotidal pattern – *neap tides of unusually small amplitude at and around equinoxes* – also is consistent with a large obliquity in Late Proterozoic time. The amplitude of fluctuations, or range, of the semidiurnal part of the lunar equilibrium tide (the principal tidal constituent, M_2) varies with $\cos^2 \delta$, where δ is the Moon's declination (Pillsbury, 1940); and maximum declination $\delta_{max} = \epsilon \pm i$, where i is the inclination of the lunar orbital plane to the ecliptic plane of 5.15°. Hence, tidal ranges at and around equinoctial neaps, when lunar declination is maximal, would be much reduced for $\epsilon \gg 23.5°$. The average equinoctial neap-tidal range of the M_2 constituent for $\epsilon = 54°$ would be ~41% that of today, for $\epsilon = 60°$ only ~30%, and for $\epsilon = 65°$ only ~21% that of today. Extended intervals of small-amplitude tides therefore would occur at and near equinoctial neaps for $\epsilon > 54°$.

It is most significant, therefore, that independent palaeogeophysical data provided by the Elatina and Reynella tidal rhythmites, which belong to the Late Proterozoic Marinoan glacial succession in South Australia, are consistent with a substantial obliquity of the ecliptic and a low palaeolatitude of deposition during Late Proterozoic glaciation.

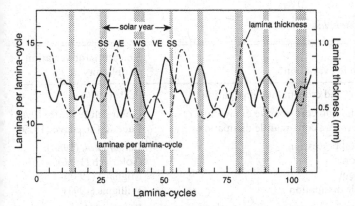

Fig. 11.10. The number of diurnal clastic laminae per fortnightly lamina-cycle counted between lamina thickness minima (neaps), for 110 lamina-cycles from the Elatina tidal rhythmites (see Fig. 11.8b; lamina-cycle number increases up-sequence). The unbroken curve (smoothed by 5-point weighted filter) reveals a strong modulation of lamina counts by a semi-annual signal of ~13 lamina-cycles; abbreviation of lamina-cycles was very marked near equinoxes. The dashed curve shows lamina thickness for the same stratigraphic interval (smoothed by 111-point filter). The shaded bands mark the stratigraphic positions of essentially unabbreviated lamina-cycles bounded by very thin, silty laminae at neaps rather than by mud drapes. VE, vernal equinox, AE, autumnal equinox; SS, summer solstice; WS, winter solstice. (Adapted from Williams, 1991.)

Inclined columnar stromatolites

Vanyo and Awramik (1982, 1985) suggested, from the apparent sinuosity of one small sample of an inclined columnar stromatolite from central Australia, that the obliquity of the ecliptic was 26.5° at ~850 Ma. In most instances, however, the origin of inclination of columnar stromatolites is unknown or can only be tentatively inferred, although current direction (Walter, 1976) and prevailing wind direction (Playford and Cockbain, 1976) appear to strongly influence such inclination. Moreover, the very high rates of stromatolite growth (up to ~10 cm/year) assumed by Vanyo and Awramik (1982, 1985) led Chivas et al. (1990) to query the use of stromatolites as possible indicators of heliotropism.

It seems unlikely that columnar stromatolites could, in fact, faithfully track the Sun throughout the year for large values of obliquity. That would require symmetrical growth, including sediment trapping, on submerged algal columns inclined up to 45° or more to the vertical (taking account of the refraction of light at the seawater surface), whose convex algal surfaces would range from near-horizontal to near-vertical. As a near-horizontal algal surface would trap settling sediment more efficiently than would a steeply inclined surface, columnar stromatolites probably would maintain a very strong vertical component of growth throughout the year, whatever the obliquity. The apparent sinuosity of certain columnar stromatolites thus could lead to large *underestimates* of palaeo-obliquity.

Alternatively, possible annual sinuosity in columnar stromatolites may reflect the annual oscillation of sea-level; this oscillation strongly influenced the paralic environment during the Late Proterozoic in South Australia and left a clear imprint on the Elatina tidal

rhythmites (Fig. 11.9a; Williams, 1988, 1989a–c, 1990, 1991). Until a widely accepted, satisfactory basis for interpreting stromatolite growth patterns is available, the suggestion of Vanyo and Awramik (1982, 1985) therefore must be viewed as unsubstantiated.

Discussion

The diverse explanations of the enigmatic Late Proterozoic glacial climate discussed here require further rigorous testing; such building and testing of hypotheses is the essence of the scientific method. Recent detailed studies of the Late Proterozoic (~650 Ma) Marinoan glacial succession in South Australia, comprising palaeomagnetic analyses that include soft-sediment field tests, time-series analysis of tidal rhythmites, detailed description and palaeoclimatic interpretation of periglacial structures, and determination of palaeowind direction for periglacial-aeolian sandstones, exemplify the types of studies from which wide inferences may be drawn. The carrying out of such studies in other continents will provide a much sharper picture of the Late Proterozoic global environment.

Palaeomagnetic studies and accurate age determinations of Late Proterozoic glaciogenic rocks may discriminate between the hypothesis of global refrigeration (which implies synchronous low-latitude glaciation) and that of a large obliquity. As pointed out by Williams (1975), broad diachronism of glaciogenic sequences might result from movement of continents across low latitudes when $\epsilon > 54°$; hence Late Proterozoic glaciogenic sequences, although deposited in low palaeolatitudes, may not be synchronous and correlative. Age determinations of Late Proterozoic glaciogenic rocks, preferably by zircon U-Pb dating of coeval volcanics, and their palaeomagnetic study therefore might distinguish between (1) global refrigeration, which predicts synchronous, correlative glacial sequences in low palaeolatitudes, and (2) glaciation in preferred low to equatorial palaeolatitudes for $\epsilon > 54°$, which could give either synchronous, or diachronous and non-correlative, glacial sequences. As noted by Li et al. (1991), if all continents were in low latitudes one could not distinguish on palaeomagnetic evidence alone between glaciation during global refrigeration and large-obliquity glaciation. The demonstration of diachronous glaciation in low palaeolatitudes would favour the large-obliquity hypothesis, although the converse does not hold. Kröner (1977) indeed concluded that the apparent diachronism of Late Proterozoic glaciations in Africa supports the large-obliquity model, and Deynoux et al. (1978) advocated diachronism of Late Proterozoic glacial sequences in Central and West Africa. Preiss (1987), however, argued that Kröner's (1977) geochronological data were insufficiently precise to demonstrate the diachronism he proposed.

Further investigations of seasonality during the Late Proterozoic could be pursued by detailed studies of varve-like deposits, rhythmites of tidal–climatic origin, and possible periglacial contraction–expansion structures, coupled with palaeomagnetic investigations into the palaeolatitudes of their formation. The large-obliquity hypothesis predicts that high latitudes were subjected to extreme seasonality, with very hot summers and severe winters, during the

Table 11.3. *Planetary obliquities (angles between equatorial and orbital planes)*

Planet	Present obliquity (°)	Range or past obliquity (°)	Mechanism for variation	Reference for variation of obliquity
Mercury	~0	<90	Tidal torque and core–mantle dissipation	Peale (1976)
Venus	177	0 to ~180	Tidal torques and core–mantle dissipation	Goldreich & Peale (1970)
		0 to ~180	Tidal torques and core–mantle dissipation	Lago & Cazenave (1979)
Earth	23.5	23 ± 1.5	Axial–orbital precession	Berger *et al.* (1984)
		~10 to 23	Tidal friction	Goldreich (1966)
		~13 to 23	Tidal friction	Mignard (1982)
		>54 to 23	Core-mantle dissipation	Williams (1993)
Moon	6.7	77 to 6.7	Tidal torque	Ward (1975)
Mars	25.2	24.4 ± 13.6	Axial–orbital precession	Ward (1979)
		27.5 ± 18.5	Axial–orbital precession (prior to Tharsis uplift)	Ward *et al.* (1979)
		25.8 ± 25.6	Axial–orbital precession	Bills (1990*a*)
		0 to ~20+	Polar ice-cap loading	Rubincam (1990)
Jupiter	3.1			
Saturn[a]	26.7	0 to ~27	Twist of orbital plane	Tremaine (1991)
Uranus	97.9	0 to ~98	Planetary impact	Safronov (1966)
		0 to ~98	Planetary impact	Korycansky *et al.* (1990)
		0 to ~98	Tidal torque	Greenberg (1974)
		0 to ~98	Twist of orbital plane	Tremaine (1991)
		0 to ~98	Axial–orbital precession	Harris & Ward (1982)
Neptune[a]	28.8	0 to ~29	Twist of orbital plane	Tremaine (1991)
Pluto	104.5	99 ± 14	Axial–orbital precession	Dobrovolskis (1989)
	118	0 to 118	Twist of orbital plane	Tremaine (1991)

Note:
[a] Krimigis (1992) gives post-Voyager 2 obliquities of Saturn and Neptune as 29.0° and 29.6°, respectively.

Late Proterozoic. The paucity of organic remains in some Late Proterozoic high-palaeolatitude deposits – for example, the carbonates and shales of the North China block (Zhang and Zhang, 1985) – may reflect a stressful, strongly seasonal milieu rather than the relatively cool climatic conditions suggested by those authors. The search for and study of high-palaeolatitude deposits of Late Proterozoic age therefore should be major objectives.

The origin of sand-wedge structures of possible periglacial origin, such as the contentious structures in the Port Askaig Formation in Scotland, should be settled by further field and laboratory studies. Sand-wedges developed in non-clayey, consolidated materials that do not expand and contract under wet-dry cycles and are not subject to soft-sediment deformation – such as the sand-wedges developed in bedrock quartzite in South Australia – can confidently be regarded as periglacial in origin.

In addition, much palaeowind data for the Late Proterozoic are required to test the prediction of non-uniformitarian zonal palaeowind directions for a large obliquity, such as low-palaeolatitude westerlies and mid-palaeolatitude easterlies. Palaeowind data would be useful only if related to apparent polar wander paths (APWP) from which north and south geographic poles can be

determined, such as the APWP for Late Proterozoic and Palaeozoic Australia (McWilliams and McElhinny, 1980; Schmidt *et al.*, 1991).

Importantly, the question of a possible mechanism for geologically significant change of obliquity also must be addressed. The concept of secular change in planetary obliquity is now well established in celestial mechanics. The obliquities of most planets have undergone or presently are subject to significant change (Table 11.3), indicating that planetary obliquity may be regarded as a potential variable. Venus, Uranus and Pluto have large obliquities that could have been acquired by several mechanisms including tidal torques, dissipative core-mantle coupling, planetary impact, resonant axial–orbital precession, and a twist of the orbital angular momentum vector. In addition, resonant axial–orbital precession causes large fluctuations in the obliquities of Mars and Pluto on 10^6-year time-scales. Axial–orbital precession also causes small variations in the Earth's obliquity ($\pm 1.5°$) about a mean value of 23°.

As noted by Gold (1966), the mean obliquity of the Earth may represent a balance between the effects of tidal friction, which increases the obliquity, and internal dissipation within the Earth, which tends to erect the axis. Dynamical calculations (Goldreich,

1966; Mignard, 1982), which assume that the modern, large value of tidal dissipation applies for much of the past, have suggested that the Earth's mean obliquity is very slowly increasing under the action of luni-solar tidal friction and that early in Earth history the obliquity was perhaps ~ 10–$15°$. Such calculations, however, do not consider geophysical processes within the Earth such as dissipative core-mantle coupling (e.g., Aoki, 1969; Aoki and Kakuta, 1971; Bills, 1990b), or take account of evidence from the geological record for much smaller past values of tidal dissipation (Williams, 1989a,b, 1990) and non-uniformitarian Precambrian climates. Possible geophysical mechanisms for geologically significant secular decrease in the Earth's mean obliquity from $54° < \epsilon < 90°$ to $\epsilon \approx 23°$ during Earth history, that incorporate all these factors as well as the widely accepted single giant impact hypothesis for the origin of the Moon (e.g., Cameron, 1986; Hartmann, 1986; Taylor, 1987), are discussed elsewhere (Williams, 1993).

Conclusions

Over the past 30 years, increasing indications of an enigmatic Late Proterozoic glacial climate – glaciation and frigid conditions near sea-level apparently in preferred low to equatorial palaeolatitudes – have emerged on several continents. Detailed palaeoclimatic and geophysical data recently have been obtained for the Adelaide Geosyncline region in South Australia, which contains some of the best-preserved and most accessible Late Proterozoic rocks in the world. These data, gained from palaeomagnetic analyses that include soft-sediment field tests, time-series analysis of tidal rhythmites, and detailed description and palaeoclimatic interpretation of periglacial structures, have confirmed that glaciation and frigid conditions indeed occurred near sea-level in low palaeolatitudes under a strongly seasonal climate. The present review of these new data and of four hypotheses that have been advanced to explain the enigmatic Late Proterozoic glacial climate – global glaciation, equatorial ice-ring system, non-axial geomagnetic field, and large obliquity – indicates that the hypothesis of a large obliquity in Late Proterozoic time must be given serious consideration. World-wide detailed investigations of Late Proterozoic rocks should aim to further test these hypotheses and better illuminate the Late Proterozoic global environment.

Acknowledgements

I thank Phil Schmidt and Brian Embleton of the CSIRO Division of Exploration Geoscience for fruitful collaborative research on the palaeolatitude of Late Proterozoic glaciation in South Australia; Larry Frakes of the University of Adelaide, Jozef Syktus of the CSIRO Division of Atmospheric Research, Bill Mitchell of the National Tidal Facility, Flinders University, and Wolfgang Preiss of the South Australian Geological Survey for helpful discussions; and Sherry Proferes for drafting. Critical reviews by Nick Eyles and another referee led to an improved paper. The work is supported by the Australian Research Council.

References

Aoki, S. (1969). Friction between mantle and core of the Earth as a cause of the secular change in obliquity. *Astronomical Journal*, **74**, 284–91.

Aoki, S. & Kakuta, C. (1971). The excess secular change in the obliquity of the ecliptic and its relation to the internal motion of the Earth. *Celestial Mechanics*, **4**, 171–81.

Berger, A., Imbrie, J., Hays, J., Kukla, G. & Saltzman, B. (eds.) (1984). *Milankovitch and climate*, Parts 1 and 2. Dordrecht: Reidel, 895 pp.

Bills, B.G. (1990a). The rigid body obliquity history of Mars. *Journal of Geophysical Research*, **95**, 14137–53.

Bills, B.G. (1990b). Obliquity histories of Earth and Mars: influence of inertial and dissipative core-mantle coupling. *Twenty-First Lunar and Planetary Science Conference, Houston, Abstracts*, pp. 81–2.

Black, R.F. (1973). Growth of patterned ground in Victoria Land, Antarctica. *Permafrost Second International Conference, Yakutsk National Research Council Publication*, 2115, 193–203. National Academy of Sciences, Washington, D.C.

Brinkman, A.W. & McGregor, J. (1979). The effect of the ring system on the solar radiation reaching the top of Saturn's atmosphere: direct radiation. *Icarus*, **38**, 479–82.

Cameron, A.G.W. (1986). The impact theory for origin of the Moon. In *Origin of the Moon*, ed. W.K. Hartmann, R.J. Phillips & G.J. Taylor, pp. 609–16. Houston: Lunar and Planetary Institute.

Chivas, A.R., Torgersen, T. & Polach, H.A. (1990). Growth rates and Holocene development of stromatolites from Shark Bay, Western Australia. *Australian Journal of Earth Sciences*, **37**, 113–21.

Chumakov, N.M. (1968). Late Precambrian glaciation of Spitsbergen. *Doklady Akademii Nauk SSSR, Earth Science Section*, **180**, 115–18.

Chumakov, N.M. & Elston, D.P. (1989). The paradox of Late Proterozoic glaciations at low latitudes. *Episodes*, **12**, 115–20.

Coats, R.P. (1981). Late Proterozoic (Adelaidean) tillites of the Adelaide Geosyncline. In *Earth's pre-Pleistocene glacial record*, ed. M.J. Hambrey & W.B. Harland. Cambridge: Cambridge University Press, 537–48.

Coats, R.P. & Preiss, W.V. (1980). Stratigraphic and geochronological reinterpretation of Late Proterozoic glaciogenic sequences in the Kimberley region, Western Australia. *Precambrian Research*, **13**, 181–208.

Connerney, J.E.P., Acuña, M.H. & Ness, N.F. (1987). The magnetic field of Uranus. *Journal of Geophysical Research*, **92**, 15329–36.

Connerney, J.E.P., Acuña, M.H. & Ness, N.F. (1991). The magnetic field of Neptune. *Journal of Geophysical Research*, **96**, 19023–42.

Crawford, A.R. & Daily, B. (1971). Probable non-synchroneity of Late Precambrian glaciations. *Nature*, **230**, 111–12.

Croll, J. (1875). *Climate and time*. London: Daldy, Isbister & Co., 577 pp.

Crowell, J.C. (1983). Ice ages recorded on Gondwanan continents. *Transactions of the Geological Society of South Africa*, **86**, 238–61.

Crowley, T.J., Baum, S.K. & Hyde, W.T. (1991). Climate model comparison of Gondwanan and Laurentide glaciations. *Journal of Geophysical Research*, **96**, 9217–26.

Crowley, T.J., Hyde, W.T. & Short, D.A. (1989). Seasonal cycle variations on the supercontinent of Pangaea. *Geology*, **17**, 457–60.

Crowley, T.J. & North, G.R. (1991). *Paleoclimatology*. New York: Oxford University Press, 339 pp.

de Boer, P.L., Oost, A.P. & Visser, M.J. (1989). The diurnal inequality of the tide as a parameter for recognizing tidal influences. *Journal of Sedimentary Petrology*, **59**, 912–21.

Deubner, F.-L. (1990). Discussion on Late Precambrian tidal rhythmites in South Australia and the history of the Earth's rotation, *Journal*, Vol. 146, p. 97–111. *Journal of the Geological Society of London*, **47**, 1083–4.

Deynoux, M. (1982). Periglacial polygonal structures and sand wedges in the Late Precambrian glacial formations of the Taoudeni Basin in Adrar of Mauritania (West Africa). *Palaeogeography, Palaeoclimatology, Palaeoecology*, **39**, 55–70.

Deynoux, M., Kocurek, G. & Proust, J.N. (1989). Late Proterozoic periglacial aeolian deposits on the West African Platform, Taoudeni Basin, western Mali. *Sedimentology*, **36**, 531–49.

Deynoux, M., Trompette, R., Clauer, N. & Sougy, J. (1978). Upper Precambrian and lowermost Palaeozoic correlations in West Africa and in the western part of Central Africa. Probable diachronism of the Late Precambrian tillite. *Geologischen Rundschau*, **67**, 615–30.

DiBona, P.A. (1991). A previously unrecognised Late Proterozoic succession: Upper Wilpena Group, northern Flinders Ranges, South Australia. *Geological Survey of South Australia Quarterly Geological Notes*, **117**, 2–9.

Dobrovolskis, A.R. (1989). Dynamics of Pluto and Charon. *Geophysical Research Letters*, **16**, 1217–20.

Dow, D.B. & Gemuts, I. (1969). Geology of the Kimberley region, Western Australia: the East Kimberley. *Geological Survey of Western Australia Bulletin*, **120**, 135 pp.

Edwards, M.B. (1975). Glacial retreat sedimentation in the Smalfjord Formation, Late Precambrian, North Norway. *Sedimentology*, **22**, 75–94.

Edwards, M.B. (1978). Glacial environments. In *Sedimentary Environments and Facies*, ed. H.G. Reading, pp. 416–38. Oxford: Blackwell.

Edwards, M.B. (1979). Late Precambrian glacial loessites from North Norway and Svalbard. *Journal of Sedimentary Petrology*, **49**, 85–91.

Embleton, B.J.J. & Giddings, J.W. (1974). Late Precambrian and Lower Palaeozoic palaeomagnetic results from South Australia and Western Australia. *Earth and Planetary Science Letters*, **22**, 355–65.

Embleton, B.J.J. & Williams, G.E. (1986). Low palaeolatitude of deposition for late Precambrian periglacial varvites in South Australia: implications for palaeoclimatology. *Earth and Planetary Science Letters*, **79**, 419–30.

Endal, A.S. & Schatten, K.H. (1982). The faint young Sun-climate paradox: continental influences. *Journal of Geophysical Research*, **87**, 7295–302.

Evans, M.E. (1976). Test of the dipolar nature of the geomagnetic field throughout Phanerozoic time. *Nature*, **262**, 676–7.

Eyles, N. & Clark, B.M. (1985). Gravity-induced soft-sediment deformation in glaciomarine sequences of the Upper Proterozoic Port Askaig Formation, Scotland. *Sedimentology*, **32**, 789–814.

Fairchild, I.J. & Hambrey, M.J. (1984). The Vendian succession of northeastern Spitsbergen: petrogenesis of a dolomite-tillite association. *Precambrian Research*, **26**, 111–67.

Fairchild, I.J., Hambrey, M.J., Spiro, B. & Jefferson, T.H. (1989). Late Proterozoic glacial carbonates in northeast Spitsbergen: new insights into the carbonate-tillite association. *Geological Magazine*, **126**, 469–90.

Frakes, L.A. (1979). *Climates throughout geologic time*. Amsterdam: Elsevier, 310 pp.

Gold, T. (1966). Long-term stability of the Earth–Moon system. In *The Earth–Moon system*, ed. B.G. Marsden & A.G.W. Cameron. New York: Plenum, 93–7.

Goldreich, P. (1966). History of the lunar orbit. *Reviews of Geophysics*, **4**, 411–39.

Goldreich, P. & Peale, S.J. (1970). The obliquity of Venus. *Astronomical Journal*, **75**, 273–84.

Gough, D.O. (1981). Solar interior structure and luminosity variations. *Solar Physics*, **74**, 21–34.

Greenberg, R. (1974). Outcomes of tidal evolution for orbits with arbitrary inclination. *Icarus*, **23**, 51–8.

Guan Baode, Wu Ruitang, Hambrey, M.J. & Geng Wuchen (1986). Glacial sediments and erosional pavements near the Cambrian-Precambrian boundary in western Henan Province, China. *Journal of the Geological Society of London*, **143**, 311–23.

Hambrey, M.J. & Harland, W.B. (eds.) (1981). *Earth's pre-Pleistocene glacial record*. Cambridge: Cambridge University Press, 1004 pp.

Harland, W.B. (1964a). Evidence of Late Precambrian glaciation and its significance. In *Problems in palaeoclimatology*, ed. A.E.M. Nairne. London: Interscience, 119–49.

Harland, W.B. (1964b). Critical evidence for a great infra-Cambrian glaciation. *Geologischen Rundschau*, **54**, 45–61.

Harland, W.B., Armstrong, R.L., Cox, A.V., Craig, L.E., Smith, A.G. & Smith, D.G. (eds) (1990). *A geologic time scale 1989*. Cambridge: Cambridge University Press, 263 pp.

Harris, A.W. & Ward, W.R. (1982). Dynamical constraints on the formation and evolution of planetary bodies. *Annual Reviews of Earth and Planetary Science*, **10**, 61–108.

Hartmann, W.K. (1986). Moon origin: the impact-trigger hypothesis. In *Origin of the Moon*, ed. W.K. Hartmann, R.J. Phillips & G.J. Taylor. Houston: Lunar and Planetary Institute, 579–608.

Hunt, B.G. (1979). The influence of the Earth's rotation rate on the general circulation of the atmosphere. *Journal of the Atmospheric Sciences*, **36**, 1392–408.

Hunt, B.G. (1982). The impact of large variations of the Earth's obliquity on the climate. *Journal of the Meteorological Society of Japan*, **60**, 309–18.

Idnurm, M. & Giddings, J.W. (1988). Australian Precambrian polar wander: a review. *Precambrian Research*, **40/41**, 61–88.

Jakosky, B.M. & Carr, M.H. (1985). Possible precipitation of ice at low latitudes of Mars during periods of high obliquity. *Nature*, **315**, 559–61.

Karte, J. (1983). Periglacial phenomena and their significance as climatic and edaphic indicators. *GeoJournal*, **7**, 329–40.

Kasting, J.F. (1989). Long-term stability of the Earth's climate. *Palaeogeography, Palaeoclimatology, Palaeoecology*, **75**, 83–95.

Kocurek, G. (1991). Interpretation of ancient eolian sand dunes. *Annual Reviews of Earth and Planetary Science*, **19**, 43–75.

Korycansky, D.G., Bodenheimer, P., Cassen, P. & Pollack, J.B. (1990). One-dimenstional calculations of a large impact on Uranus. *Icarus*, **84**, 528–41.

Krimigis, S.M. (1992). The magnetosphere of Neptune. *The Planetary Report*, **12**, 10–13.

Kröner, A. (1977). Non-synchroneity of late Precambrian glaciations in Africa. *Journal of Geology*, **85**, 289–300.

Kröner, A., McWilliams, M.O., Germs, G.J.B., Reid, A.B. & Schalk, K.E.L. (1980). Paleomagnetism of Late Precambrian to Early Paleozoic mixtite-bearing formations in Namibia (South West Africa): the Nama Group and Blaubeker Formation. *American Journal of Science*, **280**, 942–68.

Lago, B. & Cazenave, A. (1979). Possible dynamical evolution of the rotation of Venus since formation. *The Moon and the Planets*, **21**, 127–54.

Li, Yianping, Li, Yongan, Sharps, R., McWilliams, M. & Gao, Z. (1991). Sinian paleomagnetic results from the Tarim block, western China. *Precambrian Research*, **49**, 61–71.

McElhinny, M.W. (1973). *Palaeomagnetism and Plate Tectonics*. Cambridge: Cambridge University Press, 358 pp.

McWilliams, M.O. & McElhinny, M.W. (1980). Late Precambrian paleomagnetism of Australia: the Adelaide Geosyncline. *Journal of Geology*, **88**, 1–26.

Merrill, R.T. & McElhinny, M.W. (1983). *The Earth's magnetic field.* London: Academic Press, 401 pp.

Mignard, F. (1982). Long time integration of the Moon's orbit. In *Tidal friction and the Earth's rotation II*, ed. P. Brosche & J. Sünder-mann. Berlin: Springer, 67–91.

Milankovitch, M. (1930). Mathematische Klimalehre und Astronomische Theorie der Klimaschwankungen. *Handbuch der Klimatologie,* 1(A). Berlin: Gebrüder Borntraeger, 176 pp.

Ness, N.F., Acuña, M.H., Burlaga, L.F., Connerney, J.E.P., Lepping, R.P. & Neubauer, F.M. (1989). Magnetic fields at Neptune. *Science,* **246,** 1473–8.

North, G.R., Cahalan, R.F. & Coakley, J.A. (1981). Energy balance climate models. *Reviews of Geophysics and Space Physics,* **19,** 91–121.

Nystuen, J.P. (1976). Late Precambrian Moelv Tillite deposited on a discontinuity surface associated with a fossil ice wedge, Renda-len, southern Norway. *Norsk Geologisk Tidsskrift,* **56,** 29–56.

Oort, A.H. (1983). Global atmospheric circulation statistics, 1958–1973. *National Oceanographic and Atmospheric Administration Professional Paper,* **14,** 1–180.

Pariwono, J.I., Bye, J.A.T. & Lennon, G.W. (1986). Long-period variations of sea-level in Australasia. *Geophysical Journal of the Royal Astronomical Society,* **87,** 43–54.

Peale, S.J. (1976). Inferences from the dynamical history of Mercury's rotation. *Icarus,* **28,** 459–67.

Perrin, M., Elston, D.P. & Moussine-Pouchkine, A. (1988). Paleomagnetism of Proterozoic and Cambrian strata, Adrar de Mauritanie, cratonic West Africa. *Journal of Geophysical Research,* **93,** 2159–78.

Péwé, T.L. (1959). Sand-wedge polygons (tesselations) in the McMurdo Sound region, Antarctica – a progress report. *American Journal of Science,* **257,** 545–52.

Pillsbury, G.B. (1940). Tidal hydraulics. *Corps of Engineers U.S. Army Professional Paper,* **34,** 283 pp.

Piper, J.D.A. & Grant, S. (1989). A palaeomagnetic test of the axial dipole assumption and implications for continental distribution through geological time. *Physics of the Earth and Planetary Interiors,* **55,** 37–53.

Playford, P.E. & Cockbain, A.E. (1976). Modern algal stromatolites at Hamelin Pool, a hypersaline barred basin in Shark Bay, Western Australia. In *Stromatolites,* ed. M.R. Walter. Amsterdam: Elsevier, 389–411.

Plumb, K.A. (1981). Late Proterozoic (Adelaidean) tillites of the Kimber-ley-Victoria River region, Western Australia and Northern Territory. In *Earth's pre-Pleistocene glacial record,* ed. M.J. Hambrey & W.B. Harland. Cambridge: Cambridge University Press, 504–14.

Preiss, W.V. (compiler) (1987). The Adelaide Geosyncline. *Geological Survey of South Australia Bulletin,* **53,** 438 pp.

Rädler, K.-H. & Ness, N.F. (1990). The symmetry properties of planetary magnetic fields. *Journal of Geophysical Research,* **95,** 2311–18.

Rubincam, D.P. (1990). Mars: change in axial tilt due to climate? *Science,* **248,** 720–1.

Runnegar, B. (1991). Oxygen and the early evolution of the Metazoa. In *Metazoan life without oxygen,* ed. C. Bryant. London: Chapman & Hall, 65–87.

Safronov, V.S. (1966). Sizes of the largest bodies falling onto the planets during their formation. *Soviet Astronomy – AJ,* **9,** 987–91.

Schmidt, P.W., Williams, G.E. & Embleton, B.J.J. (1991). Low palaeolatitude of Late Proterozoic glaciation: early timing of remanence in haematite of the Elatina Formation, South Australia. *Earth and Planetary Science Letters,* **105,** 355–67.

Schwartz, E.J. & Symons, D.T.A. (1969). Geomagetic intensity between 100 million and 2500 million years ago. *Physics of the Earth and Planetary Interiors,* **2,** 11–18.

Sellers, W.D. (1990). The genesis of energy balance modeling and the cool sun paradox. *Palaeogeography, Palaeoclimatology, Palaeoecology,* **82,** 217–24.

Sheldon, R.P. (1984). Ice-ring origin of the Earth's atmosphere and hydrosphere and Late Proterozoic-Cambrian phosphogenesis. *Geology Survey of India Special Publication,* **17,** 17–21.

Spencer, A.M. (1971). Late Pre-Cambrian glaciation in Scotland. *Memoirs of the Geological Society of London,* **6,** 100 pp.

Spencer, A.M. (1985). Mechanisms and environments of deposition of late Precambrian geosynclinal tillites: Scotland and East Greenland. *Palaeogeography, Palaeoclimatology, Palaeoecology,* **51,** 143–57.

Sumner, D.Y., Kirschvink, J.L. & Runnegar, B.N. (1987). Soft-sediment paleomagnetic field tests of late Precambrian glaciogenic sediments (abstract). *Eos, (Transactions of the American Geophysical Union),* **68,** 1251.

Taylor, S.R. (1987). The origin of the Moon. *American Scientist,* **75,** 469–77.

Toon, O.B., Pollack, J.B., Ward, W., Burns, J.A. & Bilski, K. (1980). The astronomical theory of climatic change on Mars. *Icarus,* **44,** 552–607.

Tremaine, S. (1991). On the origin of the obliquities of the outer planets, *Icarus,* **89,** 85–92.

Van der Voo, R. (1990). The reliability of paleomagnetic data. *Tectonophysics,* **184,** 1–9.

Van Hemelrijck, E. (1982). The insolation at Pluto. *Icarus,* **52,** 560–4.

Vanyo, J.P. & Awramik, S.M. (1982). Length of day and obliquity of the ecliptic 850 Ma ago: preliminary results of a stromatolite growth model. *Geophysical Research Letters,* **9,** 1125–8.

Vanyo, J.P. & Awramik, S.M. (1985). Stromatolites and Earth-Sun-Moon dynamics. *Precambrian Research,* **29,** 121–42.

Walter, M.R. (1976). Hot-spring sediments in Yellowstone National Park. In *Stromatolites,* ed. M.R. Walter. Amsterdam: Elsevier, 489–98. Elsevier.

Ward, W.R. (1974). Climatic variations on Mars 1. Astronomical theory of insolation. *Journal of Geophysical Research,* **79,** 3375–86.

Ward, W.R. (1975). Past orientation of the lunar spin axis. *Science,* **189,** 377–9.

Ward, W.R. (1979). Present obliquity oscillations of Mars: fourth-order accuracy in orbital *e* and *I. Journal of Geophysical Research,* **84,** 237–41.

Ward, W.R., Burns, J.A. & Toon, O.B. (1979). Past obliquity oscillations of Mars: the role of the Tharsis Uplift. *Journal of Geophysical Research,* **84,** 243–59.

Washburn, A.L. (1980). *Geocryology. A survey of periglacial processes and environments.* New York: Wiley, 406 pp.

Williams, G.E. (1975). Late Precambrian glacial climate and the Earth's obliquity. *Geological Magazine,* **112,** 441–65.

Williams, G.E. (1979). Sedimentology, stable-isotope geochemistry and palaeoenvironment of dolostones capping late Precambrian glacial sequences in Australia. *Journal of the Geological Society of Australia,* **26,** 377–86.

Williams, G.E. (1986). Precambrian permafrost horizons as indicators of palaeoclimate. *Precambrian Research,* **32,** 233–42.

Williams, G.E. (1988). Cyclicity in the late Precambrian Elatina Formation, South Australia: solar or tidal signature? *Climatic Change,* **13,** 117–28.

Williams, G.E. (1989*a*). Late Precambrian tidal rhythmites in South Australia and the history of the Earth's rotation. *Journal of the Geological Society of London,* **146,** 97–111.

Williams, G.E. (1989*b*). Precambrian tidal sedimentary cycles and Earth's paleorotation. *Eos, (Transactions of the American Geophysical Union),* **70,** 33, 40–1.

Williams, G.E. (1989*c*). Tidal rhythmites: geochronometers for the ancient Earth–Moon system *Episodes,* **12,** 162–71.

Williams, G.E. (1990). Tidal rhythmites: key to the history of the Earth's rotation and the lunar orbit. *Journal of Physics of the Earth,* **38,** 475–91.

Williams, G.E. (1991). Upper Proterozoic tidal rhythmites, South Austra-
    lia: sedimentary features, deposition, and implications for the
    Earth's paleorotation. In *Clastic tidal sedimentology*, ed. D.G.
    Smith, G.E. Reinson, B.A. Zaitlin & R.A. Rahmani. *Canadian
    Society of Petroleum Geologists Memoir*, **16**, 161–178.
Williams, G.E. (1993). History of the Earth's obliquity *Earth Science
    Reviews*, **34**, 1–45.
Williams, G.E. & Tonkin, D.G. (1985). Periglacial structures and palaeocli-
    matic significance of a late Precambrian block field in the Cattle

Grid copper mine, Mount Gunson, South Australia. *Australian
    Journal of Earth Sciences*, **32**, 287–300.
Worsely, T.R. & Kidder, D.L. (1991). First-order coupling of paleogeogra-
    phy and CO_2, with global surface temperature and its latitudinal
    contrast. *Geology*, **19**, 1161–4.
Zhang, H. & Zhang, W. (1985). Palaeomagnetic data, late Precambrian
    magnetostratigraphy and tectonic evolution of eastern China.
    Precambrian Research, **29**, 65–75.

12 Isotopic signatures of carbonates associated with Sturtian (Neoproterozoic) glacial facies, central Flinders Ranges, South Australia

ADAM R. CROSSING and VICTOR A. GOSTIN

Abstract

The Appila Tillite in the southwest Flinders Ranges outcrops as a narrow, continuous unit, near the western edge of the Adelaide Geosyncline, resting disconformably on dolomitic Burra Group sediments. Interpretation of measured sections suggests the Appila Tillite represents two glacial advances, abruptly overlain by a post-glacial transgressive shale. These sections of the Tillite show condensed and commonly reworked sediments when compared with sections in the Northern Flinders Ranges, suggesting deposition close to the margin of the basin. Most clasts within this formation are of Burra Group origin with only a few erratics from the western Gawler Craton.

Dolomite is common in the Appila Tillite, either as dolostone clasts or as a diamictite matrix. Isotope analysis gives an average of -6.26 permil $\delta^{18}O$ PDB and 1.34 permil $\delta^{13}C$ PDB for the matrix, different to dolomitic clasts ($0.28\ \delta^{18}O$ and $4.39\ \delta^{13}C$). The dolomitic matrix is likely to be either detrital rock flour altered after deposition by isotopically light pore fluids, or a carbonate precursor that has undergone penecontemporaneous dolomitization. The former is favoured.

Geochemical results suggest the light $\delta^{13}C$ values of the Appila matrix were caused by input of biologically derived carbon. The variation in isotope values between sampled material from different formations (Appila Tillite Fm., Tapley Hill Fm., and Burra Group) indicate that metamorphic equilibrium has not occurred.

Introduction

The Appila Tillite (defined by Mirams, in Thompson *et al.*, 1964) in the Southern Flinders Ranges, is a local equivalent of extensive Sturtian glacial deposits found throughout the Adelaide Geosyncline (Preiss, 1987). Ten sections through this formation were measured along the Emeroo Range in the southwestern Flinders Ranges (Fig. 12.1). These sections permitted construction of a fence diagram (Fig. 12.3) used to investigate palaeoenvironmental aspects of the Appila Tillite as well as providing an accurate location for the analysed samples. Geochemical and isotopic analysis was carried out primarily to determine the origin of dolomite

within this glacial sequence. For details of precise section locations, sedimentology and geochemistry, see Crossing, (1991).

In this paper we present results from carbon and oxygen isotope work, and iron and manganese geochemistry. The results, specific to the southern Flinders Ranges, are compared with those of other studies to reveal insights into the origin of dolomite (both as clasts and matrix) within Neoproterozoic glacial deposits.

Regional stratigraphy

Within the Emeroo Range, the lowest sequence of the preglacial Burra Group is the undifferentiated Emeroo Subgroup (formerly the Emeroo Quartzite after Mawson, 1949), which displays both thickness and facies variations along the range. The lithologies suggest a fluvial environment containing coarse clastics, with silts deposited in interchannel areas (Preiss, 1987).

The overlying Mundallio Subgroup, synonymous with the Skillogalee Dolomite (Wilson, 1952), was described by Uppill (1979) who divided it into a lower pale, low energy dolomite (the Nathaltee Formation), and a dark grey upper unit (the Yadlamulka Formation) with stromatolites, dolomitic sands and silts, reworked dolomite, and black chert. Shales, and cream-coloured feldspathic quartzites, commonly with a dolomitic matrix, belong to the Undalya Quartzite (Wilson, 1952), and represent the highest beds of the Burra Group in the area.

Coats (in Thompson *et al.*, 1964) defined the succeeding Umberatana Group as containing all the glacials of the Sturtian and Marinoan (Fig. 12.2). The Appila Tillite is the lowest unit of this group in the study area, and unconformably overlies the Mundallio Subgroup (Preiss, 1987). Although no accurate age is known, Preiss (1987) suggests 750 Ma for 'Sturtian glacigenic rocks', including the Appila Tillite. The best outcrops of Appila Tillite were found within the Emeroo Range, and attempts to construct a three-dimensional picture were frustrated by poor outcrop to the east. Our sections 1–10 through this formation are shown in Fig. 12.3. They represent a condensed and dominantly reworked sequence, and were probably formed near the western margin to a Neoproterozoic basin (Crossing, 1991). The Appila Tillite is between 90 and 180 m thick, and

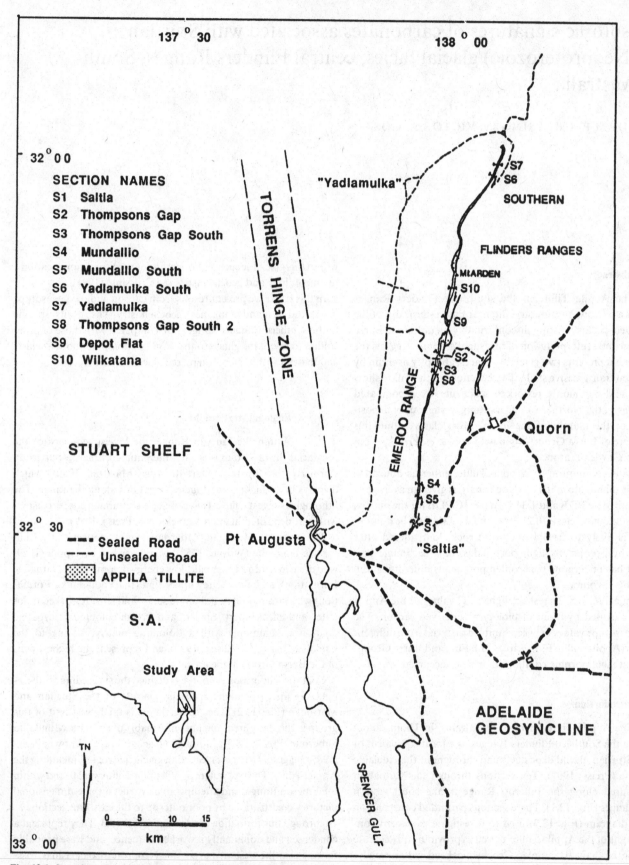

Fig. 12.1. Location map showing study area and section locations (precise coordinates in Crossing 1991).

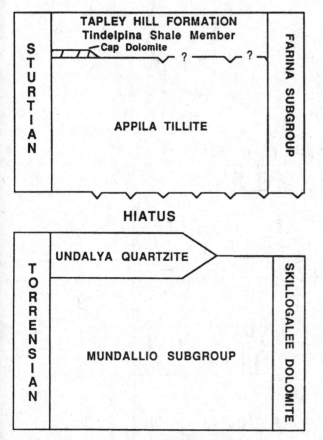

Fig. 12.2. Stratigraphic setting modified from Preiss (1987).

margin. The relative thickness of glacial unit 3 is likely to be the result of highstand conditions during an accelerated period of basin deepening, rather than a positive sea level change during glaciation. Unit 4 is interpreted to be transgressive, deposited during a period of glacial retreat and relatively high meltwater input. The division between units 3 and 4 is supported by isotopic data presented herein.

The contact between the Appila Tillite and the overlying Tapley Hill Formation has been variously described as sharply conformable or disconformable (Preiss, 1987). The basal Tindelpina Shale member of this formation in the study area generally contains basal thinly laminated dolomites grading up into alternate grey and buff weathered siltstones (Preiss, 1987). The Tindelpina Shale is best interpreted as a maximum flooding surface of a very extensive transgression that followed the Sturtian Ice Age.

Methods

A dental drill was used to extract selected carbonate from specimens collected in the field. Between 15 and 30 mg of the powdered sample (depending on the estimated carbonate purity) was reacted with 100% phosphoric acid for 16 to 30 hours at 50 °C. The CO_2 extracted was analysed for carbon and oxygen using a mass spectrometer at the University of Adelaide, treating sample fractionation in phosphoric acid as for calcite. No phosphoric acid fractionation coefficient was applied to the results, as a consistent value does not exist in the relevant papers. By leaving the values in a raw state, any coefficient may be easily applied. All measurements are in per mil relative to the PDB standard.

Nine samples and a standard (J-Do-1) were analysed for iron and manganese by dissolving dolomite powder in 50 ml of 1M hydrochloric acid for two hours at room temperature, filtering, and making up to 100 ml with deionised water. This solution was then analysed using atomic absorption spectrometry.

Isotopic and geochemical analysis

Isotopic analysis was carried out on 30 samples, and the results are displayed on Table 12.1 and Fig. 12.6.

The scatter of results in the isotope figures demonstrate that there has been no pervasive post-depositional re-equilibration; i.e. minor metamorphism responsible for cleavage development has not caused isotopic equilibration to one general signal for Appila Tillite matrix, clasts, Burra Group, and Tapley Hill Formation.

Five samples of dolomitic Burra Group from below the Appila Tillite plot with average values of -2.50 permil $\delta^{18}O$ and $+3.87$ permil $\delta^{13}C$. This approximates the area of normal marine to slightly ^{18}O-depleted (Land, 1983; Rao, 1988). Five clasts from the Appila Tillite, interpreted to have originated from the underlying Mundallio Subgroup, have average values of $+0.28$ permil $\delta^{18}O$ and $+4.39$ permil $\delta^{13}C$. This lies within the area of hypersaline to normal marine conditions (Land, 1983; Belperio, 1990).

The clasts display a slight positive shift when compared to the Burra Group sediments. That the two are not the same is unsurpris-

dominant lithologies are diamictites, sandstones, and shales containing dropstones. A dolomitic matrix is common within the diamictites and shales (see Figs 12.4 and 12.5).

Minor metamorphism has been responsible for cleavage development in many fine grained shales found within the Appila Tillite. Thin section examination of sandy dolospar from section 6 (Fig. 12.3) also reveals a cleavage fabric, with fibrous sparite growing in the pressure shadows of quartz grains (Fig. 12.4). Coarser diamictites and conglomerates display no cleavage development.

An attempt has been made to identify the fourfold division of glacial units observed in deeper eastern parts of the basin (Young and Gostin, 1988, 1990, 1991). Units 1 and 3 of this division represent glacial advances. Unit 1 was interpreted by Young and Gostin (1990) as being the result of mass flow and turbidity currents, while unit 3 represented rainout as well as mass flow. Units 2 and 4 represent glacial retreats, and were interpreted by Young and Gostin (1990) to be formed by sediment gravity flows, with some rainout, and a high meltwater input.

When these units are applied to sections 1–10 in Fig. 12.3, their condensed and reworked nature becomes apparent. Units 1 and 2 are generally thin or absent. This we attribute to non-deposition or reworking in a marginal environment. These units are more significant in distal areas of the palaeobasin. Units 3 and 4 dominate most sections; their proximal existence suggests onlapping onto the basin

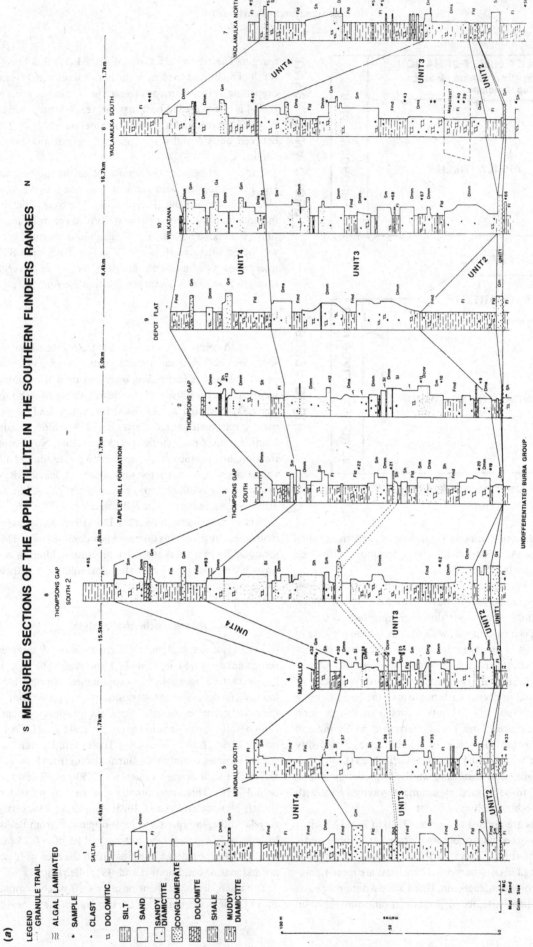

MEASURED SECTIONS OF THE APPILA TILLITE IN THE SOUTHERN FLINDERS RANGES

Fig. 12.3. (a) Measured sections of the Appila Tillite in the Southern Flinders Ranges.

(b)

D	–	–	–	Diamictite
	m			matrix supported
	c			clast supported
		m		massive
		s		stratified
		g		graded
			r	reworked
F	–	–		Mud
	m			massive
	l			laminated
		d		with dropstones
S	–	–		Sand
	m			massive
	l			laminated
	h			horizontally cross laminated
		d		with dropstones
G	–			Gravel / conglomerate
	m			massive
	s			stratified

Fig. 12.3b (*cont.*) (*b*) Lithofacies codes after Eyles and Miall (1984).

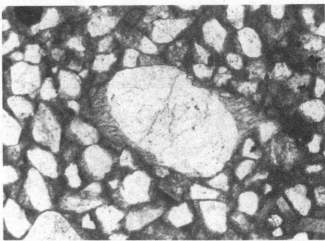

Fig. 12.4. Sandy dolospar from within the Appila Tillite, section 6 (see Fig. 12.3). Quartz grains are poorly sorted, with approximately 40% displaying undulose extinction, while the remainder are not. This suggests more than one source for detrital quartz. A cleavage fabric composed of fibrous sparite has developed in the pressure shadow of the large quartz grain that dominates the field of view. The bulk of the matrix is acicular sparite, which has replaced some grains (whose outlines may still be seen). Plane polarized light, longitudinal field of view is 4 mm.

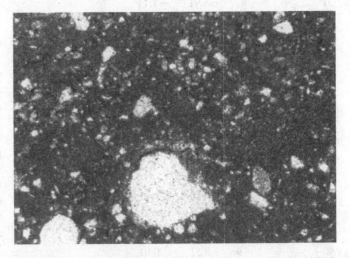

Fig. 12.5. Sandy dolomicrite from the Appila Tillite, section 2 (Fig. 12.3). In this sample the matrix is dominantly micritic, although some replacement by sparite has occurred. Detrital quartz composes ~30% of the sample and is poorly sorted, subangular to rounded. Rounded clasts of micritic and sparitic dolomite are also present; one may be seen to the right of the largest quartz clast. Plane polarized light, field of view is 2 mm.

ing, as the clasts have travelled an unknown distance (they are generally subrounded), and could have come from any level in the Burra Group, which has marine to near-shore characteristics (Uppill, 1990). Even in the limited area studied, Burra Group samples show considerable variation, supporting the field stratigraphic view that it had been eroded to various levels before deposition of the overlying strata.

Samples of dolomitic matrix from the Appila Tillite show generally more negative values, particularly of $\delta^{13}C$. Average values are -6.26 permil $\delta^{18}O$ and -1.34 permil $\delta^{13}C$. Sample 58B from a dolomite layer in section 1, (Fig. 12.3) shows a widely disparate value, although it is only a few centimetres apart from 58A in the sample rock. Inspection in hand specimen shows 58B to represent a dolomite nodule likely to have formed during later diagenesis. 58A, however, is typical of the dolomite matrix.

The Tapley Hill Formation samples from above the Appila Tillite show a large variation, but are generally more negative than the Burra Group. Averages here are -5.36 permil $\delta^{18}O$ and 0.89 permil $\delta^{13}C$.

Palaeoenvironmental and diagenetic interpretation

A considerable amount of information on the environment of deposition and early diagenetic processes within the Appila Tillite can be gleaned from the isotopic and geochemical results.

Values for the dolomitic matrix of the Appila Tillite show a major shift to lighter values when compared to the dolomitic clasts within that matrix. The matrix values are also well removed from those of the Burra Group.

As stated earlier, the clasts are interpreted to have originated from the Mundallio Subgroup, on which the Appila Tillite gener-

ally lies. At Yadlamulka South, however, the Appila Tillite overlies a thin sequence of Undalya Quartzite. Hence the unconformity at the base of the Appila Tillite represents an erosional surface extending to varying depths in the Burra Group. The slight shift in signal of the clasts to heavier $\delta^{18}O$ values when compared to the Burra Group suggests dolomite formation in normal marine to hypersaline conditions where high levels of evaporation have preferentially removed the light isotope. The clasts are hence

Table 12.1. *Stable isotope, iron and manganese values for analysed samples of Appila Tillite matrix and clasts, the underlying Burra Group, and overlying Tapley Hill Formation. (Stratigraphic position of samples shown on Fig 12.3)*

Sample	Yield (%)	^{18}O(pdb)	^{13}C(pdb)	Mn (ppm)	Fe (ppm)
CLASTS					
947–019	100	1.47	2.27		
35	89	−1.50	3.41	214	2828
43	91	−0.48	4.94	225	5026
67A	84	1.20	5.86		
67B	98	0.73	5.47		
BURRA GP					
33	84	−2.25	1.08	416	9687
50	83	−1.89	5.93	291	2346
48	85	−1.64	4.71		
49A	78	−1.41	4.43		
049B	75	−3.52	4.40		
39	71	−3.27	3.48		
40	87	−3.77	3.06		
MATRIX					
31	87	−5.91	−1.11		
45	35	−7.82	−1.30	542	10079
58A	60	−5.71	−1.10	774	16863
68	26	−4.93	−1.36	527	14100
73	11.9	−7.18	−1.46		
69	38	−3.42	−2.26		
52	23	−9.12	−0.79		
58B	18	−9.77	1.00		
74	12	−8.48	1.20		
STANDARD					
J–Do–1	98	−5.04	4.57	49	97
TAPLEY HILL					
32	31	−4.15	−0.55	1385	10768
59A	78	−4.38	2.40	573	6111
59B	74	−5.04	1.98		
56	87	−2.72	−2.00		
60	9	−8.57	0.81		
65	5	−4.94	1.25		
18	90	−8.81	3.85		

interpreted to have been transported from a source removed some distance from the site of deposition, and do not group as closely as the Burra Group samples, suggesting the source tapped by glacial activity represented a wider area containing environments grading from normal marine to hypersaline.

Clearly, Fig. 12.6 demonstrates that the dolomitic matrix of the Appila Tillite displays a signal well removed from either that of macroscopic clasts within the matrix, or the underlying Burra Group dolomites. Previous authors (e.g. Young and Gostin, 1990) have suggested that the dolomitic matrix of the Appila Tillite represents 'rock flour' derived by glacial abrasion of the underlying

units. The isotopic results in Fig. 12.6, however, show that *primary* rock flour is not the correct answer, as the isotopic shift is too great. Diagenetic alteration, or a different source for carbonate in the Appila is proposed in this paper.

Interpretation of ancient dolomites

Several problems exist in the interpretation of isotope data on dolomite. It is now accepted that carbonates and glacial material are not mutually exclusive. Nevertheless, problems remain regarding the nature of formation, deposition, and alteration of carbonates within and associated with glacial events. Some authors have suggested extreme climate variations (Williams, 1979), while others (Morris, 1978, in Williams, 1979) suggest redeposition of clastic carbonate. Young (1979) suggests Proterozoic glaciation may have been triggered by an 'anti-greenhouse' effect, caused by loss of CO_2 from the atmosphere into carbonates (and hence their common presence stratigraphically below glacially influenced strata). While dolomite formation in cold regimes is not impossible, Fairchild & Hambrey (1984) indicate that this is not necessary, and consider the dolomite to be detrital, originating from interglacial carbonates.

In addition, Land (1980) refers to the 'dolomite problem', i.e. the poorly understood conditions of formation of this mineral, and whether formation can be primary. The Precambrian contains an abundance of dolomite in relation to other carbonates; however, it is generally accepted that direct precipitation of dolomite from seawater did not occur to any significant extent. Although waters today are often saturated with respect to dolomite, a catalyst is absent. Formation of secondary dolomite generally incurs an isotopic fractionation dependent in part on the diagenetic fluid and the carbonate being replaced. δ^{18}O values shift more dramatically than δ^{13}C, as carbon is often insignificant in pore fluids. If however, dolomite forms as a syn- or early post-depositional event (generally forming micrite under low compaction or in soft sediments), the depositional signal of the precursor may be retained, or suffer only a slight shift. The micritic texture, however, still gives a primary appearance.

Another problem in interpreting the isotopic signature of dolomite centres around the fact that recrystallization may be partial or complete, in a system that is neither completely open nor closed (Land, 1980). As a result, the signal measured may be a combination of more than one process, and will not accurately record any single event.

Comparison with other Neoproterozoic dolomites

The results displayed in Figs 12.6 and 12.7 are not dissimilar to those obtained by Tucker (1982, 1983) from work on the Mid to Late Proterozoic Beck Spring Dolomite of California. This is a shallow platform dolomite (subtidal, lagoonal, tidal flats), deposited below the glacial Kingston Peak Formation. His 'depositional' components (pisolites, micrite) gave isotopic averages of −1.84 δ^{18}O and +4.21 δ^{13}C, compared with Burra Group

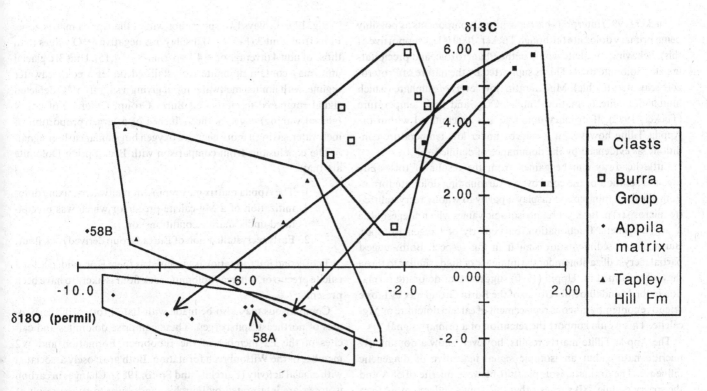

Fig. 12.6. Carbon–oxygen isotopic crossplot in permil and referenced to PDB for samples taken from the Southern Flinders Ranges. Outlier 58B is discussed in the text.

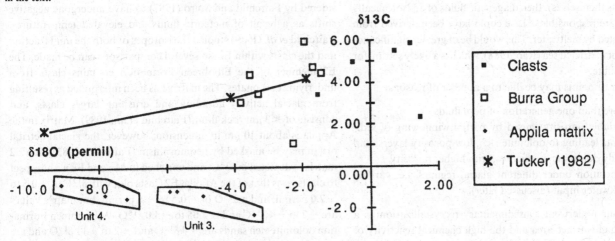

Fig. 12.7. Carbon–oxygen isotopic crossplot. Average values of micrite, fibrous dolomite, and equant dolospar (in order of decreasing values) from Tucker (1982), displayed with results from the Southern Flinders Ranges.

results of -2.5 $\delta^{18}O$ and $+3.87$ $\delta^{13}C$. The slightly more negative $\delta^{18}O$ signal for the Burra Group is indicative of a more marine rather than a marginal marine signal. Tucker's (1982) results approximate a regression (after Veizer and Hoefs, 1976) through fibrous dolomite to equant dolospar (with a value of -9.58 $\delta^{18}O$ and $+1.8$ $\delta^{13}C$), interpreted to represent early and late diagenesis. Several inferences can be drawn from the results and arguments of Tucker (1982) in relation to results from the Flinders Ranges.

Recrystallization of Appila Tillite micritic matrix is patchy, with

cathodoluminescence and thin sections showing generally minor replacement by microspar (Crossing, 1991). As a result, $\delta^{18}O$ is not as negative as Tucker's equant dolospar, while $\delta^{13}C$ values suggest a higher organic carbon input in diagenetic fluids compared with the Beck Spring Dolomite. Reduction of $\delta^{18}O$ values due to diagenesis is dependent on the nature of pore fluids. Since diagenesis of the Appila Tillite is interpreted to have occurred under low compaction or synsedimentary conditions, the seawater signature may have an important input.

Tucker (1982) interprets his depositional components as possibly being primary dolomite (although Tucker, 1990 is less supportive of this), behaving as calcite would in the Phanerozoic, and precipitating as a dolomite mud. This is supported by the nature of Proterozoic seawater; the high Mg/Ca ratio, low levels of sulphate (which inhibit dolomite formation), high pCO_2, and high temperature (Tucker, 1982). If primary dolomite was precipitated within the Appila Tillite however, it would be under low temperature conditions, as evidenced by the dominance of diamictites.

Little headway can be gained from the results of this study towards solution of the controversy surrounding dolomite formation. Clasts in thin section display a partly to wholly recrystallized (to microspar) nature, yet retain isotopic values which suggest little if any diagenetic fractionation, particularly of oxygen isotopes. Burra Group sediments are similar in this respect. Both suggest partial recrystallization under the influence of fluids similar to those they were formed in. Uppill (1990) suggests that dolomite formation in the Mundallio Sub-Group of the Burra Group was by fabric retentive contemporaneous replacement of calcian dolomite or Mg-calcite. This would support the retention of a primary signal.

The Appila Tillite matrix results, however, show a dominantly micritic nature, but an isotopic signal indicative of diagenetic influence. The dramatic isotopic shift between micrite of 58A and the sparry nodule 58B suggests that the Appila suffered more than one generation of diagenetic fluid. If the original matrix consisted of rock flour derived from abrasion of the underlying Burra Group (and the clasts themselves), then diagenetic fluids of an isotopically light nature are responsible. These could have been derived from seawater diluted by meltwater. This would be of greater significance in a shallow or restricted sea (Rao, 1988), which is a likely setting for the Appila Tillite.

The scatter of points may be due to a number of reasons:

1. More than one generation of pore fluids
2. Variable porosity caused by partial winnowing of sediments leading to dolomite rich, low porosity layers, and sandy dolomitic diamictites, with higher porosity
3. Deposition under different glacial regimes, i.e. varying meltwater input (discussed later).

If rock flour underwent synsedimentary recrystallization as a result of its high surface area and the high chemical reactivity of glacially transported dolomite (Fairchild and Spiro, 1990), the nature of seawater would be directly relevant, as it would be if dolomite formation was primary (unlikely but not disproveable). A significantly large meltwater input would be necessary to cause isotopic lightening, particularly of $\delta^{18}O$. Sea ice also preferentially takes up ^{18}O, and could enhance this effect.

Overall, synsedimentary alteration of primary rock flour would seem the most straightforward interpretation, supported by the presence of altered dolomite 'skins' on sand-sized clasts seen in thin section (Crossing, 1991). Subsequent limited penetration by pore fluids may have caused nodule growth, and formation of sparite in zones and as cavity fill. Sampling of the nodule 58B indicates that this later pore fluid was lighter than the initial pervasive fluid.

Fig. 12.6 displays two subgroups within the Appila matrix zone. Plots from unit 3 (Fig. 12.3) display less negative $\delta^{18}O$ values than those of unit 4 (average of -4.18 permil cf. -7.15). Unit 3, a glacial unit, may contain dolomite recrystallized under a cold seawater regime, with limited meltwater input giving a slightly $\delta^{18}O$ depleted signal compared to clasts and Burra Group. Dolomite of unit 4 (glacial waning) suggests the influence of a larger proportion of meltwater, giving a more negative oxygen but similar carbon signal.

The conclusions from comparison with Beck Spring Dolomite are that:

1. The Appila matrix may represent meltwater-mixing dolomitization of a Mg-calcite precursor which was precipitated under marine conditions; or
2. Early recrystallization of Burra Group derived rock flour.

The second interpretation is favoured as there is no evidence for a calcite precursor, whereas dolomitic rock flour is likely to have been present.

Comparisons may also be made with Late Proterozoic carbonates of northeast Spitsbergen. These comprise dolomites and calcites of the E2 member of the Elbobreen Formation and W2 member of the Wilsonbreen formation. Both are closely associated with glacial activity (Fairchild and Spiro, 1987). Changes in carbon isotopes are interpreted by Fairchild and Spiro to be the result of varying input of organic carbon associated with organic sediment accumulation and oceanic circulation. Oxygen isotopes are considered by Fairchild and Spiro (1987) to have undergone negative shifts as a result of meteoric fluids and elevated temperatures. Fairchild et al. (1989) studied the isotopes of both the mud fraction and the clasts within E2, so several comparisons can be made. The E2 member of the Elbobreen Formation contains clasts from underlying carbonates. The matrix has been interpreted as resulting from glacial activity abrading and crushing larger clasts, and consists of $< 3 \mu m$ rock flour (Fairchild et al., 1989). Matrix in the Appila is about 10 μm in dimension; however, the glacial detrital origin may be masked by neomorphism (Fairchild, 1983). The E2 matrix is depleted in $\delta^{13}C$, and enriched in Mn and Fe with respect to clasts, as is that of the Appila. E2 clasts plot with $\delta^{13}C$ of $+0.5$ to $+7.0$ permil, and a $\delta^{18}O$ of -0.5 to -6 permil, but matrix values are -2 to $-4 \delta^{13}C$ and $+4.65$ to $-4.0 \delta^{18}O$. Wilsonbreen Formation dolomite rich sands are $+4 \delta^{13}C$ and $+2$ to $+11 \delta^{18}O$ and are considerably different from those of the Appila. Fairchild et al. (1989) acknowledge the heavier nature of the matrix oxygen values, and interpret them as indicating a pervasive early recrystallization. Carbon depletion, and the presence of pyrite imply some input from bacterial oxidation of organic material. In comparison, the signal of the Appila matrix indicates that the E2 matrix is isotopically closer to the clasts from which it probably derived, and has undergone a shift in a different direction.

A problem acknowledged by Fairchild et al. (1989) is contamination of the sampled matrix by sand-sized carbonate clasts. This could produce a signal considerably altered in the direction of the clasts, and result in plotting in a field that appears close to hypersaline. Obviously, a hypersaline interpretation of the results

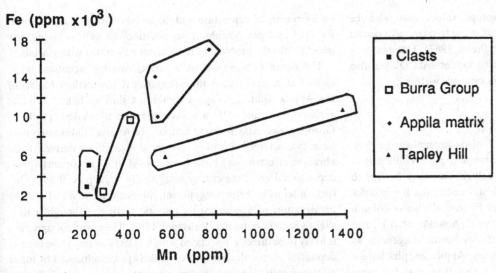

Fig. 12.8. Fe plotted against Mn for nine samples from the Southern Flinders Ranges.

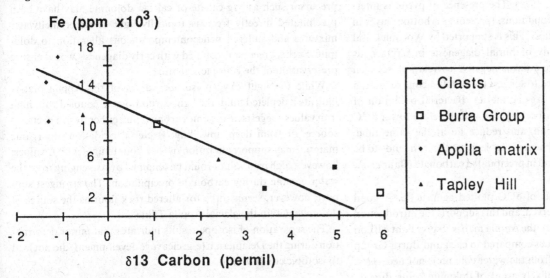

Fig. 12.9. Fe is plotted against $\delta^{13}C$ to display an approximately linear negative covariation (see text).

from E2 is unlikely, due to its glacial nature. Samples from the Appila matrix did not appear in hand specimen or thin section to have clasts so small they could not be avoided. Sodium information for the E2 member and the Appila Tillite would yield information that could answer the problem of apparent hypersalinity.

The overlying isotopic signal

Directly above the Appila Tillite lies either a thinly laminated shale or a 'cap' dolomite (Williams, 1979); both belonging to the Tapley Hill Formation. Isotopic values average −5.36 $\delta^{18}O$ and +0.89 $\delta^{13}C$. Results for the shale and for the cap dolomite do not separate, despite the differing yield and lithology. These values are enriched in ^{13}C with respect to the underlying Appila Tillite matrix, but have only slightly lower $\delta^{18}O$ values.

Williams (1979) gave values for the cap dolomite at the head of

Depot Creek as −5.5 $\delta^{18}O$ and +1.8 $\delta^{13}C$, and described the lithology as laminated 'dolomitic intrasparite and intramicrite' with ferruginous staining. Thin section description of cap dolomite from Depot Creek in Thompsons Gap agrees with this description (Crossing, 1991).

A primary or early diagenetic origin has been suggested on the basis of common micrite and a lack of volume change associated with secondary dolomitization (Williams, 1979). The coarseness of micrite (< 1 to 70 μm in Williams, 1979) is not directly suggestive of rock flour (Fairchild, 1983) but early diagenetic recrystallization may have led to a growth in crystal size. Primary dolomite formation is considered unlikely, but a syn-sedimentary or pene-contemporaneous origin would be applicable.

The lighter $\delta^{13}C$ values of the Tapley Hill, when compared to the Appila matrix suggest a microbial input, supported by the (commonly algally) laminated appearance of the cap dolomite in the study area. The lighter carbon isotope values may also be influenced by a higher freshwater input in the transgressive period following glaciation (Williams, 1979; Preiss, 1987). This may not only reduce carbon values by introducing soil derived CO_2, but also reduce oxygen values which are low in meteoric waters.

Geochemistry and the carbon signal

Nine samples were analysed for iron and manganese (Fig. 12.8). The groups show a distinct separation. Appila Tillite matrix plots with generally the highest Fe values. This may be due to reducing conditions within the sediment. In addition, if any dolomite was precipitated, it may scavenge Fe from the water column (Fairchild et al., 1989). Ricken (1986) states however, that Fe and Mn generally become enriched during carbonate diagenesis so precipitation of dolomite is not necessary. Appila samples have a maximum of about 17 000 ppm Fe, higher than that for other groups in Fig. 12.8, (short, however, of the E2 average of 30 000 ppm from Fairchild et al., 1989). The presence of pyrite is interpreted to point to anoxic conditions, (present as a bottom layer in stratified glacial water bodies). This is supported by Woronick and Land (1985) in their study of burial diagenesis in Cretaceous carbonate platforms where a linear negative correlation between $\delta^{13}C$ and Fe was interpreted to suggest the simultaneous reduction of FeO and oxidation of organic matter. Bacterial oxidation of organic rich solutions transporting iron will therefore lower $\delta^{13}C$ by an input of light carbon, and reduce Fe at the same time. Although in a reduced and hence soluble form, the iron tends to be retained on siliceous gels and in precipitated carbonate (Baur et al., 1985).

Fig. 12.9 displays results of $\delta^{13}C$ plotted against Fe. A good negative correlation is observed, and this supports the introduction of organic rich solutions into the Appila matrix to give light carbon isotopes and high iron values compared to clast and Burra Group results. Scavenging of Fe from the water column is not necessary, but could occur in the unlikely event of dolomite being directly precipitated. Organic input is likely to come from a microbially derived or bacterial source. Stromatolites form today in saline Antarctic lakes, and could be a valid organic source (Walter and Bauld, 1983).

An alternative suggestion for lower $\delta^{13}C$ and high Fe concentrations has been recently proposed by Kaufman et al. (1991) who suggest that the onset of glaciation causes upwelling in previously stratified marine basins. As a result, anoxic bottom waters, rich in unoxidised organic material and ferrous iron are brought into shelf areas, where cold, oxygen-rich surface waters precipitate iron. At the same time isotopically light organic material is oxidized, lowering the $\delta^{13}C$ signal. This argument may also be applicable to the Appila Tillite matrix; however, this process is likely to occur only at the initiation of a glacial episode.

Conclusions

Isotopic and geochemical analysis of dolomites from above, within, and below the Appila Tillite gives information on environments of deposition and diagenesis. This provides support for field and petrographic interpretation, as well as giving new insights into the problem of carbonates associated with glacials.

The Burra Group displays a normal marine signature, while clasts within the Appila Tillite are marine to hypersaline. Matrix of the Appila Tillite displays a significant shift to lighter isotopic values of $\delta^{13}C$ and $\delta^{18}O$ when compared to the underlying Burra Group, or to clasts within the matrix. The Appila Tillite matrix may have two sources. Primary dolomitic rock flour derived from abrasion of clasts may have been altered in a syn- or early postdepositional environment by isotopically light pore fluids. This early alteration before significant lithification would allow the preservation of a dominantly micritic texture, although with a slightly larger crystal size (Fairchild, 1983). The pore fluid signature is likely to be directly associated with that of seawater at the time of deposition, hence the light signal is likely to be influenced by input of isotopically light meltwater, particularly in unit 4 dolomites deposited during a glacial waning. Alternatively, a carbonate precursor such as Mg-calcite or calcian dolomite may have been precipitated directly from isotopically light seawater–meltwater mixtures and suffered penecontemporaneous alteration to dolomite. Such a precursor, coupled with early diagenesis would ensure preservation of the isotopic signature.

While the light oxygen isotope values of the Appila matrix indicate a depleted fluid, the light carbon signal, coupled with high iron values suggests an organic carbon input, either from a bacterial source, or from deep upwelling. Bacterial oxidation of organic material may support alteration of rock flour to lighter $\delta^{13}C$ values; however, high iron values could be obtained by scavenging from the water column during carbonate precipitation. The strongest support, however, seems to be for altered rock flour as the source of dolomite within the glacial Appila Tillite.

The separation of isotope results indicates that minor deformation during the Delamerian (e.g. cleavage development) did not lead to isotopic equilibration.

Acknowledgements

The encouragement and advice given by Malcolm Wallace is gratefully acknowledged. Drs Ian Fairchild and Baruch Spiro reviewed the manuscript and supplied helpful comments. Technical assistance and advice was kindly supplied by K. Turnbull. This project was partially funded by an ESSO research grant to V.A. Gostin.

References

Baur, M., Hayes, J. Studley, S. & Walter, M. (1985). Millimeter-scale variations of stable isotope abundances in carbonates from banded iron-formations in the Hamersley Group of Western Australia. Economic Geology, 80, 270–82.

Belperio, A.P. (1990). Palaeoenvironmental interpretation of the Late Proterozoic Skillogalee Dolomite in the Willouran Ranges, South Australia. In The evolution of a Late Precambrian–Early Palaeozoic rift complex: the Adelaide Geosyncline, eds., J. Jago & P. Moore, Geological Society of Australia. Special Publication 16, 85–104.

Crossing, A.R. (1991). Sedimentary environments and carbonates of the Appila Tillite in the Pt. Augusta–Hawker area, South Australia, Honours thesis, University of Adelaide.

Eyles, N. & Miall, A. (1984). Glacial facies. In *Facies Models*, ed., R. Walker, Geoscience Canada, Reprint Series 15–38.

Fairchild, I. (1983). Effects of glacial transport and neomorphism on Precambrian dolomite crystal sizes. *Nature*, **304**, 714–16.

Fairchild, I.J. & Hambrey, M.J. (1984). The Vendian succession of northeastern Spitsbergen: petrogenesis of a dolomite-tillite association. *Precambrian Research*, **26**, 111–67.

Fairchild, I.J. & Spiro, B. (1987). Petrological and isotopic implications of some contrasting Late Precambrian carbonates, NE Spitsbergen. *Sedimentology*, **34**, 973–89.

Fairchild, I.J., Hambrey, M.J., Spiro, B. & Jefferson, T.H. (1989). Late Proterozoic glacial carbonates in northeast Spitsbergen: new insights into the carbonate-tillite association. *Geological Magazine*, **126**, 469–90.

Fairchild, I.J. & Spiro, B. (1990). Carbonate minerals in glacial sediments: geochemical clues to palaeoenvironment. In *Glaciomarine environments: processes and sediments*, eds., J.A. Dowdeswell & J.D. Scourse, Geological Society of London Special Publication, **53**, 201–16.

Kaufman, A.J., Hayes, J.M., Knoll, A.H. & Germs, G.J.B. (1991). Isotopic compositions of carbonates and organic carbon from upper Proterozoic successions in Namibia: stratigraphic variation and the effects of diagenesis and metamorphism., *Precambrian Research*, **49**, 301–27.

Land, L.S. (1980). The isotope and trace element geochemistry of dolomite: the state of the art. In *Concepts and models of dolomitization*, eds., D. Zenger, J. Dunham & R. Ethington, of the Society of Economic Palaeontologists and Mineralogists Special publication **28**, 87–110.

Land, L.S. (1983). The application of stable isotopes to studies of the origin of dolomite and to problems of diagenesis of clastic sediments. In *Stable isotopes in sedimentary geology*, eds., M. Arthur, T. Anderson, I. Kaplin, J. Veizer & L. Land, Society of Economic Palaeontologists and Mineralogists Short Course 10, 4.1–4.8.

Mawson, D. (1949). Sturtian tillite of Mount Jacob and Mount Warren Hastings, north Flinders Range. *Transactions of the Royal Society of South Australia*, **72**, 244–51.

Preiss, W.V. (1987). *The Adelaide Geosyncline, Late Proterozoic stratigraphy, sedimentation, palaeontology and tectonics*. Bulletin 53, Department of Mines and Energy, Geological survey of South Australia. D.J. Woolman, Government Printer, South Australia, pp. 128–68.

Rao, C. (1988). Oxygen and carbon isotope composition of cold-water Berriedale Limestone (Lower Permian), Tasmania, Australia. *Sedimentary Geology*, **60**, 221–231.

Ricken, W. (1986). Diagenetic bedding. A model for marl-limestone alterations. In *Lecture notes in earth sciences*, eds., S. Bhattacharji, G. Friedman, H. Neugebauer & A. Seilacher, Berlin: Springer-Verlag, pp. 154–161.

Thompson, B., Coats, R., Mirams, R., Forbes, B., Dalgarno, C., & Johnson, J. (1964). Cambrian rock groups in the Adelaide Geosyncline: a new subdivision. *Quarterly Geological Notes*, The Geological Survey of South Australia, 9, 1–19.

Tucker, M.E. (1982). Precambrian dolomites: petrographic and isotopic evidence that they differ from Phanerozoic dolomites. *Geology*, **10**, 7–12.

Tucker, M.E. (1983). Diagenesis, geochemistry, and origin of a Precambrian dolomite: The Beck Spring Dolomite of eastern California. *Journal of Sedimentary Petrology*, **53**, 1097–119.

Tucker, M.E. (1990). Carbon isotopes and Precambrian–Cambrian boundary geology, South Australia: ocean basin formation, seawater chemistry and organic evolution. *Terra Nova*, **1**, 573–82.

Uppill, R.K. (1979). Stratigraphy and depositional environments of the Mundallio Subgroup (new name) in the Late Precambrian Burra Group of the Mt Lofty and Flinders Ranges., *Transactions of the Royal Society of South Australia*, **103**, 25–43.

Uppill, R.K. (1990). Sedimentology of a dolomite-magnesite-sandstone sequence in the Late Precambrian Mundallio Subgroup, South Australia. In *The evolution of a Late Precambrian–Early Palaeozoic rift complex: the Adelaide Geosyncline*, eds., J. Jago & P. Moore Geological Society of Australia Special Publication, 16, 105–128.

Veizer, J. & Hoefs, J. (1976). The nature of O^{18}/O^{16} and C^{13}/C^{12} secular trends in sedimentary carbonate rocks. *Geochimica Cosmochimica Acta*, **40**, 1387–95.

Walter, M.R. & Bauld, J. (1983). The association of sulphate evaporites, stromatolitic carbonates and glacial sediments: examples from the Proterozoic of Australia and the Cainozoic of Antarctica. *Precambrian Research*, **21**, 129–48.

Williams, G.E. (1979). Sedimentology, stable-isotope geochemistry and palaeoenvironment of dolostones capping Late Precambrian glacial sequences in Australia. *Journal of the Geological Society of Australia*, **26**, 377–86.

Wilson, A.F. (1952). The Adelaide System as developed in the Riverton–Clare Region, Northern Mount Lofty Ranges, South Australia. *Transactions of the Royal Society of South Australia*, **75**, 131–49.

Woronick, R.E. & Land, L.S. (1985). Late burial diagenesis, Lower Cretaceous Pearsall and Lower Glen Rose Formations, South Texas. In *Carbonate cements*, eds. N. Schneidermann & P.M. Harris, Society of Economic Palaeontologists and Mineralogists Special Publication, 36, 265–75.

Young, G.M. (1979). The earliest ice ages. In *The winters of the world: earth under the ice ages*, ed. B. John, Newton Abbott: David and Charles, pp. 107–30.

Young, G.M. & Gostin, V.A. (1988). Stratigraphy and sedimentology of Sturtian glacigenic deposits in the western part of the North Flinders Basin, South Australia. *Precambrian Research*, **39**, 151–70.

Young, G.M. & Gostin, V.A. (1990). Sturtian glacial deposition in the vicinity of the Yankaninna Anticline, North Flinders Basin, South Australia. *Australian Journal of Earth Sciences*, **37**, 447–58.

Young, G.M. & Gostin, V.A. (1991). Late Proterozoic (Sturtian) succession of the North Flinders Basin, South Australia; An example of temperate glaciation in an active rift setting. In *Glacial marine sedimentation; palaeoclimatic significance*: eds. J. Anderson & G. Ashley, Boulder, Colorado: Geological Society of America Special Paper, 261, 207–21.

13 Reactive carbonate in glacial systems: a preliminary synthesis of its creation, dissolution and reincarnation

IAN J. FAIRCHILD, LAWRENCE BRADBY and BARUCH SPIRO

Abstract

Subglacial transport of carbonate rock debris produces abundant fine reactive particles (rock flour) which are susceptible to dissolution. Longer transport distance and mixing with quartz increases rock flour formation, but the role of primary grain size and the quantity of the most reactive submicron-sized material is unclear. Considerations of equilibrium solubility indicate enhanced dissolution will occur in systems open to atmospheric CO_2, or where acid production by pyrite oxidation is important, or where submicron-sized crystals are abundant. Kinetic considerations emphasize the increased reactivity of fine particles with freshly exposed surfaces and lattice defects, but dissolution is often limited by the sloth of reactions involving CO_2. A number of processes can allow reprecipitation of calcite in the glacial system: ripening, warming, freezing, the common ion effect, removal of CO_2 by organic or inorganic means, input of alkalinity from organic decomposition by bacteria, evaporation, transpiration and skeletal biomineralization. The relative importance of these mechanisms has yet to be established.

Examples of contemporary processes are discussed from ongoing work in Europe and North America. Meltwaters from carbonate-rich glaciers are shown to have a wide variety of partial pressures of CO_2 reflecting their complex processes of evolution; waters become supersaturated in response to evaporation, and to some extent degassing. Regelation crusts occur on clasts and are morphologically distinct from those in vadose proglacial areas. Evidence of Holocene modification of Pleistocene glacial sediments by calcrete formation and vadose cementation of gravels is presented from other sites.

Comparison of basal ice with meltout till and refrozen meltout till at a Swiss glacier reveals a large loss of fines in the meltout tills. There is isotopic evidence for the presence of an authigenic phase in the refrozen meltout till whose apparent composition is consistent with equilibrium precipitation either by ripening, or in a series of melting–freezing cycles, or some combination of these processes.

Introduction

The chemical processes in glacial systems have received much less attention than the physical ones, but there is currently interest in the sensing of subglacial processes from studies of water quality (Sharp, 1991), and there is also an increasing awareness of the possibilities of finding a memory of depositional or early diagenetic conditions preserved in the chemistry of mineral precipitates in glacial settings (Fairchild and Spiro, 1990; Fairchild, 1993).

In comminuting rock to fine particles, glaciers create an abundance of fresh mineral surfaces which are particularly susceptible to weathering reactions in contact with dilute, acid and oxidizing meltwater (Souchez and Lemmens, 1987; Tranter et al., 1993). The phases which are most reactive during weathering (apart from rare sulphate or chloride evaporites) are calcite, dolomite and pyrite. Calcite dissolution has been shown to have a major influence in determining meltwater chemistry even where its abundance is low in relation to silicates (Simpkins and Mickelson, 1990; Tranter et al., 1993). Dolomite is less soluble, but is locally the dominant carbonate phase. Together with H^+-fixation by surface-exchange on clays, the carbonates neutralize the acidity which is derived from atmospheric sources and the oxidation of pyrite. Especially in carbonate areas, the main weathering solutes are calcium, magnesium and bicarbonate ions, together with sulphate where pyrite oxidation is important. The most common mineral to be precipitated from such meltwaters is calcite and this precipitation can be triggered by a variety of mechanisms (summarized on Fig. 13.1). The absolute and relative importance of these mechanisms are incompletely known. Dissolution and reprecipitation of carbonates in glacial sediments can also occur during subsequent non-glacial conditions (Fig. 13.1, base).

In this article we explore the processes which create and modify carbonate of various origins in glacial sediments with examples drawn from ongoing work on contemporary environments. A discussion of the characteristics and preservation potential of the different types of carbonate is given to help the interpretation of ancient glacial sediments.

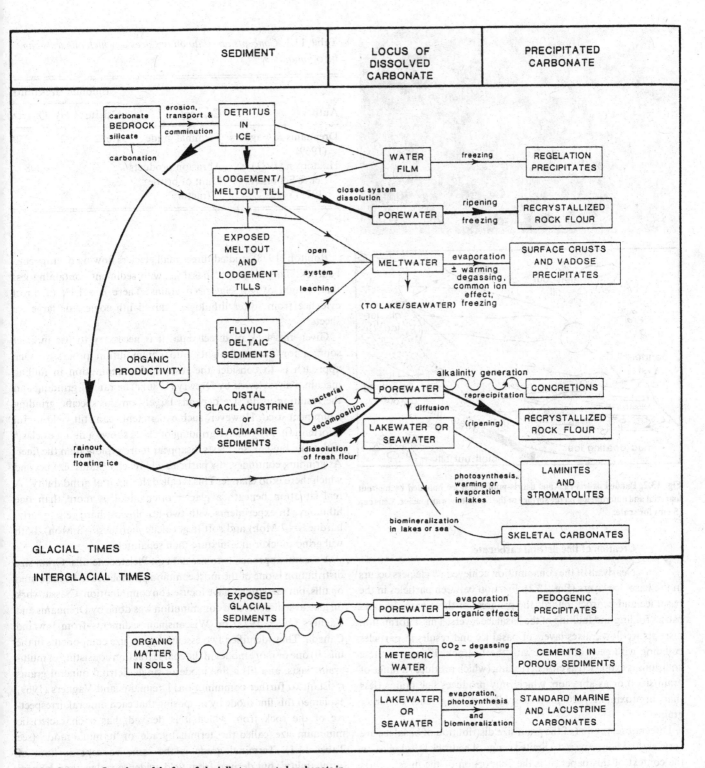

Fig. 13.1. Summary flow chart of the fate of glacially transported carbonate in both glacial and postglacial (interglacial) times. The heavy lines denote the pathways which are most likely to lead to in-situ transformation of reactive carbonate.

Fig. 13.2. **Banded debris-rich and debris-poor basal ice bounded by meltout material and outcropping on the surface of the Saskatchewan glacier. Lens cap (5 cm) for scale.**

Table 13.1. *Concepts and data on the sizes to which minerals can be comminuted*

Author	Concept	Minimum sizes (μm)		
		Calcite	Clay	Quartz
Dreimanis & Vagners (1969)	Terminal grade	1.6–2	< 2	30–250
Haldorsen (1981)	Abrasion products			3–8
Steier & Schonert (1972)	Limit of brittle fracture	3–5		1

Creation of fine detrital carbonate

Nearly all of the comminution achieved by glaciers occurs in the basal ice layers (Fig. 13.2). Friction between particles in the basal ice and bedrock retards the progress of these particles, and causes folding and faulting of the basal ice layers. This deformation disrupts ice-flow, mixes layers of basal ice and results in particles colliding with each other and with the bedrock. Energy at these collisions is expressed as either crushing (which produces a range of grain sizes) or as abrasion, which only produces fine grains (less than approximately 20 μm) as spalled-off fragments of larger grains.

Direct observations of the grain size distribution of basal ice are very few and have covered a limited range of bedrock lithologies. In the context of this paper, it is the proportion of the finest debris which is of greatest interest since, as will be amplified below, this component is the most reactive. Most glaciologists have regarded the sand and gravel fractions as crucial to the understanding of glacial systems, and have often ignored the clay-sized fraction partly due to the difficulty in working with it. This has resulted in very little published information on these grains, as is clear from Drewry's (1986) review.

Boulton (1978) studied three small glaciers flowing over igneous bedrock. They displayed basal ice with sediment containing less than weight 5% of clay-sized grains. There is a lack of direct evidence from softer lithologies, mixed lithologies, or large ice sheets.

Given this lack of direct data, it is necessary to use indirect sources for further insights into comminution processes. One approach is to consider the results of comminution in milling machines as reviewed by Drewry (1986). The rate of grinding and the resulting grain size depends largely on the specific grinding equipment used. However, such experiments can still yield useful information. Progressive grinding of clasts shows a fining of clasts with time, but coarser clasts disappear more rapidly than the finer. As grinding continues, the particles will reach a stable size beyond which there is no change. This is called the 'limit of grindability'. A real situation beneath a glacier often involves more than one lithology. In experiments with two lithologies, hard (e.g. quartz, hardness = 7 Moh) and soft (e.g. calcite, hardness = 3 Moh), both will grind quicker in a mixture than separately.

A second approach is to use data on Pleistocene tills. Grain size distribution is one of the most common parameters to be measured on tills, but few studies have focused on comminution. Classic work on glacial sediments and comminution was done by Dreimanis and Vagners (1969, 1971) on Wisconsinan sediments from lowland Canada. They identified two essential grain size components in the tills: (i) one or more modes in the coarse fraction consisting of multi-grain clasts; and (ii) a fine mode of single detrital mineral grains resistant to further comminution. Dreimanis and Vagners (1969) explained this fine mode by proposing that each mineral, irrespective of the rock from which it is derived, has a characteristic minimum size, called the 'terminal grade' or 'terminal mode' (see Table 13.1). Terminal grade is the same concept as limit of 'grindability', but derived from field evidence rather than laboratory-based milling experiments. Dreimanis and Vagners (1971) estimated that 300–500 km of ice-sheet transport was necessary to break three-quarters of the clasts in a till down to their terminal grades. They perceived source mineralogy and distance of travel (equivalent to grinding time in a milling machine) as the essential controls on grain size of till.

Haldorsen (1981) studied coarse-grained, sandstone-derived tills in Norway. She used the same approach, but identified three components. In addition to Dreimanis and Vagner's 'multi-mineral' and 'terminal-grade' modes, she identified an intermediate mode representing the separate detrital minerals released from clasts of bedrock by subglacial crushing but otherwise essentially unchanged. She conclusively showed that this monomineralic mode coincided exactly with the grain size of the bedrock. The fine 'terminal grade' was explained as the product of abrasion between the detrital sand grains. Haldorsen's (1981) study was based on a sandstone-derived till, but for tills derived from fine-grained rocks, the distinction between crushing and abrasion will be less clear. Like Boulton (1978), Haldorsen regarded the many complicated gains and losses to the subglacial system as rendering a rate of comminution (or concomitant glacier transport distance) impossible to deduce.

Slatt and Eyles (1981) came to very different conclusions on the nature of comminution. They studied the petrology and grain size of sand (not the whole range of particles) from the basal ice of 11 modern alpine glaciers, mainly flowing over igneous bedrock, with a range of crystal sizes from fine to coarse. The resulting grain sizes are non-modal, and the sand itself is enriched in lithic fragments. Slatt and Eyles (1981) interpreted this as evidence that subglacial comminution preferentially fractures grains across the mineral grains (intracrystal breakage) rather than between them (intercrystal breakage). This means that bedrock grain size is unimportant, and transport path through the glacier is the controlling factor on grain size of sediment.

A third line of approach is to consider the micron-scale mechanisms of grain breakage. This enables an assessment of the concepts of a detrital intermediate grade (Haldorsen) and of a monomineralic terminal grade (Dreimanis and Vagners; Haldorsen). All particle breakages, whether due to abrasion or crushing, are assumed to occur by brittle fracture (Rumpf, 1973). Steier and Schonert (1972) measured the compression of a single grain between two plates. For larger grains (> 100 μm), the compression causes elastic deformation and tensile stress builds up. This can be released by brittle fracture if large flaws exist for them to exploit. With small particles, even those composed of brittle materials, the compression is taken up by plastic deformation, and particles do not fracture. Steier and Schonert (1972) gives lower limits for fracturing as 3–5 μm for calcite and 1 μm for quartz.

The question posed by Slatt and Eyles still remains: what is the relative resistance to fracture of inter- and intracrystal weaknesses? The question is capable of solution. An answer would be of use to the study of reactive carbonate. If Haldorsen's (1981) concept of a mode consisting of detrital mineral grains is sound, then we would expect the finest source rocks to yield sediment most susceptible to early diagenesis. If on the other hand intracrystal breakages dominate, then we should ignore the bedrock's crystal size and examine glacial transport paths in our search for the most reactive sediment.

A second question concerns the size distribution of the products at and below the so-called 'terminal grade'. Dreimanis and Vagner's (1969) definition is somewhat vague, but it is unlikely that the values given in Table 13.1 are an absolute cut-off point. Calcite particles

less than 2 μm in size certainly exist, and being very soluble are of most importance to our work. Such particles are also the hardest to investigate. Standard grain size distribution methods, as used in our work, are not reliable for submicron-sized material. In the future, transmission electron microscopy may offer the best way forward to characterize this ultrafine material.

Chemical controls on dissolution and precipitation

One great simplification of working with terrestrial glacial systems is that the waters are dilute, so that the chemical activities of dissolved species numerically approach their concentrations in moles per litre of solution. This makes calculations of mineral stability and gas pressure in solutions much easier. Raiswell (1984) was the first to clearly expound how meltwater chemistry could be interpreted in terms of systems open or closed to the input of atmospheric carbon dioxide, and he gave a number of general relationships which followed from the principles of physical chemistry. The emphasis in the present discussion is on those chemical aspects which bear on carbonate dissolution and precipitation.

Weathering reactions

Weathering only proceeds at significant rates in the presence of dissolved acids. There is invariably some H^+ supplied from the dissociation of carbonic acid (H_2CO_3) which is created by the dissolution in water of carbon dioxide, a gas present at a concentration (partial pressure) of $10^{-3.5}$ in the atmosphere. Overall one can write carbonation reactions:

$$CO_2 + XCO_3 + H_2O \rightarrow X^{2+} + 2HCO_3^-$$

for the dissolution of a carbonate mineral and e.g.

$$2CO_2 + XSiO_3 + 3H_2O \rightarrow X^{2+} + H_4SiO_4 + 2HCO_3^-$$

for the dissolution of a silicate. The equations are written to emphasize that the carbon dioxide is consumed and that the dominant anion will be bicarbonate. Providing a calcium-containing mineral has dissolved, the solution is a potential source for a calcite precipitate, whether or not carbonate minerals were present in the weathering source. However, other factors being equal, solutions deriving from weathering of carbonate rocks tend to be more concentrated than those derived from silicates because of the higher solubility of carbonates. Hence secondary precipitates of calcite should be more abundant in areas where weathering of carbonate minerals is occurring.

Another common source of acid is through the oxidation of pyrite:

$$4FeS_2 + 15O_2 + 8H_2O \rightarrow 2Fe_2O_3 + 8SO_4^{2-} + 16H^+$$

This may be quantitatively-coupled to carbonate dissolution, thus:

$$16H^+ + 16CaCO_3 \rightarrow 16Ca^{2+} + 16HCO_3^-$$

Although other processes (e.g. exchange of solute H^+ for cations on clay mineral surfaces) may also have a role, the effects mentioned should be the dominant ones in carbonate terrains.

Equilibrium themodynamics

Calcite dissolution and precipitation are used as the quantitative examples in this section. The relevant reactions with their corresponding equilibrium constants (K) are given below (eliminating intermediate species for simplicity). The symbol $a(\ )$ denotes activity.

$$CO_2 \text{ (gas)} + H_2O \rightarrow HCO_3^- + H^+$$

$$K = \frac{a(HCO_3^-) \times a(H^+)}{p(CO_2)} = 10^{-7.69}(0\ ^\circ C) \text{ or } 10^{-7.82}\ (25\ ^\circ C)$$

$$HCO_3^- \rightarrow CO_3^{2-} + H^+$$

$$K_{II} = \frac{a(CO_3^{2-}) \times a(H^+)}{a(HCO_3^-)} = 10^{-10.63}\ (0\ ^\circ C) \text{ or } 10^{-10.33}\ (25\ ^\circ C)$$

$$CaCO_3 \text{ (calcite)} \rightarrow Ca^{2+} + CO_3^{2-}$$

$$K_{cc} = a(Ca^{2+}) \times a(CO_3^{2-})$$

Activity of the solid $CaCO_3$ is 1 and hence can be disregarded; the equilibrium constant K_{cc} is commonly referred to as the *solubility product*. Values for pK_{cc} ($-\log_{10}K_{cc}$) have recently been revised upwards from values in the 1970s literature of 8.33 and 8.41 for 0 °C and 25 °C respectively to values of 8.38 and 8.48 (Morse and Mackenzie, 1990). One reason for the slightly higher values of more recent determinations has been the elimination from the experimental materials of small amounts of fine material with enhanced solubility (see discussion below). The older values have been retained in the calculations used here for comparability with previous work (Plummer *et al.*, 1975; Raiswell, 1984).

In the general case of a solution not necessarily at equilibrium, the ionic activity product relevant to calcite is defined as:

$$IAP_{cc} = a(Ca^{2+}) \times a(CO_3^{2-})$$

The saturation index of the solution can be defined as:

$$SI_{cc} = \log_{10}(IAP_{cc}/K_{cc})$$

Negative SI_{cc} values indicate undersaturation (calcite should dissolve), a value of zero refers to saturation ($= 100\%$ saturation), and positive values refer to supersaturated solutions (calcite should precipitate).

Fig. 13.3 illustrates the increase in calcium concentration in solution during the progressive dissolution of calcite. In a closed system, that is one where carbon dioxide consumed cannot readily be replenished, the amount of dissolution is quite limited. In an open system, kept in equilibrium with carbon dioxide at a pressure of $10^{-3.5}$ atm. (as in the modern atmosphere), there is five times more dissolution of calcite before the solution becomes saturated. Depending on temperature, the saturation values are around 1 mequiv of Ca^{2+}, which corresponds to 50 mg $CaCO_3$ dissolved per litre of solution.

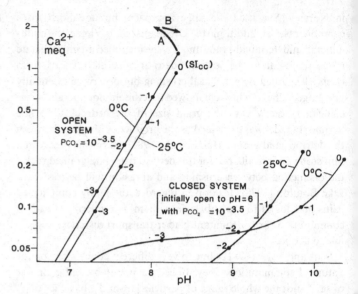

Fig. 13.3. Progressive change in dissolved calcium with pH during closed and open system calcite dissolution at 0 °C and 25 °C. Multiply meq Ca^{2+} by 50 to derive values in terms of mg $CaCO_3$ per litre of solution. The lines are labelled with values for the saturation index. Constructed from spreadsheet models using algorithms of Raiswell (1984) and a modified version of the program MIX (Plummer *et al.*, 1975). Line A indicates the pathway corresponding to additional calcite dissolution accompanying acid production (e.g. by pyrite oxidation) maintaining a saturated solution. Line B indicates an example pathway of subsequent CO_2 degassing to atmospheric values accompanied by calcite precipitation to maintain equilibrium. Constructed by modelling using MIX.

Line A on Fig. 13.3 shows the effect of a series of increments of pyrite oxidation coupled with sufficient calcite dissolution to keep the solution just saturated with calcite. The effect is a 50% increase in the amount of dissolved calcium and an increase in the partial pressure of carbon dioxide from $10^{-3.5}$ to $10^{-3.2}$. Line B illustrates the degassing of this excess CO_2 accompanied by the precipitation of some calcite (much smaller than the initial calcite dissolution).

An additional effect which is significant in glacial environments is the enhanced solubility of very small particles. There is an equilibrium thermodynamic concept of excess surface free energy which allows predictions of values of enhanced solubility by the relationship:

$$K_{cc(d)} = K_{cc(\infty)} \times \exp(6A/RT)$$

where $K_{cc(d)}$ is the solubility product for calcite of diameter d, $K_{cc(\infty)}$ is the solubility product for very large crystals, σ is the excess surface free energy per unit area, A is the surface area of the solid per mole, R is the gas constant and T the absolute temperature. Given a value of $\sigma = 85$ ergs cm^{-2} (Morse and Mackenzie, 1990), the variations of solubility with grain size are shown in Fig. 13.4. The effect is relatively small until sizes less than 0.1 μm are considered. Nevertheless the presence of small amounts of very fine-grained material may determine the calcite solubility level until that material has totally disappeared.

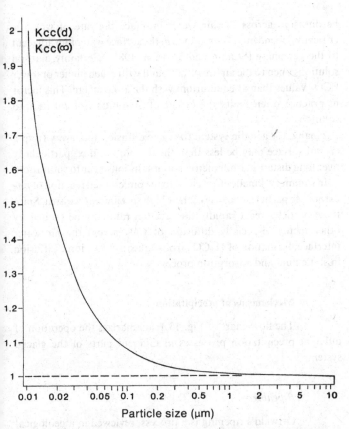

Fig. 13.4. Increase in calcite solubility with decreasing grain size (d) owing to the excess surface free energy of very small crystals. Solubility is expressed as the ratio of solubility product of calcite crystals of size d μm to solubility product of large (infinitely big) calcite crystals. Calculated from information in Morse and Mackenzie (1990).

Kinetic factors

Carbonate dissolution and precipitation are affected by other factors which are not modelled by classical equilibrium considerations. Firstly, it should be noted that solubility will increase with lattice strain and with the abundance of defects outcropping on the crystal surface. The mechanical processes of crushing and abrasion by which particles are comminuted will increase these factors, but it is not readily possible to quantify these effects. Qualitatively, it should be noted that these effects will tend to lead to more effective dissolution of glacial rock flour.

A second possible kinetic effect concerns chemical inhibition of dissolution and precipitation: an effect of great importance in marine carbonate chemistry (Morse and Mackenzie, 1990). The influence of simple anions and cations will probably be very small in (dilute) glacial waters, although there may be circumstances where organic molecules may physically cover or chemically adsorb onto mineral surfaces. However, the freshly created nature of mineral surfaces will tend to minimize this effect as far as dissolution is concerned.

A third effect concerns the rate of chemical reactions on the crystal surface which may limit the progress of dissolution (or

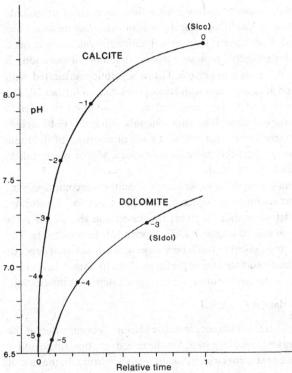

Fig. 13.5. Comparison of the dissolution kinetics of calcite and dolomite at 25 °C. Rate constants from Chou *et al.* (1989) were substituted in spreadsheets using algorithms of Raiswell (1984) in which the instantaneous dissolution flux at a series of solution compositions (at 0.1 pH intervals) was used to derive times for dissolution in each interval than relative dissolution times.

precipitation which can be regarded as the inverse process). Following Plummer *et al.* (1979), Chou *et al.* (1989) model their experimental data on calcite dissolution and precipitation in agitated solutions as the net effect of three forward and one backward reactions:

$$R = k_1 \times a(H^+) + k_2 \times a(H_2CO_3^*) + k_3 \times a(H_2O) - k_4 \times a(Ca^{2+}) \times a(CO_3^{2-})$$

where R is the reaction rate, k_1 to k_4 are rate constants, $a(\ )$ are the activities of chemical species, and $H_2CO_3^*$ refers to the sum of H_2CO_3 and dissolved CO_2.

Their results are used here to model calcite dissolution into a solution at equilibrium with atmospheric CO_2. Results show instantaneous dissolution rates diminishing from, for example, 9.3×10^{-11} mol cm^{-2} s^{-1} at pH 6.5 to 2.8×10^{-11} mol cm^{-2} sec^{-1} at pH 8.2. In Fig. 13.5 the progressive dissolution (indicated by pH and saturation index) is plotted against relative time derived from these dissolution rates in order to emphasize the decline in dissolution rates as saturation is approached.

Chou *et al.* (1989) also experimented on dolomite dissolution and modelled it using the expression:

$$R = k_1 \times a(H^+)^n + k_2 \times a(H_2CO_3^*)^n + k_3$$

Their results have been used to construct the dolomite dissolution curve of Fig. 13.5 which illustrates that in the initial stages of dissolution, dolomite dissolves more than an order of magnitude slower than calcite. Dolomite dissolution will thus be much less compared with calcite in a mixed dolomitic-calcitic sediment. Thermodynamically however, some dolomite dissolution is required to reach equilibrium. Given a solution saturated with calcite and at equilibrium with atmospheric CO_2, a further 20% of cations will need to be introduced into the solution in order to saturate for dolomite, following which the solution will be significantly supersaturated for calcite. Lower proportions of dolomite will dissolve where the calcite contains some Mg, or where calcite saturation is not reached.

The limitations of the experimental conditions become apparent when one attempts to deduce the absolute times for dissolution. Initially let us consider the time for dissolution should the rates from Chou and colleagues' (1989) experiments be applicable. The absolute times taken to reach each stage in the dissolution depends on the specific surface area of the mineral and its abundance within the solution. In the case of calcite, for a given step in the dissolution:

$$dm_{Ca} = R \times dt \times A$$

where dm_{Ca} is the difference in molar calcium concentration at the beginning and end of the step, R is the rate, dt the time taken and A the surface area across which dissolution is occurring. In the case of a 1% suspension of calcite grains in turbulent water, 54 g of calcite would be in contact with each litre of solution. Given a specific surface area (surface area per unit mass of crystal) of $S\ m^2\ g^{-1}$, each litre of solution would thus be in contact with $5.4 \times S \times 10^5\ cm^2$ surface area of calcite. Specific surface area and crystal size are inversely proportional: e.g. for 2 μm rhombohedra, S is around 1.1 $m^2\ g^{-1}$. The time taken to approach saturation is then directly proportional to crystal size: around 15 seconds for 1 μm crystals, 150 seconds for 10 μm crystals.

However the experiments of Chou et al. (1989) were conducted in situations where ionic supply was not a limiting factor, PCO_2 being allowed to fall as dissolution progressed. Other workers using fixed PCO_2 (e.g. Plummer et al., 1979) considered situations of high ratio of fluid volume to crystal surface areas where again there is no supply problem. Lasaga (1984) commented that the formation of H_2CO_3 from aqueous CO_2 could limit mineral dissolution and this is evaluated below. Given this same situation of a 1% suspension of 2 μm calcite grains in turbulent water, each litre of solution would thus be in contact with $5.4 \times S \times 10^5\ cm^2$ surface area of calcite, and the instantaneous dissolution flux would be $1.5 \times 10^{-10}\ mol\ cm^{-2}\ sec^{-1}$ (Chou et al., 1989, pH = 6) corresponding to a flux per unit volume of $8 \times 10^{-5}\ mol\ l^{-1}\ s^{-1}$. Under these conditions, use of the MIX modelling program indicates that there would be half a mole demand for aqueous CO_2 for each mole of $CaCO_3$ dissolved, so the CO_2 flux required is $4 \times 10^{-5}\ mol\ l^{-1}\ s^{-1}$. Using rate constants given in Lasaga (1984, for 25 °C and an ionic strength of 0.5) and for a solution initially in equilibrium with atmospheric CO_2, the maximum rate of formation of H_2CO_3 from aqueous CO_2 can be calculated as $4.37 \times 10^{-7}\ mol\ l^{-1}\ s^{-1}$. The supply of H_2CO_3 at the onset of dissolution is thus around two orders of magnitude less than that required by the dissolution flux.

In non-turbulent solution, the ultimate limiting factor could well

be diffusion across the air–water interface the rate of which is critically dependent on conditions in the surface water film adjacent to the gas phase (Stumm and Morgan, 1981). Certainly, natural solutions open to the air are often found with much higher or lower PCO_2 values than at equilibrium with the atmosphere. This factor in practice often limits the extent of dissolution of calcite into solution.

Finally, for a static system the ionic diffusion flux away from a crystal surface may be less than the dissolution flux, particularly over long distances (millimetres) or in solutions near to saturation.

In summary, kinetic effects lead to the preferential reaction of *fine* carbonate particles because of their high specific surface area. Since these particles react rapidly the reaction rate may be limited by other factors especially diffusion of CO_2 across the air–water interface, formation of H_2CO_3 from aqueous CO_2, ionic diffusion in static fluid and adsorption processes.

Mechanisms of precipitation

The flow chart of Fig. 13.1 summarizes the operation of different precipitation processes in different parts of the glacial system.

Ripening

Ostwald's ripening is a process, reviewed in a geological context by Morse and Casey (1988), in which small particles tend to dissolve and reprecipitate as overgrowths on larger crystals in a saturated solution. This tendency arises from the higher chemical potential (lower stability) of curved rather than flat interfaces, a phenomenon that becomes more significant in irregularly shaped or very small particles with high specific surface area. The rate of ripening is the mass dissolved or precipitated per unit time. The lower the temperature, the smaller the crystal size has to be for ripening to be significant. Attention is increasingly focusing on diagenetic applications (e.g. to opaline silica (Williams et al., 1985; burial diagenesis of clays, Eberl et al., 1990). Gregg et al. (1992) have evidence for the action of ripening in coarsening crystal size in Holocene dolomitic crusts at earth surface temperatures. Previously Fairchild et al. (1989) and Fairchild and Spiro (1990) had invoked early diagenetic ripening of glacial rock flour as an explanation for the chemistry of dolomitic matrices of certain Precambrian glacimarine sediments. These observations stimulated our current work.

A major problem concerns lack of knowledge of the rate of ripening at earth surface temperatures. Given that it is essentially a dissolution–reprecipitation phenomenon, ripening ought to occur much more rapidly in calcite than dolomite and there is some experimental information on calcite.

Ostwald's ripening was utilized by Lorens (1981) to carry out experimental determinations of trace element partition coefficients into calcite at slow growth rates. Doping with the radioactive species ^{45}Ca allowed a very sensitive determination of overall reaction rate which was found to be on average 3.7×10^{-11} mol $mg^{-1}\ min^{-1}$ (0.23% of the total material per day). Rates were constant for a given run over periods of between 100 minutes and 4 days. From SEM study Lorens (1981) estimated the specific surface area of the calcite (S) to be $0.8 \, m^2 \, g^{-1}$ which would correspond to a crystal size around 3 μm.

For faster growth rates stimulated by other mechanisms Lorens (1981) found a linear relationship between rate and $(\Omega - 1)$ where Ω is a saturation parameter defined as the ionic concentration product (corrected for free ions) divided by the concentration product at equilibrium. In the dilute solutions used Ω approximates to 10^{SI} as used in this paper. In the four slowest experiments ($\Omega < = 1.29$) the relationship was:

$$R = 15(\Omega - 1)$$

where R is the reaction rate. If this relationship between crystal growth rate and supersaturation is extrapolated to the ripening experiment, Ω is found to be 1.0025.

Saturations around $\Omega = 1 (SI = 0)$ can be thought of as a transition zone between a field of precipitation and a field of dissolution. Ripening is essentially a combination of dissolution and precipitation proceeding simultaneously. Also, dissolution and precipitation can be thought of as analogous processes symmetrically disposed about the bulk saturation level: hence they can each be described in theory by the same form of rate equation (Plummer et al., 1979; although experiments indicate variable fit to the theory depending on PCO_2). In the context of subglacially transported sediments an undersaturated static solution interacting with a calcite rock flour would gradually increase in its saturation level until at a bulk saturation level of around 1 some reprecipitation would start to occur. Ω would continue to rise as long as fine crystals were present with enhanced solubilities. The value of 1.0025 thus represents the level of Ω where the dissolution and precipitation rates are equal. Since for the precipitation process $\Omega - 1 = 0.0025$, dissolution should be driven by a similar difference between saturation for fine crystals and the actual solution composition, i.e. the saturation value for the finer crystals is 1.005 times the value for the bulk material.

This result can now be related with the theoretical increase in free energy of finer crystals related to their high surface area (Fig. 13.4). Given a bulk crystal size of 3 μm, ripening could be driven by a component higher in solubility by a factor of 1.005: this corresponds to a crystal size of around 1 μm.

A number of assumptions have been made in this analysis. On the one hand growth at very low supersaturations is likely to be slower than the linear extrapolation from higher rates assumed, and growth would also be slowed by any adsorbed impurities. If these factors were important the size of the finer material being dissolved would have to be rather smaller than 1 μm. Lattice strain and defects would have the opposite effect however, in that they enhance solubility and hence rate of reaction.

Given these uncertainties, directed experimental work is required and we have some work in progress experimentally ripening naturally and artificially comminuted materials. Initial results using $\delta^{18}O$ as a monitor of reaction progress indicate that chalk crushed in a Tema® mill in contact with isotopically depleted water showed no detectable ripening over a period of 50 days, indicative of at least

two orders of magnitude slower reaction than in Lorens' (1981) work. This may simply reflect the disintegration of the chalk into its primary micron-sized calcite crystals without crushing.

In summary, ripening is a process that can only occur in saturated solutions in contact with material of extremely fine crystal size. Given that the finest materials dissolve most readily in (flowing) water, ripening is therefore most likely to occur in the pore waters of subglacially transported sediments freshly released from glacial ice. The extent to which ripening occurs will be critically dependent on the crystal size distribution (abundance of the finest material) and the extent of lattice defects in the crystals.

Warming

The extent to which warming can promote precipitation is indicated by Fig. 13.3. Given an already saturated solution, warming could in principle result in the precipitation of a moderate fraction of the dissolved carbonate. In the terrestrial environment Swett (1974) argued that the warming of solutions drawn up by capillary effects above permafrost could be an agent in the precipitation of calcareous crusts on the upper surface of pebbles and cobbles. Otherwise, appreciable warming is likely only to take place in closed pools where it may have a much less significant effect than evaporation. Evaporation is also likely to be more important than warming in stimulating carbonate precipitation in formerly glacimarine environments following deglaciation.

Freezing

Experimental work by Hallet (1976) confirmed the potential importance of freezing in carbonate precipitation. Very little of the solute is trapped in the ice lattice, although the extent of supersaturation of the remaining solution depends on the extent to which carbon dioxide, which is also concentrated in the solution, remains dispersed or forms bubbles. Although precipitation only occurred in the final stages of Hallet's experiments, the initial solutions were undersaturated, and precipitation throughout the course of freezing would be expected if the initial fluid were saturated (Sharp et al., 1990).

Freezing effects occur in response to regelation, i.e. pressure changes along bedrock surfaces in the subglacial environment where the carbonate precipitates are well known, but it should also occur at advancing permafrost fronts and close to the ground surface, where they are not well distinguished from other precipitation mechanisms (e.g. Sletten, 1978).

Common ion effect

Supersaturation for calcite may be achieved by the dissolution of minerals which have an ion in common with calcite (e.g. gypsum or, as mentioned earlier, dolomite). This was the mechanism concluded by Atkinson (1983) to account for the bulk of the cave carbonates in the Castleguard area beneath the Columbia Icefield of western Alberta, Canada.

Changes in CO_2 and alkalinity

Partial pressures of CO_2 will be increased by acid input to a solution. The case of acid introduction through pyrite oxidation has already been discussed, arrow B on Fig. 13.3. illustrating the potential for a small amount of carbonate precipitation if the solution degasses to atmospheric CO_2 pressure. This was inferred to be a subordinate precipitation mechanism in the Castleguard area previously mentioned (Atkinson, 1983).

Clark and Hansel (1989) describe 20–30 cm cemented sediment at the interface of lodgement till with underlying sand at a locality in Illinois. They attribute cementation to degassing of CO_2 from subglacial water saturated with CO_2 by regelation processes as it passed into the sands. However, since such cementation has not been documented in a modern context, it might be post-glacial.

Very high CO_2 pressures result from the oxidation of organic matter in soils, but in glacial conditions organic oxidation is not likely to be important, except possibly if quantities of fresh organic matter were incorporated in tills. Interglacial soil formation on calcareous tills will yield carbonate-rich waters with much potential for carbonate precipitation in caverns or porous lithologies where degassing is possible.

Photosynthesis provides another form of CO_2-removal, although microbial photosynthesis is the only likely form in glacial settings. Lawrence and Hendy (1989) for instance have documented photosynthetically triggered calcite precipitation in the water column of a lake in the Dry Valley region of Antarctica, and calcitic stromatolites are known there also (e.g. Wharton et al., 1982). Photosynthesis was also invoked by Sharp et al. (1990) as a possible triggering agent for some subglacial precipitates with heavy $\delta^{13}C$ values.

Microbial decomposition of organic matter raises alkalinity significantly and can trigger early diagenetic carbonate concretion formation in glacilacustrine and glacimarine muds (e.g. Lamothe et al., 1983) and tills with derived organic matter (Fairchild and Spiro, 1990).

Evaporation and transpiration

The arid Dry Valley region of Antarctica includes some excellent examples of evaporites forming in lakes, pools and as subaerial crusts (e.g. Burton, 1981 and review by Fairchild et al., 1989 and Fairchild and Spiro, 1990). The waters are isotopically enriched in deuterium and especially $\delta^{18}O$ compared with their starting compositions.

Evaporation is arguably the major cause of the precipitation of crusts of calcite near to the sediment surface in tundra areas (e.g. Swett, 1974), although freezing could also be important (Sletten, 1978). In areas such as interior Alaska which display a continental climate and hence are forested despite subzero annual average temperatures, significant fine carbonate precipitation in the soil zone can be attributed principally to transpiration through the trees (Marion et al., 1991).

Overall, the effects of evaporation could only be significant in glacial areas in small water bodies (or larger bodies in very arid conditions), or in soil zones. However, the potential for carbonate dissolution and reprecipitation in soil zones in post-glacial conditions with warmer temperatures and more extensive biotas is clearly larger. The climatic limit of lithified calcretes is not clear, but is likely to be determined principally by summer temperatures or aridity.

Skeletal biomineralization

The ultimate removal mechanism for much carbonate in glacial lakes and seas is in the form of skeletal ostracods, foraminifera, molluscs etc. This precipitation normally occurs from undersaturated waters. Shells may be found dispersed in diamicts (e.g. Eyles and Schwarz, 1991) or concentrated as lag deposits (Eyles and Lagoe, 1989) or in areas receiving little glacial sediment (Domack, 1988).

Recent examples

Some illustrations of present-day phenomena are given here from ongoing work. Two sites frequently referred to in the text are the Tsanfleuron glacier of western Switzerland (location map in Sharp et al., 1990) and the Saskatchewan glacier of the Canadian Rockies (Ford, 1983).

Carbonate saturations of proglacial waters

Study of meltwater chemistry gives some indication of their history (Raiswell, 1984) and potential for carbonate dissolution or precipitation. Fig. 13.6 gives an overview to illustrate the complexity of the system in which carbonate saturation is approached across a wide range of CO_2 pressures. These data are from alpine glaciers at altitudes of 1800 m (Saskatchewan glacier) to 2400 m (Tsanfleuron glacier) where the atmospheric CO_2 pressure is closer to $10^{-3.6}$ than $10^{-3.5}$. Open-system weathering would lead to a horizontal distribution of points at around $pP_{CO_2} = 3.6$, but this is not the dominant feature. A series of points define a diagonal line from lower left to upper right which is approximately the course of closed-system weathering, but these include ices where CO_2 may have also have been excluded from the ice during the freezing process. Whilst many data points lie in the upper (CO_2-depleted) part of the diagram reflecting at least some carbonate dissolution out of contact with the atmosphere, others near the base require more acid than could be supplied from atmospheric sources. These latter points are restricted to the Saskatchewan glacier area and the north flank of the Tsanfleuron glacier, both areas where clasts bear pyrite, oxidation of which could have supplied the additional acid source. Some of these waters, with alkalinities of up to 200 mg $CaCO_3$ per litre, are carbonate-saturated, presumably by coupled calcite dissolution. Calculations indicate that CO_2-degassing could lead to moderate oversaturation (to around $SI_{cc} = 0.7$).

In most cases, the surface waters and melted ice and snow samples would be capable of continued dissolution of suspended or

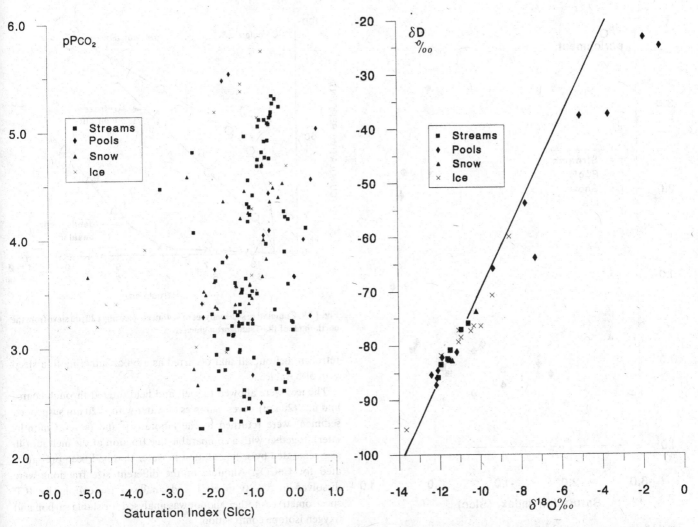

Fig. 13.6. Data on CO_2 pressure (expressed as pP_{CO_2}) plotted against saturation index for meltwaters and (melted) surface snow and ice samples in glacial areas on carbonate bedrock. Data are derived from primary measurements of fluid temperature, pH, alkalinity and cation concentrations. The biggest source of error is likely to be from pH determinations; error of ± 0.1 pH unit gives an error of ± 0.1 pP_{CO_2}. Most results were calculated from unpublished data of Roger (1978) on the Tsanfleuron glacier, supplemented by further observations there in 1990 and 1991, and at the Marmolada glacier in 1990 and the Saskatchewan glacier in 1991.

Fig. 13.7. Istopic compositions of meltwaters, ice and snow samples from the Tsanfleuron and Marmolada areas, 1990, plotted as δD versus $\delta^{18}O$ (parts per thousand versus SMOW). Note the adherence of most points to the meteoric water line ($\delta D = 8\delta^{18}O + 10$).

settled sediment, but the near-saturation achieved in low-discharge waters away from areas of snow and ice-melt indicates that sediment porewaters are at or near saturation. This could allow ripening processes to proceed.

Supersaturation is found in certain pools, the causative agent for which can be demonstrated by isotopic data. Fig. 13.7 illustrates the tendency for water samples to follow the meteoric water line which typefies atmospheric precipitation. Many pools are heavy in both deuterium and oxygen indicative of significant water supply from summer rainfall rather than ice and snowmelt. Positive deviations of $\delta^{18}O$ (with small δD enrichments) from the meteoric water line characterize evaporative processes, which lead to evaporation

trajectories with a slope of around 5.5. As a simple index of evaporation, the difference between the $\delta^{18}O$ value of the water and its appropriate value for the meteoric water line (calculated from δD) has been determined. In Fig. 13.8, this is plotted against saturation and it is seen that all of the evaporated waters are also mildly supersaturated. Evaporation can thus have significant effects in an Alpine glacial environment. Additionally, one of the data points, with a moderate $\delta^{18}O$ enrichment, came from a small pool with actively photosynthesizing microbiol mats and pH of 9.4. Here CO_2 withdrawal by photosynthesis is clearly important. We thus predict that carbonate precipitates could form in ephemeral pools. Fig. 13.8 illustrates that supersaturation can also be reached by water bodies experiencing no significant isotopic enrichment from evaporation; this is principally dependent on sufficient contact with carbonate rock flour coupled either with the common ion effect, degassing or warming in particular cases.

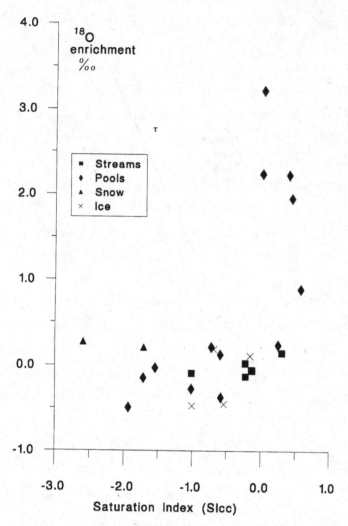

Fig. 13.8. Heavy-oxygen enrichment plotted against saturation index for 1990 water and ice samples from the Tsanfleuron and Marmolada areas. The enrichment equals the horizontal distance of the data point from the meteoric water line of Fig. 13.7.

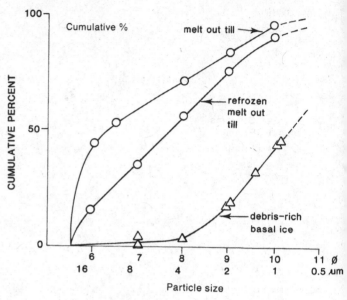

Fig. 13.9. Cumulative size analyses of sediments passing a 20 μm sieve from the north flank of the Tsanfleuron glacier.

refrozen meltout till that occurred as a block adhering to a steep wall of basal ice.

The ices were allowed to melt and field-filtered through coarse and fine (20 μm) sieves. Samples of water with <20 μm suspended sediment were returned to the laboratory and (several months later), together with a comparable fine fraction of the meltout till, size-separated by sedimentation in water that had been pre-saturated for $CaCO_3$. Aliquots of the different size fractions were dissolved in 5% (v/v) HCl and analysed for cations by ICP spectrometry. Another aliquot was analysed for stable carbon and oxygen isotope composition.

Grain-size distributions were determined by traditional setting methods in solutions pre-saturated for calcite. The three samples of <20 μm sediment (Fig. 13.9) display a notably fine-skewed distribution for the debris-rich basal ice with only 50% of the sediment being greater than 1 μm in size. Conversely the other samples showed a more even relationship between abundance and φ size fractions, with cumulative percentages approaching 100 at the 1 μm level. Insoluble residue determinations are in the range of 70–80% for the basal ice and increase from around 60–65% in the <16 and <8 μm sizes of the meltout materials to 95% and 80% in the <1 μm fraction of the meltout till and refrozen meltout tills respectively.

The data show that much more fine sediment is present in the basal ice as compared with the meltout materials. Assuming that the sample of debris-rich ice is representative (an assumption which may not be valid), this suggests the action of at least two processes: dissolution and removal in suspension. The finest material has enhanced solubility and dissolution kinetics as discussed earlier. Melting, and later passage of water through the resulting meltout till will enhance this selective dissolution effect. However, even total carbonate dissolution would be inadequate to account for the difference between the debris-rich basal ice and the meltout tills. Quantitative removal of both carbonate and siliciclastic material in

Processes in the ice-marginal meltout zone

To study the diagenetic changes involving reactive carbonate in diamicts, we have tried to locate a field setting with a single source lithology, and where the age of diamicts and their processes of formation is well constrained. In practice, this proves extremely difficult!

We are currently investigating material from the north flank of the Tsanfleuron glacier, close to the locality studied by Hallet *et al.* (1978), where detrital material appears to be exclusively from Hauterivian (Cretaceous) muddy limestones of the Wildhorn nappe, outcrops of which bound this flank of the glacier. The ice margin is characterized by complexly deformed basal ice reflecting lateral transport of subglacially transported debris. Hallet *et al.* (1978) argued that at least part of this basal ice formed during regelation and was in contact with a calcite-saturated water. We have sampled basal ice, adjacent meltout till, and a sample of

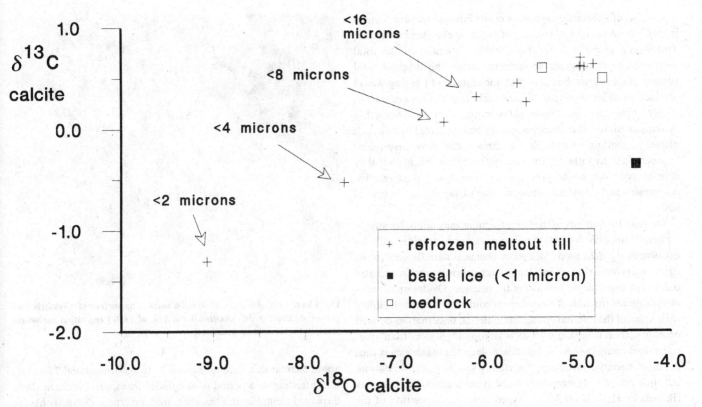

Fig. 13.10. Carbon–oxygen isotopic cross plot (parts per thousand versus PDB) of carbonates from the north flank of the Tsanfleuron glacier. Sieved coarser fractions of the refrozen meltout till cluster with local Cretaceous (Hauterivian limestone) bedrock composition. Finer fractions obtained by sedimentation methods are increasingly negative on a linear trend which is interpreted as reflecting a mixture of finer authigenic carbonate with the coarser clastic carbonate.

suspension in meltwaters is also required, and this is consistent with the typically turbid nature of the meltwater at the site. Carbonate is also prone to further dissolution in turbulent meltwater. Thus ice margins are prone to export a high proportion of the most reactive fine carbonate in suspension and solution. Despite this, we have evidence for the formation of some new calcite in this environment, as is explained below.

Isotopic data obtained to date is presented in Fig. 13.10, and chiefly relate to different size fractions of the refrozen meltout till. The analysis of medium silt to sand-sized fractions (separated by laboratory sieving) yielded very homogeneous compositions, similar to the composition of two samples of bedrock from Hauterivian outcrops. The finest fractions obtained by settling methods, and which therefore contained a mixture of sizes, show a progressive decrease in both $\delta^{13}C$ and $\delta^{18}O$ along a linear trend. A $<1\ \mu m$ fraction from basal ice showed some ^{13}C-depletion compared with coarser clasts, but lacked the ^{18}O-depletion characterizing the refrozen meltout till. Isotopic analysis of the ice in the refrozen meltout till yielded a $\delta^{18}O$ compositions of $-10.3‰_{SMOW}$.

ICP data on separated size fractions from the three samples showed no clear trends in Mg, Mn and Sr in the three sediments. There is a general increase in Fe with decreasing grain size in the three samples, attaining values of 20 000–40 000 ppm in the <2 and $<1\ \mu m$ fractions.

The linear trend of the isotopic data for the refrozen meltout till is consistent with a physical mixture of two components, a detrital material, whose composition is well-constrained with $\delta^{18}O$ around $-5‰_{PDB}$, and an ^{18}O-depleted authigenic component. The latter is at least as isotopically light as the lightest sample analysed ($-9.1‰_{PDB}$).

There are at least two possible origins for the authigenic carbonate. Firstly, it could have formed largely by ripening. Calcite of composition $-9.1‰_{PDB}$ would precipitate in equilibrium with water of composition $-10.3‰_{SMOW}$ at a temperature of around 10 °C, which is well within the range of surface summer temperatures. Alternatively the authigenic component could have formed largely by freezing. There are fractionation effects when water freezes which results in residual water depleted in ^{16}O. Modelling using algorithms of Jouzel and Souchez (1982) which assume isotopic equilibrium indicates that the carbonate precipitated during progressive freezing of water with isotopic composition of $-10.3‰_{SMOW}$ would become progressively lighter in ^{18}O, reaching a composition of $-9.1‰_{PDB}$ when around 50% of the water had frozen. This composition also lies within the range (-5 to

− 12‰$_{PDB}$) of subglacial carbonate crusts inferred to have formed from different degrees of freezing of basal ice elsewhere under the Tsanfleuron glacier (Sharp *et al.*, 1990). Estimates of the total weight of the authigenic component versus the original total volume of ice suggest however that one episode of freezing would produce an order of magnitude less carbonate than is present, so multiple episodes of freezing would be required, even assuming that no redissolution of the authigenic component occurred during each phase of melting. Although we cannot currently distinguish between these hypotheses, they are both comparable in that they refer to processes which progressively remove the most reactive component and create an authigenic one of slightly larger crystal size.

The high Fe contents of the finest fractions may not be located in carbonate since finely crystalline iron oxides or hydroxides will also dissolve in the acid used. Meltwater chemical data suggest pyrite oxidation is occurring where Hauterivian debris is present and iron oxide is an expected by-product of the reaction. Oxidation of some of the organic fraction of the sediment could account for the lower $\delta^{13}C$ value of the inferred authigenic material, since marine-derived organic carbon is typically 25–28‰ isotopically lighter than associated carbonate carbon. A bacterial role in the diagenesis is thus possible; bacterial involvement certainly speeds pyrite oxidation, but their action in carbonate-buffered systems generally is unclear (Bierens de Haan, 1991). The oxygen isotope composition of the < 1 μm sample from the basal ice does not evidence any authigenic component, although the carbon isotope value is lighter than expected for detrital material. The reason for this discrepancy is unknown.

Although this work is as yet incomplete, the identification of a dispersed authigenic fine-grained carbonate component in a glacial medium is intriguing, and clearly warrants continuing investigation of diamicts in neoglacial and Pleistocene settings.

Crusts on clasts: regelation versus vadose zone phenomena

Regelation is the process by which ice melts in response to high pressure on the upflow side of obstacles and refreezes on the lee side. Refreezing is associated with carbonate precipitation and there is an extensive literature on the fibrous and micritic crusts that develop on bedrock surfaces (e.g. Sharp *et al.*, 1990). In certain Alpine tills, Souchez and Lemmens (1985) observed carbonate encrusted pebbles associated with calcite-lined fissures and hypothesized that these features related to regelation in the basal ice layer, rather than to subsequent pedogenic processes. This awaits confirmation by detailed description. Here we offer some new observations on pebble coatings.

At the Saskatchewan glacier thrusting brings abundant debris-rich basal ice to the glacier surface and the debris accumulates by meltout. Fig. 13.11 illustrates part of the surface of a striated boulder where cream-coloured micritic crusts occur in small depressions on the surface; crusts also cover much of the lee face of the boulder. A random sample of 50 cobbles and pebbles larger than

Fig. 13.11. Part of surface of striated boulder on surface of Saskatchewan glacier illustrating the localized formation of (pale) regelation carbonate precipitates in leeside depressions.

about 5 cm at this locality showed 31 to be striated and 7 faceted, thus attesting to a period of subglacial transport. Nineteen clasts displayed submillimetre thickness micritic crusts, distinctly harder and of different colour from nearby clastic matrix. Crusts were nucleated more abundantly on limestone (10 of 15) than mixed limestone-dolostone (7 of 22) or dolostone (2 of 13). Since many clasts were either not striated or else striated in several directions, the relationship of crust distribution to that of leeside cavities was not very obvious, crusts tending to occur in crevices in various orientations on clast surfaces.

A contrasting form of crust was found on clasts at the surface of fluted moraines that had been exposed for up to a few decades in front of the Saskatchewan glacier. Typically, the precipitate occurs as a creamy-white skin concentrated either on pebble tops (Fig. 13.12) or on their lower surfaces just above where embedded in diamict. Examination of 50 clasts (representing all the clasts > 2 cm in a 60 × 30 cm area) at a moraine crest showed 24 to bear crusts (14 of 19 on limestone; 9 of 15 on mixed limestone-dolostone and 1 of 16 on dolomite). Rather than filling depressions, the crusts formed a submillimetre smooth skin over larger areas of the clast surfaces, tapering gradually at the edges. Both evaporation and freezing are probably involved in their formation, the role of freezing being particularly clear for nearby occurrences of similar material within the cracks of frost-shattered clasts. Incipient examples of crusts on clast tops have also been seen on fluvial gravels near the Tsanfleuron glacier.

Both these types of crust could be confused with subsurface calcrete crusts, a point that is taken up in the discussion. Here we confirm that regelation crusts on clasts exist, but that they tend to cover only a small part of clast surfaces, being concentrated in depressions. Whilst some could have developed on bedrock surfaces prior to erosion, others occur on different surfaces and seem to

Fig. 13.12. Surface carbonate crusts on the crests of limestone pebbles collected from the moraine surface in front of the Saskatchewan glacier.

Fig. 13.13. Late glacial (*c.* 18000 BP) fluvial gravels, from Somersham, Cambridgeshire (West, 1991) exhibiting two periods of calcrete formation. A contemporaneous calcrete crust is reworked as intraclasts (along the horizon of the black lens cap) into succeeding fluvial deposits. Modern white carbonate precipitates around black roots occur in centre right area.

Fig. 13.14. Collection of pebbles taken from the Newburg pit, Wisconsin, USA illustrating the preferential cementation of sandy matrix beneath clasts where water drips concentrate.

Fig. 13.15. Advanced cementation of openwork gravels (same site as Fig. 13.14) with clear dripstone morphology above lens cap and complex pattern of hanging sheets (right).

have formed since erosion. Like the regelation types, crusts on clasts exposed in the proglacial zone prefer to nucleate on limestone, but cover more of the clast surface. There is no simple relationship to gravity unlike other vadose zone phenomena.

Post-glacial transformations

Carbonate rock flour in glacial deposits exposed postglacially on the land surface is highly susceptible to modification. On the one hand, the warmer climate encourages plant growth and resulting decay in organic matter leads to high CO_2 pressures in the soil zone which enhances carbonate dissolution. Conversely, evaporation is intensified, increasing the likelihood of soil carbonate precipitation.

We have noted conspicuous carbonate precipitation associated with modern roots in near surface tills in the North American Great Lakes region and East Anglia (UK), areas which would normally be climatically marginal for soil carbonate precipitation. In one case (Fig. 13.13) this Holocene carbonate is seen to merge spatially into a temporally separate epsiode of calcrete crust formation in the original late-glacial fluvial sediments. The timing of this earlier period of crust formation is fixed by the reworking of the crust as intraclasts in the fluvial sediments (Fig. 13.13).

In the shallow subsurface, dissolution of carbonate by percolating waters would tend to lead to the selective removal of finer calcite versus coarser calcite, and versus dolomite. Here there is also considerable scope for reprecipitation of calcite in caverns or porous sediments where CO_2-degassing can occur.

A post-glacial, rather than glacial origin of calcite cement can be deduced by geochemical or textural evidence. For example, presumed subglacial sands and gravels beneath lodgement till at Blackhall Rocks, County Durham, UK, are cemented by calcite with isotopic composition $\delta^{13}C = -10.28$ and $\delta^{18}O = -5.44$,

which points to Holocene precipitation by degassing of waters enriched in light carbon from the soil zone.

Another example is from drumlin sands and gravels in Wisconsin, where patchy calcite cementation occurs. Incipient cementation is concentrated beneath pebbles (Fig. 13.14) and more advanced stages are characterized by dripstone morphologies (Fig. 13.15), both features typical of vadose diagenesis (a subglacial origin can certainly be ruled out). This vadose diagenesis most probably occurred post-glacially when enhanced opportunities for CO_2-degassing were present. These conclusions are consistent with those of Aber (1979) who recognized similar styles of vadose cementation in both late-glacial and postglacial conglomerates on the Appalachian Plateau of New York State.

Application to glacial sedimentary sections

Carbonate ranges from sparse to dominant in ancient glacial sediments, but we are some way from fully utilizing the information it can potentially yield. This section is to provide some guidelines. The morphology of secondary carbonate precipitates gives information on diagenetic environments through which a glacial sediment has passed, whilst their chemical and isotopic composition can yield depositional or diagenetic information (Fairchild and Spiro, 1990).

Crusts on clasts

In principle crusts would form subglacially, proglacially, pedogenically following climatic warming or, in older sequences, during burial diagenesis or metamorphism.

Regelation crusts are restricted to clasts which have not been abraded following subglacial transport, that is they will only occur in diamicts. They may be micritic or fibrously sparitic and can incorporate sediment grains (e.g. Sharp et al., 1990). Our observations suggest a concentration in crevices on clast surfaces.

Proglacial crusts can develop on clasts at the land surface, but since such surfaces are liable to erosion they probably have a low preservation potential. Crusts attached to angular fragments formed by frost-shattering could be distinctive.

Circumgranular micritic or sparitic crusts are a very common phenomenon in both pedogenic and groundwater calcretes in response to crystal growth force in strongly supersaturated solutions (e.g. the groundwater calcretes of periglacial salt lakes, Fairchild et al., 1989). Crusts will tend to be thicker and more continuous than those mentioned above, especially under warmer climatic conditions, and may be less discriminating as to the lithology of the host.

Burial diagenetic reaction rinds will tend to occur on all sides of clasts, whilst tectonic pressure fringe-type fibrous growths are characteristically oriented parallel to the cleavage trace (Fairchild, 1985).

Diamict matrix and muds

Whilst the sand and gravel fraction of diamicts can readily be identified as unaltered detritus, in carbonate-rich diamicts the matrix could have a complex composite origin. We have shown that the glacially transported rock flour component is liable to dissolution, the degree of which is critically dependent on the amount of fluid that has interacted with the sediment. In relatively closed

system situations, ripening, or repeated freezing and thawing could lead to recrystallization of the finest matrix. There is also a limited role of precipitation forced by the common ion effect or following CO_2-loss of sulphate-rich waters. Evaporation could lead to localized micrite formation in surface sediments, e.g. in muds within fluvioglacial deposits.

We have identified an authigenic component in a modern meltout till, but have yet to investigate in the same detail its counterparts in the historical or geological record. Nor yet do we have data on lodgement till. Quaternary evidence for ripening of glacimarine deposits is as yet sparse, although an abstract by Clasen et al. (1991) describes the special role of dissolution of ice-rafted carbonate in allowing authigenic carbonate precipitation in North Atlantic deep-sea sediments of Pleistocene age. Although not yet proven, we are working at the possibility of using an authigenic component of diamict matrix to distinguish marine and non-marine glacial sediments.

Sorted sands and gravels

The processes observed in modern glacial systems do not suggest a rapid cementation of sand and gravel facies within glacigenic successions in glacial climates. Therefore, there is a high probability that calcite-cemented gravels reflect post-glacial activity or, in the older deposits, burial processes. An exception may be the development of precipitates in the near-surface zone of outwash plains subject to evaporation and freezing processes.

Acknowledgements

Numerous collaborators and other friends have helped us in gathering these data, particularly Roland Souchez and Jean-Louis Tison (Free University, Brussels), Martin Sharp, Bryn Hubbard, Richard West and Phil Gibbard (University of Cambridge), Mike Hambrey (Liverpool John Moores University), David Mickelson (University of Wisconsin, Madison) and Nick Eyles, Jo Boyce and Mike Gipp (University of Toronto, Scarborough). This work is supported by NERC grant GR3/7687 to IJF.

References

Aber, J.S. (1979). Glacial conglomerates of the Appalachian Plateau, New York. Quaternary Research, 11, 185–96.
Atkinson, T.C. (1983). Growth mechanism of speleothems in Castleguard Cave, Columbia Icefield, Alberta, Canada. Arctic and Alpine Research, 15, 523–36.
Bierens de Haan, S. (1991). A review of the rate of pyrite oxidation in aqueous systems at low temperature. Earth Science Reviews, 31, 1–10.
Boulton, G.S. (1978). Boulder shapes and grain size distributions of debris as indicators of transport paths through a glacier and till genisis. Sedimentology, 25, 773–99.
Burton, R. (1981). Chemistry, physics and evolution of Antarctic saline lakes. Hydrobiologica, 82, 339–62.
Chou, L., Garrels, R.M. & Wollast, R. (1989). Comparative study of the kinetics and mechanisms of dissolution of carbonate minerals.

Chemical Geology, **78**, 269–82.

Clark, P.U. & Hansel, A.K. (1989). Clast ploughing, lodgement and glacier sliding over a soft glacier bed. *Boreas*, **18**, 201–7.

Clasen, S., Jantschik, R. & Meischner, D. (1991). Authigenic dolomite in the North Atlantic deep sea. In *Dolomieu Conference on Carbonate Platforms and Dolomitization Abstracts*, Ortisei, Italy, p. 53.

Domack, E.W. (1988). Biogenic facies in the Antarctic glacimarine environment: basis for a polar glacimarine summary. *Palaeogeography, Palaeoclimatology, Palaeoecology*, **63**, 357–62.

Dreimanis, A. & Vagners, U.J. (1969). Lithologic relation of till to bedrock. In *Quaternary geology and climate*, ed. H.E. Wright Jr, Royal Society of Canada, Ottawa, 11–49.

Dreimanis, A. & Vagners, U.J. (1971). Bimodal distribution of rock and mineral fragments in basal tills. In *Till: a symposium*, ed. R.P. Goldthwait, Columbus: Ohio State University Press, pp. 237–50.

Drewry, D.J. (1986). *Glacial geologic processes*. London: Edward Arnold, 176 pp.

Eberl, D.D., Srodon, J., Krakle, M., Jdyler, B.E. & Peterman, D.E. (1990). Ostwald ripening of clays and metamorphic minerals. *Science*, **248**, 474–7.

Eyles, N. & Lagoe, M.B. (1989). Sedimentology of shell-rich deposits (coquinas) in the glaciomarine upper Cenozoic Yakatage Formation, Middleton Island, Alaska. *Geological Society of America Bulletin*, **101**, 129–42.

Eyles, N. & Schwarz, H.P. (1991). Stable isotope record of the last glacial cycle from lacustrine ostracodes. *Geology*, **19**, 257–60.

Fairchild, I.J. (1985). Petrography and carbonate chemistry of some Dalradian dolomitic metasediments: preservation of diagenetic textures. *Journal of the Geological Society of London*, **142**, 167–85.

Fairchild, I.J., Hambrey, M.J., Jefferson, T.H. and Spiro, B. (1989). Late Proterozoic glacial carbonates in NE Spitsbergen: new insights into the carbonate-tillite association. *Geological Magazine*, **126**, 469–90.

Fairchild, I.J. & Spiro, B. (1990). Carbonate minerals in glacial sediments: geochemical clues to palaeoenvironment. In *Glacimarine environments: processes and sediments*, ed., J.D. Scourse & J.A. Dowdeswell. Geological Society of London Special Publication, **53**, 241–56.

Fairchild, I.J. (1993). Balmy shores and icy wastes: the paradox of carbonates associated with glacial deposits in Neoproterozoic times. *Sedimentology Review*, **1**, 1–16.

Ford, D.C. (1983). The physiography of the Castleguard Karst and Columbia Icefields area, Alberta, Canada. *Arctic and Alpine Research*, 15, 427–436.

Gregg, J.M., Howard, S.A. and Mazzullo, S.J. (1992). Early diagenetic recrystallization of Holocene (<3000 years old) peritidal dolomites, Ambergris Cay, Belize. *Sedimentology*, **39**, 143–60.

Haldorsen, S. (1981). Grain-size distribution and its relation to glacial crushing and abrasion. *Boreas*, **10**, 91–105.

Hallet, B. (1976). Deposits formed by subglacial precipitation of $CaCO_3$. *Geological Society of America Bulletin*, **87**, 1003–15.

Hallet, B., Lorrain, R. & Souchez, R. (1978). The composition of basal ice from a glacier sliding over limestones. *Geological Society of America Bulletin*, **89**, 314–20.

Jouzel, J. & Souchez, R.A. (1982). Melting–refreezing at the glacier sole and the isotopic composition of the ice. *Journal of Glaciology*, **28**, 35–42.

Lamothe, M., Hillaire-Marcel, C. & Pagé, P. (1983). Découverte de concrétions calcaires striées dans le till de Gentilly, basses-terres du Saint-Laurent, Québec. *Canadian Journal of Earth Sciences*, **20**, 500–505.

Lasage, A.C. (1984). Chemical kinetics of water-rock interactions. *Journal of Geophysical Research*, **89**, B6, 4009–25.

Lawrence, M.J.F. & Hendy, C.H. (1989). Carbonate deposition and Ross Sea ice advance, Fryxell Basin, Taylor Valley, Antarctica. *New Zealand Journal of Geology and Geophysics*, **32**, 267–77.

Lorens, R.B. (1981). Sr, Cd, Mn and Co distribution coefficients in calcite as a function of calcite precipitation rate. *Geochimica Cosmochimica Acta*, **45**, 553–61.

Marion, G.M., Introne, D.S. & Van Cleve, K. (1991). The stable isotope geochemistry of $CaCO_3$ in the Tanana River floodplain of interior Alaska, U.S.A.: composition and mechanisms of formation. *Chemical Geology (Isotope Geoscience Section)*, **86**, 97–110.

Morse, J.W. & Casey, W.H. (1988). Ostwald processes and mineral parageneses in sediments. *American Journal of Science*, **288**, 537–60.

Morse, J.W. & Mackenzie, F.T. (1990). *Geochemistry of sedimentary carbonates*, Amsterdam: Elsevier, 707 p.

Plummer, L.N., Parkhurst, D.L. & Kosiur, D.R. (1975). MIX2, a computer program for modeling chemical reactions in natural waters. *United States Geological Survey, Water Resources Investigations Report*, 61, 75 pp.

Plummer, L.N., Parkhurst, D.L. & Wigley, T.M.L. (1979). Critical review of the kinetics of calcite dissolution and precipitation. In *Chemical modelling in aqueous systems: speciation, sorption, solubility, and kinetics*, ed., E.A. Jenne. American chemical society symposium series, 93, 537–73.

Raiswell, R. (1984). Chemical models of solute acquisition in glacial melt waters. *Journal of Glaciology*, **30**, 49–57.

Roger, M. (1978). Contribution à l'étude des processus géomorphologiques en bordure d'un glacier alpine sur substrat calcaire. Université libre de Bruxelles, unpublished Ph.D. thesis.

Rumpf, H. (1973). Physical aspects of comminution and new formulation of a Law of Comminution. *Powder Technology*, **7**, 145–59.

Sharp, M. (1991). Hydrological inferences from meltwater quality data: the unfilled potential. British Hydrological Society 3rd National Hydrology Symposium, Southampton, 5.1–5.7.

Sharp, M., Tison, J.-L. & Fierens, G. (1990). Geochemistry of subglacial calcites: implications for the hydrology of the basal water film. *Arctic and Alpine Research*, **22**, 141–52.

Simpkins, W.W. & Mickelson, D.M. (1990). Groundwater flow systems and geochemistry near the margin of the Burroughs Glacier. In *Proceedings of the Second Glacier Bay Science Symposium*, ed., A.M. Milner & J.D. Wood, National Park Service, Anchorage, Alaska, pp. 26–33

Slatt, R.M. & Eyles, N. (1981). Petrology of glacial sand: implications for the origin and mechanical durability of lithic fragments *Sedimentology*, **28**, 171–83.

Sletten, R.S. (1978). The formation of pedogenic carbonates on Svalbard: the influence of cold temperatures and freezing. In *Permafrost Fifth International Conference Proceedings, Volume 1*, ed., K. Sesseset, Tapir, Trondheim, pp. 467–72.

Souchez, R.A. & Lemmens, M. (1985). Subglacial carbonate deposition: an isotopic study of a present-day case. *Palaeogeography, Palaeoclimatology, Palaeoecology*, **51**, 357–64.

Souchez, R.A. & Lemmens, M. (1987). Solutes. In *Glacio-fluvial sediment transfer*, ed., A.M. Gurnell & M.J. Clark. New York: Wiley, pp. 285–303.

Steier, K. & Schonert, K. (1972). Verformung und Bruchphanomene unter Druckbeanspruchung von sehr kleinen kornen aus Kalkstein, Quartz und Polystyrol. *Europ. Symp Zerkleinen*, Cannes.

Stumm, W. & Morgan, J.J. (1981). *Aquatic chemistry*. New York: Wiley, 780 pp.

Swett, K. (1974). Calcrete crusts in an Arctic permafrost environment. *American Journal of Science*, **274**, 1059–63.

Tranter, M., Brown, G., Raiswell, R., Sharp, M. & Gurnell, A. (1993). A conceptual model of solute acquisition by Alpine meltwaters. *Journal of Glaciology*, **39**.

West, R.G. (1991). Somersham, Cambridgeshire (TL 375795). In *Central East Anglia and the Fen Basin*, ed., S.G. Lewis, C.A. Whiteman, &

D.R. Bridgland, London: Quaternary Research Association, Field Guide, pp. 120–2.

Wharton, R.A., Parker, B.C., Simmons, G.M., Seaburg, K.G. & Love, F.G. (1982). Biogenic calcite structures forming in Lake Fryxell, Antarctica. *Nature*, **295**, 403–5.

Williams, L.A., Parks, G.A. & Crerar, D.A. (1985). Silica diagenesis, I. Solubility controls. *Journal of Sedimentary Petrology*, **55**, 301–11.

14 A Permian argillaceous syn- to post-glacial foreland sequence in the Karoo Basin, South Africa

JOHAN N.J. VISSER

Abstract

The basal part of the thick mudrock sequence of the Ecca Group overlying the glacigenic Dwyka Group (late Carboniferous to Artinskian in age) shows evidence of periglacial conditions during deposition. The dark-coloured to black mudrocks which attain a thickness of up to 250 m, contain organic-rich horizons, siltstone and fine-grained sandstone interbeds, marine fossils, and carbonate, siliceous and phosphate concretions, lenses and beds. Deposition was by suspension settling of mud, turbidity current activity and minor fall-out of airborne volcanic ash in a large sea with episodic anoxic bottom conditions. The climate probably varied from subpolar to temperate. The syn- to post-glacial mudrock sedimentation in the foreland basin was controlled by the rapid collapse of the marine ice sheet, basin tectonics (subsidence and rising foreland arc), oceanic circulation, relative sea-level change and a cool to temperate climate.

Introduction

The present Karoo Basin formed when the Gondwana supercontinent broke up during the Late Jurassic to Early Cretaceous. The beds of the Karoo Supergroup consist of remnants of upper Carboniferous to lower Jurassic strata which covered extensive parts of southwestern Gondwana. In southern Africa, Karoo beds also occur in several smaller, mostly fault-controlled, basins towards the north, but this paper deals exclusively with the post-glacial sequence in the main Karoo Basin, which covers an area of just over 600000 km² (Fig. 14.1). The glacigenic Dwyka Group which occurs at the base of the Karoo Supergroup, is conformably overlain by a predominantly mudrock sequence (Ecca Group) up to 3000 m thick (Tankard *et al.*, 1982).

This study considers the basal part of the Ecca Group. The Prince Albert and the overlying Whitehill Formations form the basal Ecca in the western and central parts of the Karoo Basin, and the Pietermaritzburg Formation forms the basal Ecca in the eastern Karoo Basin. Reliable stratigraphic data on the basal part of the Ecca Group in the southeastern part of the basin could not be

obtained because of the effects of the Mesozoic folding. The Whitehill Formation has been explored, without success, as a possible source of hydrocarbons since the early 1970s (Anderson and McLachlan, 1979; Cole and McLachlan, 1991).

Spatial and age relationships of the Prince Albert and Whitehill Formations

The early to late Permian depositional basin of the Prince Albert and Whitehill Formations was a further development of the Permo-Carboniferous Dwyka (Sowegon) basin (Visser, 1989). This feature consisted of an elongated, north-trending foreland basin which was bounded on the east by a continental highland and on the west by a tectonic arc formed by the subduction of the palaeo-Pacific plate underneath the Gondwana plate. The basin had an open end towards the paleo-north (the present direction toward South America).

The white-weathering Whitehill Formation which is a synchronous marker horizon (Oelofsen, 1987), is present only in the western and central parts of the basin. Toward the east and northeast it loses its character, so that the mudrocks of the underlying Prince Albert Formation become indistinguishable from the upper mudrocks. Therefore beyond the lateral cutoff (Fig. 14.1) the argillaceous Pietermaritzburg Formation overlies the glacigenic beds (Fig. 14.2). The Prince Albert and Whitehill Formations can be traced northward into the Warmbad Basin. The upper diamictite units of the Dwyka Group show lateral facies changes to the basal mudrocks of the Ecca Group in the west and north (Figs 14.3 and 14.4; Visser, 1982).

The age of the Prince Albert and Whitehill Formations is based on palynology, their correlation with formations in the Paraná Basin of Brazil, the correlation of the Whitehill Formation with the coal-bearing beds of the Vryheid Formation of the Ecca Group, and the presence of *Eurydesma* fossils near the top of the glacigenic Dwyka Group in Namibia. Based on these data Visser (1990) suggested a Permian age; possibly Artinskian to Kungurian for the Prince Albert Formation and late Kungurian to early Ufimian for the Whitehill Formation.

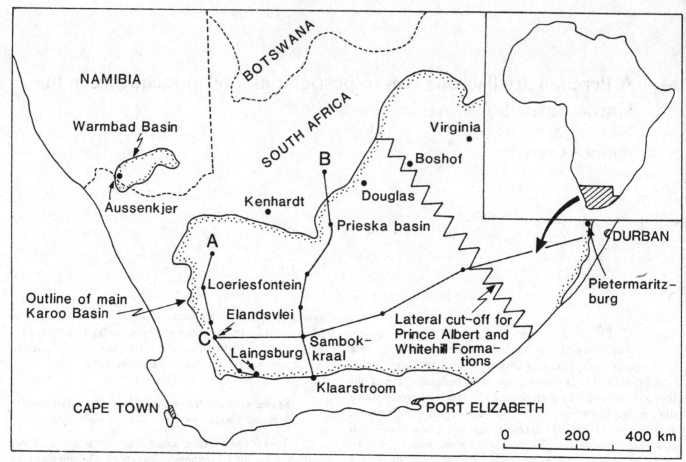

Fig. 14.1. Location of the main Karoo Basin in southern Africa. A, B and C represent section lines.

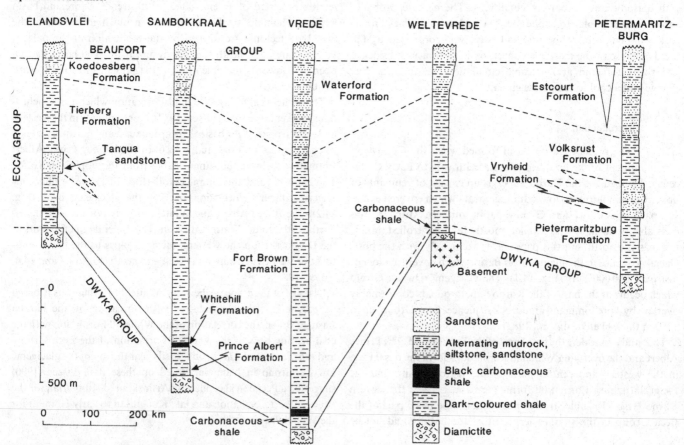

Fig. 14.2. East–west stratigraphic section through the Ecca Group in the Karoo Basin. See Fig. 14.1 for the location of section C.

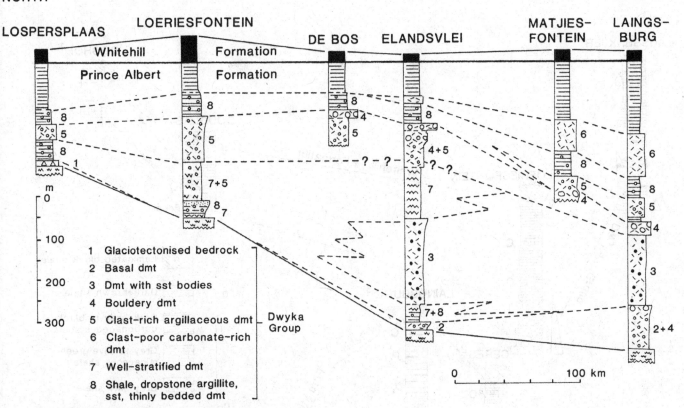

NORTH

SOUTH

LOSPERSPLAAS   LOERIESFONTEIN   DE BOS   ELANDSVLEI   MATJIES-FONTEIN   LAINGS-BURG

Whitehill   Formation

Prince Albert   Formation

m
0

100

200

300

1 Glaciotectonised bedrock
2 Basal dmt
3 Dmt with sst bodies
4 Bouldery dmt
5 Clast-rich argillaceous dmt
6 Clast-poor carbonate-rich dmt
7 Well-stratified dmt
8 Shale, dropstone argillite, sst, thinly bedded dmt

Dwyka Group

0   100 km

Fig. 14.3. North–south stratigraphic section of the Dwyka Group and basal Ecca Group along the western margin of the Karoo Basin. See Fig. 14.1 for the location of section A. dmt, diamictite; sst, sandstone.

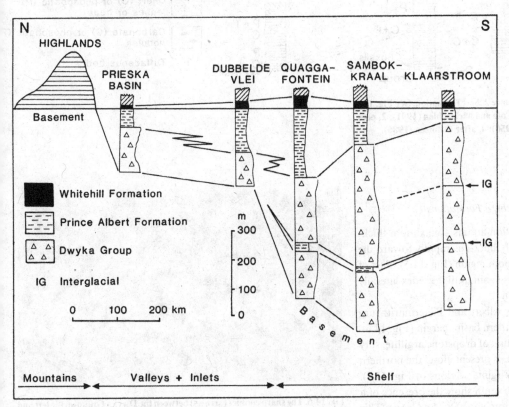

N   S

HIGHLANDS

PRIESKA BASIN   DUBBELDE VLEI   QUAGGA-FONTEIN   SAMBOK-KRAAL   KLAARSTROOM

Basement

Whitehill Formation

Prince Albert Formation

Dwyka Group

IG   Interglacial

m
300

200

100

0

Basement

0   100   200 km

IG

IG

Mountains   Valleys + Inlets   Shelf

Fig. 14.4. North–south stratigraphic section of the Dwyka Group and basal Ecca Group in the central part of the Karoo Basin. See Fig. 14.1 for the location of section B.

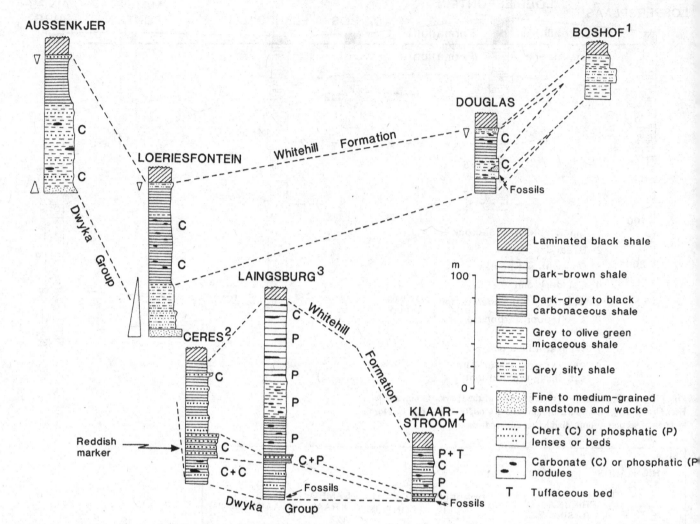

Fig. 14.5. Stratigraphic sections of the Prince Albert Formation. See Fig. 14.1 for the location of the sections. 1, after Cole and McLachlan (1991); 2, pers. comm., J.C. Loock; 3, after Strydom (1950); 4, after Oelofsen (1986).

Lithology

Prince Albert/Pietermaritzburg Formations

The Prince Albert Formation attains a maximum thickness of just under 200 m in the south, but thins rapidly toward the northeast as well as in the Klaarstroom region (Fig. 14.5). East of Klaarstroom the formation thickens again until it grades laterally into the Pietermaritzburg Formation.

The basal contacts of the Prince Albert and Pietermaritzburg Formations are sharp along the southern basin margin (Fig. 14.6), but a pronounced transition, consisting of dropstone argillite, silty rhythmite, sandstone and diamictite, is present along the northern basin margin (Loeriesfontein and Virginia sections in Fig. 14.7). Detailed stratigraphic sections in the south show the presence of a thin unit of bedded diamictite at the top of the Dwyka Group. This is overlain by a persistent thin carbonaceous shale which weathers white and contains marine fossils, chert and phosphorite lenses, and

Fig. 14.6. The sharp contact (arrows) between the Dwyka Group on the left and the Prince Albert Formation on the right at Laingsburg. The white-weathering carbonaceous shale at the base of the Prince Albert Formation can be clearly seen. RM, reddish marker.

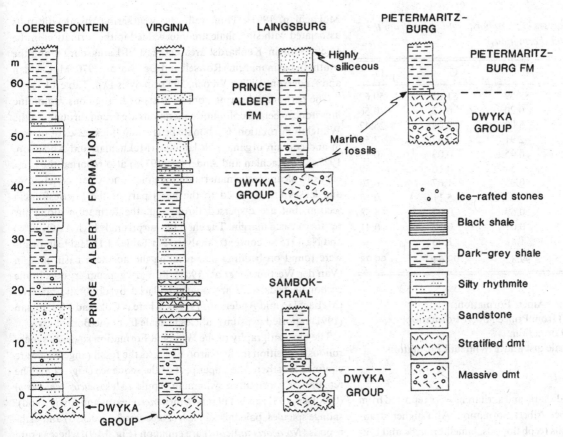

Fig. 14.7. The Dwyka–Ecca transition in the Karoo Basin. Sections at Loeriesfontein and Virginia represent valley facies, whereas the other sections are from the shelf facies. Section at Pietermaritzburg after Von Brunn and Gravenor (1983).

very sparsely distributed dropstones (Fig. 14.6 and 14.7). The upper contact with the Whitehill Formation is sharp, except in the north where it constitutes a thin coarsening-upward sequence from carbonaceous mudrock to silty shale (Fig. 14.5).

The northern outcrops of the Prince Albert Formation are characterized by the presence of greyish to olive-green, micaceous and grey, silty shale in addition to the normal dark-grey to black carbonaceous shale. The formation shows a facies change towards the northeast and northwest where siltstones and sandstones are present (e.g. northern sections in Fig. 14.5). Cross-bedding and slump structures are present in the sandstones, and parallel and ripple lamination in the silty shales. Concretions and lenticular bodies of brownish carbonate are found primarily in the dark-coloured shales.

The southern outcrops of the Prince Albert Formation consist predominantly of dark-grey, carbonaceous, pyrite-bearing, splintery shale (Fig. 14.5). Lenticular units of grey, greenish and dark-brown mudrock, locally micaceous, are developed near Laingsburg and at the eastern end of the outcrop belt (cf. Wright, 1969). At Laingsburg claystone breccia and east of Klaarstroom, massive, greenish-grey wackes are interbedded near the base of the formation. Viljoen (1990) recognised widely distributed volcanigenic material within the southern outcrops of the formation, whereas

discrete tuffaceous layers are present near the top of the formation between Laingsburg and Klaarstroom.

The basal part of the Pietermaritzburg Formation, which is in total up to 400 m thick (Du Toit, 1956), consists of dark-grey, fissile shale with occasional lenses and concretions of impure carbonate and iron-rich material.

Bluish-grey chert nodules and lenses are widely distributed in the dark-grey shales of the Prince Albert Formation. Near the base of the sequence a well-bedded chert-rich zone which weathers reddish, can be traced along the southern basin margin to a position north of Port Elizabeth (Fig. 14.5). A chemical analysis of chert from this zone shows a SiO_2 content of almost 89%, an Al_2O_3 content of 5% and a Fe_2O_3 content of about 3% (analysis B in Table 14.1). Phosphatic nodules, up to 0.3 m across and consisting of amorphous collophanite and carbonaceous material (Strydom, 1950), lenticular masses, up to 10 m long and 40 cm thick, and thin beds consisting primarily of francolite (Bühmann et al., 1989) are distributed throughout the formation. The P_2O_5 content of the rock varies between 30 and 35% (analysis A in Table 14.1). Carbonate concretions and lenticular bodies, commonly with a brown encrustation, occur in the formation in the west and along the northern basin margin.

The remains of a fossil shark (Oelofsen, 1986) and sponge

Table 14.1. *Chemical analyses of phosphorite and cherts from the Prince Albert and Whitehill Formations*

	A	B	C	D
SiO_2	14.92	88.59	81.05	50.62
TiO_2	0.12	0.20	0.12	0.56
Al_2O_3	4.30	5.09	9.04	25.42
Fe_2O_3	1.54	2.91	0.67	7.58
MnO	1.24	0.05	0.03	0.15
MgO	0.91	0.59	0.10	2.59
CaO	41.04	0.05	1.67	6.30
Na_2O	0.49	0.23	5.11	0.57
K_2O	0.53	0.69	0.07	5.58
P_2O_5	30.42	0.01	0.12	0.41
LOI	4.34	1.41	1.66	—
	99.85	99.82	99.64	99.78

Notes:

A, phosphorite from Prince Albert Formation;
B, chert (reddish marker) from Prince Albert Formation;
C, chert from Whitehill Formation;
D, Tuff interbedded in shale just above Whitehill Formation (after Wickens, 1984).

spicules, foraminifera, radiolaria and acritarchs were reported from near the base of the Prince Albert Formation. At a higher stratigraphic level marine fossils (cephalopods, lamellibranchs and brachiopods), as well as fossil wood and palaeoniscoid fish remains and coprolites were recorded (McLachlan and Anderson, 1973). Fish trails and arthropod trackways are also present (Anderson, 1981). Strydom (1950) also noted the presence of radiolaria in phosphatic nodules distributed throughout the formation in the Laingsburg–Klaarstroom area.

Whitehill Formation

The Whitehill Formation, which has a maximum thickness of about 80 m, pinches out as a stratigraphic unit towards the east and east-northeast (Fig. 14.1) due to facies changes (cf. Fig. 14.2 and 14.8). The lower contact of the formation is sharp and well-defined, except in the Boshof section where the definition of the formation becomes vague because of an increase of grey, silty shale in the unit. The upper contact is commonly transitional along the northern basin margin, where silty beds are interbedded with the black carbonaceous shales (Fig. 14.8).

Along the western outcrop belt the Whitehill Formation consists of a basal and an upper shale horizon separated by a carbonate-bearing zone. Interbedded silty shale is also common along the northwestern margin (Fig. 14.8). Zones of grey, silty shale also become prominent toward the pinch-out of the formation (cf. Boshof section in Fig. 14.8). Black, carbonaceous, pyrite-bearing shale constitutes between 75 and 100% of the formation. The shale is very thinly laminated and contains up to 14% carbonaceous material, mostly amorphous kerogen (Du Toit, 1956; Cole and

McLachlan, 1991). Thin, yellowish-weathering rhyodacitic tuffs associated with silty shale are interbedded in the formation in the Loeriesfontein–Kenhardst area and east of Laingsburg along the southern basin margin (Rowsell and De Swardt, 1976; McLachlan and Anderson, 1977a; Viljoen, 1990; analysis D in Table 14.1).

The carbonate-bearing zone consists of ferruginous, dolomitic limestone, whereas dolomitic concretions are found throughout the Whitehill Formation. Carbonate beds attain thicknesses of up to 1 m and contain organic-rich laminae (McLachlan and Anderson, 1977a). McLachlan and Anderson (1977a) also reported the presence of volcaniclastic material in the dolostone. Chert concretions and lenses are confined to the basal part of the Loeriesfontein section, but are dispersed throughout the formation along the southern basin margin. The chert has surprisingly high Al_2O_3 (9%) and Na_2O (5%) contents (analysis C in Table 14.1). Halite imprints were found on bedding planes along the northern basin margin (Van der Westhuizen *et al.*, 1981). Primary gypsum crystals in the carbonate beds were apparently replaced by dolomitic material (McLachlan and Anderson, 1977a), whereas Cole and McLachlan (1991) reported gypsum lenticles in shale from the Boshof area.

The biostratigraphy of the Whitehill Formation suggests synchronous deposition in the Karoo Basin, as the fossil range zones are confined mostly to the upper part of the sequence (Fig. 14.8). The remains of a primitive swimming reptile (*Mesosaurus tenuidens*) (Oelofsen, 1981) and of plants (*Glossopteris* leaves and fossil wood), sponge spicules, paleoniscoid fish (*Palaeoniscus capensis*) and arthropods (*Notocaris tapscotti*) are common (Fig. 14.9), whereas rare fossil insect wings were also reported (McLachlan and Anderson, 1977b). Worm trails were found in grey, silty shale near the middle of the formation at Loeriesfontein. Anderson (1975) reported the presence of limulid trackways at the top of the formation in the Aussenkjer area.

Palaeoclimatic reconstruction

The last glacigenic deposits of the Dwyka Group along the southern margin of the Karoo Basin were deposited under subpolar climatic conditions with the production of small amounts of meltwater, an increase in relative sea-level and highly increased depositional rates (Visser, 1989). After the collapse of the marine ice sheet cold to temperate glacial conditions were maintained on the highlands along the paleo-eastern basin margin. The basin in which the Prince Albert and Whitehill Formations were deposited, was situated between paleolatitudes 50° and 70° (Smith *et al.* 1981) implying a cold to temperate climate. Further evidence of the climatic conditions can be inferred from the presence of sparsely distributed dropstones in the basal part of the Prince Albert Formation, the fossil remains and time stratigraphic correlation with events in a more proximal setting to the basin margins.

The upper part of the Prince Albert Formation correlates with the lower part of the coal-bearing Vryheid Formation of the highlands, whereas the upper part of the latter correlates with the Whitehill Formation (cf. Cole and McLachlan, 1991). The sedimentary model for the Prince Albert Formation thus comprises the

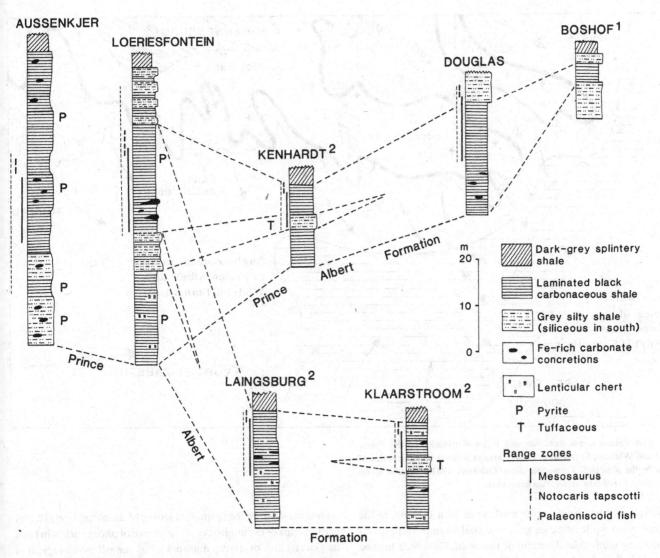

Fig. 14.8. Stratigraphic sections of the Whitehill Formation. See Fig. 14.1 for the location of the sections. 1, after Cole and McLachlan (1991); 2, after Oelofsen (1981).

Fig. 14.9. Fossils from the Whitehill Formation from near Loeriesfontein. (a) Primitive free-swimming reptile (*Mesosaurus tenuidens*). (b) Arthropod (*Notocaris tapscotti*).

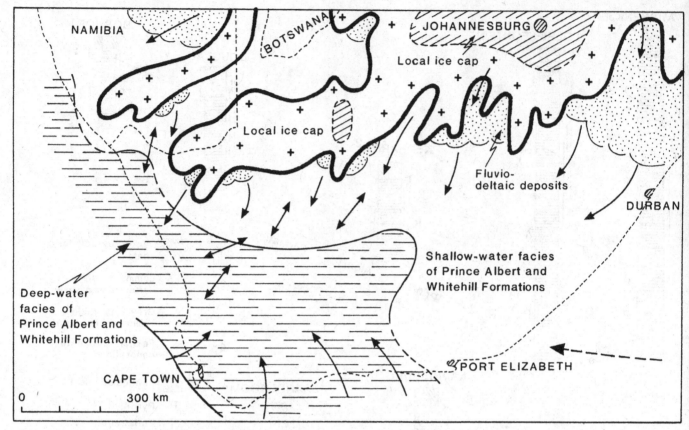

Fig. 14.10. Palaeocurrent directions and facies distribution of the Prince Albert and Whitehill Formations. Double arrows indicate the orientation of fossils in the Whitehill Formation (after Oelofsen, 1981). Broken arrow represents inferred flow along main basin axis.

deposition of a prism of marine mud, some of it organic, in the foreland basin while small ice caps and coal swamps were maintained along the highland margin of the basin. Coal beds formed under glaciofluvial and glaciolacustrine conditions (LeBlanc Smith and Eriksson, 1979). The dropstones within the lower 30 m of the Prince Albert Formation, derived from iceberg rafting, suggest cool to temperate water conditions with tidewater glaciers in the valleys along the paleo-eastern basin margin.

The shark remains, brachiopods and microfossils suggest temperate to cool water temperatures hospitable to marine life (cf. Bühmann *et al.*, 1989). The restricted faunal species in the Whitehill Formation also supports cool conditions. Disseminated phosphorite was reported from the upper layers of the glacigenic Dwyka Group (Bühmann *et al.*, 1989). The deposition of phosphorite is controlled by temperature, organic productivity levels in the water column, oceanic circulation, marine transgression and oceanic water composition (cf. Cook, 1984). However, its formation in the Dwyka and throughout the Prince Albert Formation indicates continuous cool conditions during mudrock deposition. The major element geochemistry of the basal mudrocks of the Whitehill Formation indicates the presence of relatively unweathered material, suggestive of cold conditions with possible ice caps in the source areas of the muds (Visser and Young, 1990).

Thus considering the latitudinal setting of the basin, lateral facies relationships of the mudrocks, the presence of phosphorites in both the glacials and overlying mudrocks and the palaeontology, it is believed that overall cool to temperate climatic conditions prevailed during deposition of the Prince Albert and Whitehill Formations.

Depositional model for the mudrocks

Visser (1989) suggested deposition of the upper part of the Dwyka Group by a marine ice sheet. In the black shales overlying the glacial deposits along the southern basin margin, the remains of a shark and the presence of sponge spicules, foraminifera and acritarchs are strong evidence for marine conditions after retreat of the ice (cf. Oelofsen, 1987). The presence of marine fossils at higher stratigraphic levels and the *Rb/K ratio* of the shales (Visser and Young, 1990) suggest marine conditions during deposition of the Prince Albert and at least the lower part of the Whitehill Formations. The *Rb/K ratio* of the upper Whitehill shales and the paleontology suggest brackish conditions during deposition (Visser and Young, 1990; Oelofsen, 1981). Water depths in the basin varied from shallow to moderate (< 400 m) (Visser and Loock, 1978) with maximum water depth along the foredeep in the southwestern and western parts of the basin (Fig. 14.10).

A major source area was located to the paleo-east where glacial outwash, reworking of glacial deposits and erosion of the bedrock highlands supplied sediment (Fig. 14.10). The recognition of inter-bedded claystone breccia and wacke in the Prince Albert Formation suggests that sediment was also supplied from the paleo-west and probably along the basin axis. Rhyodacitic volcanism to the paleo-west of the basin was a minor sediment source.

Sedimentation of the mudrocks took place by suspension settling of mud (e.g. well-laminated shales) and mud turbidites (thinly bedded mudrocks and silty rhythmites). The black organic-rich muds accumulated under anoxic, reducing conditions as shown by the rarity of bioturbation and the presence of pyrite. The black shales contain a high proportion of amorphous kerogen which represents entrained plant algae, bacterially altered land plant cuticle and pollen, and possible benthic cyanobacterial mats (Cole and McLachlan, 1991). The tuffaceous shales (K-bentonites of Viljoen, 1990) represent airborne volcanic ash.

The coarser-grained silt and sand within the mudrock sequence were deposited by tractional bottom currents in a shallow environment as shown by the cross-bedding and ripple lamination. However, some of the coarser-grained units in the Loeriesfontein and Aussenkjer sections may have formed by rapid suspension fall out from large inflows of sediment-laden water as shown by the presence of slumping and parallel lamination (cf. Domack, 1983).

The cherts show a general lack of sedimentary structures with only parallel micro-laminations, some of them organic-rich, being locally present. Major element geochemistry studies of bedded cherts by Iijima et al. (1985) suggest that the less alumina-rich types ($MnO/Al_2O_3 = 0.01$ for the Prince Albert cherts) represent slow suspension fall-out of siliceous skeletal material. This is also supported by the higher MnO and Fe_2O_3 contents of such cherts. Those cherts with high alumina contents ($MnO/Al_2O_3 = 0.003$ for the Whitehill cherts) formed as siliceous turbidites. These results, however, contradict the interpretations based on the mudrock sedimentology in that the Whitehill muds accumulated under shallower conditions than those of the Prince Albert Formation. The presence of carbon-rich microlaminations within the Whitehill cherts also argues against a turbidite origin. It is suggested that all cherts in the mudrock sequence formed by slow suspension fall-out of siliceous skeletal material under anoxic bottom conditions, but in the case of the Whitehill Formation contamination of the siliceous sediment also by the suspension fall out of clay and plant debris took place.

The phosphatic material in the mudrocks represents direct precipitation of possible 'francolite' at the water–sediment interface in restricted anoxic microenvironments (cf. Strydom, 1950; Büh-mann et al., 1989). The high phosphorus concentrations in the cold oxygen-poor bottom waters were derived from the degradation of organic material.

The presence of marine micro-organisms in the carbonate bodies interbedded in the Prince Albert and Whitehill mudrocks may indicate brief periods of sediment starvation in the basin (A.H. Bouma, pers. comm.; cf. Scotchman, 1991). During these periods biogenetic sedimentation predominated and solution of the skeletal material contributed to the secondary precipitation of carbonate. The dolomitic limestones of the shallower facies, containing car-bon-rich microlaminations and casts of gypsum crystals, also formed under anoxic bottom conditions (McLachlan and Ander-son, 1977a; cf. Sweeney et al., 1987). The lack of evidence of shallow-water sedimentary structures suggests a possible stratified water body.

Deposition of the gypsum and halite crystals reflects greatly increased salinities with the formation of brines during quiet shallow-water conditions (Van der Westhuizen et al., 1981; cf. Eugster, 1985). Early diagenetic processes caused the formation of the concretions, lenses and irregular bodies consisting of dolomite, dolomitic limestone, phosphorite and chert interbedded in the mudrock sequence.

Conclusions

The abrupt change from glacigenic deposition to open marine mud sedimentation is attributed to self-destructive collapse of the marine ice sheet after a relative sea-level rise (Visser, 1991). Shales overlie the diamictites without the presence of an intervening shallow-water facies over about 75% of the Karoo Basin, indicating very rapid inundation of the area during a major transgression.

The thin (< 4 m) black shale unit overlying the glacials (Fig. 14.7), is a condensed marine section deposited during the rapid transgression that followed the collapse of the marine section of the ice sheet (Visser, 1991). The sedimentation rate was low, there was a lack of meltwater-derived terrigenous sediment, water temperature was low (cf. Bühmann et al., 1989) and the preservation rate of marine organic material during anoxic bottom conditions was high. These conditions correspond well with a subpolar climatic setting (cf. Domack, 1988).

Although the climate ameliorated to cool to temperate and sea-level remained relatively high during deposition of the Prince Albert Formation, small ice caps were maintained on the highlands (Fig. 14.10). Along the paleo-eastern basin margin abundant terrestrial vegetation was present, meltwater input was high and deltas built out from the valleys. Deltas probably also formed in the palaeo-west. The Pietermaritzburg and upper Prince Albert shales were the distal equivalents of these deltaic deposits (cf. Tavener-Smith, 1985). Delta progradation may have been partly controlled by post-glacial isostatic rebound of the highlands in the paleo-east, whereas the rising foreland basin margin in the paleo-west was a major sediment source for the foredeep. The different lithofacies in the Prince Albert Formation along the southern margin of the Karoo Basin represent basin floor mud fans entering the basin from the foreland margin. Acidic volcanism along part of the foreland arc supplied windblown volcaniclastic material to the basin. The mud deposits have a high ratio of terrestrial to marine carbon and bottom conditions probably varied between oxygenated and anaer-obic. In the deeper section of the basin most of the mud was deposited by distal mud turbidites, whereas siliceous ooze and phosphatic sediment accumulated during quiescent periods. Silty mud, silt and carbonate deposits formed shoreward.

Hälbich *et al.* (1983) reported a major compressional event, dated at 258 Ma, in the foreland basin. This compression occurred during deposition of the Whitehill Formation and produced an incipient retro-arc fold-thrust belt with an adjacent foredeep in which the deep-water facies of the Whitehill Formation was deposited. This tectonic event had a profound influence on the configuration of the Ecca Basin restricting marine circulation within the basin. A lowering in relative sea-level (global event?) had already started during deposition of the upper part of the Prince Albert Formation as suggested by the progradation of coal-forming deltas along the paleo-eastern basin margin. The subsiding foredeep formed a topographic low on the sea floor furthermore inhibiting bottom circulation and allowing stagnant water to persist (cf. Hallam and Bradshaw, 1979). The sea-level lowering and increased tectonic activity in the foreland basin resulted in a greater influx of clastic and terrigenous organic material. The increased organic component created a strong drain on the oxygen content of the oceanic water column (cf. Leggett, 1980) thereby resulting in an oxygen deficient water mass, which generated conditions favourable for the preservation of organic carbon and in many cases banished benthic fauna. The high volume of suspended clastic material resulted in increased sedimentation rates which also contributed significantly to the preservation of organic material.

Shoreward a shallow-water fine-grained sandstone and siltstone facies was deposited together with grey, silty and black mudrocks (Fig. 14.8). These deposits accumulated largely under oxygenated conditions as shown by the worm trails and limulid trackways. During periods of restricted clastic sediment supply dolomitic limestones were also deposited in the shallow regions (cf. Oelofsen, 1987). About halfway through the deposition of the Whitehill Formation the basin became almost completely isolated from the ocean thereby largely reducing the marine influence. This resulted in brackish conditions during the last depositional stage of the Whitehill muds (Visser and Young, 1990). This was an ideal setting for aquatic fauna to invade the basin despite the continuation of anoxic bottom conditions.

Subsidence of the foreland ramp associated with a relative rise in sea-level abruptly terminated Whitehill and coal deposition and dark-grey mud was laid down as a blanket deposit over the entire area. After this depositional event all evidence of glacial influence in the Karoo Basin ceased.

It can be concluded that post- and synglacial mudrock deposition was controlled by the rapid collapse of the marine section of the Gondwana Ice Sheet, tectonics in the foreland basin, a cool to temperate climate and, in the case of the distinct Whitehill Formation, a restricted oceanic circulation. Climate and eustatic change, however, directly controlled all organic input into the basin.

Acknowledgements

Financial assistance for the study of the Prince Albert and Whitehill Formations by the Foundation for Research Development (FRD) and the University of the Orange Free State is gratefully acknowledged. The manuscript benefited from the comments of L.A. Krissek (Ohio State University) and an anonymous referee. Mr J.C. Loock is also thanked for fruitful discussions on and for the photographs of the fossils in the Whitehill Formation.

References

Anderson, A.M. (1975). Limulid trackways in the late Palaeozoic Ecca sediments and their palaeoenvironmental significance. *South African Journal of Science*, **71**, 249–51.

Anderson, A.M. (1981). The *Umfolozia* arthropod trackways in the Permian Dwyka and Ecca Series of South Africa. *Journal of Paleontology*, **55**, 84–108.

Anderson, A.M. & McLachlan, I.R. (1979). The oil-shale potential of the early Permian White Band Formation in southern Africa. In *Some sedimentary basins and associated ore deposits of South Africa*, ed., A.M. Anderson & W.J. van Biljon. Special Publication of the Geological Society of South Africa, 6, 83–9.

Bühmann, D., Bühmann, C. & Von Brunn, V. (1989). Glaciogenic banded phosphorites from Permian sedimentary rocks. *Economic Geology*, **84**, 741–9.

Cole, D.I. & McLachlan, I.R. (1991). Oil potential of the Permian Whitehill Shale Formation in the main Karoo Basin, South Africa. In *Gondwana Seven, Proceedings*, ed., H. Ulbrich & A. Rocha Campos. Instituto de Geociências, Universidade de Sao Paulo, 379–90.

Cook, P.J. (1984). Spatial and temporal controls on the formation of phosphate deposits – a review. In *Phosphate minerals*, ed. J.O. Nriagu & P.B. Moore. New York: Springer-Verlag, pp. 242–74.

Domack, E.W. (1983). Facies of late Pleistocene glacial-marine sediments on Whidbey Island, Washington. In *Glacial-marine sedimentation*, ed., B.F. Molnia. New York: Plenum Press, pp. 535–70.

Domack, E.W. (1988). Biogenic facies in the Antarctic glacimarine environments: basis for a polar glacimarine summary. *Palaeogeography, Palaeoclimatology, Palaeoecology*, **63**, 357–72.

Du Toit, A.L. (1956). *Geology of South Africa*. Edinburgh: Oliver and Boyd, 611 pp.

Eugster, H.P. (1985). Oil shales, evaporites and ore deposits. *Geochimica et Cosmochimica Acta*, **49**, 619–35.

Hälbich, I.W., Fitch, F.J. & Miller, J.A. (1983). Dating the Cape Orogeny. In *Geodynamics of the Cape Fold Belt*, ed. A.P.G. Söhnge & I.W. Hälbich. Special Publication of the Geological Society of South Africa, 12, 149–64.

Hallam, A. & Bradshaw, M.J. (1979). Bituminous shales and oolitic ironstones as indicators of transgressions and regressions. *Journal of the Geological Society of London*, **136**, 157–64.

Iijima, A., Matsumoto, R. & Tada, R. (1985). Mechanism of sedimentation of rhythmically bedded chert. *Sedimentary Geology*, **41**, 221–33.

LeBlanc Smith, G. & Eriksson, K.A. (1979). A fluvioglacial and glaciolacustrine deltaic depositional model for Permo-Carboniferous coals of the northeastern Karoo Basin, South Africa. *Palaeogeography, Palaeoclimatology, Palaeoecology*, **27**, 67–84.

Leggett, J.K. (1980). British Lower Palaeozoic black shales and their palaeo-oceanographic significance. *Journal of the Geological Society of London*, **137**, 139–56.

McLachlan, I.R. & Anderson, A.M. (1973). A review of the evidence for marine conditions in southern Africa during Dwyka times. *Palaeontologia Africana*, **15**, 37–64.

McLachlan, I.R. & Anderson, A.M. (1977a). Carbonates, 'stromatolites' and tuffs in the lower Permian White Band Formation. *South African Journal of Science*, **73**, 92–4.

McLachlan, I.R. & Anderson, A.M. (1977b). Fossil insect wings from the early Permian White Band Formation, South Africa. *Palaeontologia Africana*, **20**, 83–6.

Oelofsen, B.W. (1981). An anatomical and systematic study of the Family Mesosauridae (Reptilia; Proganosauria) with special reference to its associated fauna and palaeoecological environment in the Whitehill Sea. Unpublished Ph.D. Thesis, University of Stellenbosch, 163 pp.

Oelofsen, B.W. (1986). A fossil shark *Neurocranium* from the Permo-Carboniferous (lowermost Ecca Formation) of South Africa. In *Indo-Pacific fish biology: proceedings of the Second International Conference on Indo-Pacific Fishes*, ed., T. Uyeno, R. Arai, T. Taniuchi & K. Matsuura. Tokyo: Ichthyological Society of Japan, pp. 107–24.

Oelofsen, B.W. (1987). The biostratigraphy and fossils of the Whitehill and Irati Shale Formations of the Karoo and Paraná Basins. In *Gondwana Six: Stratigraphy, sedimentology, and paleontology*, ed., G.D. McKenzie. Geophysical Monograph of the American Geophysical Union, 41, 131–8.

Rowsell, D.M. & De Swardt, A.M.J. (1976). Diagenesis in Cape and Karoo sediments, South Africa, and its bearing on their hydrocarbon potential. *Transactions of the Geological Society of South Africa*, **79**, 81–145.

Scotchman, I.C. (1991). The geochemistry of concretions from the Kimmeridge Clay Formation of southern and eastern England. *Sedimentology*, **38**, 79–106.

Smith, A.G., Hurley, A.M. & Briden, J.C. (1981). *Phanerozoic paleocontinental world maps*. Cambridge: Cambridge University Press, 102 pp.

Strydom, H.C. (1950). The geology and chemistry of the Laingsburg Phosphorites. *Annals of the University of Stellenbosch*, **26A**, 267–85.

Sweeney, M., Turner, P. & Vaughan, D.J. (1987). The Marl Slate: a model for the precipitation of calcite, dolomite and sulphides in a newly formed anoxic sea. *Sedimentology*, **34**, 31–48.

Tankard, A.J., Jackson, M.P.A., Eriksson, K.A., Hobday, D.K., Hunter, D.R. & Minter, W.E.L. (1982). *Crustal evolution of southern Africa*. New York: Springer-Verlag, 523 pp.

Tavener-Smith, R. (1985). The lowest sedimentary cycle in the Vryheid Formation (Ecca) of the Umfolozi Game Reserve: a constructive, wave-dominated delta front? *South African Journal of Science*, **81**, 469–74.

Van der Westhuizen, W.A., Loock, J.C. & Strydom, D. (1981). Halite imprints in the Whitehill Formation, Ecca Group, Carnarvon District. *Annals of the Geological Survey of South Africa*, **15**, 43–6.

Viljoen, J.H.A. (1990). K-bentonites in the Ecca Group of the south and central Karoo Basin. *Abstracts of the 23rd Congress of the Geological Society of South Africa, Cape Town*, 576–9.

Visser, J.N.J. (1982). Implications of a diachronous contact between the Dwyka Formation and the Ecca Group in the Karoo Basin. *South African Journal of Science*, **78**, 249–51.

Visser, J.N.J. (1989). The Permo-Carboniferous Dwyka Formation of southern Africa: deposition by a predominantly subpolar marine ice sheet. *Palaeogeography, Palaeoclimatology, Palaeoecology*, **70**, 377–91.

Visser, J.N.J. (1990). The age of the late Palaeozoic glacigene deposits in southern Africa. *South African Journal of Geology*, **93**, 366–75.

Visser, J.N.J. (1991). Self-destructive collapse of the Permo-Carboniferous marine ice sheet in the Karoo Basin: Evidence from the southern Karoo. *South African Journal of Geology*, **94**, 255–62.

Visser, J.N.J. & Loock, J.C. (1978). Water depth in the main Karoo Basin, South Africa, during Ecca (Permian) sedimentation. *Transactions of the Geological Society of South Africa*, **81**, 185–91.

Visser, J.N.J. & Young, G.M. (1990). Major element geochemistry and paleoclimatology of the Permo-Carboniferous glacigene Dwyka Formation and post-glacial mudrocks in southern Africa. *Palaeogeography, Palaeoclimatology, Palaeoecology*, **81**, 49–57.

Von Brunn, V. & Gravenor, C.P. (1983). A model for late Dwyka glaciomarine sedimentation in the eastern Karoo Basin. *Transactions of the Geological Society of South Africa*, **86**, 199–209.

Wickens, H. de V. (1984). Die stratigrafie en sedimentologie van die Groep Ecca wes van Sutherland. Unpublished M.Sc. Thesis, University of Port Elizabeth, 86 pp.

Wright, A.B. (1969). The geology of a portion of northwestern Albany. Unpublished M.Sc. Thesis, University of Rhodes, Grahamstown, 114 pp.

15 A palaeoenvironmental study of black mudrock in the glacigenic Dwyka Group from the Boshof – Hertzogville region, northern part of the Karoo Basin, South Africa

DOUGLAS I. COLE and ANGUS D.M. CHRISTIE

Abstract

Thin horizons (< 0.5 m) of black mudrock are present in the Dwyka Group (< 25 m thick), which overlies the floor of a palaeovalley near the northern margin of the Karoo Basin, South Africa. This palaeovalley was excavated by southward-moving glaciers from the mid-Namurian until the late Sakmarian. The Dwyka Group was deposited during deglaciation (late Sakmarian–early Artinskian), when a tidewater glacier retreated northwards up the valley.

The black mudrock forms a minor component of an argillaceous sequence up to 10 m thick that was deposited in the northern part of the palaeovalley in a subaqueous proglacial environment proximal to the glacier ice-front. The underlying lithofacies are dominated by diamictite that represent debris flow and minor debris rain deposits. Overlying strata comprise sandstone (± 2 m thick), which is attributed to sedimentation on an ice-proximal, prograding, subaqueous outwash fan, and a thin (< 0.3 m) conglomerate-diamictite couplet that represents sediment gravity flows associated with failure of unstable material, when rapid glacial melting caused a sudden rise in water-level.

Presence of *Botryococcus braunii* in the black mudrock indicates freshwater that was derived from the nearby melting glacier and also small icebergs. The marine setting of the Karoo Basin during the Dwyka glaciation implies a brackish-water environment in the palaeovalley due to mixing with freshwater near the glacier ice-front.

Total organic carbon (TOC) contents of the black mudrock average 1.88% ($n = 27$) and much of the organic material comprises finely macerated inertinite. Fossilized plant fragments, exinite and sporinite are also present and a local terrestrial floral source plus a freshwater algal contribution, is envisaged for the organic matter. It is proposed that genesis of the black mudrock was primarily controlled by an amelioration of the climate, which led to a substantial increase in organic productivity, principally in the form of floral and algal growth. The genesis was also enhanced by an anoxic or dysaerobic substrate, evidence for which is shown by: (1) lack of bioturbation and (2) preservation of carbonaceous plant fragments and pyrite nodules.

Introduction

Drilling in the Boshof–Hertzogville region (Fig. 15.1) showed the presence of black mudrock overlying diamictite in the glacigenic Dwyka Group. It was thought that factors such as eustatic sea-level, climatic and organic productivity changes associated with the glaciation may have influenced the genesis of this black mudrock, and a primary aim of this study was to assess these factors and determine the palaeoenvironment of the black mudrock.

The depositional environment of the Dwyka Group in the Boshof–Hertzogville region is described elsewhere by Cole (1991). The black mudrock occurs as thin (< 0.5 m) beds, which are found within an argillaceous sequence up to 10 m thick. Data was obtained solely from boreholes (Fig. 15.2), and the TOC contents and organic composition of samples of black mudrock were determined in order to gain a better understanding of the palaeoenvironment.

The Dwyka Group in the study area (Fig. 15.1) is thin (< 25 m) and was mainly deposited in the lowermost part of a palaeovalley (Fig. 15.2; Cole, 1991), which was situated on the southern slope of glaciated highlands, referred to as the Cargonian Highlands (Visser, 1986, 1987). The strata were laid down during the final phase (Early Permian) of the Dwyka glaciation by a northward-retreating tidewater glacier (Cole, 1991). The study area is located on the northern margin of the Karoo Basin (Fig. 15.1), which was a huge back-arc basin that extended southwards to the palaeo-Pacific margin of Gondwana during and after the Dwyka glaciation from mid-Carboniferous until Early Permian times (Visser, 1987, 1990).

Geology

The strata in the Boshof and Hertzogville areas mostly consist of Permian Ecca Group mudrock and Dwyka Group diamictite, sandstone and mudrock. The Precambrian basement consists of Early Proterozoic andesitic lava and quartz prophyry (Ventersdorp Supergroup) and Archaean granite, the latter forming small inliers. The Dwyka Group unconformably overlies the basement and is in turn conformably overlain by the Ecca Group.

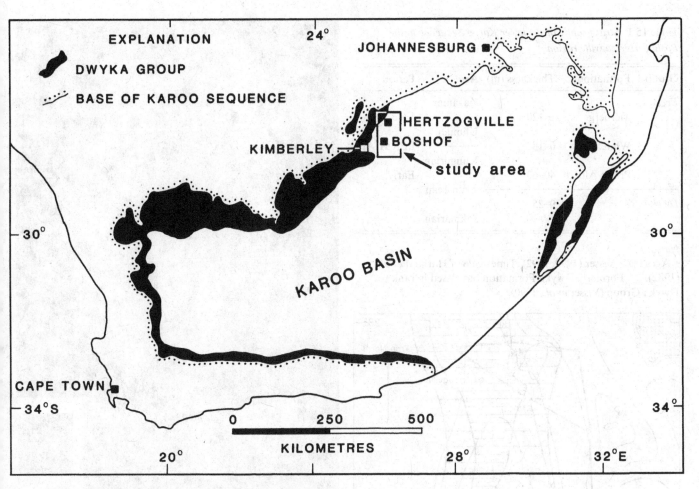

Fig. 15.1. Plan of Karoo Basin showing outcrop of Dwyka Group and location of study area.

The Ecca Group consists of three formations which, in ascending stratigraphic order, are the Prince Albert, Whitehill and Tierberg (Table 15.1; SACS, 1980, Fig. 7.3.3; Cole and McLachlan, 1991). Around basement highs these formations overlap the Dwyka Group and rest unconformably on Precambrian rocks.

The Ecca and Dwyka groups form the lower part of the Karoo Sequence (SACS, 1980, p. 535) and, in the study area (Fig. 15.1), have a thickness of at least 200 metres and a regional dip of less than one degree southeast. They are intruded by Middle Jurassic dolerite dykes and sills, which normally form between 25 and 50% of the succession.

Depositional environment of the Dwyka Group

A synthesis of Cole's (1991) publication is given in order to show the disposition of the black mudrock within the complete glacial sequence. The Dwyka Group is divided into a maximum of five lithofacies associations, with each association reflecting a specific set of glacigenic deposits as summarized in Table 15.2.

Glacial deposition was preceded by a long period of erosion (mid-Namurian–late Sakmarian; Visser, 1990, 1992) by cold-based gla-

ciers, which excavated a valley (Hertzogville Valley) into Precambrian basement (Fig. 15.2). Deposition occurred during the deglaciation stage (late Sakmarian–early Artinskian; Table 15.1), when a tidewater glacier (Powell, 1984) retreated northwards up the valley. The initial deposits are represented by a basal diamictite association (<7.3 m thick) that consists of clast-rich, argillaceous, massive diamictite overlain in places by thin (0.2 m) beds of clast-poor, flow-banded diamictite. The clasts mimic the composition of the underlying bedrock and are faceted or bullet-shaped in places, thus inferring a glacial origin. The basal diamictite is thought to represent subaqueous debris flows derived from unstable lateral moraines situated on the valley shoulders (Cole, 1991). These were deposited after the glacier had retreated to a position just north of Hertzogville (Fig. 15.2). Stabilization of the tidewater ice front led to the deposition of rhythmites in a subaqueous, proglacial, ice-proximal environment. These are confined to the northern part of the Hertzogville Valley and consist of suspension-sedimented muds interbedded with thin sands that were deposited from sediment-charged, freshwater underflows originating from subglacial melt-water streams (cf. Domack, 1984). Scattered lonestones are interpreted as dropstones emplaced from rapidly melting small icebergs

Table 15.1. *Stratigraphy of the lower Karoo Sequence in the Boshof–Hertzogville region*

Group	Formation	Thickness (m)	Age[a]	Period
Ecca			Kazanian	
	Tierberg	±120	Ufimian	Late
	Whitehill	0–18		PERMIAN
	Prince Albert	40–60	Kungurian	Early
			Artinskian	
Dwyka[b]		0–25	Sakmarian	

Notes:
[a] Ages from Visser (1990, 1992). Time-scale of Harland *et al.* (1982).   [b] Formerly Dwyka Formation but raised in rank to Dwyka Group (Visser *et al.*, 1990).

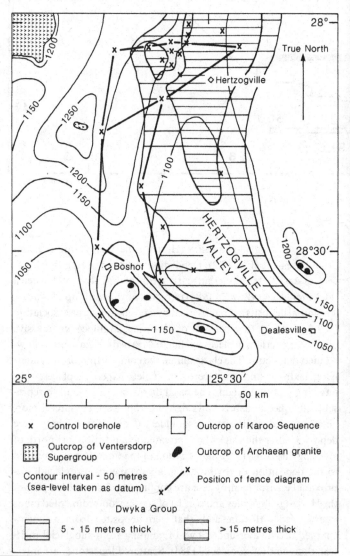

Fig. 15.2. Topographic contour map of the pre-Dwyka surface and thickness of the Dwyka Group in the Boshof–Hertzogville region.

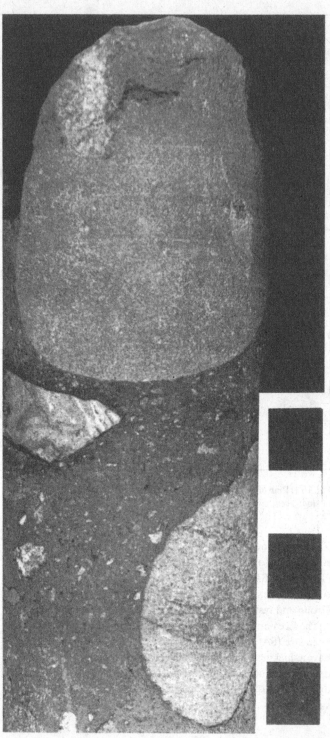

Fig. 15.3. Clast-poor, massive diamictite having a polymictic clast composition. Scale is in cm.

Table 15.2. *Lithofacies interpretation of the Dwyka Group: summary*

Lithofacies association	Lithofacies	Interpretation
Upper rudite	Diamictite, conglomerate	Debris and turbidity flows
Sandstone	Sandstone, minor mudrock and conglomerate	Proglacial traction and turbidity flows on subaqueous outwash fans
Mudrock	Mudrock with scattered lonestones	Proglacial suspension muds with iceberg-rafted dropstones
Diamictite-mudrock	Clast-poor diamictite, rhythmite with lonestones	Proglacial debris flows, iceberg-rafted debris rain, suspension muds and sandy underflows
Basal diamictite	Clast-rich diamictite	Proglacial debris flows

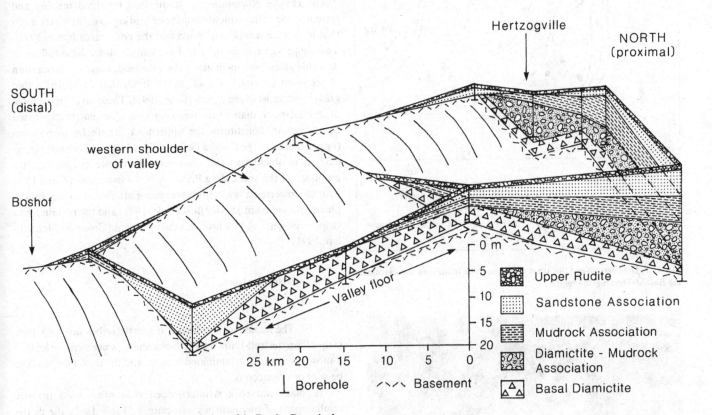

Fig. 15.4. Fence diagram of lithofacies associations of the Dwyka Group in the Hertzogville Valley.

in front of the glacier (Powell, 1984). The rhythmites form the basal unit of the diamictite–mudrock association (Table 15.2) and the overlying diamictite frequently shows soft-sediment deformation at its base. This diamictite is clast-poor, massive, and has a polymictic clast composition (Fig. 15.3). It is interpreted as a debris-flow and/or debris-rain deposit that was laid down when the tidewater glacier resumed its northward retreat (Cole, 1991).

The succeeding mudrock association (Table 15.2) is confined to the northern part of the Hertzogville Valley (Figs. 15.4 and 15.5) and includes thin beds of black mudrock, which are the focus of this paper. The association comprises laminated and massive mudrock, which may contain lonestones and arenaceous or rudaceous laminae (Fig. 15.6). Thin (<0.4 m) beds of rhythmite, sandstone, conglomerate (Fig. 15.7) and diamictite occur sporadically.

The mudrock is interpreted as a suspension deposit that formed by rapid flocculation of fine detritus as freshwater plumes from the ice-front mixed with basinal saline water. The lonestones again represent iceberg-rafted material that was dropped in a proglacial environment close to the ice-front. The arenaceous and rudaceous laminae, rhythmites, sandstones and some conglomerates (Fig. 15.7) are interpreted as turbidity flow deposits, whereas the diamictites and matrix-supported conglomerates represent debris-flow

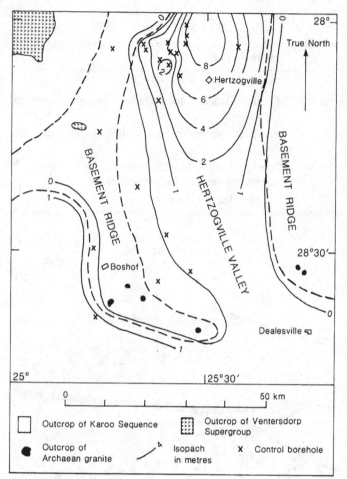

Fig. 15.5. Distribution and thickness of the mudrock lithofacies association in the Boshof–Hertzogville region.

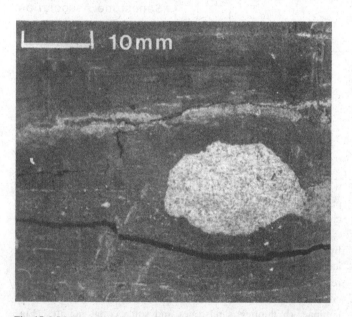

Fig. 15.6. Mudrock containing very fine-grained sandstone lamina and an ice-rafted dropstone composed of andesitic lava.

deposits (Cole, 1991). These flows were most likely triggered by sediment slumping, usually near the ice-front, and iceberg grounding on the valley floor (Powell, 1981; Gilbert, 1983; Miall, 1983). Continued stabilization of the tidewater ice front a few tens of kilometres north of Hertzogville, resulted in the progradation of sandy sediments (sandstone association) across the mudrock (cf. Mustard and Donaldson, 1987) in the northern part of the valley (Fig. 15.4). The sandstone association represents deposition on the distal portions of subaqueous outwash fans (Boulton and Deynoux, 1981) by sediment gravity flow and subaqueous traction currents (Cole, 1991). The distal southern part of the valley was sediment starved, except for a sandy basin slope fan at the foot of the western ridge of the valley near Boshof (Fig. 15.4; Cole, 1991).

The Dwyka glaciation was terminated by rapid melting and retreat of the valley (tidewater) glacier leading to rising water levels that may have completely inundated the entire area (Cole, 1991). The rising water is thought to have caused downslope failure of unstable glacial sediment along the valley sides and the generation of sediment gravity flows (cf. Eyles, 1987) that deposited a thin gravel over most of the region (Cole, 1991). This gravel, in the form of a clast-poor diamictite overlying clast- or matrix-supported conglomerate, constitutes the uppermost lithofacies association (upper rudite) of the Dwyka Group (Fig. 15.4) and does not exceed 0.32 m in thickness. The clast-poor diamictite grades up into mudrock of the succeeding Prince Albert Formation (Table 15.1) with the lowermost few centimetres being attributed to the waning phase of a sediment gravity flow (Cole, 1991) and the remainder to suspension-settling in a prodelta environment (Cole and McLachlan, 1991).

Black mudrock

Description

The name black mudrock better describes this rock-type rather than the well-known term 'black shale', since massive texture is more common than laminated texture, and the rock-type tends to be silty or arenaceous.

The black mudrock is interbedded with other units (mostly argillaceous) of the mudrock association (Table 15.2), the distribution and thickness of which are shown in Fig. 15.5. The colours (Goddard et al., 1963) are black (N1) or greyish black (N2), in contrast to the dark grey (N3) and medium-dark grey (N4) mudrock that predominates in the mudrock association. The beds average 0.2 m in thickness and have both sharp planar and gradational contacts. Irregular upper contacts are also present, where sandstone or conglomerate beds overlie the black mudrock (Fig. 15.7). In localized areas on the shoulders of the Hertzogville Valley (Fig. 15.4), black mudrock rests directly on Precambrian basement with a sharp contact (Fig. 15.8).

In places, the black mudrock contains arenaceous laminae and scattered lonestones, which are predominantly granule-size and rarely small pebble. The mudrock is apparently undisturbed by biogenic activity and is usually micaceous. It is occasionally carbo-

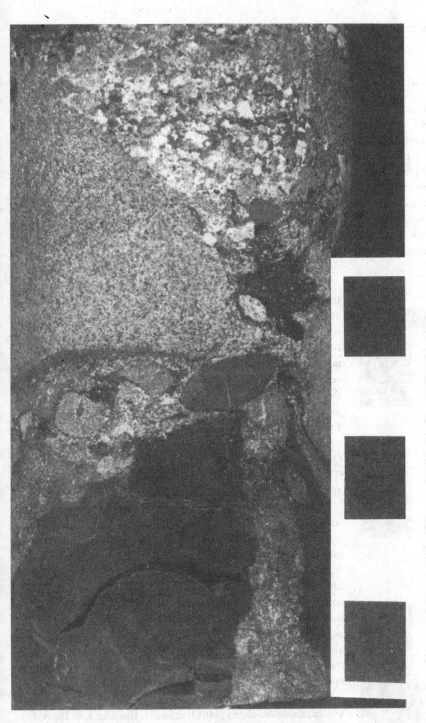

Fig. 15.7. Clast-supported, massive conglomerate discordantly overlying black mudrock. The conglomerate is interpreted as a high-density, turbidity-flow deposit. Note the injection of sandy conglomeratic material into the underlying mudrock, probably as a result of dewatering (cf. Visser *et al.*, 1984). Scale is in cm.

naceous, and fossilized plant debris and pyrite nodules were noted in some of the borehole cores.

Total organic carbon and organic composition

The total organic carbon (TOC) contents of 27 core samples of black mudrock retrieved from the mudrock association and, for comparison, six samples of mudrock retrieved from the other lithofacies associations (Table 15.2) and basal part of the Prince Albert Formation were determined. The adverse, contact metamorphic effects of the dolerite dykes and sills on the organic matter were primarily avoided by sampling core from a distance of at least the thickness of the intrusion away from its contact with the sedimentary strata (Perregaard and Schiener, 1979; Cole and McLachlan, 1991). The results of the TOC analysis are summarized in Table 15.3.

Table 15.3. *Summary of total organic carbon (TOC) contents*

	TOC (%)		
Stratigraphic unit	Mean	Range	Number of samples
Prince Albert Formation	3.20	3.0–3.4	2
Sandstone association	1.46	0.6–2.3	2
Mudrock association	1.88	0.1–4.3	27
Diamictite-mudrock association	0.25	0.2–0.3	2

Note that only black mudrock was sampled from the mudrock association, whereas greyish-coloured mudrock was sampled from the other units.

TOC contents of selected black mudrock horizons from within the mudrock association range between 0.1 and 4.3% with an average of 1.88%. There is no discernible vertical trend in carbon content within the mudrock association nor differences in the contents of samples taken from near the valley floor with those taken from above the valley shoulder (Figs. 15.4 and 15.5). The contents are generally higher than those of mudrocks sampled from the other lithofacies associations of the Dwyka Group, but the mean value is lower than that of the two samples taken from the Prince Albert Formation (Table 15.3). Compared with other black mudrocks, the TOC values are lower, e.g. the Early Permian Quamby Mudstone Group in Tasmania averages about 8% TOC (Domack, 1988) and the Permian Kupferschiefer of Europe averages 6% TOC (Wedepohl *et al.*, 1978).

Microscopically, the organic matter in the black mudrock consists of inertodetrinite (fragments of inertinite up to 20 μm in size but more commonly between 5 and 8 μm), with exinite making up just a small proportion and vitrodetrinite being rarely present. Although much of the organic material could only be classified as amorphous kerogen (5–10 μm in size), a single *Botryococcus braunii* speciment 150 μm in length was observed. It is possible that the fluorescence observed in some samples may be associated with this organism. A small quantity of sporinite (10–20 μm in length) was also observed. Pyrite occurs in concentrations of less than 0.5% in the form of rounded to subrounded grains rarely greater than 5 μm in diameter. Rare nodules up to 6 mm in diameter are present.

Macroscopically, the only organic matter present consists of unidentifiable fossilized plant fragments. Elsewhere, in the northeastern part of the Karoo Basin (Fig. 15.1), fragments of the tree floras *Gangamopteris*, *Glossopteris* and *Cordaites* have been reported from the coal-bearing Vryheid Formation (Du Toit, 1929; Plumstead, 1964), a unit which is laterally equivalent to the Prince Albert and Whitehill formations of the study area (Table 15.1; Cole and McLachlan, 1991; Visser, 1992).

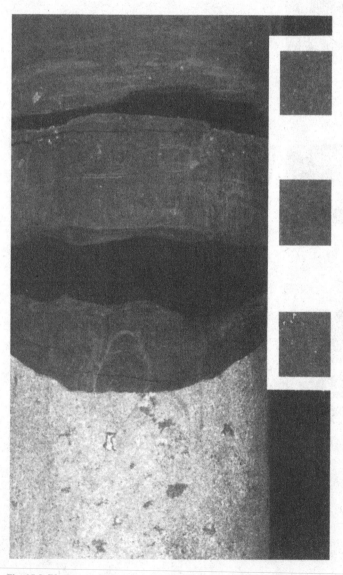

Fig. 15.8. Black mudrock resting unconformably on andesitic lava of the Early Proterozoic Ventersdorp Supergroup. Scale is in cm.

Palaeoenvironment

Lithofacies analysis has indicated that the black mudrock is the product of pelagic mud settling in a proglacial environment

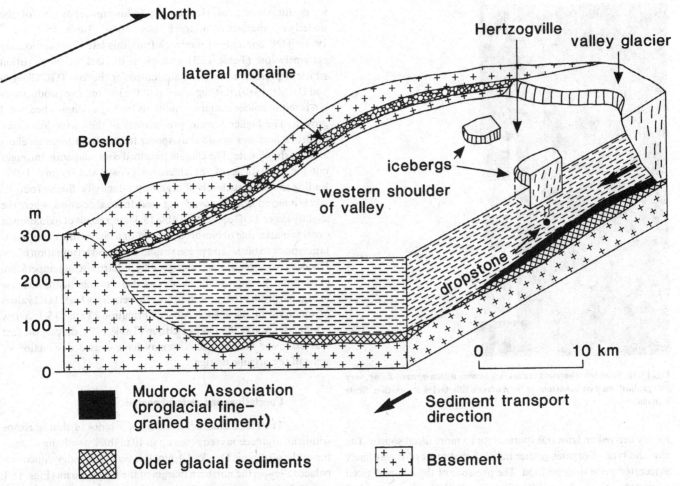

Fig. 15.9. Palaeogeography and depositional environment of the mudrock lithofacies association (including black mudrock horizons) in the Boshof–Hertzogville region.

proximal to the ice-front of a tidewater (valley) glacier (Cole, 1991). Its concentration in the northern segment of the Hertzogville Valley (Fig. 15.5) is related to the tidewater glacial regime, whereby the bulk of the sediments accumulates close to the ice-front as a result of: (1) freshwater plumes mixing with basinal saline water leading to flocculation of muds and (2) rapid melting of small icebergs with a concomitant emplacement of coarser detritus (Powell, 1984). Thin (<1 m) units of black mudrock also occur in the more distal parts of the valley and, locally, on the lower shoulders (Fig. 15.8).

The ice-front was probably located a few tens of kilometres north of Hertzogville and the glacier may have been about 150 m thick, using the estimates of Visser and Loock (1988) for a Dwyka glacier in an adjacent valley 65 kilometres west of Boshof (Fig. 15.1). Since the present pre-Dwyka topography (Fig. 15.2) largely reflects the original glacial landscape (Cole, 1991), it is apparent that the upper shoulders of the Hertzogville Valley would have projected above the ice and that the water depth south of the ice-front was between 100 m for the upper and 200 m for the lower part of the valley. Such depths are consistent with those given by Visser (1987, 1989) for the

Dwyka Group elsewhere in the basin. The above findings are illustrated in a palaeogeographic sketch (Fig. 15.9).

The absence of bioturbation, together with the presence of carbonaceous plant fragments and pyrite nodules suggest that the substrate of the black mudrock horizons was anoxic or dysaerobic (Savrda and Bottjer, 1989). Apart from the presence of minor turbidite deposits (Fig. 15.7), rare debris-flow deposits and scattered ice-rafted erratics (Fig. 15.6), the relatively slow sedimentation rate of the suspension-deposited muds would have been conducive for the deposition and preservation of organic material. The greyish-coloured mudrocks, which form the bulk of the mudrock association, are bioturbated in places, and unbranched sand-filled horizontal burrows occupying mudstone laminae in thin sandstone beds (Fig. 15.10), were noted in two borehole cores. This biogenic activity implies oxidation of the substrate and degradation of some of the organic matter with a consequent lowering of the TOC content (Savrda and Bottjer, 1989).

TOC contents of the black mudrock suggest a moderate level of organic productivity. The shape of the valley and its aspect with respect to the basin makes it likely that most organic matter was

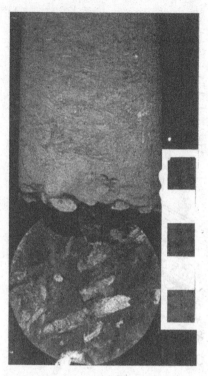

Fig. 15.10. Sand-filled burrows in mudrock lamina within upward-fining, very fine-grained, massive sandstone of the mudrock lithofacies association. Scale is in cm.

locally derived and not transported from a more distal source. The size and type of organic matter indicate that most of it was finely macerated prior to deposition. The presence of the colonial green alga *Botryococcus braunii* (Cook *et al.*, 1981) implies that algal growth took place in freshwater close to the melting ice-front north of Hertzogville (Fig. 15.9) but its contribution to the TOC of the black mudrock is unknown. The water in the Hertzogville Valley would have become progressively more saline away from the ice-front, as a result of mixing of fresh glacial meltwater with basinal marine water (Visser, 1989). The occurrence of fossilized plant fragments, exinite and sporinite indicate a floral origin. Floral growth probably took place on the adjacent ice-free shoulders of the valley (Fig. 15.9) under cool temperate conditions (Visser and Young, 1990). Elsewhere, in the northeastern part of the Karoo Basin, the climate was sufficiently mild to allow the growth of conifer-dominated gymnosperms that are found in the upper part of the Dwyka Group (Falcon, 1986). The floral material was either windblown to the valley depository and/or transported by the glacier. Its contribution to the TOC of the black mudrock is again difficult to assess.

The black mudrock horizons accumulated during a long period of ice-front stabilization and were disrupted by episodes of biogenic activity and, eventually, by progradation of sandy detritus (sandstone association) under more vigorous and turbid conditions. Eustatic sea-level changes are thought to have had a negligible effect on the genesis of the black mudrock. Only a small rise in water-level is envisaged during the retreat of the tidewater glacier to

its position north of Hertzogville following deposition of the underlying diamictite-mudrock association (Table 15.2; Cole, 1991). TOC contents of mudrock from this latter association are extremely low (Table 15.3) and, given the lack of bioturbation which, if present, could have accounted for the low TOC (Savrda and Bottjer, 1989), it is suggested that the low organic productivity is a result of colder climatic conditions leading to diminished floral growth. The higher organic productivity of the succeeding black mudrock horizons would thus appear to be related to an amelioration of the climate. The climate remained cool temperate throughout the final stages of the glaciation (Visser and Young, 1990), leading to moderate levels of organic productivity. Such productivity is inferred from the overlying sandstone association, where the slightly lower TOC contents (Table 15.3) are a result of oxidation of organic matter due to bioturbation and/or a higher rate of sedimentation prior to burial and preservation. The rapid termination of the Dwyka glaciation was accompanied by a sharp rise in temperature (Visser and Young, 1990). There was also a simultaneous increase in organic productivity, as shown by the relatively high TOC values of mudrock in the Prince Albert Formation (Table 15.3) a few centimetres above the Dwyka Group (Table 15.1), despite the fact that extensive biogenic activity may have caused destruction of some of the organic matter.

Conclusions and regional analysis

The black mudrock consists of a series of thin horizons within an argillaceous sequence up to 10 m thick overlying diamictite of the glacigenic Dwyka Group. It is very locally developed in a palaeovalley on the northern margin of the Karoo Basin (Figs. 15.1 and 15.2) and was deposited subaqueously in a proglacial environment proximal to the ice-front of a northward-retreating tidewater glacier during Early Permian (early Artinskian) times. The organic-rich muds formed as a result of increased organic productivity associated with an amelioration of the climate, when the valley (tidewater) glacier retreated to a position several tens of kilometres north of Hertzogville (Fig. 15.9). The organic matter, most of which is finely-macerated, was probably derived from both flora that colonized the ice-free valley shoulders, and freshwater algae (*Botryococcus braunii*).

This is the first reported occurrence of black mudrock from the Dwyka Group of the Karoo Basin. However, in the southern part of the basin, a thin (± 5 m) unit of black shale, which contains marine fossils, occurs at the base of the Prince Albert Formation (Table 15.1) immediately above diamictite of the Dwyka Group (Visser, 1991, 1992). This unit is thought to be late Sakmarian in age and was deposited after collapse of the marine ice sheet, when shelf subsidence caused a rise in sea level (Visser, 1991).

In the southern and central parts of the Karoo Basin, the Dwyka Group consists predominantly of diamictite, which was deposited by rain-out from the decoupled marine ice sheet (Visser, 1986, 1991). It thus differs from the valley facies sequence that characterizes the northern part of the Karoo Basin (including the study area). This sequence comprises diamictite, mudrock, rhythmite, sandstone, and conglomerate that were laid down by tidewater glaciers (Visser, 1986, 1989), which persisted until the early Artinskian (Visser, 1990). The valley facies sequence of the Dwyka Group is, in places, overlain by a marine organic-rich unit, which occupies the basal part of the Prince Albert Formation. This unit can be traced diachronously from Douglas, some 150 kilometres southwest of Boshof (Fig. 15.1), to the southern part of the Karoo Basin. It is also present in the eastern part of the basin (Natal), where phosphatic material is abundant (Bühmann et al., 1989). The moderate TOC content of the basal Prince Albert Formation (Table 15.3) in the study area, indicates that this part of the stratigraphic sequence is probably a lateral equivalent of the organic-rich unit to the southwest at Douglas (Fig. 15.1). Since Visser and Young (1990) showed that a rapid amelioration of the climate took place after the Dwyka glaciation, it is suggested that this was the primary cause of the increase in organic productivity that characterizes the basal part of the Prince Albert Formation. Sediment starvation in the basin, during which biogenetic sedimentation predominated, has also been proposed (Visser, 1991). The organic-rich unit would thus be similar to the Early Permian, algal-rich, Quamby Mudstone Group in Tasmania, which also overlies glacigenic diamictite associated with a marine ice-shelf (Domack, 1988).

Acknowledgements

The authors are most grateful to the Geological Survey of South Africa for providing borehole core and analysing for total organic carbon. The Industrial Development Corporation and Anglo-American Corporation are kindly thanked for making available several borehole cores and/or logs.

References

Boulton, G.S. & Deynoux, M. (1981). Sedimentation in glacial environments and the identification of tills and tillites in ancient sedimentary sequences. Precambrian Research, 15, 397–422.

Bühmann, D., Bühmann, C. & Von Brunn, V. (1989). Glaciogenic banded phosphorites from Permian sedimentary rocks, Economic Geology, 84, 741–50.

Cole, D.I. (1991). Depositional environment of the Dwyka Group in the Boshof–Hertzogville region, Orange Free State. South African Journal of Geology, 94, 272–87.

Cole, D.I. & McLachlan, I.R. (1991). Oil potential of the Permian Whitehill Shale Formation in the main Karoo basin, South Africa. In Gondwana Seven Proceedings, ed., H. Ulbrich & A.C. Rocha Campos. Sao Paulo: Instituto de Geociências, Universidade de Sao Paulo, pp. 379–390.

Cook, A.C., Hutton, A.C. & Sherwood, N.R. (1981). Classification of oil shales. Bulletin Centre Recherche Exploration – Production, Elf-Aquitaine, 5, 353–81.

Domack, E.W. (1984). Rhythmically bedded glaciomarine sediments on Whidbey Island, Washington. Journal of Sedimentary Petrology, 54, 589–602.

Domack, E.W. (1988). Biogenic facies in the Antarctic glacimarine environment: basis for a polar glacimarine summary. Palaeogeography Palaeoclimatology Palaeoecology, 63, 357–72.

Du Toit, A.L. (1929). A short review of the Karoo fossil flora. International Geological Congress, 15th Session, 1929, South Africa, Proceedings, pp. 239–251.

Eyles, N. (1987). Late Pleistocene debris-flow deposits in large glacial lakes in British Columbia and Alaska. Sedimentary Geology, 53, 33–71.

Falcon, R.M.S. (1986). A brief review of the origin, formation, and distribution of coal in Southern Africa. In Mineral Deposits of Southern Africa, Volumes 1 and 2, ed., C.R. Anhaeusser & S. Maske. Johannesburg: Geological Society of South Africa, pp. 1879–98.

Gilbert, R. (1983). Sedimentary processes of Canadian Arctic fjords. Sedimentary Geology, 36, 147–75.

Goddard, E.N., Trask, P.D., De Ford, R.K., Rove, O.N., Singewald, J.T., (Jnr) & Overbeck, R.M. (1963). Rock-Color Chart. New York: Geological Society of America.

Harland, W.B., Cox, A.V., Llewellyn, P.G., Pickton, C.A.G., Smith, A.G. & Walter, S.R. (1982). A Geological Time Scale. Cambridge: Cambridge University Press, 131 pp.

Miall, A.D. (1983). Glaciomarine sedimentation in the Gowganda Formation (Huronian), northern Ontario. Journal of Sedimentary Petrology, 53, 477–91.

Mustard, P.S. & Donaldson, J.A. (1987). Early Proterozoic ice-proximal glacio-marine deposition: the lower Gowganda Formation at Cobalt, Ontario, Canada. Geological Society of America Bulletin, 98, 373–87.

Perregaard, J. & Schiener, E.J. (1979). Thermal alteration of sedimentary organic matter by a basalt intrusive (Kimmeridgian shales, Milne Land, east Greenland). Chemical Geology, 26, 331–43.

Plumstead, E.P. (1964). Gondwana floras, geochronology and glaciation in South Africa. International Geological Congress, 22nd Session, 1964, India, pp. 303–319.

Powell, R.D. (1981). A model for sedimentation by tidewater glaciers. Annals of Glaciology, 2, 129–34.

Powell, R.D. (1984). Glacimarine processes and inductive lithofacies modelling of ice shelf and tidewater glacier sediments based on Quaternary examples. Marine Geology, 57, 1–52.

SACS: South African Committee for Stratigraphy (1980). Stratigraphy of South Africa. Part 1: Lithostratigraphy of the Republic of South Africa, South West Africa/Namibia, and the Republics of Bophuthatswana, Transkei and Venda, (Compiler L.E. Kent). Geological Survey of South Africa Handbook, No. 8, 690 pp.

Savrda, C.E. & Bottjer, D.J. (1989). Trace-fossil model for reconstructing oxygenation histories of ancient marine bottom waters: application to Upper Cretaceous Niobrara Formation, Colorado. Palaeogeography Palaeoclimatology Palaeoecology, 74, 49–74.

Visser, J.N.J., Colliston, W.P. & Terblanche, J.C. (1984). The origin of soft-sediment deformation structures in Permo-Carboniferous glacial and proglacial beds, South Africa. Journal of Sedimentary Petrology, 54, 1183–96.

Visser, J.N.J. (1986). Lateral lithofacies relationships in the glacigene Dwyka Formation in the western and central parts of the Karoo Basin. Geological Society of South Africa Transactions, 89, 373–83.

Visser, J.N.J. (1987). The palaeogeography of part of southwestern Gondwana during the Permo-Carboniferous glaciation. Palaeogeography Palaeoclimatology Palaeoecology, 61, 205–19.

Visser, J.N.J. & Loock, J.C. (1988). Sedimentary facies of the Dwyka Formation associated with the Nooitgedacht glacial pavements, Barkly West District. South African Journal of Geology, 91, 38–48.

Visser, J.N.J. (1989). The Permo-Carboniferous Dwyka Formation of Southern Africa: deposition by a predominantly subpolar marine ice sheet. Palaeogeograph Palaeoclimatology Palaeoecology, 70, 377–91.

Visser, J.N.J. (1990). The age of the late Palaeozoic glacigene deposits in southern Africa. South African Journal of Geology, 93, 366–75.

Visser, J.N.J., Von Brunn, V. & Johnson, M.R. (1990). Dwyka Group. In:

Catalogue of South African Lithostratigraphic Units, ed., M.R. Johnson. South African Committee for Stratigraphy, No. 2, Government Printer, Pretoria, 15–17.

Visser, J.N.J. & Young, G.M. (1990). Major element geochemistry and palaeoclimatology of the Permo-Carboniferous glacigene Dwyka Formation and post-glacial mudrocks in southern Africa. *Palaeogeography Palaeoclimatology Palaeoecology*, **81**, 49–57.

Visser, J.N.J. (1991). Self-destructive collapse of the Permo-Carboniferous marine ice sheet in the Karoo Basin: evidence from the southern Karoo. *South African Journal of Geology*, **94**, 255–62.

Visser, J.N.J. (1992). Late Carboniferous – Permian sea-level changes in the Karoo Basin. *Abstracts, Geological Society of South Africa 24th Congress*, Bloemfontein, 453–5.

Wedepohl, K.H., Delevaux, M.H. & Doe, B.R. (1978). The potential source of lead in the Permian Kupferschiefer bed of Europe and some selected Paleozoic mineral deposits in the Federal Republic of Germany. *Contributions to Mineralogy and Petrology*, **65**, 273–81.

16 Late Paleozoic post-glacial inland sea filled by fine-grained turbidites: Mackellar Formation, Central Transantarctic Mountains

MOLLY F. MILLER and JAMES W. COLLINSON

Abstract

Sandstones and shales of the Lower Permian Mackellar Formation conformably overlie glacigenic sediments and are conformably overlain by braided stream deposits. They record the filling of a large fresh to brackish water post-glacial inland sea by fine-grained turbidites (silt, fine sand) deposited in prograding channel-overbank systems. Upward-coarsening sequences 10 to 25 m thick are capped by broadly cross-cutting channels a few metres deep and hundreds of metres wide. Sediment-laden streams entering the inland sea from the adjacent braidplain delta delivered sediment directly to the turbidite systems; Gilbert-type deltas and distributary mouth bar deposits are rare or absent.

Fine-grained submarine fans and turbidite systems typically lack upward-coarsening trends and are characterized by a few stable channels. The Mackellar turbidites illustrate that upward-coarsening sequences recording progradation by channel-overbank complexes form in fine-grained turbidite systems and demonstrate that these systems can be characterized by numerous shifting channels.

Fine-grained turbidite systems adjacent to and conformably overlain by sandy braidplain delta deposits may be characteristic of the filling of large post-glacial lakes or inland seas. Continental glaciation produces large quantities of poorly sorted sediment. After glaciation, streams reworking the glacial deposits transport fine-grained sediment to the inland sea, leaving behind the coarser sand on the braidplain.

Introduction

Melting of Permo-Carboniferous ice sheets in Gondwana formed lakes and caused widespread marine incursions onto the continent in response to sea-level rise (e.g. LeBlanc Smith and Eriksson, 1979; Tankard et al., 1982; Banks and Clark, 1987; Domack, 1988; Martini and Rocha Campos, 1991; Redfern, 1991). Marine salinity conditions in many areas are indicated by the presence of brachiopods and bivalves and other components of the marine Eurydesma and Levipustula zone faunas (e.g. McLachlan and Anderson, 1973; Runnegar and Campbell, 1976; Lopez Gamundi, 1989). Large marine embayments formed inland seas of variable salinity that extended thousands of kilometres onto the continent, analogous to the present-day Baltic Sea (Barrett et al., 1986; Oelofsen, 1987; Collinson and Miller, 1991.)

Salinity conditions, sedimentary processes and depositional environments at the distal margins of these broad inland seas are not well understood, in spite of their importance in interpreting response to sea-level fluctuation (e.g. Lopez Gamundi, 1989). The purpose of this paper is to describe the processes and environments that characterize the filling of a large post-glacial inland sea.

The Lower Permian Mackellar Formation and its lateral equivalents were deposited in a post-glacial inland sea that extended from the present-day Nimrod Glacier in the Transantarctic Mountains to the Ellsworth Mountains (Fig. 16.1; Collinson and Miller, 1991; Collinson et al., 1992). During the austral summer 1985–6, we investigated 12 outcrops of the Mackellar Formation over a 24 000 km² area in the region of the Beardmore and Nimrod Glaciers (Figs. 16.1 and 16.2). The Mackellar Formation is underlain by the glacigenic Pagoda Formation and overlain by the braided stream deposits of the Fairchild Formation; Lindsay (1968a) and Barrett (1968) provided the stratigraphic framework (Table 16.1).

Extensive vertical and lateral exposure of the predominantly fine-grained deposits of the Mackellar Formation facilitated recognition of large-scale features. This information was integrated with analysis of facies and paleocurrent data and interpretation of salinity conditions to reconstruct the filling history of this post-glacial inland sea. The Mackellar Formation primarily records deposition of fine-grained turbidites in channel-overbank systems fed directly by streams traversing a braidplain delta that eventually prograded over the turbidite systems. Similar processes may have controlled sedimentation at the distal margins of post-glacial inland seas extant at diverse times in the geologic past.

Stratigraphic setting

The Upper Carboniferous to Triassic stratigraphic sequence in the Transantarctic Mountains is similar to that of other parts of Gondwana that bordered the paleo-Pacific margin (Tankard et al., 1982; McKenzie, 1987; deCastro, 1988; Ulbrich and Rocha Campos, 1991). The Permian Glossopteris and Triassic

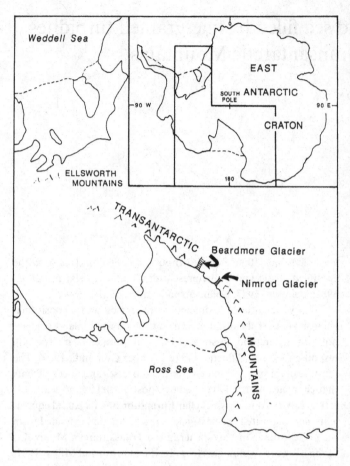

Fig. 16.1. Location of study area in region of Beardmore and Nimrod Glaciers, Central Transantarctic Mountains. Ellsworth Mountains block has been rotated clockwise since the Permian; Permian position shown in Fig. 16.13.

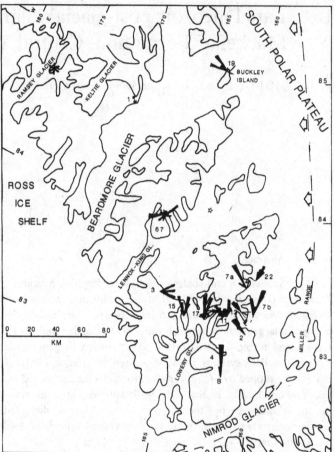

Fig. 16.2. Location of outcrops and paleocurrent data. Numbers indicate number of readings and also identify localities. Detailed locality information is given in the Appendix. *, location along Ramsey Glacier illustrated in Figs. 16.10 and 16.11. Data at (4) B collected by T. Horner. Dashed lined is inferred shoreline; open arrows point toward sea.

Dicroidium floras as well as palynomorph floras and the Early Triassic *Lystrosaurus* and *Cynognathus* vertebrate faunas tie Antarctica biostratigraphically to other Gondwana continents (Hammer, 1990; Taylor and Taylor, 1990). The sequence consists of Permo-Carboniferous glacial deposits overlain by marine or lacustrine shale capped by coal-bearing fluvial deposits of Permo-Triassic age. During the initial break-up of Gondwana during the Jurassic, diabase dikes and sills intruded the sequence and tholeitic basalts were extruded locally.

In Antarctica the thickest (2.5 km) and most complete section of the Gondwana sequence, including the Permian units, is located in the Beardmore Glacier region of the central Transantarctic Mountains (Fig. 16.1) where the strata are nearly flat lying. The sequence records continental glaciation followed by sediment infilling of a post-glacial basin (Table 16.1; Elliot, 1975; Barrett *et al.*, 1986). Diamictite, sandstone, and shale of the Pagoda Formation were deposited in diverse glacial environments with evidence of several glacial advances and retreats (Miller, 1989). The region was flooded after glaciation, and the Mackellar Formation and Fairchild Formation record the filling of this post-glacial sea. Fine- to coarse-

grained sandstones of the Fairchild Formation were deposited in braided streams flowing over the extensive outwash plain of the receding continental ice sheet (Isbell and Collinson, 1988, 1991). Such a braided outwash plain adjacent to a lake or sea has been termed a 'braid delta' or 'braidplain delta' (McPherson *et al.*, 1988; Nemec, 1991a). With time this braidplain delta prograded over the turbidite deposits.

The Mackellar Formation apparently pinches out north of the Nimrod Glacier (Figs. 16.1 and 16.2; Laird *et al.*, 1971). Its lateral equivalent is sandstone, similar to that of the Fairchild Formation. This suggests that a shoreline of the Mackellar inland sea initially was located in the Nimrod Glacier region.

In the Beardmore Glacier area the Mackellar Formation consists of 60 to 140 m of shale, siltstone and fine-grained sandstone, with subordinate diamictite. Interbedded sandstone and shale predominates, with sandstone: shale ratios varying from very low to nearly one.

The Mackellar Formation does not contain laterally extensive lithologic subunits that can be correlated between outcrops. Broad ranges and low diversity of the microflora within the Mackellar

Table 16.1. *Stratigraphic and depositional interpretation of Permian strata in the Beardmore Glacier area. (From Barrett et al., 1986; Isbell & Collinson, 1988, 1991; Miller, 1989)*

Formation	Rock Type (fossil)	Thickness (m)	Inferred Environment
Buckley Fm	Sandstone, shale, coal (*Glossopteris*)	750	Braided stream, flood plain
Fairchild Fm	Sandstone	250	Braided stream
Mackellar Fm	Shale, fine-grained sandstone	100	Inland sea
Pagoda Fm	Diamictite	200	Terrestrial glacial, proglacial

Formation preclude biostratigraphic subdivision (Kyle and Schopf, 1982). Although high-resolution correlation of beds within the formation is not possible, similarities in facies and sequences allow interpretation of the depositional processes and reconstruction of the hierarchy of these processes.

Contact relations between the Mackellar Formation and underlying Pagoda Formation and overlying Fairchild Formation generally are conformable, but suggest abrupt changes in environmental conditions. The upper part of the Pagoda Formation commonly is shale that is massive or disturbed by soft-sediment deformation and contains dropstones. In contrast, shale of the Mackellar Formation lacks dropstones and characteristically is well laminated. The transition between massive Pagoda shale and well-laminated Mackellar shale occurs in a meter or less, but the contact is not along a single plane. The contact with the overlying fluvial deposits of the Fairchild Formation typically is gradational. Sandstone in the lowermost Fairchild Formation resembles sandstone of the upper Mackellar Formation in grain size, sedimentary structures, bed thickness, and lateral continuity of beds. Interbedded sandstone and shale, the dominant facies of the Mackellar Formation, occurs within the lowermost Fairchild Formation. At a few localities, the Fairchild/Mackellar contact is erosional, probably reflecting channeling by the braided streams that deposited sands of the Fairchild Formation.

Facies

Six facies are recognized in the Mackellar Formation, but it is dominated by the interbedded sandstone and shale facies (Fig. 16.3). Facies characteristics and interpretation are discussed briefly below and summarized in Table 16.2

Shale facies

Description

Some of the laminated shale and siltstone of this facies consists of laminae that are uniformly 0.5 mm to 3 mm thick grouped into siltstone/shale couplets (Fig. 16.4). Within each

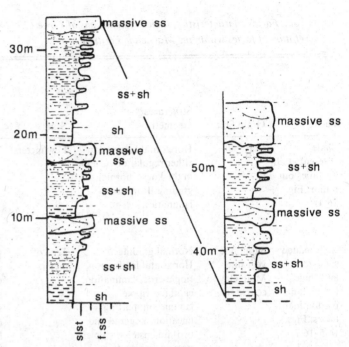

Fig. 16.3. Measured section of Mackellar Formation at Locality 17 (Fig. 16.2) showing vertical sequence and relative abundance of facies. sh, shale facies; ss + sh, interbedded sandstone and shale facies; massive ss, massive sandstone facies.

couplet, contact between the siltstone and overlying shale is gradational, but the upper and lower contact of the couplets are abrupt. In less regularly laminated shales, siltstone/shale couplets are not well defined, laminae thickness is variable (0.5 mm to 10 mm), and some siltstones are ripple cross-laminated.

Interpretation

Fine-grained sediments of the shale facies were deposited from suspension and from low density turbidity currents (Lowe, 1982; Pickering *et al.*, 1986) in relatively quiet water conditions that occur in diverse environments (e.g. tidal flat, deep or anoxic marine, proglacial lake). Regularly laminated shales of the Mackellar Formation closely resemble those deposited from suspension in glacial lakes from interflows or overflows generated when entering stream water is less dense than part or all of the lake water (Ashley *et al.*, 1985). Cross-laminated siltstone layers in less regularly laminated shale layers are similar to those deposited by underflows (low density turbidity currents) as described by Ashley *et al.* (1985).

Interbedded sandstone and shale facies

Description

Shale and siltstones are interbedded with beds of very fine- to fine-grained sandstones that range from a few mm to 100 cm thick but typically are less than 30 cm thick. Sandstones are characterized by: (1) abrupt, commonly erosional, bases that are irregular or smooth; (2) normal grading, with a slight upward decrease in sand grain size and upward increase in fine-grained

Table 16.2. *Facies characteristics, facies designations of Lowe (1982), Pickering* et al. *(1986) and Mutti & Ricci Lucchi (1972), and interpretation of facies within the Mackellar Formation*

Facies	Grain Size	Structures/ Geometry	Scale	Trends/ relationships	Interpretation	L = Lowe (1982) P = Pickering *et al.* (1986) M = Mutti & Ricci Lucchi (1972)
Shale (Present at all localities but minor; Fig. 16.4)	Clay, silt	Horizontal lamination either regular or variable in thickness; normal grading. Rare cross-lamination	Few meters	Most common at base of upward-coarsening sequences	Deposition mainly from suspension and low density turbidity currents; (inter-overflows and underflows of Ashley *et al.*, 1985)	L = $T_{e,d}$ P = D2.3; E2.1; E2.2 M = G,E
Interbedded sandstone and shale. (Abundant, predominant facies; Figs. 16.5–16.7)	Sand; fine to very fine Mud: clay to coarse silt	Normal grading. Horizontal lamination, ripple-cross-lamination, climbing-ripple lamination; parting lineation; linguoid and two-dimensional ripples, starved ripples, trace fossils. Soft sediment deformation, convolute bedding, flame structures, small normal faults. Laterally discontinuous in 100m; channel fills common.	Tens of meters	Gradationally overlies shale. Has hierarchy of upward-coarsening sequences. Largest scale are capped by massive sandstone facies.	Turbidites $T_{b,c-e}$. deposited from high and low density turbity currents	L = T_t, T_{b-d} P = C2.2; C2.3 M = E
Massive sandstone (Present at all localities; Fig. 16.7)	Fine sand (minor medium sand)	Primarily massive, upward into planar lamination and climbing ripple lamination. Fills broad channels < 5 m deep, hundreds of meters wide.	Few meters	Fills series of broad cross-cutting channels capped upward-coarsening sequences (10–25 m) within interbedded sandstone and shale facies.	Deposited from high and low (minor) density turbidity currents.	L = S_3 P = B1.1 M = B
Burrowed sandstone (locality 7a)	Fine sand	Bioturbation; ripple cross-lamination; oscillation ripple marks	10 m	Near top of Mackellar Fm surrounded by interbedded sandstone and shale facies	Deposited above wave base, removed from main loci of deposition	NA
Large-scale cross-stratified sandstone (Localities 11, 13; Fig. 16.8)	Fine sand, subordinate silt, clay	(a) Giant-scale (10 m) cross-stratification (Locality 13) (b) Cross-stratification; normally graded foresets (sand, silt, clay) with climbing-ripple and ripple cross-lamination	10 m	At top of Mackellar Fm; overlies interbedded sandstone and shale or massive sandstone facies and grades upward into cross-bedded sandstone of Fairchild Fm.	(a) Possible Gilbert-type delta. (b) Episodic deposition on sloping (e.g. bar) surface from waning currents.	NA
Diamictite (Localities 7b, 22; Fig. 16.9)	Variable; clasts a few cm to 3 m in diameter	Fills broad shallow channel at one locality.	Few meters	Fills channel or extends laterally > 1 km. Abruptly overlies inter-bedded sandstone and shale facies.	Mudflow and debris flow deposit locally derived from remanent highs of glacial diamictite	NA

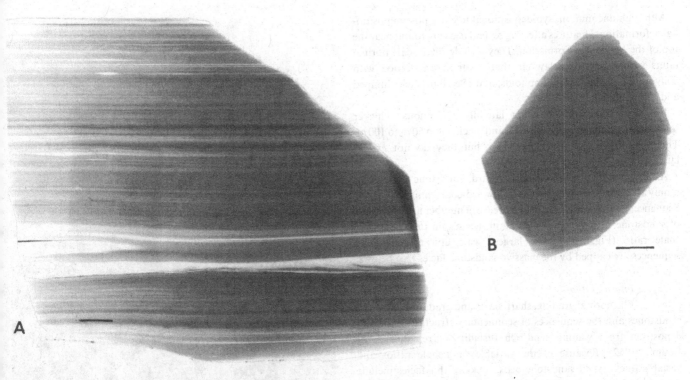

Fig. 16.4. Radiographs of shale; coarse-grained layers are light (Locality 7b, Fig. 16.2). (a) Irregularly laminated shale with cross-laminated siltsone. (b) Regularly laminated shale. Bar in both photographs = 1 cm.

Fig. 16.5. Structures within interbedded sandstone and shale facies showing deposition from waning currents. (a) Small-scale ripple cross-lamination draped with siltstone. Ruler (15 cm long) at left for scale (Locality 67, Fig. 16.2). (b) Oblique view of ripple cross-laminated sandstone bounded by thin shale beds. Note upward increase in angle of climb of climbing-ripple lamination. Base of the sandstone bed is nearly a vertical section; top of the sandstone approaches a bedding plane view. Brunton compass for scale (Locality 19, Fig. 16.2).

matrix; and (3) gradational contact with the overlying shale. Thin beds (< 15 cm thick) are horizontally or ripple-cross laminated with some starved ripples (Fig. 16.5a). In thicker beds the sandstone displays parting lineation, horizontal laminations, ripple cross-lamination, and, rarely, large-scale low-angle cross-stratification. Climbing ripple-lamination, convolute lamination and load structures are common and flame structures are rare (Fig. 16.5b). Ripple marks and trace fossils are common on bedding planes of both thick and thin sandstones. Three-dimensional current ripples, particu-

larly linguoid ripples, are most common, followed by two-dimensional current ripples. Trace fossils are endostratal trails that are typically preserved on bedding surfaces with linguoid ripples.

The vertical sequence of structures within sandstone varies, but the following vertical sequence is typical: (1) basal small-scale ripple cross-lamination; (2) climbing-ripple lamination with lee, but not stoss-side, preserved; (3) climbing-ripple lamination with both stoss and lee-side laminae preserved (Type B); (4) draped lamination of Ashley et al. (1982); and (5) siltstone with small starved ripples.

Although uncommon, syn-depositional to early post-depositional deformational features affecting several beds are found near the top of the Mackellar Formation. They include small scale normal faults with offsets of a few cm that occur in association with convolute bedding, rare slump folds, and one 10 m thick slumped and rotated block.

Individual sandstones are not laterally continuous. Thicker sandstones are channel fills that thin and pinch out in 50 m to 100 m. Thin sandstones are more sheet-like, but they do not extend laterally more than tens of meters.

Although individual beds fine upward, sandstone beds commonly are arranged in a hierarchy of upward-coarsening sequences. Sequences coarsen upward by an increase in number and thickness of sandstone beds with a concomitant increase in the sandstone: shale ratio (Fig. 16.6). The largest scale upward-coarsening sequences are capped by the massive sandstone facies (see below).

Interpretation

The normal grading, sharp bases and gradational tops of sandstones and the sequences of sedimentary structures indicates deposition from waning sand-rich turbidity currents ($T_{b,c-e}$ of Bouma, 1962). Because of the variable sand:shale ratios and variable thickness of sandstone beds, rocks of this facies include those in facies C2.2 and C2.3 of Pickering et al., (1986) and T_t and T_{b-d} of Lowe (1982) deposited primarily by low density turbidity currents. The lenticular shape of some sandstones suggests that turbidity currents were confined to channels or record small channeled splays. Lack of lateral continuity of other sandstones on a scale greater than tens of meters also suggests that the turbidity currents did not spread over a wide area.

Massive sandstone facies

Description

This facies consists primarily of structureless fine-grained sandstone with subordinate medium-grained sandstone. Beds 0.5 m to 3 m thick have erosive bases with flute marks and flame structures. Horizontal stratification is common in the upper 50 cm of the thicker of these beds, overlain by small-scale ripple cross-lamination or climbing ripple-lamination and siltstone or shale. Each massive bed is a broad (100 m to 200 m-wide) channel-fill sandstone. The massive sandstone facies at a single locality typically comprises two or more of these cross-cutting channel fills (Fig. 16.7).

Interpretation

The massive structureless sandstones are interpreted as having been rapidly deposited by high-density turbidity currents (T_a of Bouma, 1962; S_3 of Lowe, 1982; B1.1 of Pickering et al., 1986). The horizontal lamination, ripple cross-lamination and climbing ripple-lamination overlying the massive sandstones were deposited from low density flows that probably were genetically related to the high-density turbidity currents (Middleton and Hampton, 1973; Lowe, 1982). The high-density turbidity currents

Fig. 16.6. Small-scale upward-coarsening sequence within interbedded sandstone and shale facies, recording progradation of small channels and associated overbank deposits. Thickest sandstone at top pinches out within 50 meters. Hammer at right for scale (Locality 67, Fig. 16.2).

were confined to channels, as indicated by the ubiquitous occurrence of the massive sandstone facies as channel-fill deposits.

Burrowed sandstone facies

Description

A 10 m thick burrowed, fine-grained sandstone unit occurs near the top of the Mackellar Formation at one locality. Oscillation ripples on bedding planes of beds 3 to 5 cm thick are discernible, although horizontal to oblique endostratal trails (unlined burrows) 0.5 to 1 cm in diameter are abundant. In vertical section, cross-lamination is visible, although it has been disrupted by burrowing. This corresponds to an ichnofabric index to 3 to 4 (Droser and Bottjer, 1987).

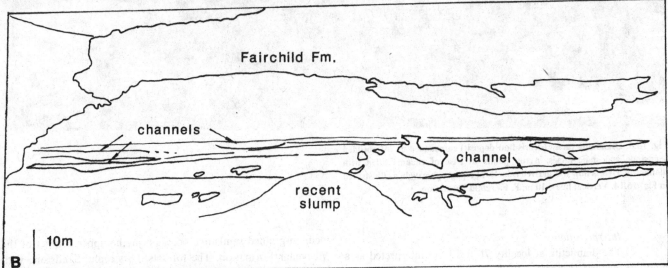

Fig. 16.7. Two upward-coarsening sequences consisting of interbedded sandstone and shale facies capped by massive sandstone facies. Note cross-cutting
of channels filled with massive sandstone; these are analogous to channels
illustrated in Fig. 16.14.

Interpretation

The burrowed sandstone was deposited above wave base
and reworked by wave processes. The ratio of rate of infaunal
destruction of physical sedimentary structures to rate of production
by physical processes was higher than during deposition of any
other facies. Extensive disruption by biological activity may have
been caused by a large population of infaunal animals, a smaller
population of very active animals, a low rate of physical reworking,
or a combination of these factors. In any case, deposition occurred
in a regime that was more affected by wave activity and that
supported a more prolific infauna than that of other facies. Possibilities include, but are not limited to, a well-oxygenated shallow-
water environment marginal to the major locus of deposition.

Diamictite facies

Description

Coarse-grained deposits are rare in the Mackellar Formation, and this facies occurs at only two localities. At one locality (7b,
Fig. 16.2) the diamictite is 6.6 m thick and extends laterally for more
than 1 km. Clasts include boulders up to 90 cm in diameter, which
are composed of graywacke, granite, gneiss, mudstone, limestone,
and large (tens of meters in longest dimension) contorted blocks of
sandstone and mudstone. Elsewhere (locality 22, Figs. 16.2 and
16.8) the clast composition is similar, but the diamictite facies is
slightly thinner (5 m) and pinches out over several hundred meters.

Fig. 16.8. Diamictite facies (debris flow deposit) surrounded by interbedded sandstone and shale facies. Arrow points to base of channel filled with diamictite. Debris flow was derived from local topographic high, as illustrated in Fig. 16.14. Vertical line = 10 m; F, Fairchild Formation.

Interpretation

The diamictite at locality 7b has been interpreted as a mudflow deposit (Lindsay, 1968b). Similarly, the diamictite at locality 22 is a mudflow or debris flow deposit (Middleton and Hampton, 1973; Lowe, 1982). Compositionally the clasts resemble those in the underlying Pagoda Formation that are interpreted to be derived from local basement rocks. The large contorted blocks of sandstone and diamictite probably were derived from the Pagoda Formation, in which the predominant rock types are sandstone and diamictite that lithologically resemble the rocks in the contorted blocks within the Mackellar mudflow deposit.

The geometry of the diamictite at locality 22 indicates that the debris flow was restricted to a channel. At its other occurrence at locality 7, the diamictite facies has a more sheet-like geometry, suggesting that it was not confined to a channel.

Large-scale cross-stratified sandstone facies

Description

This facies consists of two types of large-scale cross-stratified sandstone, each of which is found at a single locality. At locality 13 (Fig. 16.2), giant-scale (10 m) cross-stratified fine- to medium-grained sandstone occurs near the upper contact of the Mackellar Formation. The foresets dip steeply. Sandstone with similar scale cross-stratification is visible on photomosaics of the same stratigraphic interval across a valley from this outcrop; here the sandstone extends laterally for hundreds of meters.

At locality 11 (Figs 16.2 and 16.9), large-scale cross-strata composed of upward-fining foresets (very fine-grained sandstone to shale) occur near the top of the Mackellar Formation. The foresets are dominantly very fine-grained sandstone with ripple cross-lamination and climbing ripple-lamination. Two dimensional current ripples and trace fossils are exposed on bedding planes.

Interpretation

The giant scale cross-stratified sandstone at locality 13 may record progradation of a Gilbert-type delta; the lateral continuity of this facies across the valley is consistent with this interpretation. Alternatively, the sandstone could record the lateral infilling of a migrating channel. However, the foresets dip more steeply than those of epsilon cross-stratified units, and are not sigmoidal in vertical section (Collinson and Thompson, 1982).

The origin of the large-scale cross-stratified sandstone at locality 11 is unclear. Upward-fining within each foreset implies deposition

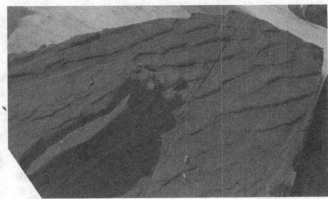

Fig. 16.9. (*a*) **Large-scale cross-stratification, with upward-fining sequences within foresets. Total thickness of outcrop is 4 m.** (*b*) **Current ripples on bedding plane surface at top of A (Locality 11, Fig. 16.2).**

by episodic waning currents. Unlike epsilon cross-stratification which commonly occurs as a single set, this is composed of several cross-cutting sets. Low density turbidity currents probably deposited these sediments on an inclined surface in or marginal to larger-scale channels that are not clearly exposed at this outcrop.

Large-scale channels

In addition to numerous small channels, channels on the scale of a few meters deep and hundreds of meters wide are common in the Mackellar Formation. Broadly cross-cutting channels filled with fine-grained sandstone of the massive sandstone facies are most abundant (Fig. 16.7; see 'massive sandstone facies' above). Other channel types occur at single localities, including the channel filled with diamictite interpreted as a debris-flow deposit (Fig. 16.8; see 'diamictite facies' above).

Two distinctive channel forms in an inaccessible outcrop that is continuous for over a kilometer were seen from a helicopter. This southernmost observed outcrop of the Mackellar Formation is 80 km from the nearest measured section. It is farther from the inferred original shoreline of Mackellar inland sea (Fig. 16.2) than other outcrops, and therefore may provide a glimpse at more distal deposits. Here a series of broadly cross-cutting channels occur in a zone 20 to 30 m thick in the middle of the Mackellar Formation (Fig. 16.10). Channel-fill ranges from cross-stratified sandstone to interbedded sandstone and shale with diverse sandstone:shale ratios. Some of these channels were cut, abandoned and filled with very fine-grained sediment. Most, however, were cut and subsequently filled with turbidites, implying that they functioned as long-lived conduits for sediment.

The zone of broadly cross-cutting channels is overlain and underlain by interbedded sandstone and shale that resemble turbidites occurring elsewhere in the Mackellar Formation, but appear to be more laterally continuous.

A unique channel filled with two series of gently inclined beds was

observed about a kilometer from the broadly cross-cutting channels (Fig. 16.10), within approximately the same stratigraphic horizon (Fig. 16.11). Individually, except for their greater lateral continuity, the inclined beds resemble the typical Mackellar turbidites. From a distance the inclined beds in the upper series appear similar to lateral accretion surfaces. If they are lateral accretion surfaces, it implies that turbidity currents deposit fine-grained sediment from suspension on inclined surfaces.

The importance of the abundance and diversity of channels in the Mackellar Formation is threefold. First, they demonstrate the key role of channels in the development and maintenance of the Mackellar turbidite systems. Second, they corroborate Bouma's (1988) observation that channels and channel-fill processes within turbidite systems are complex. Third, the subtlety of many of these channels as seen in large exposures suggests that similar channels go undetected in smaller, less well-exposed outcrops of turbidites.

Facies associations and depositional environments

Facies associations

The most common vertical sequence of facies consists of the shale facies followed upward by the interbedded sandstone and shale facies and the massive sandstone facies (Fig. 16.3). Two or more of these upward-coarsening sequences are present in the Mackellar Formation at most outcrops, the upper of which commonly is conformably overlain by trough cross-bedded sandstones of the Fairchild Formation. The diamictite facies is overlain and underlain by the interbedded sandstone and shale facies. Where present, the large-scale cross-stratified sandstone facies and the burrowed sandstone facies occur at the top of the Mackellar Formation, conformably underlying the Fairchild Formation.

The massive sandstone facies passes laterally into the interbedded sandstone and shale facies, as does the diamictite facies. Lateral facies relationships between the large-scale cross-bedded sandstone facies and the shale facies were not observed.

Fig. 16.10. Series of broadly cross-cutting channels; arrows point to the bases of some. n indicates cross-stratified sandstone that is cut by a channel filled with interbedded sandstone and shale. Note apparent lateral continuity of turbidites above channel zone. Vertical line in upper left center = approximately 10 m. Photo taken from helicopter at location marked by *, Fig. 16.2.

Depositional environments

Turbidite facies

Vertical and lateral associations of facies within the Mackellar Formation, combined with interpretation of most of the facies as deposited by turbidity currents (see above) indicate deposition in fine-grained turbidite systems. During periods of quiescence, shales and siltstones were deposited. Episodic turbidity currents deposited sandstones of the interbedded sandstone and shale facies. Sandstones of this facies are not laterally continuous and many fill small channels, suggesting deposition in channel and in overbank channel-splay systems. This is consistent with their similarity to Facies E turbidites of Mutti and Ricci-Lucchi (1972) interpreted as overbank deposits and with presence of features (climbing ripple-lamination, convolute lamination) indicative of deposition in a channel-levee complex (Walker, 1985). The massive sandstone facies resembles Facies B channel deposits, commonly associated with Facies E to form channel-overbank sequences

(Mutti, 1977). Upward-coarsening on both small scale (several meters) within the interbedded sandstone and shale facies and large scale (10 to 25 m) represented by upward-coarsening within the interbedded sandstone and shale facies capped by the massive sandstone facies records progradation of both small-scale and large-scale channel-overbank sequences.

In most submarine fan models based on ancient turbidite sequences, upward-coarsening sequences of laterally extensive turbidites (Facies C and D or Mutti & Ricci-Lucchi, 1972) record progradation and aggradation of depositional lobes that form distal to the channelized portion of the fan, typically in the lower fan, but also in the mid-fan (e.g. Mutti and Ricci-Lucchi, 1972; Walker, 1978; Mutti and Normark, 1987; Shannmugam and Moiola, 1988). Although turbidites of the Mackellar Formation resemble lobe deposits in comprising upward-coarsening sequences, they lack the characteristic lateral continuity of turbidites forming depositional lobes. Mackellar turbidites were deposited in channel-overbank systems rather than in unchannelized

Fig. 16.11. Arrow points to top of channel-fill consisting of two series of gently inclined beds (probably turbidites), that are possible lateral accretion surfaces. Scoured base of channel visible to left of corner. Vertical bar at lower right = approximately 100 m. Photo taken from helicopter at location marked by *, Fig. 16.2.

lobes, and thus the Mackellar turbidite system is not compatible with most ancient submarine fan depositional models.

Recent work on both modern and ancient fine-grained submarine fan deposits and turbidite systems suggests that fine-grained systems differ fundamentally from coarser-grained systems on which submarine fan models have been based. Major differences are in lack of lobe deposits and prevalence of channel-overbank deposits throughout the system (Melvin, 1986; Shanmugam et al., 1988). Characteristics of the interbedded sandstone and shale facies and massive sandstone facies of the Mackellar Formation are consistent with deposition in such a fine-grained turbidite system.

Non-turbidite facies

The position of the large-scale cross-bedded sandstone facies and the burrowed sandstone facies near the top of the Mackellar Formation and their gradational contact with the overlying trough cross-bedded sandstones of the Fairchild Formation, recording braided stream deposition, indicate that these facies were deposited in the transition zone between the turbidite system and the braided outwash plain that supplied the sediment. Deposition of the burrowed sandstone facies occurred above wave base, as indicated by presence of symmetrical ripples, but sufficiently protected from the main locus of deposition to maintain an infauna whose activity overshadowed physical reworking.

The large-scale cross-stratified facies includes a rare occurrence of a Gilbert-type delta, with foresets reaching a maximum of 10 m. Elsewhere the foresets of the large-scale cross-stratified facies are smaller and consist of upward-fining couplets of sandstone and shale, indicating episodic deposition. Most commonly, the braided stream deposits of the Fairchild Formation gradationally overlie the massive sandstone facies or the interbedded sandstone and shale facies. This suggests that sediment typically was delivered directly to the turbidite system by streams flowing off the outwash plain rather than stored near the shoreline in deltaic complexes.

Two lines of evidence suggest that the diamictite facies was deposited by locally derived debris flows off topographic highs on the Mackellar sea floor rather than by debris flows generated at or near the shoreline. First, the braided outwash plain adjacent to the shoreline of the Mackellar inland sea, recorded by sandstones of the overlying Fairchild Formation, lacked large clasts and thus could not have been a source for the diamictite. Second, glacial deposits of the Pagoda Formation underlying the Mackellar Formation do include clasts resembling those in the diamictite facies and thus were a potential source. Oversteepened mounds of glacial debris could have been flooded by the Mackellar sea, and surrounded by turbidites before spawning debris flows, the deposits of which subsequently were enclosed by turbidity current deposits.

Paleocurrents

Orientation of ripple-cross lamination, climbing ripple-lamination, three dimensional current ripples and parting lineation indicate a predominantly north-to-south paleocurrent direction and also suggest an eastward component in the southern part of the area (Fig. 16.2). These data are consistent with currents directed basinward away from the inferred shoreline of the Mackellar inland sea (Fig. 16.2) and with the postulated both northern and southern source areas for clay minerals in shale of the Mackellar Formation (Krissek and Horner, 1988; 1991).

Paleosalinity

Biogenic structures, carbon:sulphur ratios, and comparison with contemporaneous post-glacial marine deposits all suggest that the Mackellar Formation was deposited in fresh to brackish water.

Biogenic structures

Biogenic structures give three lines of evidence that the part of the Mackellar inland sea that covered the present day Beardmore – Nimrod Glacier region was fresh to brackish. First, two morphologically distinctive trace fossils resemble those from nonmarine rocks. One is a small (<0.5 cm in diameter) looped endostratal trail (Fig. 16.12a) that is identical to an unnamed possible millipede trace from (Devonian) fluvial deposits of the Catskill deltaic complex of New York (Bridge and Gordon, 1985). The ichnogenus *Isopodichnus* includes these and a variety of other small bilobed endostratal trails produced by diverse arthropods and found in both non-marine and marine deposits (Fig. 16.12b; Miller, 1979; Bromley and Asgard, 1979; Pollard, 1988). Second, an unnamed endostratal trail that is one of the few trace fossils found in the Mackellar Formation, also occurs in overbank pond facies of the younger fluvial Buckley Formation (Table 16.1) and in upper Paleozoic freshwater deposits of the Seaham Formation in New South Wales, Australia (samples collected by J.W. Collinson).

A third line of ichnological evidence suggestive of fresh to brackish water depositional conditions is provided by the facts that trace fossil diversity is low (five or fewer types), trace fossils are not abundant, and, with the exception of the single occurrence of the burrowed sandstone facies, bioturbation is minimal. Paleozoic non-marine rocks typically are not bioturbated (Miller, 1984), and

modern lakes are characterized by low diversity of infaunal animals, particularly at profundal depths (Ricker, 1952; Reid and Wood, 1976). Decreases in benthic faunal diversity and biomass parallels decrease in salinity within the Baltic Sea (Elmgren, 1978). The Mackellar inland sea may have had a similar gradient in faunal diversity. The laterally equivalent Polarstar Formation in the Ellsworth Mountains, interpreted to have been closer to and had more free interchange with the paleo-Pacific Ocean, has a more diverse ichnofauna than the Mackellar Formation, including the trace fossils *Phycodes* and *Rhizocorallium*, that are absent from the Mackellar Formation. The complex feeding trace *Zoophycos* occurs in many marine shales and turbidite sequences deposited under dyaerobic conditions and has been identified as indicative of low-oxygen marine depositional environments (Frey and Pemberton, 1984). Its absence from the Mackellar Formation is evidence, albeit negative, that low-oxygen conditions did not exert strong control on the ichnofauna.

Carbon:sulfur ratios

Ten samples from the Mackellar Formation were selected for determination of carbon:sulfur ratios. Of these, only five had sufficient organic matter (> aproximately 1% total organic carbon) to yield meaningful results. All had high carbon:sulfur ratios (range 86–202; mean = 130; R.A. Berner, 1986, pers. comm.) that are within the ranges of both modern and ancient freshwater sediments (Berner and Raiswell, 1984, Figs. 16.2 and 16.3). The method does not discriminate between fresh and slightly brackish water. Marine waters contain abundant sulfate, which would provide detectable sulfur in some form regardless of post-depositional alteration (R.A. Berner, 1986, pers. comm.). A non-marine environment is consistent with, although not required by, the predominance of vitrinite and inertinite, rarity of amorphous organic matter, and absence of alganite that was observed during visual kerogen typing (W.G. Dow, 1987, pers. comm.).

The effects of contact metamorphism precluded both kerogen typing by Rock-Eval analysis and the use of other mineralogical and geochemical techniques (e.g. Curtis, 1964; Tourtelot, 1964; Veizer and Domovic, 1974).

Comparison with Upper Paleozoic post-glacial marine sequences

Permo-Carboniferous glaciomarine or post-glacial marine sediments in Tasmania, other parts of Australia, southern Africa, and Argentina are characterized by the presence of marine fossils. A post-glacial marine transgression covered much of Australia, resulting in widespread distribution of the cold-water *Eurydesma* fauna (Runnegar and Campbell, 1976). In southern Africa, mudrocks interbedded with and overlying the glacigenic Dwyka Formation contain acritarchs, radiolarians, and forams, as well as marine macrofossils including gastropods, cephalopods, crinoids, and echinoids (Visser, 1987). Permian post-glacial mudrocks in Tasmania record diverse marine environments analogous to those

Fig. 16.12. Trace fossils (all approximately 3 mm wide) on upper bedding surfaces. (*a*) Large arrow points to *Isopodichnus*; note lateral change in morphology. Small arrow points to looped trail. Both traces possibly produced by same animal moving at different depths within sediment (Locality 11, Fig. 16.2). (*b*) *Isopodichnus* (Locality 67, Fig. 16.2).

presently surrounding Antarctica (Domack, 1988). Basinal mud-rocks contain ice-rafted clasts with elements of the shallow-water *Eurydesma* fauna and include bryozoan-rich turbidites derived from nearshore environments. Beds several meters thick within the mudstone are dominated by *Tasmanites*, a planktonic marine alga. Melting of sea ice may have stimulated '*Tasmanites* blooms' analogous to the diatom blooms that occur off the coast of Antarctica, providing a rich source for the biogenic sediment accumulating at depths below 500 m (Domack, 1988).

The Mackellar Formation lacks all elements of the diverse macro- and microflora and fauna that occur in Permo-Carboniferous glaciomarine and post-glacial marine sediments, suggesting that it was not deposited under marine conditions.

Summary

No single piece of data yields conclusive evidence that the Mackellar Formation was deposited under non-marine conditions. However, disparate lines of evidence each suggest non-marine depositional conditions. These include: high carbon:sulfur ratios, presence of non-marine trace fossils, low level of bioturbation, absence of marine trace fossils even those found in turbidites and low oxygen deposits, and absence of marine macro and microfossils that occur in other Permo-Carboniferous post-glacial marine units. The combination of these facts is stronger than any one of its components, and supports the interpretation of the Mackellar Formation as deposited under fresh to brackish water conditions.

Water depth

Limited information suggests that the Mackellar inland sea in the present-day Beardmore Glacier area was relatively shallow. Interpretation of water depth is constrained only by the following facts: (1) the inland sea was underlain by continental rather than oceanic crust, (2) thicknesses of upward-coarsening sequences require a minimum depth of 25 m, and (3) the unit is underlain and overlain by rocks deposited on land, and is less than 150 m thick throughout the study area. While not conclusive, all of these factors suggest deposition in relatively shallow water, probably no more than several hundred metres deep at the maximum, and possibly shallower.

Paleogeography

Thicknesses of Lower Permian rocks vary little to the south through the central Transantarctic Mountains. The Mackellar Formation and its lateral equivalents are typically 100 to 200 m thick (Long, 1965; Aitchison *et al.*, 1984). In contrast, the post-glacial shale in the Ellsworth Mountains, the Polarstar Formation, is at least 1000 m thick (Collinson *et al.*, 1992). In the Ellsworth Mountains, sediment-fill occurred in a more basinal setting where subsidence was greater and where shale accumulated for a longer period of time. Sandstones in the Polarstar Formation are volcaniclastic, and were derived from a different source area than the quartzo-feldspathic sandstones in the Mackellar Formation and

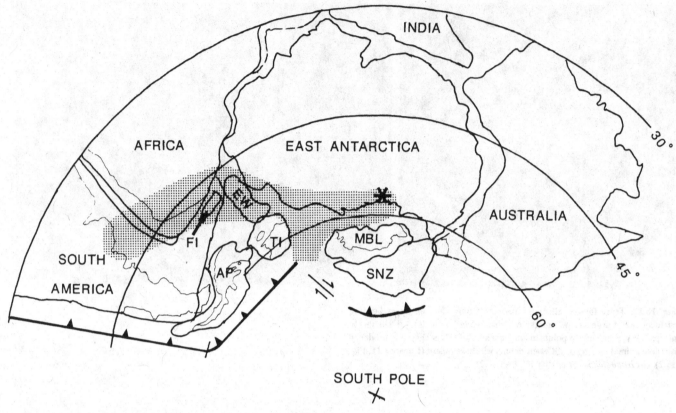

Fig. 16.13. Reconstruction of Gondwana for 230 Ma from Grunow *et al.*
(1991) modified to show position of study area (*) relative to inland sea that
may have extended over entire stippled area. MBL, Marie Byrd Land; SNZ,
South Island, New Zealand; TI, Thurston Island; EW, Ellsworth Mountain
block; FI, Falkland Islands; AP, Antarctic Peninsula.

elsewhere in the Transantarctic Mountains (Frisch & Miller, 1991;
Collinson *et al.*, in press). These data suggest that post-glacial shale
deposition from the present-day Nimrod Glacier to the Ellsworth
Mountains occurred in an extensional back-arc basin, bounded on
one side by the East Antarctic Craton and on the other side by
several allochthonous crustal blocks. During the Permo-Triassic
the paleo-Pacific margin was a convergent continental margin
bordered by calc-alkaline volcanics arcs (Fig. 16.13; Collinson,
1990).

A recent reconstruction based on paleomagnetic data shows a
major gap in continental crust between the Marie Byrd Land
(MBL)/Southern New Zealand (SZN) blocks and the Thurston
Island (TI)/Antarctic Peninsula (AP) blocks (Fig. 16.13; Grunow *et
al.*, 1991). This major gap may represent an arm of the paleo-Pacific
Ocean that extended into the continental basin associated with the
Mackellar inland sea. The basin may have stretched beyond the
Ellsworth Mountains to connect with the Karoo Basin of South
Africa (Fig. 16.13; Visser, 1987; 1989).

Model for filling of the Mackellar post-glacial inland sea

Extensive cliff-face exposure of the Mackellar Formation
allows reconstruction of the facies relationships that record the
processes by which a large post-glacial inland sea was filled and the

depositional environments in which these processes were acting
(Fig. 16.4). Interpretation of facies and facies relationships identifies key features of the depositional regime:

1. The Mackellar inland sea was bordered by a braided
   outwash plain formed in front of a receding continental
   glacier.
2. Deposition within the Mackellar inland sea was dominated by turbidity currents carrying fine-grained sediment
   (primarily clay to fine sand; subordinate medium sand).
3. The turbidite systems extended into the basin by progradation of channel-overbank systems rather than by progradation of lobes comprised of sheet-like sand bodies.
4. Braided streams of the outwash plain were linked directly
   or nearly directly to the turbidite systems.

Any model for filling of the post-glacial inland sea must be
compatible with these constraints and must integrate aspects of
both deep-sea and deltaic depositional models.

Basinal processes and environments

Well-described turbidite facies of Mutti and Ricci-Lucchi
(1972), Lowe (1982) and Pickering *et al.* (1986) accurately characterize the shale facies, interbedded sandstone and shale facies, and

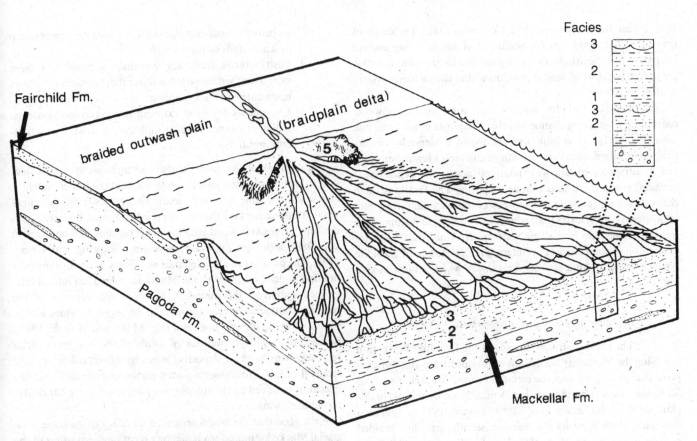

Fig. 16.14. Depositional setting of Mackellar Formation showing major facies and relationships to braided stream deposits of overlying Fairchild Formation and to glacigenic deposits of underlying Pagoda Formation. Only one of the many postulated feeder–turbidity systems is shown. Not to scale; widest channels are several hundred meters wide. Key to facies: 1, shale; 2, interbedded sandstone and shale; 3, massive sandstone; 4, large-scale cross-stratified sandstone; 5, burrowed sandstone. Burrowed sandstone and cross-stratified sandstone facies are typically absent. Not shown is diamictite facies deposited from debris flows off topographic highs of Pagoda glacigenic deposits, such as illustrated at left of diagram.

massive sandstone facies that dominate the Mackellar Formation. Associations of the standard facies are keys to interpretation of subenvironments within submarine fan facies models (Mutti & Richi-Lucchi, 1972; Walker, 1978). These classical submarine fan models are based on ancient turbidite systems that contained fine to very coarse sediment or on submarine fans at the sand-rich end-point of Stow's (1985) ternary classification of submarine fans based on grain size. According to classical and sand-rich radial submarine fan models, upward-coarsening and upward-thickening laterally continuous turbidites (Facies C and D of Mutti and Ricci-Lucchi, 1972) record progradation of lobes in the lower or mid fan. The Mackellar turbidites differ in lacking laterally continuous lobe deposits. On this basis, as well as because of the shallow-water setting, Mackellar turbidites would not be considered submarine fan deposits by Shanmugam and Moiola (1988), regardless of salinity conditions. The Mackellar turbidite systems resemble fan deltas (the coarse-grained fan endmember of Stow, 1985) in their direct links between feeder and fan channels and their occurrence in shallow water (Fig. 16.14; Reading and Orton, 1991), but differ in lacking sediment coarser than medium sand.

The Mackellar turbidites more closely resemble those of fine-grained submarine fans (Stow, 1985), but there are still significant differences. Like the Mackellar turbidite systems, ancient fine-grained submarine fans are dominated by channel-overbank deposits (e.g. Cheema et al., 1977; Siemers, 1978; Braithwaite et al., 1988). However, they commonly do not display upward-coarsening and upward-thickening sequences (e.g. Melvin, 1986) as do Mackellar turbidites. Modern fine-grained submarine fans are large, elongate, and gently sloping and characterized by a single (or few) sinuous channel(s) bounded by high levees (Bouma et al., 1985; Damuth et al., 1988; Shanmugam et al., 1988). The resulting deposits consist of unconnected linear channel sand bodies surrounded by mud (e.g. Surlyk, 1987; Shanmugam and Moiola, 1991) that contrast markedly with the broadly cross-cutting channels of the Mackellar turbidite systems.

Submarine fan models assume introduction of sediment to turbidite systems through a single point source (e.g. a major river). However, sediment may also move downslope unchannelized to form slope aprons (e.g. Ineson, 1989) or be delivered by multiple feeder channels of approximately equal size (e.g. Chan and Dott,

1983; Heller and Dickenson, 1985; Collinson, 1986). Turbidites of the (Carboniferous) Mam Tor Sandstone of northern England are inferred to have been linked to delta plain distributary channels that controlled patterns of sand deposition within the turbidite system (Collinson, 1986).

The Mackellar turbidite systems were comprised of upward-coarsening prograding channel-overbank deposits of mud and fine sand (Fig. 16.14). Overbank deposits display a hierarchy of partially channelized splay deposits. Major channels a few meters deep and hundreds of meters wide migrated across the depositional surfaces, eroding previous channel deposits. The fact that these channels are on the scale of hundreds of meters wide rather than kilometers wide indicates that sediment was delivered by numerous sources rather than by a single major river. Thus, the Mackellar Formation is comprised of an interdigitating mosaic of turbidite systems fed by multiple sources, only one of which is illustrated in Fig. 16.14.

Shoreline processes and environments

The braided stream deposits of the Fairchild Formation overlying the Mackellar Formation were deposited on an outwash plain that prograded over the turbidites as the Mackellar sea was filled; such coastal braidplains are termed 'braid deltas' or 'braidplain deltas' (McPherson et al., 1988; Nemec, 1991a). Subaerial braidplain delta deposits are characteristically sheet-like braided stream deposits such as those of the Fairchild Formation (Isbell and Collinson, 1988, 1991). Typically, subaqueous braidplain delta deposits display large-scale cross-stratification attributable to Gilbert-type deltas, formed as coarse-grained sediment is dumped into quiet, relatively shallow water (McPherson et al., 1988). Large-scale cross-stratification characteristic of Gilbert-type deltas occurs at only one outcrop of the Mackellar Formation. Also, distributary mouth bar deposits that form instead of Gilbert-type deltas in response to low slope or other factors (Postma, 1991) are absent.

We suggest that sediment entering the Mackellar sea was funneled directly into the turbidite systems rather than deposited at the shoreline in Gilbert-type deltas or in distributary mouth bars. The cold, sediment-laden water probably formed underflows as it entered the Mackellar sea. Continuing downslope these could, under some conditions, expand into turbidity currents (see Nemec, 1991b for discussion of processes). This would link streams on the braidplain delta with channels in the turbidite systems, and is consistent both with the observed close and conformable relationship between turbidite sandstones and overlying braided stream deposits and with the paucity of Gilbert-type deltas and lack of bar deposits.

Model

Components of the model for the filling of the Mackellar post-glacial inland sea are as follows:

1. Fine-grained turbidite systems prograded by a series of upward-coarsening channel-overbank sequences capped by broadly cross-cutting channels a few (<5) meters deep and hundreds of meters wide.
2. Distributaries traversing a subaerial braidplain delta introduced sediment-laden water that formed underflows upon entering the inland sea.
3. Underflows delivered sediment to the turbidite systems; sediment was not stored in Gilbert-type deltas or distributary mouth bars.

The paradox of the model is that the turbidite system is fine-grained (primarily mud to fine sand), whereas the feeder system is coarse-grained (fine to coarse sand). This differs from most fine-grained submarine fans that are supplied with mud by large rivers draining extensively weathered low-relief terrains (Shanmugam et al., 1988), although there are exceptions (Reading and Orton, 1991). The paradox can be explained as follows. First, although the braidplain delta adjacent to the Mackellar inland sea lacked fine-grained sediment, the abundant poorly-sorted deposits of the receding continental glacier throughout the region provided a large source of mud, silt and fine sand (e.g. Miller and Waugh, 1991). Second, the relatively well-sorted sands of the braidplain delta represent winnowed lag deposits formed by stream reworking and removal of fines from poorly sorted glacial sediments. Third, the fines thus removed by the streams were delivered to the Mackellar turbidite systems.

We suggest that the model presented for filling of the large postglacial Mackellar inland sea based on exceptional exposures of the fine-grained Mackellar Formation can be applied to other postglacial inland sea or large lake deposits that are less well exposed. The model may be widely applicable because continental glaciation produces large volumes of poorly sorted sediment. Through reworking this sediment will be separated into fine-grained components, eventually deposited as turbidites in the inland sea or lake, and coarser-grained components that remain on the braid delta as lag deposits. As the inland sea/lake is filled, the fine-grained turbidites are gradually covered by the prograding coarse-grained braidplain delta.

Conclusions

1. Shale, siltstone, and fine-grained sandstone of the Mackellar Formation were deposited near the margin of a fresh to brackish water, shallow inland sea. The sea may have been connected to the paleo-Pacific Ocean through gaps in crustal blocks, and may have extended beyond the Karoo Basin of South Africa. In the Central Transantarctic Mountains the sea was bordered by a braided outwash plain.

2. The Mackellar Formation is dominated by fine-grained turbidites that were deposited in prograding channel-overbank sequences.

3. Abundance and diversity of large-scale channels illustrate that channel-forming and filling processes in fine-grained turbidite systems are complex. Many channels are broad (hundreds of meters wide), shallow (few meters deep), and cross-cutting.

4. The fine-grained turbidite systems were adjacent to and pro-graded by coarser-grained (fine to coarse sand) braidplain delta deposits. The braided streams traversing the subaerial braidplain delta delivered sediment directly to the turbidite systems, rather than depositing it at the shoreline in Gilbert-type deltas or distributary mouth bars.

5. A vertical sequence comprised of fine-grained turbidites covered by coarser braided stream deposits may be typical for the fill of large post-glacial lakes or inland seas. Streams remove fine sediments from the poorly sorted glacial debris and transport them to the shoreline where they may enter turbidite systems through underflows. Coarser sediments are left on the subaerial braidplain delta as winnowed lag deposits; with time these sands prograde over the turbidite systems.

Appendix. Outcrop locations

Brief descriptions of outcrop locations below are summarized from Frisch (1987). USGS Antarctica 1:250 000 Reconnaissance Series quadrangle names are given in parentheses. Outcrop numbers refer to locations on Fig. 16.2.

Outcrop number	Location
4(b)	82°52′S, 160°55′ (Nimrod Glacier). North-facing slope of peak (4000 m), 6 km west of Mount Markham
2	83°13′S, 160°28′E (Mount Rabot). East-facing slope north of north–south trending ridge, 4.0 km south of Mount Counts
11	83°18.9′S, 161°49′E. (Mount Rabot). Northwest-facing slope at the head of the helm Glacier
17	83°15.9′S, 162°52′E. (Mount Rabot). East-facing slope 10.4 km northwest of Bengaard Peak
15	83°18.3′, 164°23′E. (Mount Rabot). East-facing slope in the southern section of Bunker Cum, the westward extension of the Clarkson Peak ridge, 2 km northwest of Clarkson Peak
3	83°23.6′S, 165°49′E (Mount Elizabeth and Mount Kathleen)
7a	83°30.4′S, 160°49′E (Mount Rabot). North-facing slope at the northern end of the ridge extending north-northwest of Mt Weeks
7b	83°17.8′S, 160°38′E (Mount Rabot). Southwest-facing slope 3.6 km northwest of Mount Angier in the Moore Mountains
13	83°19.9′, 160°41′E. (Mount Rabot) North-facing slope 2.9 km northwest of Mt Angier
22	83°31.5S, 160°47′E. (Mount Rabot). South-southwest-facing slope of the dry valley on the southern section of Mount Weeks Massif
67	83°54.5′S; 166°34′E (Mount Elizabeth and Mount Kathleen). Northwest-facing slope 7 km north of Mount Mackellar, near the head of Tillite Glacier
19	85°01.1′S, 164°09′E. (Buckley Island and Plunket Point). Southwest-facing slope of Mount Bowers
1	84°40.9′S, 170°36′E. (The Cloudmaker). Southwest-facing slope at the western edge of Mount Deakin

Acknowledgements

We wish to thank R. Frisch, L. Krissek, T. Horner, J. Isbell, B. Lord, J. Miller, and B. Waugh for their help in the field and for many fruitful discussions. J. Isbell, P.G. DeCelles, R.L. Phillips, J. May, and R. Winn made helpful comments on earlier versions of this manuscript. This study was supported by NSF grants DPP84-18445 to M.F. Miller and J.M.G. Miller, and DPP84-18354 to J.W. Collinson.

References

Aitchison, J.C., Bradshaw, M.A. & Newman, J. (1988). Lithofacies and origin of the Buckeye Formation: Late Paleozoic glacial and glaciomarine sediments, Ohio Range, Transantarctic Mountains, Antarctica. *Palaeogeography, Palaeoclimatology, Palaeoecology*, **64**, 93–104.

Ashley, G.M., Southard, J.B., & Boothroyd, J.C. (1982). Deposition of climbing-ripple beds: a flume simulation: *Sedimentology*, **29**, 67–79.

Ashley, G.M. & Smith, N.D. (1985). *Glacial sedimentary environments*. Society of Economic Paleontologists and Mineralogists Short Course No. 16 Lecture Notes, 246 pp.

Banks, M.R. & Clarke, M.J. (1987). Changes in the geography of the Tasmania basin in the late Paleozoic. In *Gondwana Six: Stratigraphy, sedimentology, and paleontology*, ed., G.D. McKenzie, Geophysical Monography 41, American Geophysical Union, 1–14.

Barrett, P.J. (1968). The post-glacial Permian and Triassic Beacon Rocks in the Beardmore Glacier area, Central Transantarctic Mountains, Antarctica. Ph.D. dissertation, Ohio State University, 144 pp.

Barrett, P.J., Elliot, D.H. & Lindsay, J.F. (1986). The Beacon Supergroup (Devonian–Triassic) and Ferrar Group (Jurassic) in the Beardmore Glacier area, Antarctica. In *Geology of the Central Transantarctic Mountains*, ed., M.D. Turner & J.F. Splettstoesser. Antarctic Research Series, American Geophysical Union, 36, 339–429.

Berner, R.A. & Raiswell, R. (1984). C/S method for distinguishing freshwater from marine rocks. *Geology*, **12**, 365–8.

Bouma, A.H. (1962). *Sedimentology of some flysch deposits*. (Amsterdam: Elsevier, 168 pp.

Bouma, A.H., Stelting, C.E. & Coleman, J.M. (1985). Mississippi Fan, Gulf of Mexico. In *Submarine fans and related turbidite systems*, ed., A.H. Bouma, W.R. Normark, & N.E. Barnes. New York: Springer-Verlag, 137–50.

Bouma, A.H. (1988). Migratory and braided channels in Eocene submarine fan deposits in the French Maritime Alps (abstract). *American Association of Petroleum Geologists Bulletin*, **72**, 990.

Braithwaite, P., Armentrout, J.M., Beeman, C.E. & Malecek, S.J. (1988). East breaks block 160 field, offshore Texas: a model for deep-water deposition of sand. Twentieth Annual Offshore Technology Conference 1988 Proceedings 2, 145–56.

Bridge, J.S. & Gordon, E.A. (1985). Quantitative interpretation of ancient river systems in the Oneonta Formation, Catskill Magnafacies. In *The Catskill Delta*, ed., D.L. Woodrow & W.D. Sevon, Geological Society of America Special Paper 201, 163–81.

Bromley, R.G. & Asgaard, V. (1979). Triassic freshwater ichnocoenoses from Carlsbert Fjord, East Greenland. *Palaeogeography, Palaeoclimatology, Palaeoecology,* **23**, 39–80.

Chan, M.A. & Dott, R.H., Jr. (1983). Shelf and deep-sea sedimentation in Eocene forearc basin, western Oregon – fan or non-fan? *American Association Petroleum Geologists Bulletin,* **67**, 2100–16.

Cheema, M.R., Donaldson, A.C., Heald, M.T. & Renton, J.J. (1977). Gas producing submarine-fan deposits of Catskill delta complex in north-central West Virginia (abstract). *American Association of Petroleum Geologists Bulletin,* **61**, 775–6.

Collinson, J.D. & Thompson, D.B. (1982). *Sedimentary structures.* London: George Allen and Unwin, 194 pp.

Collinson, J.D. (1986). Comment on 'submarine ramp facies model for delta-fed, sandrich turbidite systems.' *American Association of Petroleum Geologists Bulletin,* **70**, 1742–3.

Collinson, J.W. (1990). Depositional setting of Late Carboniferous to Triassic biota in the Transantarctic basin. In *Antarctic paleobiology: its role in the reconstruction of Gondwana,* ed., T.N. Taylor & E.L. Taylor. New York: Springer-Verlag, pp. 1–14.

Collinson, J.W. & Miller, M.F. (1991). Comparison of Lower Permian postglacial black shale sequences in the Ellsworth and Transantarctic Mountains, Antarctica. In *Gondwana Seven Proceedings,* ed., H. Ulbrich & A.C. Rocha Campos, Instituto de Geosciencias, Universidade de Sao Paulo, 217–31.

Collinson, J.W., Vavra, C.L. & Zawiskie, J.M. (1992). Sedimentology of the Polarstar Formation, Permian, Ellsworth Mountains, Antarctica. In *Geology and paleontology of the Ellsworth Mountains, Antarctica,* ed., G.F. Webers, C. Craddock & J.F. Splettstoesser. Geological Society of America Memoir, 63–79.

Curtis, C.D. (1964). Studies on the use of boron as a paleoenvironmental indicator: *Geochimica Cosmochimica Acta.,* **28**, 1125–37.

Damuth, J.E., Flood, R.D., Kowsmann, R.O., Belderson, R.H. & Gorini, M.A. (1988). Anatomy and growth pattern of Amazon deep-sea fan revealed by long-range side-scan sonar (Gloria) and high resolution seismic studies. *American Association of Petroleum Geologists Bulletin,* **72**, 885–911.

de Castro, J.C. (1988). Sedimentology, stratigraphy and paleontology of the Gondwana sequence of the Paraná Basin. Field excursion AZ Guidebook, Seventh Gondwana Symposium. Petrobras, Rio de Janeiro, 29 pp.

Domack, E.W. (1988). Biogenic facies in the Antarctic glacimarine environment: basis for a polar glacimarine summary. *Palaeogeography, Palaeoclimatology, Palaeoecology,* **63**, 357–372.

Droser, M.L. & Bottjer, D.J. (1987). Development of ichnofabric indices for strata deposited in high-energy nearshore terrigenous clastic environments. In *New concepts in the use of biogenic sedimentary structures for paleoenvironmental interpretation,* ed., D.J. Bottjer. Los Angeles: Pacific Section, Society of Economic Paleontologists and Mineralogists, pp. 29–33.

Elliot, D.H. (1975). Gondwana basins of Antarctica. In *Gondwana geology,* ed., K.S.W. Campbell. Canberra: Australia National University Press, pp. 493–536.

Elmgren, R. (1978). Structure and dynamics of Baltic benthos communities, with particular reference to the relationship between macro- and meiofauna: *Kieler Meeresforschungen,* **4**, 1–22.

Frey, R.W. & Pemberton, S.G. (1984). Trace fossil facies models. In *Facies models,* 2nd edition, ed., R.G. Walker. Geoscience Canada Reprint Series 1, 189–207.

Frisch, R.S. & Miller, M.F. (1991). Provenance and tectonic implications of sandstones within the Permian Mackellar Formation, Central Transantarctic Mountains. In *Geological evolution of Antarctica,* ed., M.R.A. Thomson, J.A. Crame, J.W. Thomson. Cambridge: Cambridge University Press, pp. 219–23.

Frisch, R.S. (1987). Permian sandstones of the Mackellar Formation (Beacon Supergroup), Central Transantarctic Mountains: provenance and tectonic implications, Unpublished Master's Thesis, Vanderbilt University, Nashville, Tennessee, 144 pp.

Grunow, A.M., Kent, D.V. & Dalziel, I.W.D. (1991). New paleomagnetic data from Thurston Island: implications for the tectonics of West Antarctic and Weddell Sea Opening. *Journal of Geophysical Research,* **96**, 17 935–54.

Hammer, W.R. (1990). Triassic terrestrial vertebrate faunas of Antarctica. In *Antarctic paleobiology: its role in the reconstruction of Gondwana,* ed. T.N. Taylor & E.L. Taylor, New York: Springer-Verlag, pp. 42–50.

Heller, P.L. & Dickinson, W.R. (1985). Submarine ramp facies model for delta-fed, sand-rich turbidite systems. *American Association of Petroleum Geologists Bulletin,* **69**, 960–76.

Ineson, J.R. (1989). Coarse-grained submarine fan and slope apron deposits in a Cretaceous back-arc basin, Antarctica. *Sedimentology,* **36**, 793–819.

Isbell, J.L. & Collinson, J.W. (1988). Fluvial architecture of the Fairchild and Buckley Formations (Permian), Beardmore Glacier area. *Antarctic Journal of the United States,* **23**, 3–5.

Isbell, J.L. & Collinson, J.W. (1991). Sedimentological significance of fluvial cycles in the Permian of the central Transantarctic Mountains. In *Gondwana Seven Proceedings,* ed., H. Ulbrich & A.C. Rocha Campos. Instituto de Geosciencias, Universidade de Sao Paulo, 189–99.

Krissek, L.A. & Horner, T.C. (1988). Geochemical record of provenance in fine-grained Permian clastics, central Transantarctic Mountains. *Antarctic Journal of the United States,* **23**, 19–21.

Krissek, L.A. & Horner, T.C. (1991). Clay mineralogy and provenance of fine-grained Permian clastics, Central Transantarctic Mountains. In *Geological evolution of antarctica,* ed., M.R.A. Thomson, J.A. Crame & J.W. Thomson. Cambridge: Cambridge University Press, pp. 209–13.

Kyle, R.A. & Schopf, J.M. (1982). Permian and Triassic palynostratigraphy of the Victoria Group, Transantarctic Mountains. In *Antarctic geoscience,* ed., C. Craddock. Madison: University of Wisconsin Press, pp. 649–59.

Laird, M.G., Mansergh, G.D. & Chappell, J.M.A. (1971). Geology of the central Nimrod Glacier area, Antarctica: *New Zealand Journal Geology and Geophysics,* **14**, 427–68.

LeBlanc Smith, C. & Eriksson, K.A. (1979). A fluvioglacial and glaciolacustrue deltaic and depositional model for Permo-Carboniferous coals of the northwestern Karoo Basin, South Africa. *Palaeogeography, Palaeoclimatology, Palaeoecology,* **27**, 67–84.

Lindsay, J.F. (1968a). Stratigraphy and sedimentation of the lower Beacon rocks of the Queen Alexandra, Queen Elizabeth and Holland Ranges, Antarctica, Ph.D. Dissertation, Ohio State University, 300 pp.

Lindsay, J.F. (1968b). The development of clast fabric in mudflows. *Journal of Sedimentary Petrology,* **38**, 1242–53.

Long, W.E. (1965). Stratigraphy of the Ohio Range, Antarctica. In *Geology and paleontology of the Antarctic,* ed., J.B. Hadley, American Geophysical Union, Antarctic Research Series 6, 71–116.

Lopez Gamundi, O.R. (1989). Post-glacial transgressions in Late Paleozoic basin of western Argentina: a record of glacioeustactic sea level rise. *Palaeogeography, Palaeoclimatology, Palaeoecology,* **71**, 257–70.

Lowe, D.R. (1982). Sediment gravity flows. II. Depositional models with special reference to the deposits of high density turbidity currents. *Journal of Sedimentary Petrology,* **52**, 279–97.

Martini, I.P. & Rocha Campos, A.C. (1991). Lower Gondwana coal sequences in the Paraná Basin, Brazil. In *Gondwana Seven Proceedings,* ed., H. Ulbrich & A.C. Rocha Campos, Instituto de Geosciencias, Universidade de Sao Paulo, pp. 317–29.

McKenzie, G.D. (ed.) (1987). *Gondwana Six: Stratigraphy, Sedimentology, and Paleontology.* Geophysical Monograph 41. American Geophysical Union, 250 pp.

McLachlan, I.R. & Anderson, A.M. (1973). A review of the evidence for marine conditions in southern Africa during Dwyka times. *Palaeontologica Africana*, **15**, 37–64.

McPherson, J.G., Shanmugam, G. & Moola, R.J. (1988). Fan deltas and braid deltas: conceptual problems. In *Fan deltas; sedimentology and tectonic settings*, eds., W. Nemec & R.J. Steel, London: Blackie and Son, pp. 14–22.

Melvin, J.B. (1986). Upper Carboniferous fine-grained turbidite sandstones from southwest England: a model for growth in an ancient delta fed subsea fan. *Journal of Sedimentary Petrology*, **56**, 19–34.

Middleton, G.V. & Hampton, M.A. (1973). Sediment gravity flows: mechanics of flow and deposition. In *Turbidites and deep-water sedimentation*, eds., G.V. Middleton & A.H. Bouma. Pacific Section, Society of Sedimentary Paleontologists and Mineralogists, pp. 1–38.

Miller, J.M.G. (1989). Glacial advance and retreat sequences in a Permo-Carboniferous section, Central Transantarctic Mountains. *Sedimentology*, **36**, 419–30.

Miller, J.M.G. & Waugh, B.J. (1991). Permo-Carboniferous glacial sedimentation in the central Transantarctic Mountains and its paleotectonic implications (extended abstract). In *Geological evolution of Antarctica*, eds., M.R.A. Thomson, J.A. Crame & J.W. Thomson, Cambridge: Cambridge University Press, pp. 205–8.

Miller, M.F. (1979). Paleoenvironmental distribution of trace fossils in the Catskill deltaic complex, New York State. *Palaeogeography, Palaeoclimatology, Palaeoecology*, **28**, 117–41.

Miller, M.F. (1984). Distribution of biogenic structures in Paleozoic non-marine and marine-margin sequences. *Journal of Paleontology*, **58**, 550–70.

Mutti, E. & Ricci Lucchi, F. (1972). Turbidites of the northern Apennines: introduction to facies analysis (English translation by T.H. Nilsen, 1978). *International Geology Review*, **20**, 125–66.

Mutti, E. (1977). Distinctive thin-bedded turbidite facies and related depositional environments in the Eocene Hecho Group (south-central Pyrenees, Spain), *Sedimentology*, **24**, 107–31.

Mutti, E. & Normark, W.R. (1987). Comparing examples of modern and ancient turbidite systems: problems and concepts. In *Marine clastic sedimentology: concepts and case studies*, ed., J.R. Leggett & G.G. Zuffa, London: Graham and Trotman, 1–37.

Nemec, W. (1991*a*). Deltas – remarks on terminology and classification. In *Coarse-grained deltas*, eds., A. Colella & D.B. Prior, International Association of Sedimentologists Special Publication 10. Oxford: Blackwell Scientific Publications, pp. 3–12.

Nemec, W. (1991*b*). Aspects of sediment movement on steep delta slopes. In *Coarse-Grained Deltas*, eds., A. Colella & D.B. Prior, International Association of Sedimentologists Special Publication 10. Oxford: Blackwell Scientific Publications, pp. 29–73.

Oelofsen, B.W. (1987). The biostratigraphy and fossils of the Whitehill and Irati Shale Formations of the Karoo and Paraná Basins. In *Gondwana Six: stratigraphy, sedimentology and paleontology*, ed., G.D. McKenzie, Geophysical Monograph 41. American Geophysical Union, 131–8.

Pickering, K.T., Stow, D.A.V., Watson, M.P. & Hiscott, R.N. (1986). Deep-water facies, processes and models: review and classification scheme for modern and ancient sediments. *Earth Science Reviews*, **23**, 75–174.

Pollard, J.E. (1988). Trace fossils in coal-bearing sequences. *Journal of the Geological Society of London*, **145**, 339–50.

Postma, G. (1991). Depositional architecture and facies of river and fan deltas: a synthesis. In *Coarse-grained deltas*. eds., A. Colella & D.B. Prior, International Association of Sedimentologists Special Publication 10, Oxford: Blackwell Scientific Publications, p. 13–27.

Reading, H.G. & Orton, G.J. (1991). Sediment calibre: a control on facies

models with special reference to deep-sea depositional systems. In *Controversies in modern geology: evolution of geological theories in sedimentology, earth history and tectonics*, ed., D.W. Muller, J.A. McKenzie & H. Weissert, London: Academic Press, pp. 85–111.

Redfern, J. (1991). Subsurface facies analysis of Permo-Carboniferous glacigenic sediments, Canning Basin, western Australia. In *Gondwana Seven Proceedings*, ed., H. Ulbrich & A.C. Rocha Campos. Instituto de Geosciencias, Universidade de Sao Paulo, pp. 349–364.

Reid, G.K. & Wood, R.D. (1976). *Ecology of inland waters, 2nd edition*. New York: Van Nostrand, 485 pp.

Ricker, W.E. (1952). The benthos of Cultos Lake. *Journal of the Fisheries Research Board of Canada*, **9**, 204–12.

Runnegar, B. & Campbell, K.S.W. (1976). Late Paleozoic faunas of Australia. *Earth Science Reviews*, **12**, 235–57.

Shanmugam, G. & Moiola, R.J. (1988). Submarine fans: characteristics, models, classification, and reservoir potential. *Earth Science Reviews*, **24**, 383–428.

Shanmugam, G., Moiola, R.J., McPherson, J.G. & O'Connell, S. (1988). Comparison of turbidite facies association in modern turbidite facies associations in modern passive-margin Mississippi fan with ancient active-margin fans. *Sedimentary Geology*, **58**, 63–77.

Shanmugam, G., & Moiola, R.J. (1991). Types of submarine fan lobes: models and implications. *American Association of Petroleum Geologists Bulletin*, **75**, 156–79.

Siemers, C.T. (1978). Submarine fan deposition of the Woodbine–Eagle Ford interval (Upper Cretaceous), Tyler County, Texas. *Gulf Coast Association of Geological Societies Transactions*, **28**, 493–533.

Stow, D.A.V. (1985). Deep-sea clastics: where are we and where are we going? In *Sedimentology: recent developments and applied aspects*. Eds. P.J. Brenchley & B.P.J. Williams. Special Publication of the Geological Society, 18, 67–93.

Surlyk, F. (1987). Slope and deep shelf gully sandstones, Upper Jurassic, East Greenland. *American Association Petroleum Geologists Bulletin*, **71**, 464–75.

Tankard, A.J., Eriksson, K.A., Hunter, D.R., Jackson, M.P.A., Hobday, D.K. & Minter, W.E.L. (1982). *Crustal evolution of Southern Africa: 3.8 billion years of earth history*. New York: Springer-Verlag, 523 pp.

Taylor, T.N. & Taylor, E.L. (1990). *Antarctic paleobiology: its role in the reconstruction of Gondwana*. New York: Springer-Verlag, 261 pp.

Tourtelot, H.A. (1964). Minor element composition and organic carbon content of marine and non-marine shales of Late Cretaceous age in the western interior of the United States. *Geochimica et Cosmochimica Acta*, **20**, 1579–604.

Ulbrich, H. & Rocha Campos A.C. (eds) (1991). *Gondwana Seven Proceedings*: Instituto de Geosciencias, Universidade de Sao Paulo, 714 pp.

Veizer, J. & Demovic, R. (1974). Strontium as a tool in facies analysis: *Journal of Sedimentary Petrology*, **44**, 93–115.

Visser, J.N.J. (1987). The paleogeography of part of southwestern Gondwana during the Permo-Carboniferous glaciation. *Palaeogeography, Palaeoclimatology, Palaeoecology*, **61**, 205–19.

Visser, J.N.J. (1989). The Permo-Carboniferous Dwyka formation of southern Africa: deposition by a predominantly subpolar marine ice sheet. *Palaeogeography, Palaeoclimatology, Palaeoecology*, **70**, 377–91.

Walker, R.G. (1978). Deep-water sandstone facies and ancient submarine fans: models for exploration for stratigraphic traps. *American Association of Petroleum Geologists Bulletin*, **52**, 932–66.

Walker, R.G. (1985). Cretaceous Wheeler Gorge conglomerates, California: a possible channel-levee complex. *Journal of Sedimentary Petrology*, **55**, 279–90.

17 Ice scouring structures in Late Paleozoic rhythmites, Paraná Basin, Brazil

ANTONIO C. ROCHA-CAMPOS, PAULO R. DOS SANTOS and JOSÉ R. CANUTO

Abstract

Series of long, sinuous and subparallel furrows on bedding planes of rhythmites of the Itararé Subgroup (Early Permian), cropping out in the State of Santa Catarina, Brazil are interpreted as iceberg scours. Troughs are associated with abundant ice-rafted clasts, pellets and mounds of dumped debris melted out from floating icebergs, and lenses of debris probably released from grounded icebergs. Rhythmites were deposited in a sizeable, relatively deep fresh water body, at least in part as varves, during the last episode of deglaciation of the late Paleozoic glaciation. Iceberg scours are common in modern and Pleistocene glaciated shelves, but are not known from the ancient record despite selective preservation of glaciomarine rocks. This paper presents the first detailed description of iceberg scours from pre-Pleistocene glacial strata.

Introduction

Ice scours caused by the interaction of floating ice masses with basin floor sediments are widely reported from the floors of the Pleistocene lakes and are a marked feature of cold climate shallow marine shelves (e.g. Belderson and Wilson, 1973; Harrison and Jolleymore, 1974; Dredge, 1982; Hélie, 1983; Barnes et al., 1984; Thomas and Connell, 1985; Woodworth-Lynas et al., 1985; Eyles and Clark, 1988).

There is much microgeomorphological data regarding iceberg scours available from seismic and sonar surveys, and submarine diving observations (Barnes et al., 1973; Woodworth-Lynas et al., 1985), but little structural data to aid interpretation of these structures in the pre-Pleistocene strata. With exception of the brief mentions of Rattigan (1967) and von Brunn (1977), no fossil ice scour structures have been identified from the Earth's pre-Pleistocene glacial record despite the widespread, selective preservation of glaciomarine rocks.

We describe here series of long furrows exposed in plan and section developed on bedding planes of late Paleozoic rhythmites of the Paraná Basin in southeastern Brazil, interpreted as ice scour structures. These rhythmites belong to the upper part (Rio do Sul Formation) of the Late Carboniferous–Early Permian Itararé Subgroup (Castro, 1988) and represent facies associated with the last event of the Gondwana glaciation in the area.

The excellent exposures allow the detailed examination of the morphology and structures associated with the furrows and thus furnish valuable information for the identification of ice scour features in the pre-Pleistocene glacial record.

Stratigraphic setting

Ice scour features described herein have been found on bedding planes of rhythmites cropping out at a quarry operated by Mineração Sul Brasil located some 300 m north of km 161 + 100 m of Road BR-470, near Trombudo Central, in central-eastern Santa Catarina (Rocha-Campos et al., 1990; Santos et al., 1992). (Fig. 17.1).

Stratigraphic relations of the rhythmites with other facies of the upper part of the Rio do Sul Formation are synthetically depicted in Fig. 17.2 (Castro, 1988). In the section exposed between km 158 and 162 of road BR-470, the rhythmites crop out with a total thickness of about 50 m. They seem to intercalate with lenticular, sigmoidal delta front sands and interdigitate with and are covered by the upper marine shale (Castro, 1988). This lithology is partly contemporaneous with, and may represent the prodelta muds of the prograding deltaic sands of the basal part of the Rio Bonito Formation (Triunfo Member), that locally contains thin coal layers. The geological evolution of the area is interpreted as involving episodes of advance and retreat of the ice at the margin of the Rio do Sul sub-basin accompanied by changes in sea-level, resulting in alternating regressive and transgressive episodes. Advances of the glacier are documented by deposition of a lower basal tillite (lodgement) and the influx of thick turbidite sands, sand-shale rhythmites and shale (deep and fringe lobe deposits), and retreats by the deposition of glaciomarine and marine shale.

The rhythmites exposed at the quarry faces, about 4 m thick, are defined by cyclic repetition of couplets 3.5 to 8.5 cm thick composed of a basal light-coloured silt layer, in sharp contact with an upper dark clay thinner lamina of relatively constant thickness (up to 4 mm). Bright layers are composed of several millimetric to centi-

Fig. 17.1. Location of structures described in the Paraná Basin, Brazil.

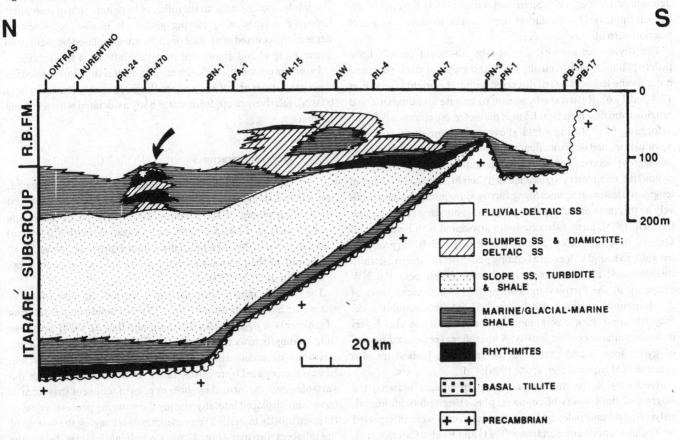

Fig. 17.2. Stratigraphic panel of the Rio do Sul Formation from Lontras to well PB-17. Arrow points to location of rhythmites studied at BR-470. R.B.FM., Rio Bonito Formation (Triunfo Member). (Adapted from Castro, 1988.)

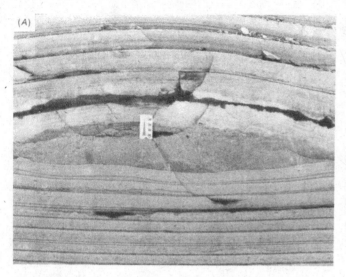

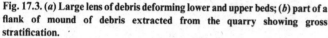

Fig. 17.3. (*a*) Large lens of debris deforming lower and upper beds; (*b*) part of a flank of mound of debris extracted from the quarry showing gross stratification.

metric thick layers deposited by a combination of different dispersive mechanisms including turbidity currents (graded layers) and debris flows (Ashley, 1975; Smith and Ashley, 1985). Except for rare current ripples and some starved ripple marks no other evidence of traction currents has been seen.

The rhythmites are characterized by abundant rafted clasts, pellets (Thomas and Connell, 1985) and chips of dark clay, and mounds and lenses of debris dispersed within the bright layers (Figs 17.3 and 17.6). Rafted clasts, mostly of granite and quartzite and ranging from millimeters to 1.5 m in diameter, are commonly found deforming to rupturing strata above and below them and were obviously melted out from floating icebergs. Mounds (planar base) and lenses (biconvex) of non-sorted mixtures of sand, silt, clay and diamictite, millimeters to decimeters in height, and up to meters in length, the latter sometimes filling furrows, are interpreted respectively as products of iceberg dumping and grounding (Thomas and Connell, 1985). Larger structures are associated with bending and faulting of layers below and/or above (Fig. 17.3). A group of irregular to round bulges on a bedding plane, decimetric to metric in diameter and up to decimeters high, was found near the NW extremity of the furrows and correspond to a concentration of dump mounds or lenses of debris melted out from grounded ice. Invertebrate trails are commonly seen on top of the clay layers indicating intense benthic activity. A microflora (pollen and spores) of gymnosperms and pteridophytes and algal bodies are also preserved (Marques-Toigo *et al.*, 1989).

Regularity of the rhythmites, the sharp contact between the bright and dark layers of couplets, plus other sedimentological, palynological and paleomagnetic characteristics were interpreted by Rocha-Campos and Sundaram (1981) and Rocha-Campos *et al.* (1981) to be indicative of seasonal control of the deposition of rhythmites of the Trombudo Central area as varves. They also pointed out that close association of the rhythmites and the upper

dark shale, indicative of marine conditions, and salinity constraints for the formation of varves would require deposition of the rhythmites associated with the influx of large amounts of meltwater from the neighbouring calving glacier. It is also possible that deposition occurred in an inlet or embayment isolated by oscillation in the sea-level and dominated by deltaic inflows of fresh water.

From the evidence above, the environment setting interpreted for the rhythmites is one of a reasonably large, probably coastal or near coastal, relatively deep, fresh water body associated with a calving ice margin.

Ice scour structures

Ice scour structures form series of subparallel, elongated, sinuous and shallow depressions or furrows extensively developed on bedding planes of the rhythmites well exposed by quarrying operations (Fig. 17.4). Furrows are of variable width (20–50 cm) and depth (up to 20 cm) along a maximum exposed length of 70–80 m. In spite of their sinuous aspect, furrows have a consistent gross orientation of 60°N–70°W.

The furrows, mostly asymmetrical in section, are either open V-shaped or have flat bottoms (Fig. 17.5). No bulldozing, fracturing of sediments or jigger marks (Reimnitz and Barnes, 1974) have been noted along furrows that have smooth sides, rarely exhibiting fine striations or crenulations parallel to their axes. Troughs are, however, separated by relatively raised, convex areas, also long and variable in width, formed by slices or piles of sediment that seem to have been displaced laterally, during the scouring process, and also thrusted mostly towards SW, sometimes overhanging the margin of the adjacent furrows (Fig. 17.6). Underlying laminae below the troughs at both their margins are cut by small reverse faults dipping towards and away from the center of the structures, diminishing in size downwards. Sections cut parallel to the axes of the troughs also

Fig. 17.4. General view (looking northwest) of deformed bedding planes of the rhythmites.

Fig. 17.5. Furrows and associated features on bedding planes of the rhythmites. Note flat bottom trough at center with overhanging thrusted mass of sediments on the left (foreground) and sharp cut right flank near hammer.

show reverse faults and recumbent folds indicating a sense of movement towards the northwest, coincident with the direction of the furrows.

Along the total deformed thickness of less than 1 m, starting from 2 or 3 different bedding planes where the troughs are more deeply impressed, deformation seems to attenuate downwards, suggesting repetition of the process. Beds above and below the deformed zone are undisturbed.[1]

Discussion

The following mechanisms might account for the origin of the deformations and must be analysed before picturing a comprehensive model for their origin: (1) tectonic movements; (2) slumping; (3) grounding of ice.

The fact that the structures represent a localized intrastratal occurrence does not favour the hypothesis of a tectonic origin. In addition, there is no record of tectonic compressional phases affecting the Paraná Basin during the Paleozoic.

[1] Another furrowed surface was recently found at another quarry face, some 6 m below the main deformed zone. Except for their smaller dimensions and different orientation the troughs are identical to the ones described here.

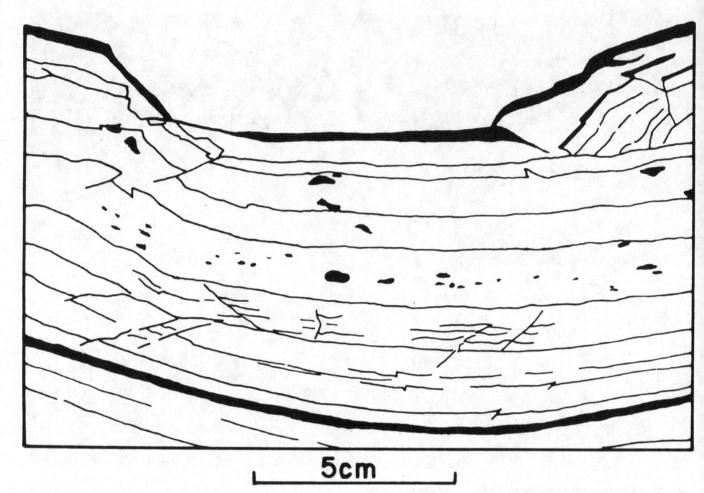

5cm

Fig. 17.6. Diagram of a section of a small trough showing morphology and associated structures.

Folding and other deformations caused by intrastratal gravity slumping along a bedding plane or 'décollement' surface are relatively common structures associated with rhythmites, including glaciolacustrine varves (Pickrill and Irwin, 1983) and are well exposed in a nearby quarry of the Itararé rhythmites. The geometry of the troughs, however, do not exhibit any distinctive characteristics of slumped strata, that are in general structurally more complex.

The sedimentary and environmental context described above strongly suggest the origin of the troughs associated with the former presence of grounded ice over the bottom of the basin of deposition, under three possible circumstances, namely: (a) glacier advance; (b) lake/sea ice; and (c) icebergs.

Structures similar in shape to the troughs, generated by deformation of overriding ice flow have been described in the Pleistocene and pre-Pleistocene literature as fluted (e.g. Boulton, 1976) or furrowed surfaces (von Brunn, 1977; Tomazelli and Soliani, 1982). However, such structures are usually associated with the lodgement and deformation of basal till and may show evidence of shear, as

faults, folds or glaciotectonic features (Hicock and Dreimanis, 1985). In the present case the occurrence would be expected of intercalated deposit (tillite) resting upon an unconformable surface, as well as of shear structures in the underlying overpressured rhythmites.

Either seasonal lake/sea ice or icebergs may be responsible for the deformation structures. One argument against the hypothesis of lake/sea ice is the abundance of particles and debris rafted and dumped from icebergs or released from grounded ice associated with the furrows, which suggests the frequent presence of icebergs in the water body (Thomas and Connell, 1985). In addition, seasonal ice ridges are usually formed in the nearshore zone of fast ice and their gouging activity decreases sharply in offshore direction. The relatively deep water depth suggested for the present site indicates that seasonal ice masses are unlikely to have been a significant influence on the basin floor (Barnes et al., 1984; Fischbein, 1987). Moreover, in the case of one ice ridge we would not expect a series of parallel troughs as occurs in the study area.

Thus, the sedimentary and paleoenvironmental context and the

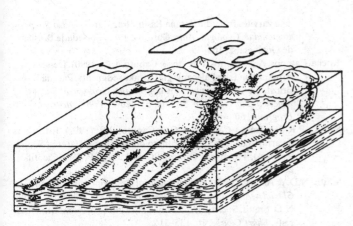

Fig. 17.7. Proposed model for the formation of the scour structures. Larger arrow indicates sense of displacement of iceberg; smaller arrows denote lateral oscillation and rotation.

set of characteristics shown by the troughs, particularly orientation and association with abundant rafted clasts, and mounds and lenses of debris dumped or melted out from ice are highly suggestive of an origin through grounding of a single iceberg with some irregularities at the base of the ice.

Fig. 17.7 shows a model for the evolution of the scour structures comprising a single iceberg with an irregular base. The iceberg displaced on bedding planes of the rhythmites towards the northwest probably in response to wind coupling combined with lateral thrusting action to the southwest. The trajectory of the iceberg apparently involved some slight rotation and oscillation, producing the sinuous, meandering pattern of the troughs and lateral piling-up of sediment. The berg finally came to rest, melted and discharged its contained debris which is preserved as mounds or lenses at the far end of the furrows.

Comparison of the structures described here and examples of ice-gouging from the Recent and Pleistocene glaciomarine and glaciolacustrine record, however, shows several differences that deserve discussion. First of all, troughs in the Itararé rhythmites are relatively small-scale features when compared with modern and Pleistocene scours on shelves and glaciolacustrine lakes, which are generally larger (e.g. Barnes, 1987; Barnes et al., 1984; Belderson and Wilson, 1973; Reimnitz and Barnes, 1974; Dredge, 1982; Hélie, 1983; Woodworth-Lynas et al., 1985, and pers. comm., 1990). We suggest, as a possible explanation, that relatively small icebergs just touching and sliding on the soft clay bottom could be capable of producing less intense and smaller deformation. One may speculate that interference among successive troughs in present day shelves might actually cause obliteration of smaller features, selectively preserving the bigger ones, that are associated with stronger deformation of the substratum.

The idea that ice scouring is necessarily a multiple event that generates always highly reworked and deformed piles of sediment, as suggested by examples on present-day marine shelves may, therefore, be not always the case. Interpretation of the Itararé

structures as iceberg scours, as discussed here, might indicate rather that paleoscours may occur as small-scale features that necessitate good outcrops, such as the Itararé one, to identify. We hope that data presented here stimulate detailed structural descriptions of other pre-Pleistocene sections exhibiting similar features.

Acknowledgements

Support for this research was provided to A.C. Rocha-Campos by the Fundação de Amparo à Pesquisa do Estado de São Paulo (FAPESP), Brazil (Proc. 89/3092-2; 91/0546-2), and to P.R. dos Santos by the Conselho Nacional de Desenvolvimento Científico e Tecnológico (CNPq), Brazil (Proc. 203090-89.6). A first draft of this paper was written during the tenure of the CNPq research fellowship awarded to P.R. dos Santos at the Glaciated Basin Research Group, University of Toronto, Scarborough Campus. We thank N. Eyles for many useful comments and to Christopher Woodworth-Lynas for suggestions and bibliographic help.

References

Ashley, G.M. (1975). Rhythmic sedimentation in Glacial Lake Hitchcock, Massachusetts-Connecticut. In *Glaciofluvial and glaciolacustrine sedimentation*, eds., A.V. Jopling & B.C. McDonald. Society of Economic Paleontologists and Mineralogists Special Publication, 23, 304–20.

Barnes, P.W. (1987). Morphologic studies of the Wilkes Land Continental Shelf, Antarctica – glacial and iceberg effects. In *The Antarctic continental margin: geology and geophysics of offshore Wilkes Land*, eds., S.L. Eittreim & M.A. Hampton, Circum-Pacific Council for Energy and Mineral Resources, Houston, Earth Science Series, 5A, 175–94.

Barnes, P.W., Rearic, D.M. & Reimnitz, E. (1984). Ice gouge characteristics and processes. In *The Alaskan Beaufort Sea-ecosystems and environments*, eds., P.M. Barnes, D.M. Schell & E. Reimnitz. New York: Academic Press, pp. 159–84.

Barnes, P.W., Reimnitz, E., Gustafson, C.W. & Larsen, B.R. (1973). U.S.G.S. marine studies in the Beaufort Sea off Northern Alaska, 1970–1977. *U.S. Geological Survey Open-File Report*, 561, 11 pp.

Belderson, R.H. & Wilson, J.B. (1973). Iceberg plough marks in the vicinity of the Norwegian Trough. *Norsk Geologisk Tidsskrift*, 53, 323–8.

Boulton, G.S. (1976). The origin of glacially fluted surface: observations and theory. *Journal of Glaciology*, 17, 287–309.

Castro, J.C. de (1988). Stratigraphic framework of Rio do Sul/Rio Bonito Fms. In *Sedimentology, stratigraphy and paleontology of the Gondwana sequence of the Paraná Basin; Guide Book, Excursion A2.* Seventh Gondwana Symposium, São Paulo, 1988, Petróleo Brasileiro S/A, Rio de Janeiro, 1–8.

Dredge, L.A. (1982). Relict ice scour marks and late phases of Lake Agassiz in northern Manitoba. *Canadian Journal of Earth Sciences*, 19, 1079–87.

Eyles, N. & Clark, B.M. (1988). Storm-influenced deltas and ice scouring in late Pleistocene glacial lake. *Geological Society of America Bulletin*, 100, 793–809.

Fischbein, S.A. (1987). Analysis and interpretation of ice-deformed sediments from Harrison Bay, Alaska. *U.S. Geological Survey, Open-File Report*, 262, 73 pp.

Harrison, I.M. & Jolleymore, R.G. (1974). Iceberg furrow marks on the continental shelf northeast of Belle Island, Newfoundland. *Canadian Journal of Earth Sciences*, 11, 43–52.

Hélie, R.G. (1983). Relict iceberg scours, King William Island, Northwest Territories. *Geological Survey of Canada Current Research, Part B*, 83-1B, 415–18.

Hicock, S.R. & Dreimanis, A. (1985). Glaciotectonic structures as useful ice-movement indicators in glacial deposits: four Canadian case studies. *Canadian Journal of Earth Sciences*, **22**, 339–46.

Marques-Toigo, M., Dias-Fabricio, M.E., Guerra-Sommer, Cazzulo-Klepzig, M. & Piccoli, A.E.M. (1989). Afloramento da área de Trombudo Central, Permiano Inferior, Santa Catarina: palinologia, icnologia e sedimentologia. In *XI Congresso Brasileiro de Paleontologia, Resumos das Comunicações*, Curitiba, 1989, Sociedade Brasileira de Paleontologia, 43–4.

Pickrill, R.A. & Irwin, J. (1983). Sedimentation in a deep glacier-fed lake: Lake Tekapo, New Zealand. *Sedimentology*, **30**, 63–75.

Rattigan, J.H. (1967). Depositional, soft sediment and post-consolidation structures in a Paleozoic aqueoglacial sequence. *Journal of the Geological Society of Australia*, **14**, 5–18.

Reimnitz, E. & Barnes, P.W. (1974). Sea ice as a geological agent on the Beaufort Sea shelf of Alaska. In *The coast and shelf of the Beaufort Sea*, eds., J.C. Rudd & J.E. Sater. Arlington: Arctic Institute of North America, pp. 301–53.

Rocha-Campos, A.C. & Sundaram, D. (1981). Geological and palynological observations on Late Paleozoic varvites from the Itararé Subgroup, Paraná Basin, Brazil. In *Anais do II Congresso Latino-Americano de Paleontologia*, Porto Alegre, 1981, ed., Y.T. Sanguinetti. Porto Alegre: Universidade Federal do Rio Grande do Sul, 1, 257–75.

Rocha-Campos, A.C., Ernesto, M. & Sundaram, D. (1981). Geological, palynological and paleomagnetic investigations on Late Paleo-zoic varvites from the Paraná Basin, Brazil. In *Terceiro Simpósio Regional de Geologia, Atas*, Curitiba, 1981. Sociedade Brasileira de Geologia, 2, 162–75.

Rocha-Campos, A.C., Santos, P.R. dos & Canuto, J.R. (1990). Ice-scouring structures in Late Paleozoic glacial lake sediments, Paraná Basin, Brazil. In *25th Annual Meeting of the Geological Society of America, Northeastern Section*, Syracuse, 1990, *Abstracts with Programs*, 66.

Santos, P.R.dos, Rocha-Campos, A.C. & Canuto, P.R. (1992). Estruturas de arrasto de icebergs em ritmito do Subgrupo Itararé (Neopaleozóico), Trombudo Central, SC. Bol.lG-USP, Série Científica, 23, 1–18.

Smith, N.D. & Ashley, G.M. (1985). Proglacial lacustrine environments. In *Glacial sedimentary environments*, eds., G.M. Ashley, J. Shaw, N.D. Smith. Society of Economic Paleontologists and Mineralogists, *Short Course*, 16, 135–216.

Thomas, G.S.P. & Connell, R.J. (1985). Iceberg drop, dump and grounding structures from Pleistocene glaciolacustrine sediments, Scotland. *Journal of Sedimentary Petrology*, **55**, 243–9.

Tomazelli, L.J. & Soliani Jr., E. (1982). Evidências de atividade glacial no Paleozóico Superior do Rio Grande do Sul, Brasil. In *Anais do XXXII Congresso Brasileiro de Geologia*, Salvador, 1982, Sociedade Brasileira de Geologia, 4, 1378–1391.

von Brunn, V. (1977). A furrowed infratillite pavement in the Dwyka Group of northern Natal. *Transactions of the Geological Society of South Africa*, **80**, 125–30.

Woodworth-Lynas, C.M.T., Simms. A. & Rendell, C.M. (1985). Iceberg grounding and scouring on the Labrador continental shelf. *Cold Regions Science & Technology*, **10**, 163–86.

18 Soft-sediment striated surfaces and massive diamicton facies produced by floating ice

C.M.T. WOODWORTH-LYNAS and J.A. DOWDESWELL

Abstract

Soft-sediment striated surfaces, commonly associated with massive diamictites, have been described from many pre-Quaternary glacial sedimentary sequences around the world. A number of authors interpret these associations as evidence for a subglacial origin. However, soft-sediment striated surfaces are formed by the mechanical scouring action of seabed- (or lakebed-) touching free-floating ice masses over large areas of modern high latitude continental shelves, and in glacial lakes and in lakes subject to seasonal freeze-up. Prolonged reworking of seabed and lakebed sediments by scouring ice keels can cause homogenization of pre-existing stratification to form massive diamictons that may be described as iceberg turbates, *sea-ice turbates* or *lake-ice turbates*. Evidence from previous work on soft-sediment striated surfaces by other authors is presented and reviewed in the context of possible origin by scouring ice keels.

In the modern East Greenland glacimarine environment massive diamicton has been deposited during the Holocene by the combined action of iceberg-rafting and suspension settling from subglacial and seasonal terrestrial meltwater runoff. During deposition continuous scouring by deep-drafted icebergs (up to 550 m) has resulted in mechanical mixing of the diamicton. The present surface of this iceberg turbate is characterized morphologically by the typical hummocky topography of criss-crossing iceberg scour marks. We show that Quaternary ice keel scour processes have affected an order of magnitude more of the seafloor than have subglacial processes, and suggest that this order of magnitude difference probably can be applied to pre-Quaternary glacial marine and lacustrine environments. Consequently the effects of ice keel scour should be more frequently preserved than features of ice shelf grounding zone processes. Field work directed at reappraising pre-Quaternary soft-sediment striated surfaces associated with diamictites now may be appropriate.

Introduction

Many pre-Quaternary glacial sedimentary sequences contain bedding plane surfaces that display subparallel ridges, grooves and striations. Ridges and associated grooves have relief of a few centimetres to a few tens of centimetres, and striations have relief of a few millimetres. These subparallel features are generally thought to have been formed by the mechanical action of laterally moving ice in contact with underlying unlithified soft sediment. Throughout this paper we refer to ice-formed, subparallel ridges, grooves and striations on bedding planes in unconsolidated, unfrozen sediments as *soft-sediment striated surfaces*. These surfaces should not be confused with striated boulder pavements which have different origins (e.g. Eyles, 1988, in press).

Soft-sediment striated surfaces have been described from a number of Proterozoic and Phanerozoic glacial sedimentary sequences, for example in Mauritania (Deynoux, 1980), Algeria (Beuf *et al.*, 1971), Namibia (Kroner and Rankama, 1972; Martin, 1965), South Africa (Savage, 1972; Visser and Hall, 1984; Visser, 1985; Visser and Loock, 1988; von Brunn and Marshall, 1989; Visser, 1990), Saudi Arabia (Vaslet, 1990), Australia (O'Brien and Christie-Blick, 1992) and Antarctica (Aitchison *et al.*, 1988; Miller, 1989). Nearly all reported soft-sediment striated surfaces are interpreted to have been formed by the mechanical action of continental to grounded, or nearly neutrally-buoyant, glaciers and ice sheets (e.g. Beuf *et al.*, 1971; Visser, 1989; O'Brien and Christie-Blick, 1992). Soft-sediment striated surfaces may be formed by the mechanical action of ice in three different environmental settings, including: (1) the development of subglacial flutes, grooves and striations at the base of continental glaciers or fully grounded tidewater glaciers; (2) the action of nearly neutrally buoyant ice at its grounding line, the base of which just grazes the sediment surface as it decouples from the bed to form a floating ice shelf or ice tongue; (3) the action of scouring by the keels of free-floating glacial and non-glacial ice masses. Each of these different mechanisms is discussed below.

In the geological record, soft-sediment striated surfaces are generally associated with the upper surface of diamictites (lithified diamictons). Diamictites are poorly sorted, massive to weakly stratified sediments, and are often found in association with the former presence of glaciers. The depositional and post-depositional environment of massive, structureless diamictites is often difficult to define, and discussion of origin often includes not only the sedimentology of the unit, but also that of bounding surfaces and adjacent units (Eyles *et al.*, 1983).

This paper commences with a brief overview of the three principal types of soft-sediment striated surfaces: (1) subglacial flutes, (2) ice sheet grounding line striations, and (3) surfaces formed by free-floating ice masses. We then address three main points: first, the formation and morphology of ice keel scour marks. Ice keel scour marks are described and discussed with respect to the origin of soft-sediment striated surfaces found in the geological record. It is proposed that some soft-sediment striated surfaces are likely to be the product of scour by freely floating ice of either glacier, sea- or lake-ice origin. Second, evidence is presented on the nature of modern massive diamictons formed in East Greenland glacimarine environments, and it is suggested that the diamictons are produced by iceberg-rafting of debris and prolonged reworking by the scouring keels of deep-drafted icebergs. Third, drawing on the first two points, it is suggested that the formation of some massive diamictites and associated soft-sediment striated surfaces can be explained as a product of ice keel scouring and turbation.

Origins of soft-sediment striated surfaces

Subglacial flutes

Striating and fluting mechanisms may operate at the sole of an actively sliding glacier where deformable sediments are present (e.g. Dyson, 1952; Hoppe and Schytt, 1953; Baranowski, 1970; Paul and Evans, 1974; Boulton, 1976, 1982). Glacial flutes originate on the lee sides of rigid subglacial obstructions, frequently boulders, that project at least 0.3–0.5 m above the lodgement till surface (Boulton, 1976). Flutes, which may extend up to 1 km in length, may be formed when deformable subglacial sediment is intruded into cavities which open up in the base of the ice and propagate from the lee sides of obstructions (Boulton, 1976) or by deposition of debris from the glacier sole into cavities on the lee side of rigid obstructions (e.g. Boulton, 1982). The crest lines of glacial flutes intersect the lee sides of initiating obstructions in an upward-concave till wedge so that the downstream height of the flute is always less than the height of the obstruction. Flute width is normally slightly less than boulder width (Boulton, 1976). Subglacial flutes are also deflected around boulders in their path, and this is explained by Boulton (1976) as the result of disturbance of ice flow lines. Solheim et al. (1990) described a striking example of a Late Pleistocene fluted surface imprinted on an area of about 4000 km² of the northern Barents Sea, that was formed when grounded glacier ice flowed across the area.

Ice sheet grounding line striations

It has been suggested that the action of nearly neutrally buoyant ice at its grounding line, the base of which just grazes the sediment surface as it decouples from the bed to form a floating ice shelf or ice tongue, may produce striations in the form of ridges and grooves as a result of irregularities in the ice sheet base (e.g. O'Brien and Christie-Blick, 1992). Soft-sediment striated surfaces may develop if the grounding line position fluctuates through time. It should be noted that, although a grounding 'line' is usually referred

to in the literature, it should more correctly be viewed as a grounding 'zone' because tidal fluctuations can cause relatively large migrations of the line of grounding where the bed slopes only gently (Echelmeyer et al., 1991). Ridges, grooves and striae in this grounding zone are likely to be laterally extensive and highly consistent in orientation. Soft-sediment striated surfaces formed in this situation should have no marginal berms, because the sediment surface is in contact with laterally extensive overlying ice.

Large floating ice shelves, which are dynamically a part of the parent grounded ice sheet, occur around much of the periphery of Antarctica today, but are not found in the contemporary Northern Hemisphere. Floating ice tongues, fed by fast-flowing outlet glaciers are, however, present in a number of locations around Greenland (e.g. Olesen and Reeh, 1969; Echelmeyer et al., 1991). A major difference between these ice tongues and Antarctic ice shelves is one of scale. Whereas the Greenland ice tongues are 5–10 km in width, the major ice shelves of Antarctica are hundreds of kilometres wide at the grounding line. Floating ice shelves require a number of specific conditions in which to form, including a limitation to relatively deep water due to buoyancy considerations, a high ice velocity which feeds mass from large interior basins, and lateral pinning points (Syvitski, 1991). These conditions are met only rarely in the northern Polar regions today. Even where ice advanced to the continental shelf break during Northern Hemisphere full-glacial conditions, ice shelves would not necessarily have formed because of the very high rates of mass loss by iceberg calving likely to be associated with such exposed locations.

Surfaces formed by free-floating ice masses

The contemporary phenomenon of scour by icebergs and seasonal floating ice in the Earth's polar and sub-polar oceans is well documented (cf. the extensive bibliography by Goodwin et al., 1985). The occurrence of ice keel scour marks, modern scouring rates and the process of scouring is described and quantified in a number of papers (e.g. Geonautics Limited, 1985; Hodgson et al., 1988; Lewis and Woodworth-Lynas, 1990; Woodworth-Lynas et al., 1991; Dowdeswell et al., 1993). On the present day Antarctic continental shelf, icebergs probably scour the seabed in water depths as great 500 m (Barnes and Lien, 1988). In the northern hemisphere, icebergs are known to scour in water depths up to 550 m on the East Greenland continental shelf (Dowdeswell et al., 1993), and to 230 m on the eastern Canadian continental shelf (Hotzel and Miller, 1983). In contrast to such extremely deep iceberg keels, scouring by sea-ice pressure ridges is generally restricted to water depths of less than 60 m, and often to much smaller depths (Reimnitz et al., 1984). In large polar to temperate freshwater bodies, present-day scouring by seasonally occurring ice generally takes place during the spring breakup. Examples include Great Slave Lake (Weber, 1958) and Lake Erie, where ice keels scour in water depths up to 25 m (Grass, 1984, 1985).

The planiform pattern of relict Quaternary lake ice and iceberg scour marks has been reported from large areas of Canada and the northern United States. Examples include Lake Agassiz (Horberg, 1951; Clayton et al., 1975; Dredge, 1982; Mollard, 1983; Woodworth-Lynas and Guigné, 1990), Lake Ojibway (Dionne, 1977) and Lake Iroquois (Gilbert *et al.*, 1992), and also southern Norway where iceberg scouring occurred during a very large Preboreal glacial outburst flood (Longva and Bakkejord, 1990; Longva and Thoresen, 1991). Similar patterns, formed by icebergs during deglaciation, also have been identified on the modern lake floor of Lake Superior (Berkson and Clay, 1973). Marine ice keel scour marks are seen also on subaerially exposed seafloor areas on King William Island (Hélie, 1983; Woodworth-Lynas *et al.*, 1986), Victoria Island (Sharpe, pers. comm., 1987), and on Amund Ringnes Island and Ellef Ringnes Island (Hodgson, 1982) in the Canadian Arctic, and on Coats Island (Aylsworth and Shilts, 1987; Josenhans and Zevenhuizen, 1990) and Mansel Island (Josenhans, pers. comm., 1990) in Hudson Bay.

The nature of glacimarine and glacilacustrine sediments provides important contextual evidence for identification of ice keel scour marks. A common association is that of massive diamict facies and ice keel scour marks produced by the combined action of melt-out and fall, through the water column, of ice-rafted debris and continuous mechanical deformation of the seafloor by scouring keels. Prolonged reworking of marine sediments by the action of grounding and scouring iceberg keels will produce a variety of diamict facies termed iceberg turbates (Vorren *et al.*, 1983). The more general term ice keel turbate may be used to describe facies reworked by scouring icebergs, seasonal ice, or by some combination of both when the nature of the scouring agent cannot be determined (Barnes and Lien, 1988). We propose the additional, specific terms *sea-ice turbate* and *lake-ice turbate* to describe marine and lacustrine facies, respectively, when it can be established that sediments have been reworked by the keels of floating ice of non-glacial origin.

Scour by free-floating ice masses: ice keel scour mark morphology and identification criteria

Ice keel scour marks observed in modern oceanic environments are generally curvilinear trough-like features (Fig. 18.1). The trough is generally incised into the seabed, and may range in depth from as little as a few centimetres to as much as 25 m, and in width from less than 1 m to as much as 250 m (e.g. Lien, 1981). In fine-grained sediments, the flat-bottomed trough is generally characterized as a soft-sediment striated surface, on which ridges, grooves and striations are oriented parallel to the trend of the ice keel scour mark. Ridges and grooves commonly range in relief and width from a few millimetres to 0.3 m (Hodgson *et al.*, 1988; Woodworth-Lynas *et al.*, 1991). The striated surface is formed by moulding of seabed material as it passes from beneath the ice at the trailing edge of the keel (Woodworth-Lynas *et al.*, 1991). Ridges, grooves and striations are formed by the moulding of sediment in cracks, fissures and other irregularities at the trailing edge of the scouring keel. Ridges that initiate from lodged boulders exhibit crest-lines that are at the same elevation as the tops of the initiating boulders. Ridge width is the same as boulder width, and ridge shape may mirror the cross-sectional shape of the initiating boulder (Woodworth-Lynas *et al.*, 1991). In general ridges and grooves are always subparallel and are

not deflected around boulders. In places on the inner flanks of ice keel scour mark berms, ridges, grooves and striations may be developed at angles up to 45° to those within the ice keel scour mark trough. These disjunct features, restricted to the sides of the trough, probably form as sediment is extruded laterally at the free-surfaces on either side of the scouring keel (Hodgson *et al.*, 1988). In places the ridges, grooves and striations may be punctuated by small pits that are the result of dissolution of blocks of ice broken from, and pressed into the seabed by the scouring keel (Woodworth-Lynas *et al.*, 1991). The ice keel scour mark trough is nearly always flanked by two parallel ridges, or berms, of sediment. The positive relief of the berms is a function of some combination of scour-induced seabed heave and deposition of material excavated from the trough by, and displaced to both sides of the scouring keel (Woodworth-Lynas *et al.*, 1991).

Criteria for identifying ice keel scour marks in stratigraphic section have been presented by Woodworth-Lynas (in press). The criteria are based on data from: (1) visual observations of the surface morphology of modern, marine iceberg scour marks from submersibles (Woodworth-Lynas *et al.*, 1991); (2) cross-sections of relict Late Pleistocene iceberg scour marks preserved in clay sediments of glacial Lake Agassiz, southern Manitoba (Woodworth-Lynas and Guigné, 1990); (3) sandy, lacustrine delta-front sediments at Scarborough Bluffs, Toronto, Ontario (Eyles and Clark, 1988); and (4) silty, glacial outburst flood-related sediments near Romerike in southern Norway (Longva and Bakkejord, 1990; Longva and Thoresen, 1991).

In general, ice keel scour marks may be distinguished from subglacial flutes, and other features interpreted as striations formed in soft sediments, by a number of criteria. These include the following (Fig. 18.2):

1. The presence of one or two (depending on outcrop extent) co-linear berms
2. A trough or region, between the berms, that is generally below the level of the seabed or lakebed bedding surface
3. Restriction of a soft-sediment striated surface to the area between the berms
4. Boulder-initiated ridges that mirror the shape of the boulders
5. The occurrence of ice dissolution voids that truncate the soft-sediment striated surface
6. The presence of flat-topped mounds in topographically low areas within the ice keel scour mark trough
7. The nature of associated sediments (e.g. marine, lacustrine or subglacial).

Examples of ice keel scour marks and soft-sediment striated surfaces in ancient glacial sediments

Background

With a few notable exceptions (Martin, 1965; Rattigan, 1967; Kröner and Rankama, 1972; Powell and Gostin, 1990; Rocha-Campos *et al.*, this volume) neither ice keel scour marks nor ice keel turbates have been documented in the rock record (e.g.

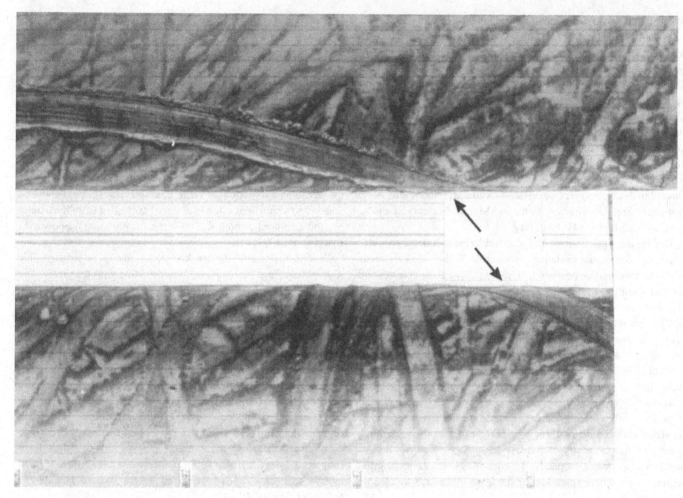

Fig. 18.1. Sidescan sonograph of a typical modern continental shelf seafloor that has been affected by scouring icebergs. The intersecting, typically curvilinear, parallel-sided features are segments of scour marks, in this case up to 50 m wide and several hundreds of metres long. The new scour mark (shown by arrows) displays two positive-relief berms (up to 5 m above the undisturbed seabed). The soft-sediment striated surface occurs in the negative-relief trough between the two berms. Striated surfaces in the troughs of the older features illustrated here have been either covered by post-scour sediments or eroded through the action of bioturbation (a process that will not have affected Proterozoic scour marks) and subsequent winnowing by currents. This example from the Labrador Sea, water depth 120 m.

Woodworth-Lynas, 1988). There may be several reasons for this. First, criteria for the visual recognition of ice keel scour mark morphology and ice keel turbate sedimentology in rocks of glacia-quatic origin had not been established. Secondly, such sediments may not be preserved in the geological record. However, their presence over large areas of modern continental shelves (e.g. Vorren *et al.*, 1983; Dowdeswell *et al.*, in press) and Quaternary glacial lake beds (e.g. Woodworth-Lynas and Guigné, 1990), and the likely relatively good preservation potential of deposits formed in these areas, makes this an unlikely proposition. A more likely explanation is that the sediments concerned have simply not been considered in the context of a sedimentary environment dominated by ice keel scouring and iceberg rafting of debris.

Examples

Some of the Proterozoic and Phanerozoic soft-sediment striated surfaces described below are discussed fully by Woodworth-Lynas (in press). As noted above, very few ice keel scour marks have been described from pre-Quaternary sedimentary rocks, and most of these are small-scale features (generally less than 1–2 m wide). Ice keel scour marks 30–40 cm long and 10–20 cm wide have been identified on bedding surfaces of the Kuibis Series quartzite within the Late Precambrian Nama System of Namibia (Martin, 1965; Kröner and Rankama, 1972) (Fig. 18.3). Deynoux (1980) described soft-sediment striated surfaces of Precambrian age in Mauritania and interpreted them to have been formed by the

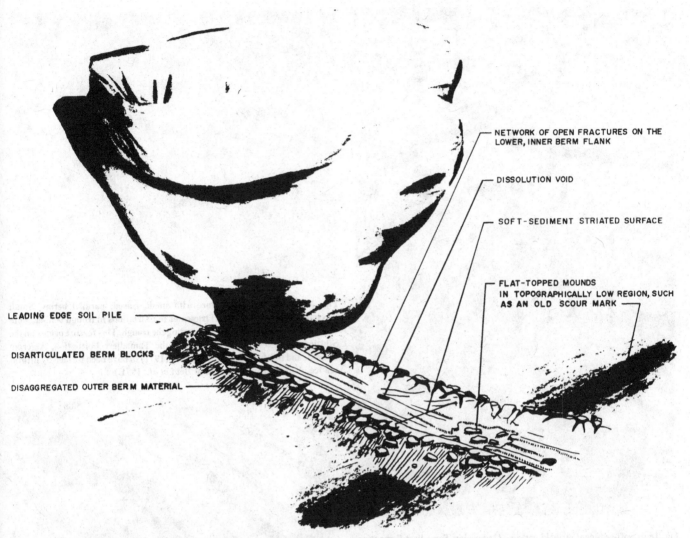

NETWORK OF OPEN FRACTURES ON THE LOWER, INNER BERM FLANK

DISSOLUTION VOID

SOFT-SEDIMENT STRIATED SURFACE

FLAT-TOPPED MOUNDS IN TOPOGRAPHICALLY LOW REGION, SUCH AS AN OLD SCOUR MARK

LEADING EDGE SOIL PILE

DISARTICULATED BERM BLOCKS

DISAGGREGATED OUTER BERM MATERIAL

Fig. 18.2. Diagram to summarize the salient features of a typical continental shelf scour mark. Typical water depth would be in the range 100–550 m, and the scour mark would be in the range 10–100 m wide (exceptionally up to 250 m wide) and on the order of 1–2 m deep (exceptionally 10–15 m deep).

Fig. 18.3. Short (< 1 m), randomly-oriented, soft-sediment furrows on a bedding surface of Kuibis Series quartzite, Nama System, Namibia. These features were interpreted by Martin (1965) to have been formed by grounding ice floes. (Taken from Martin (1965), Plate XI, figure 1.)

Fig. 18.4. Soft-sediment striated surface, Ordovician Tamadjert Formation, Algerian Sahara. Note unstriated margin on left side and gradual amplitude change and lateral migration of ridges across adjacent grooves – features characteristic of ridges and grooves formed at the trailing edge of scouring ice keels. Liftoff of the graving tools is implied where striations are developed on either side of but not within small (few cm wide) depressions along the left margin. (From Beuf *et al.*, 1971, frontispiece of Chapter 3.)

Fig. 18.5. Double-sided feature with small, raised, parallel berms. Small surcharge pile (tape measure) truncates fine striations in trough behind. Note ridge in background obliquely truncating the trough. This feature occurs on the Silurian transgression surface, above the Tamadjert Formation, Algerian Sahara, and is interpreted by Beuf *et al.* (1971) to have been formed by littoral sea ice. (Taken from Figure 62 of Beuf *et al.*, 1971.)

Fig. 18.6. Soft-sediment striated surface from the Upper Ordovician Sarah Formation, Saudi Arabia, and described by Vaslet (1990) as part of a 'glacial floor'. We suggest that the distorted striations may have been caused by breakout of a small piece of ice from the trailing edge of a scouring ice keel. (Taken from Figure 8b of Vaslet, 1990.)

action of drifting ice. Regularly-spaced 'chatter mark' features, superimposed roughly perpendicularly to the striations are similar to those formed in modern tidal estuarine environments by wave-induced, oscillating, scouring ice floes described by Dionne (1972, 1988) and to larger-scale features observed in some modern iceberg scour mark troughs (Bass and Woodworth-Lynas, 1988). From the Algerian Sahara Desert, Beuf *et al.* (1971) illustrated and described several soft-sediment striated surfaces exposed in over 1200 km of outcrop from the Upper Ordovician Tamadjert Formation. In general they interpreted these surfaces to have been generated by clastic material embedded in the sole of an active terrestrial glacier. One such surface with ridges and grooves that change shape and transgress longitudinally, interpreted to reflect dynamic change of irregularities in the bottom topography of the ice, is illustrated in

Fig. 18.4. A possible ice keel scour mark about 1 m wide is described by Beuf *et al.* (1971) from the Silurian-age transgression surface (Fig. 18.5). Soft-sediment striated surfaces of Upper Ordovician-age are found in glacial paleo-valley deposits of the Sarah Formation, Saudi Arabia (Vaslet, 1990). Here, distortions of the striations may suggest breakout of small ice blocks from the trailing edge of the scribing ice mass (Fig. 18.6). Powell and Gostin (1990) described ice keel scour marks 1–2 m wide in Permian glacimarine sediments in Australia (Fig. 18.7).

Fig. 18.7. Small-scale ice scour mark (approximately 1 m wide) on bedding plane surface of a bioturbated glaciomarine diamictite, Permian Pebbley Beach Formation, Australia. The trough is filled with sandstone thus obscuring the soft-sediment striated surface that may occur at the base of the incised trough. The contrast in grain size and sorting between the diamictite and sandstone in the trough is similar to observations from large-scale, Pleistocene scour marks exposed on King William Island, Northwest Territories, Canada described by Woodworth-Lynas et al. (1986). (Photograph kindly provided by R. Powell, Northern Illinois University.)

Fig. 18.8. Downward-displaced varved bedding beneath a groove within a large-scale (several metres wide) scour mark trough in the Permian Rio do Sul Formation, Paraná Basin, Brazil. Note upwarping on either side. (Photograph kindly provided by A.C. Rocha-Campos, Universidade de Sao Paulo (see also Rocha-Campos et al. this volume).)

There is only one locality where large-scale, pre-Quaternary ice keel scour marks have been identified, and these are Permian-age features in Brazil (Rocha-Campos et al., 1990; Rocha-Campos et al., this volume). Here, multiple subparallel small furrows (up to 50 cm wide and 20 cm deep) on a bedding plane surface are interpreted to have been formed by a single scouring event (Rocha-Campos et al., this volume). This represents the only documented example of a feature that is comparable in size to modern iceberg and sea ice pressure ridge scour marks that are prolific over large areas of high latitude continental shelves (Fig. 18.8).

Ridges, grooves and striations on a bedding plane exposure of Carboniferous-age Dwyka Formation pebbly quartz-rich sandy diamictite are shown in Fig. 18.9. Savage (1972), who described

these features, noted that slumping of ridges across adjacent grooves indicated that soft-sediment slumping must have occurred immediately after the grooves were formed, and acknowledged difficulty in explaining how the marks would be preserved if they had formed beneath an ice sheet. Visser (1990) examined the same Dwyka Formation surface from a location near that described by Savage (1972), and suggested that associated erosional troughs and channels could have been formed by scouring icebergs, although he preferred an interpretation of their origin due to subglacial melt-water flow. From another location in Dwyka Formation sandstone, Visser (1985) described current ripples oriented normal to striations on a 'very small' soft-sediment striated surface, an association that could be interpreted in terms of ice keel scouring by a current-driven mass of free-floating ice.

Soft-sediment striated surfaces are also common in the lower part of the Buckeye Formation in the Transantarctic Mountains, where they are associated with dropstone-bearing diamictites (Aitchison et al., 1988). In parts of the Buckeye Formation, hummocky stratification is interpreted to indicate periods of relatively shallow open-water conditions during which sediments were affected by storm-wave action. An example of a soft-sediment striated surface from the Buckeye Formation that shows an abrupt margin beyond which no striations are evident is given in Fig. 18.10. Although Aitchison et al. (1988) state that the striated surfaces are of subglacial origin, they acknowledge the possibility that such surfaces may have been formed from action by 'the grounding of floating ice such as bottom scraping berg ice'.

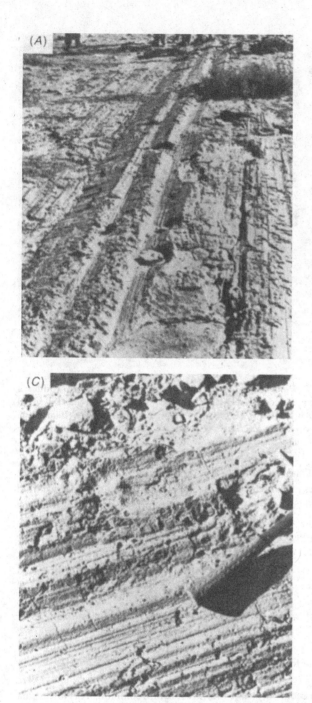

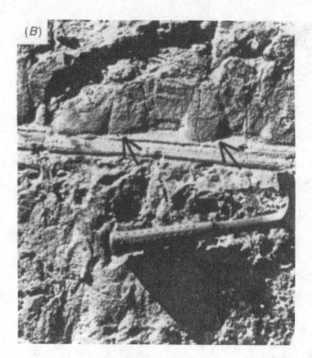

Fig. 18.9. Soft-sediment striated surface on a sandstone bedding plane, Late Carboniferous Dwyka Formation, South Africa. (*a*) General view of the striated surface, (*b*) striations partially masked by slumping of adjacent ridge (arrows), (*c*) striations above hammer handle partially masked by adjacent material (Reproduced from Savage, 1972, figures 2, 3 and 4.)

Rattigan (1967) described soft-sediment striated surfaces in siltstone laminites of the Carboniferous-age Kuttung Facies, New South Wales, Australia. In some exposures this author notes that both current-ripple crests and the axial surfaces of localized, intraformational, overturned slump folds are approximately normal to associated striae and concludes that this association may be an indication of formation by grounding ice floes. Elsewhere the occurrence of dropstones lodged in the troughs of broad, striated depressions is interpreted by Rattigan (1967) to be the result of grounded glacial ice (in this context, presumably ice that is structur-

ally part of a coherent, floating ice sheet). The striae within these trough-like surfaces occur in two or more sets of varying orientation and are overlain by dropstone-bearing layers (Rattigan, 1967). However, the morphological association of lodged stones on variably oriented striated surfaces within broad troughs, stratigraphically overlain by dropstone-bearing layers may be viewed as consistent with a re-interpretation of the features as ice keel scour marks that formed in open-water conditions with wind- and current-driven free-floating scouring ice masses moving in different directions.

Fig. 18.10. Soft-sediment striated surface from the Permian Buckeye Formation, Transantarctic Mountains, Antarctica. Note the margin (to the right of the standing figure) separating striated from non-striated surface: possibly the margin of an ice scour mark. (Taken from figure 5 of Aitchison et al., 1988.)

O'Brien and Christie-Blick (1992) described in some detail soft-sediment striated surfaces from the Lower Paleozoic Carolyn Formation in the Grant Range of Western Australia. They ruled out formation of the surfaces by free-floating scouring ice masses because of the spatial and stratigraphic consistency of ridge and groove orientation, preferring instead an interpretation of formation by episodic minor advances of a glacial grounding line.

Iceberg scouring and massive diamicton formation in the modern glacimarine environment of East Greenland

Ice keel turbates, formed through the mechanical disturbance of unconsolidated seabed sediments by icebergs, have been recognized in Quaternary glacimarine sediments from acoustic profiles, sidescan sonographs and sediment cores (e.g. Vorren et al., 1983; Josenhans et al., 1986; Barnes et al., 1987). However, of particular importance in terms of the interpretation of soft-sediment striated surfaces is the association between iceberg scouring, iceberg rafting of sediments and the formation of massive diamicton (Dowdeswell et al., in press). Therefore, in the geological record the juxtaposition of soft-sediment striated surfaces, interpreted to be the result of iceberg scouring, and massive diamicton facies can be explained as a product of closely related processes.

Icebergs and glacimarine sediments in East Greenland

Surficial glacimarine sediments in the fjord system of Scoresby Sund (area 13 700 km²) and on the adjacent East Greenland continental shelf are predominantly massive diamictons, comprised of poorly sorted sandy, pebbly muds (Dowdeswell et al., in press). Acoustic reflection-profile records show that more than 30 000 km² of the seafloor of the fjord and adjacent shelf, in water depths of less than 500–600 m, is characterized by irregular topography, and surficial sediments show little internal acoustic structure (Dowdeswell et al., 1993). Irregular seafloor topography is formed by numerous iceberg scour marks, which may have trough-to-berm crest relief of up to 10 m, and berm-crest to berm-crest apparent widths up to 20 m. Iceberg scour marks are most prolific in water depths of between 300–400 m, becoming less profuse with increasing depth until beyond about 550 m where they are absent (Fig. 18.11).

Large numbers of tabular icebergs, calved from the marine termini of Daugaard-Jensen Gletscher and several other fast-flowing outlet glaciers of the Greenland Ice Sheet have been observed drifting in the Scoresby Sund fjord system and on the adjacent continental shelf (Dowdeswell et al., 1992). Measurements of iceberg freeboards, using sextant and radar from a ship, allow the calculation of iceberg keel depths, and show that many icebergs within Scoresby Sund have drafts up to 550 m (Dowdsewell et al., 1992). Observations of stationary icebergs from sequential Landsat satellite imagery also demonstrate that significant numbers of icebergs are aground, implying that iceberg scouring has taken place. Scouring of the sea floor by modern icebergs is occurring, therefore, but the iceberg scour mark population probably represents a cumulative record of superimposed forms dating from the Late Weichselian (Dowdeswell et al., 1993).

Origin of massive diamicton and ice keel turbate

Acoustic profiles from Scoresby Sund fjord show that between 5 and 15 m of surficial seabed sediments overlie Jurassic and Cretaceous bedrock. Coring at 16 sites in the fjord and on the shelf beyond the fjord mouth (Fig. 18.12) retrieved a total of 39 m of core (Marienfeld, 1991; Dowdeswell et al., in press). Four distinct lithofacies were identified in the following proportions, averaged for all cores: (1) 89% massive diamicton, (2) 4% gravel or coarse sand in lenses or thin beds, (3) 4% laminated fine sand to clay, (4) 3% crudely layered, poorly-sorted sediments with large clasts. A radiograph of the massive diamicton facies is shown in Fig. 18.13. The longest core retrieved a column of sediment 5.4 m in length, but did not penetrate the base of the diamicton. On the adjacent continental shelf the acoustically internally structureless surficial sediment facies thicken to several tens of metres. AMS radiocarbon dates on benthic foraminifera from fjord cores show that the diamicton is of Holocene age (< 10 000 BP), and accumulated at rates of between 0.1 and 0.3 m/1000 years. Glacier ice had retreated from the outer part of the Scoresby Sund fjord system before the diamicton was deposited. The diamicton is interpreted to be glacimarine in origin, the bulk of which was deposited probably by iceberg-rafting and subsequent melt-out of included debris (Dowdeswell and Murray, 1990; Dowdeswell et al., in press). Also there is probably a contribution to Holocene sedimentation by suspension settling of fine material carried by meltwater from tidewater glaciers and subaerial rivers at the fjord margins.

The two dominant processes of iceberg-rafting and iceberg scouring during the Holocene have produced the massive diamicton facies, or an iceberg turbate (Vorren et al., 1983), that characterizes about 10 000 km² of the fjord system and the adjacent East

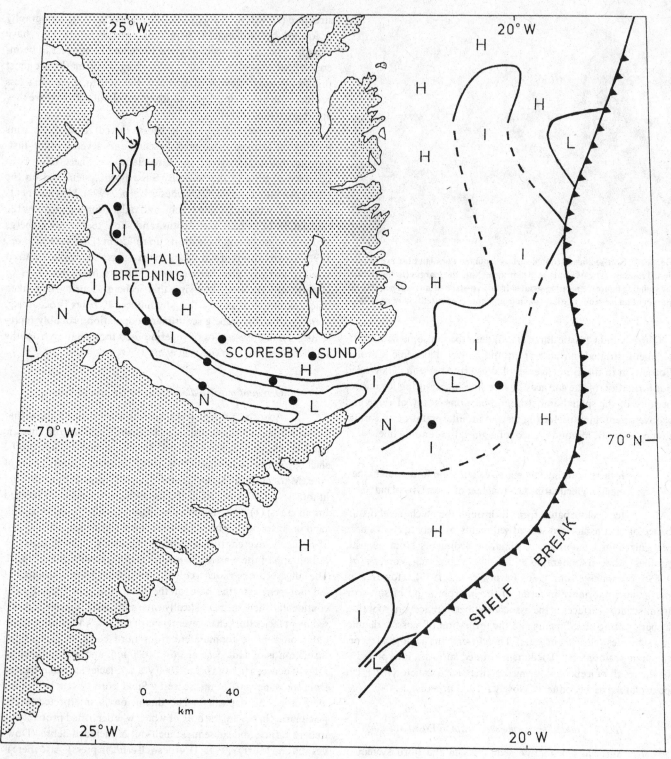

Fig. 18.11. Map of Scoresby Sund and the East Greenland continental shelf showing areas and intensity of iceberg scouring. H is high intensity, I is intermediate and L is low. N represents areas where scour marks are absent. The shelf break is at about 500 m. The map is based on acoustic data. (From Dowdeswell *et al.*, in press.)

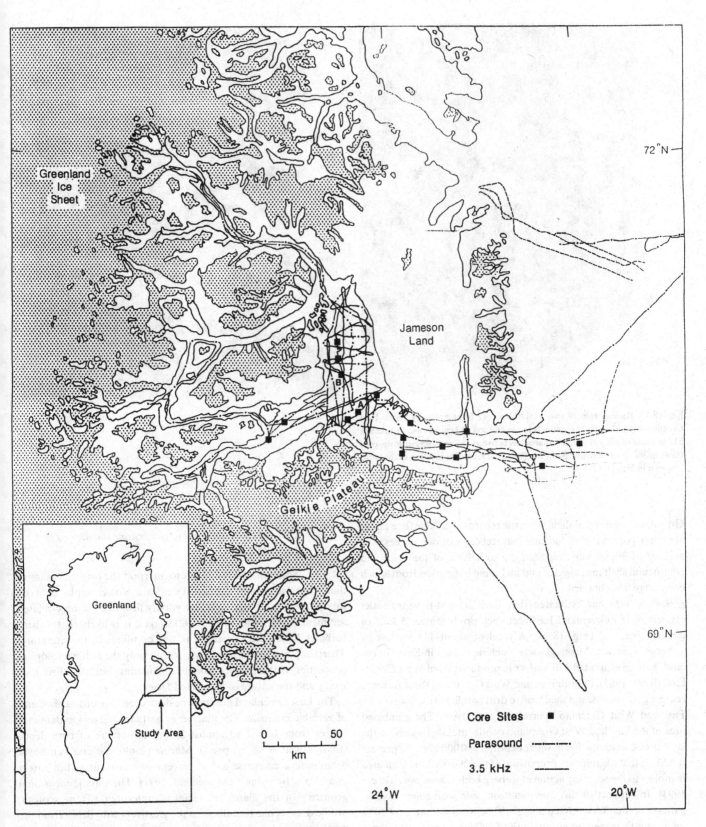

Fig. 18.12. Map of Scoresby Sund and the East Greenland continental shelf showing track lines of acoustic data collection and core locations. Areas covered by glacier ice are shaded. Parts of the cores labelled A and B are shown in Fig. 18.13.

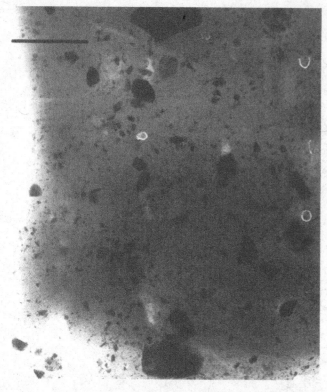
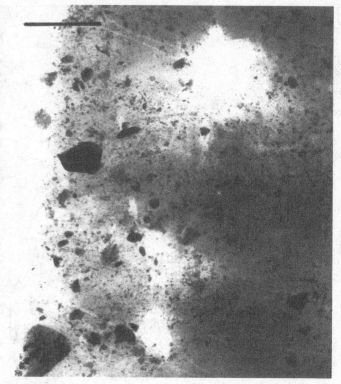

Fig. 18.13. Radiographs of two sections of core from Scoresby Sund, East Greenland. (*a*) Massive diamicton between 200 and 210 cm down-core taken in 512 m water depth. (*b*) Massive diamicton between 155 and 165 cm down-core taken in 386 m water depth. Scale bars are 2 cm long. The location of the cores is shown in Fig. 18.12.

Greenland continental shelf. Acoustic records, available for a larger area than core material, indicate that iceberg scouring occurs over an area of 30 000 km², representing over 90% of the fjord and continental shelf investigated and including large areas from which cores were not obtained.

Reeh (1985) has estimated that over 20 fast-flowing outlet glaciers of the Greenland Ice Sheet each produce over 5 km³ of icebergs per year (Fig. 18.14). A total of about 115 km³/yr of icebergs is produced from glaciers reaching the sea in East Greenland, and an estimated 170 km³/yr is produced from West Greenland (Reeh, 1985). In both East and West Greenland those icebergs reaching the continental shelf tend to drift parallel to the coast in the East and West Greenland currents, respectively. The combined area of the East and West Greenland continental shelves above the shelf break, at about 500 m water depth, is 500 000 km². At present it is likely that a significant proportion of this 500 000 km² shelf area is subject to the scouring action of iceberg keels (Dowdeswell *et al.*, 1993). In support of this interpretation, side scan sonar investigations of the West Greenland shelf between 64° and 69°30′N indicate little iceberg scouring south of 69°N, but intensive iceberg scouring north of this latitude associated with the northerly drift of icebergs from major iceberg-producing outlet glaciers (Brett and Zarudzki, 1979).

Criteria for distinguishing glacimarine diamictons and related facies

Sometimes it is difficult to interpret the origin of massive diamicton facies, and confusion can exist, for example, between interpreting basal tills, sediment gravity flows and iceberg-related sediments. Eyles and Eyles (1992) suggest it is likely that iceberg turbates have been described as tills or tillites in the literature. Therefore, it is useful to summarize not only the sedimentological properties of the East Greenland diamictons, but also their geometry and the nature of associated facies.

The East Greenland glacimarine diamictons are unstratified and of variable grainsize. The number of particles > 2 mm in diameter ranges from 1 or 2 to almost 20 per centimetre of core depth (Dowdeswell *et al.*, in press). Marine planktonic and benthonic foraminifera comprise a few percent of most cores, but rarely exceed 5% by volume (Marienfeld, 1991). The three-dimensional geometry of the diamicton facies, as interpreted from acoustic profile data, is one that thickens progressively with depth from the northern margins of the fjord to its deeper, southern side (from 5 to 15 m), and thickens out onto the continental shelf (to several tens of metres). The lack of core penetration beyond 5.4 m means that there is little direct evidence of the physical nature of underlying related

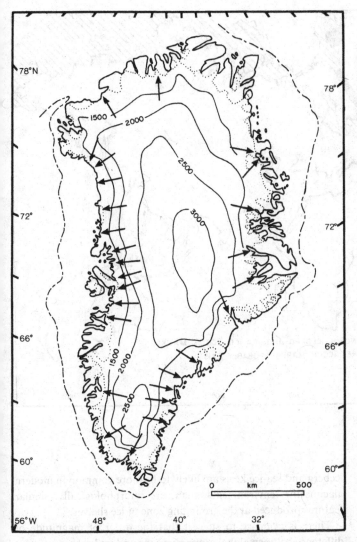

Fig. 18.14. Map of the Greenland Ice Sheet (elevations in metres) and continental shelf (after Dowdeswell et al., 1993). Fast-flowing outlet glaciers with an estimated ice discharge exceeding 5 km³/yr are shown by arrows (after Reeh, 1985). The 500 m isobath (dashed line) indicates the location of the shelf break.

continental shelves both eustatic and isostatic effects will lead to a series of sea-level fluctuations during glacial–interglacial cycles, and any site will be first more and then less proximal to glacier-influenced sources of sediment as glacial ice masses grow and decay (Boulton, 1990). The Late Cenozoic Yakataga Formation in Alaska is well exposed in a sequence that is up to 5 km in stratigraphic thickness, and has been discussed in the context of proximal and distal glacier-influenced sources by Eyles and Lagoe (1990) and Eyles and Eyles (1992). These authors interpret fossiliferous diamicton units, each up to 100 m thick, to have been formed by iceberg-rafting of coarse material and suspension settling of meltwater-derived fines. The diamictons contain a series of boulder pavements (e.g. Eyles, 1988, in press), shell-rich beds or coquinas, and laterally extensive marine muds. Ice-rafted clasts are found throughout the formation. Boulder lags and coquinas are interpreted to record shallow water, and the mud facies are linked to deep-water deposition when ice volume was reduced (Eyles and Lagoe, 1990). The Yakataga Formation thus provides key information on facies that may be associated with glacimarine diamictons, and suggests that much facies variability can be explained by the interplay of glacier advance–retreat cycles and sea-level changes during glacial–interglacial cycles.

Discussion

Scouring of sediments by floating ice keels is a phenomenon occurring today over vast areas of high latitude and subpolar continental shelves and lake beds (Figs. 18.15 and 18.16). Ice keel scouring results in the reworking of large volumes of sediment. For example, a conservative estimate of the volume of sediment reworked by scouring icebergs on the Labrador continental shelf each year is 0.03 km³ (Lewis et al., 1989), or as much as 300 km³ in the last 10 000 years. Although much of this sediment is repeatedly reworked by numerous iceberg scour events, the latter figure is likely to be a conservative estimate because it is based on modern iceberg flux rates which are probably significantly smaller than during the deglaciation phase of glacial–interglacial cycles.

Prolonged exposure to scouring by many keels can modify sediments to the point of almost total homogenization of any pre-existing sedimentary structures (e.g. Vorren et al., 1983, Norwegian shelf; Josenhans et al., 1986, Labrador shelf). In Lake Agassiz, for example, the most recent period of iceberg scouring probably lasted for as little as 1200 years (9900–8700 BP) and resulted in severe reworking of primary layering (Woodworth-Lynas and Guigné, 1990). Ice keel scour marks also may be present in marine sediments far removed from glaciers and ice sheets (Eyles et al., 1985), as a product of iceberg drift far from the parent ice mass or through the presence of seasonal pack ice unassociated with continental glaciation. There are a number of modern analogues where ice keel scoured areas are considerable distances from glaciers and ice sheets. These include the eastern Canadian continental shelf south of Hudson Strait (Lewis and Blasco, 1990) that is scoured presently by icebergs that have drifted from Greenland outlet glaciers; the Beaufort Sea (Lewis and Blasco, 1990) which is affected by first and

facies. However, thin beds of gravel and coarse sand (facies 2) found in several cores exhibit non-gradational upper and lower contacts with the surrounding diamicton. These coarse layers are interpreted as dump deposits, formed when material was released into the water column from the surfaces of overturning icebergs. The distribution of thin Quaternary deposits in the fjord suggests that ice advance under full glacial conditions removed much of the pre-existing sediments and re-deposited them on the outer shelf and on a major fan system beyond the continental shelf-slope break. Observed seaward thickening of the acoustically defined massive diamicton may support this interpretation.

It is difficult to provide detailed information on the lateral and vertical changes in physical properties associated with diamicton facies using short cores from modern marine settings. However, on

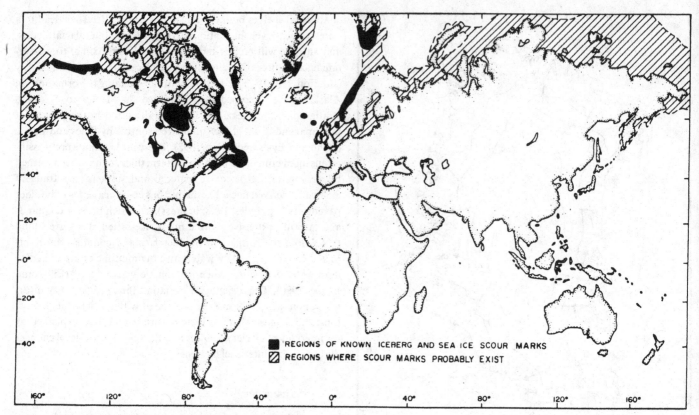

Fig. 18.15. Map showing distribution of scour marks for northern and mid-latitudes (based on numerous published works and on interpolation).

multi-year sea-ice pressure ridge keels and, rarely, by tabular icebergs that originate far to the northeast in the Canadian Arctic islands; the Okhutsk Sea, east of Sakhalin Island (Astafiev *et al.*, 1991) which is affected by first year pressure ridge keels, and Lake Erie (Grass, 1984; 1985) the floor of which is scoured by winter lake-ice pressure ridges.

Estimates of the areal extent of tidewater glacier grounding zones in the modern Polar regions indicate that they probably affect significantly smaller areas of the seafloor compared with the widespread action of scouring icebergs and seasonal pack-ice. In Antarctica today, floating ice shelves comprise almost 15 000 km of the coastline (Drewry *et al.*, 1982). Around the modern Greenland Ice Sheet, floating ice tongues associated with fast-flowing outlet glaciers form a few hundred kilometres of the coast at most. Making the assumption that ice shelf width does not vary by more than 10–20% between grounding zone and terminus, this gives a total global length of modern ice shelf grounding zones of less than 16 000 km. By assuming the grounding zone is two kilometres wide, we obtain a value of 32 000 km² affected by grounding zone processes. Even allowing for a migration of all grounding zones by 100 km during the Holocene yields an affected seafloor area of only 1 600 000 km². This figure is likely to be an overestimate, but it is still an order of magnitude smaller than the area of modern continental shelves affected by iceberg and sea-ice scouring (Figs. 18.15 and 18.16). Therefore, soft-sediment striated surfaces produced by scouring

iceberg and sea-ice keels are likely to be more common in modern glacimarine environments than are morphologically similar features produced at the grounding zone of ice shelves.

There is evidence to suggest that the order of magnitude of difference between global grounding zone and ice keel scour extent can be applied to the pre-Quaternary geological record, subject, of course, to possible differences in the preservation potential of soft-sediment striated surfaces formed by floating ice keels and grounding zone processes respectively. First, there is increasing evidence that where the margins of Quaternary ice sheets were marine-based (that is, grounded below sea-level), deglaciation was very rapid and associated with very large discharges of icebergs (e.g. Hughes, 1988; Eyles and McCabe, 1989). Second, major pulses of iceberg-rafting associated with instability in the Wisconsinan Laurentide Ice Sheet are recorded as layers of relatively coarse, carbonate-rich debris deposited across the North Atlantic up to 3000 km from source areas in eastern Canada (Heinrich, 1988; Andrews and Tedesco, 1992; Bond *et al.*, 1992). These Heinrich layers are observed at six stratigraphic intervals between 70 000 and 14 000 BP (Bond *et al.*, 1992). During each of these major pulses large numbers of icebergs would have been available to scour the continental shelves.

Soft-sediment striated surfaces associated with diamictites are known from the Earth's pre-Quaternary rock record. The diamictites may be interpreted as lodgement tills deposited beneath a grounded glacier or ice sheet (e.g. Boulton and Deynoux, 1981;

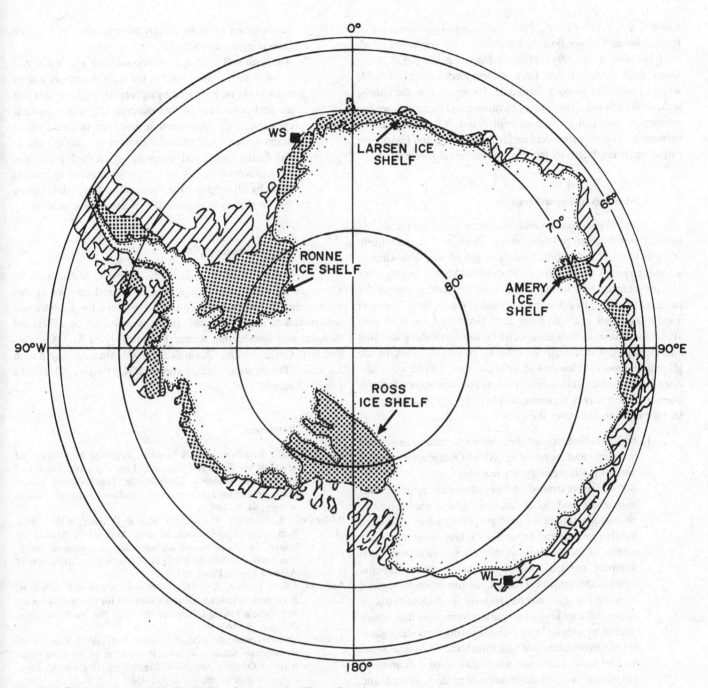

Fig. 18.16. Map showing probable distribution of scour marks (diagonal stripes) around Antarctica (generally above the 500 m contour). WS and WL are the Weddell Sea and Wilkes Land scour mark study areas respectively, reported in Barnes and Lien (1988). Distribution is interpolated based on maximum water depths from which scour marks have been found around Antarctica (reported by Barnes and Lien, 1988).

Lawson, 1981; Eyles *et al.*, 1983), or as rain-out deposits that formed beneath either floating ice shelves (e.g. Anderson *et al.*, 1991; Hambrey *et al.*, 1991) or from icebergs (e.g. Eyles *et al.*, 1985; Dowdeswell *et al.*, in press). Each of these mechanisms is a viable way of producing massive diamictite. However, it is the role of icebergs in both sediment delivery by rafting, and by ice keel/seabed interaction, that can provide an explanation for the association between ice keel scour-marked surfaces and diamicton facies over extensive areas affected by modern glacimarine sedimentation.

Summary and conclusions

In this paper we have shown that ice keel scour marks and ice keel turbates are probably more common than previously thought in the Earth's pre-Pleistocene glacial and cold-climate sedimentary rock record. We have presented specific examples of soft-sediment striated surfaces that could be reinterpreted in the context of scour by free-floating ice-masses (Woodworth-Lynas, in press). Using an example from East Greenland we have also attempted to demonstrate an association between iceberg scouring and rafting and the origin of massive diamicton facies in the glacimarine record (Dowdeswell *et al.*, in press). Field work directed at reappraising soft-sediment striated surfaces associated with diamictites now may be appropriate. The following points summarize the main conclusions of the study.

1. Soft-sediment striated surfaces are formed in unlithified sediments and are preserved in both Quaternary and more ancient glacial sedimentary sequences.
2. Soft-sediment striated surfaces often have been ascribed to origins by subglacial fluting, subglacial abrasion or by processes acting at ice shelf grounding zones. Although modern marine and lacustrine ice keel scour marks are similar in morphology to features reported from soft-sediment striated surfaces associated with diamictite facies in the geological record, the formation by ice keel scouring has seldom been proposed for such surfaces.
3. We describe ice keel scour marks and massive diamictons formed by iceberg-rafting and scouring in modern glacimarine environments of East Greenland. We suggest that ice keel scour marks and associated massive diamictons are present over significant areas of modern, iceberg- and sea-ice-influenced continental shelves, and also in modern, Quaternary, and pre-Quaternary lakes influenced by icebergs and/or winter lake-ice. Around Greenland alone, a significant proportion of the 500 000 km² continental shelf may be subject to iceberg scouring today.
4. Globally, scouring by ice keels is a process that probably affects an order of magnitude more of modern glacimarine and high-latitude oceanic seafloor than ice shelf grounding zone processes. We suggest that this order of magnitude of difference should apply also to pre-Quaternary glacimarine and glacilacustrine environments, and that consequently the effects of ice keel scour should

be preserved more frequently than features of ice shelf grounding zone processes.
5. Although soft-sediment striated surfaces and associated massive diamictites found in the pre-Quaternary glacial record may be produced by iceberg-rafting of debris and ice keel reworking of this material this process is not exclusive. Our interpretation does not preclude (i) an origin for some soft-sediment striated surfaces by subglacial fluting, subglacial abrasion or ice shelf grounding zone processes or (ii) an origin for massive diamictite facies by till lodgement beneath active terrestrial glaciers and ice shelves or as a result of mass wasting processes.

Acknowledgements

We would like to thank N. Eyles, J. Miller and M. Deynoux for their assistance with, and critical reviews of this manuscript. Support for Woodworth-Lynas for the analysis and interpretation included in this paper is provided by a Natural Sciences and Engineering Research Council (NSERC) (Canada) Strategic Grant entitled 'Quantification of seabed damage due to ice scour'. This paper is a contribution to IGCP Project 260 (Earth's Glacial Record).

References

Aitcheson, J.C., Bradshaw, M.A. & Newman, J. (1988). Lithofacies and origin of the Buckeye Formation: Late Paleozoic glacial and glaciomarine sediments, Ohio Range, Transantarctic Mountains, Antarctica. *Palaeogeography, Palaeoclimatology, Palaeoecology*, **64**, 93–104.

Anderson, J.B., Kennedy, D.S., Smith, M.J. & Domack, E.W. (1991). Sedimentary facies associated with Antarctica's floating ice masses. In *Glacial marine sedimentation: paleoclimatic significance*, eds., J.B. Anderson & G.M. Ashley, Geological Society of America Special Paper, 261, 1–25.

Andrews, J.T. & Tedesco, K. (1992). Detrital carbonate-rich sediments, northwestern Labrador Sea: implications for ice-sheet dynamics and iceberg rafting (Heinrich) events in the North Atlantic. *Geology*, 20, 1087–90.

Astafiev, V.N., Polomoshnov, A.M. & Truskov, P.A. (1991). *Stamukhi on the northern Sakhalin offshore*. Proceedings of the First International Offshore and Polar Engineering Conference, Edinburgh, August 11–16, vol. II, 462–466.

Aylsworth, J.M. & Shilts, W.W. (1987). Surficial geology, Coats Island, District of Keewatin, Northwest Territories. Scale 1:250 000. Geological Survey of Canada Map 1633A.

Baranowski, S. (1970). The origin of fluted moraine at the fronts of contemporary glaciers. *Geografiska Annaler*, **52A**, 68–75.

Barnes, P.W. & Lien, R. (1988). Icebergs rework shelf sediments to 500 m off Antarctica. *Geology*, **16**, 1130–3.

Barnes, P.W., Asbury, J.L., Rearic, D.M. & Ross, C.R. (1987). Ice erosion of a sea-floor knickpoint at the inner edge of the stamukhi zone, Beaufort Sea, Alaska. *Marine Geology*, **76**, 207–22.

Bass, D.W. & Woodworth-Lynas, C.M.T. (1988). Iceberg crater marks on the sea floor, Labrador Shelf. *Marine Geology*, 79, 243–60.

Berkson, J.M. & Clay, C.S. (1973). Microphysiography and possible iceberg grooves on the floor of western Lake Superior. *Geological Society of America Bulletin*, **84**, 1315–28.

Beuf, S., Biju-Duval, B., de Charpal, O., Rognon, P., Gariel, O. & Bennacef,

A. (1971). *Les grés du paleozoique inferieur au Sahara: sédimentation et discontinuités évolution structurale d'un craton*. Paris: Editions Technip, 464 pp.

Bond, G., Heinrich, H., Broecker, W., Labeyrie, L., McManus, J., Andrews, J. *et al.* (1992). Evidence for massive discharges of icebergs into the North Atlantic ocean during the last glacial period. *Nature*, **360**, 245–9.

Boulton, G.S. (1976). The origin of glacially fluted surfaces: observations and theory. *Journal of Glaciology*, **17**, 287–309.

Boulton, G.S. (1982). Subglacial processes and the development of glacial bedforms. In *Glacial, glacio-fluvial and glacio-lacustrine systems*, eds., R. Davidson-Arnott, R. Nickling & D.D. Fahey, Norwich: Geo Books (in association with Geomorphology Symposium, University of Guelph) Geographical Publication No. 6, 1–31.

Boulton, G.S. (1990). Sedimentary and sea level changes during glacial cycles and their control on glacimarine facies architecture. In *Glacimarine environments: processes and sediments*, eds., J.A. Dowdeswell & J.D. Scourse, Geological Society of London Special Publication No. 53, 121–37.

Boulton, G.S. & Deynoux, M. (1981). Sedimentation in glacial environments and the identification of tills and tillites in ancient sedimentary sequences. *Precambrian Research*, 15, 397–420.

Brett, C.P. & Zarudzki, E.F.K. (1979). Project Westmar: a shallow marine geophysical survey on the West Greenland continental shelf. Grønlands Geologiske Undersøgelse, Rapport Nr. 87: 27 pp.

Clayton, L., Laird, W.M., Klassen, R.W., & Kupsch, W.D. (1975). Intersecting minor lineations on Lake Agassiz plain. *Journal of Geology*, **73**: 652–6.

Deynoux, M. (1980). Les formations glaciaires du Précambrien Terminal et de la fin de l'Ordovicien en Afrique de l'Ouest: deux examples de glaciation d'inlandis sur un plateforme stable. *Trauaux des Laboratoires des Sciences de la Terre, St.-Jérome, Marseille*, (B), **17**, 554 pp.

Dionne, J.C. (1972). Ribbed grooves and tracks in mud tidal flats of cold regions. *Journal of Sedimentary Petrology*, **41**, 848–51.

Dionne, J.C. (1977). Relict iceberg furrows on the floor of glacial Lake Ojibway, Quebec and Ontario. *Maritime Sediments*, **13**, 79–81.

Dionne, J.C. (1988). Characteristic features of modern tidal flats in cold regions. In Tide-influenced sedimentary environments and facies, eds., P.L. de Boer, *et al.*, Dordrecht: Reidel pp. 301–32.

Dowdeswell, J.A., Whittington, R.J. & Marienfeld, P. (in press) The origin of massive diamicton facies by iceberg rafting and scouring, Scoresby Sund, East Greenland. *Sedimentology*.

Dowdeswell, J.A., Villinger, H., Whittington, R.J. & Marienfeld, P. (1993). Iceberg scouring in Scoresby Sund and on the East Greenland continental shelf. *Marine Geology*, **111**, 37–53.

Dowdeswell, J.A., Whittington, R.J. & Hodgkins, R. (1992). The sizes, frequencies, and freeboards of East Greenland icebergs observed using ship radar and sextant. *Journal of Geophysical Research*, **97** (C3), 3515–28.

Dowdeswell, J.A. & Murray, T. (1990). Modelling rates of sedimentation from icebergs. In *Glacimarine environments: processes and sediments*, eds., J.A. Dowdeswell & J.D. Scourse, Geological Society of America Special Publication, 53, 121–37.

Dredge, L.A. (1982). Relict ice scour marks and late phases of Lake Agassiz in northernmost Manitoba. *Canadian Journal of Earth Sciences*, 19, 1079–87.

Drewry, D.J., Jordan, S.R. & Jankowski, E. (1982). Measured properties of the Antarctic Ice Sheet: surface configuration, ice thickness, volume and bedrock characteristics. *Annals of Glaciology*, **3**, 83–91.

Dyson, J.L. (1952). Ice-ridged moraines and their relation to glaciers. *American Journal of Science*, **250**, 204–11.

Echelmeyer, K., Clarke, T.S. & Harrison, W.D. (1991). Surficial glaciology of Jackobshavns Isbrae, West Greenland: Part I. Surface morphology. *Journal of Glaciology*, **37**, 368–82.

Eyles, C.H. (1988). A model for striated boulder pavement formation on glaciated, shallow-marine shelves: an example from the Yakataga Formation, Alaska. *Journal of Sedimentary Petrology*, **58**, 62–71.

Eyles, C.H. (In press). Intertidal boulder pavements in the northeastern Gulf of Alaska and their geological significance. *Sedimentary Geology*.

Eyles, C.H., Eyles, N. & Miall, A.D. (1985). Models of glaciomarine sedimentation and their application to the interpretation of ancient glacial sequences. *Palaeogeography, Palaeoclimatology, Palaeoecology*, **51**, 15–84.

Eyles, C.H. & Lagoe, M.B. (1990). Sedimentation and facies geometries on a temperate glacially influenced continental shelf: the Yakataga Formation, Middleton Island, Alaska. In *Glacimarine environments: processes and sediments*, eds., J.A. Dowdeswell & J.D. Scourse, Geological Society of London Special Publication No. 53, 363–86.

Eyles, N. & Clark, B.M. (1988). Storm-influenced deltas and ice scouring in a late Pleistocene glacial lake. *Geological Society of America Bulletin*, **100**, 793–809.

Eyles, N. & Eyles, C.H. (1992). Glacial depositional systems. In *Facies models: response to sea level change*, eds., R.G. Walker & N.P. James. Geological Association of Canada, 73–100.

Eyles, N., Eyles, C.H. & Miall, A.D. (1983). Lithofacies types and vertical profile models; an alternative approach to the description and environmental interpretation of glacial diamict and diamictic sequences. *Sedimentology*, **30**, 393–410.

Eyles, N. & McCabe, A.M. (1989). The Late Devensian (< 22000 BP) Irish Sea Basin: the sedimentary record of a collapsed ice sheet margin. *Quaternary Science Reviews*, **8**, 307–51.

Geonautics Limited. (1985). Design of an iceberg scour repetitive mapping network for the Canadian east coast. Environmental Studies Revolving Funds Report No. 043, Ottawa, 50 pp.

Gilbert, R., Handford, K.J. & Shaw, J. (1992). Ice scours in the sediments of glacial Lake Iroquois, Prince Edward County, eastern Ontario. *Géographie physique et Quaternaire*, **46**, 189–94.

Goodwin, C.R., Finley, J.C. & Howard, L.M. (1985). Ice scour bibliography. Environmental Studies Revolving Funds Report No. 010, Ottawa: 99 pp.

Grass, J.D. (1984). Ice scour and ice ridging studies in Lake Erie. IAHR Ice Symposium, Hamburg, 33–43.

Grass, J. (1985). Lake Erie cable crossing: ice scour study. In *Workshop on ice scouring, 15–19 February, 1982*. National Research Council of Canada, Associate Committee on Geotechnical Research, technical memorandum No. 136: 1–16.

Hambrey, M.J., Ehrmann, W.U. & Larsen, B. (1991). Cenozoic glacial record of the Prydz Bay continental shelf, East Antarctica. In *Proceedings of the Ocean Drilling Program, Scientific Results*, eds. J. Barron & B. Larsen, 119, 77–131.

Heinrich, H. (1988). Origin and consequences of cyclic ice rafting in the Northeast Atlantic Ocean during the past 130000 years. *Quaternary Research*, **29**, 143–52.

Hélie, R.G. (1983). Relict iceberg scours, King William Island, Northwest Territories. Geological Survey of Canada, Paper 83-1B, 415–17.

Hodgson, D.A. (1982). Surficial materials and geomorphological processes, western Sverdrup and adjacent islands, District of Franklin. Geological Survey of Canada, Paper 81–9, 37 pp.

Hodgson, G.J., Lever, J.H., Woodworth-Lynas, C.M.T. & Lewis, C.F.M. (1988). The dynamics of iceberg grounding and scouring (DIGS) experiment and repetitive mapping of the eastern Canadian continental shelf. Environmental Studies Research Funds Report No. 094, Ottawa, 316 pp.

Hoppe, G. & Schytt, V. (1953). Some observations on fluted moraine surfaces. *Geografiska Annaler*, **XXXV**, 105–15.

Horberg, L. (1951). Intersecting minor ridges and periglacial features in the Lake Agassiz basin, North Dakota. *Journal of Geology*, **59**, 1–18.

Hotzel, I.S. & Miller, J.D. (1983). Icebergs: their physical dimensions and the presentation and application of measured data. *Annals of Glaciology*, **4**, 116–23.

Hughes, T. (1988). Ice dynamics and deglaciation models when ice sheets collapsed. In *North America and adjacent oceans during the last deglaciation*, eds., W.F. Ruddiman & H.E. Wright Jr., Geology of North America, K3, Boulder: Geological Society of America, pp. 183–220.

Josenhans, H.W., Zevenhuizen, J. & Klassen, R.A. (1986). The Quaternary geology of the Labrador Shelf. *Canadian Journal of Earth Sciences*, **23**, 1190–213.

Josenhans, H.W. & J. Zevenhuizen. (1990). Dynamics of the Laurentide ice sheet in Hudson Bay, Canada. *Marine Geology*, **92**, 1–26.

Kröner, A. & Rankama, K. (1972). Late Precambrian glaciogenic sedimentary rocks in southern Africa: a compilation with definitions and correlations. Precambrian Research Unit, University of Cape Town, Bulletin II.

Lawson, D.E. (1981). Distinguishing characteristics of diamictons at the margin of the Matanuska Glacier, Alaska. *Annals of Glaciology*, **2**, 78–84.

Lewis, C.F.M. & Blasco, S.M. (1990). Character and distribution of sea-ice and iceberg scours. In Ice scouring and the design of offshore pipelines, proceedings of an invited workshop, April 18–19, eds., J.I. Clark, I. Konuk, F. Poorooshasb, J. Whittick & C. Woodworth-Lynas, Calgary, Alberta. Canada Oil and Gas Lands Administration and Centre for Cold Ocean Resources Engineering, pp. 57–101.

Lewis, C.F.M. & Woodworth-Lynas, C.M.T. (1990). Ice scour. In *Geology of the continental margin of eastern Canada*, eds., M.J. Keen & G.L. Williams, Geology of Canada, No. 2. Geological Survey of Canada, pp. 785–92.

Lewis, C.F.M., Josenhans, H.W., Simms, A., Sonnichsen, G.V. & Woodworth-Lynas, C.M.T. (1989). The role of seabed disturbance by icebergs in mixing and dispersing sediment on the Labrador shelf: a high latitude continental margin. In program with abstracts, Canadian Continental Shelf Symposium (C2S3), Bedford Institute of Oceanography, Dartmouth, October 2–7.

Lien, R.L. (1981). Sea bed features in the Blaaenga area, Weddell Sea, Antarctica. in *Proceedings of 6th International Conference on Port and Ocean Engineering under Arctic Conditions*. Quebec, July 23–31, 2, 706–16.

Longva, O. & Bakkejord, K.J. (1990). Iceberg deformation and erosion in soft sediments, southeast Norway. *Marine Geology*, **92**, 87–104.

Longva, O. & Thoresen, M.K. (1991). Iceberg scours, iceberg gravity craters and current erosion marks from a gigantic Preboreal flood in southeastern Norway. *Boreas*, **20**, 47–62.

Marienfeld, P. (1991). Holozäne sedimentationsentwicklung im Scoresby Sund, east Greenland. *Boreas*, **21**, 169–86.

Martin, H. (1965). The Precambrian geology of South West Africa and Namaqualand. Precambrian Research Unit, University of Cape Town, 159 pp.

Miller, J.M.G. (1989). Glacial advance and retreat sequences in a Permo-Carboniferous section, central Transantarctic Mountains. *Sedimentology*, **36**, 419–30.

Mollard, J.D. (1983). The origin of reticulate and orbicular patterns on the floor of the Lake Agassiz basin. Geological Association of Canada Special Paper No. 26, 355–374.

O'Brien, P.E. & Christie-Blick, N. (1992). Glacially grooved surfaces in the Grant Group, Grant Range, Canning Basin and the extent of Late Palaeozoic Pilbara ice sheets. *Journal of Australian Geology and Geophysics*, **14**, 87–92.

Olesen, O.B. & Reeh, N. (1969). Preliminary report on glacier observations in Nordvestfjord, East Greenland. Grønlands Geologiske Undersøgelse, Rapport, 21, 41–53.

Paul, M.A. & Evans, H. (1974). Observations on the internal structure and origin of some flutes in glacio-fluvial sediments, Blomstrand-breen, north-west Spitsbergen. *Journal of Glaciology*, **13**, 393–400.

Powell, R.D. & Gostin, V.A. (1991). A glacially influenced, storm-dominated continental shelf system on the Permian Australian-Gondwana margin. Abstract of Papers, 13th International Sedimentological Congress, Nottingham, U.K., August 26–31, pp. 435–6.

Rattigan, J.H. (1967). Depositional, soft sediment and post-consolidation structures in a Palaeozoic aqueoglacial sequence. *Journal of the Geological Society of Australia*, **14**, p.

Reeh, N. (1985). Greenland Ice-Sheet mass balance and sea-level change. In Glaciers, ice sheets and sea level: effect of a CO_2-induced climatic change. Report DOE/EV/60235-1, 155–171, Washington, D.C.: U.S. Department of Energy.

Reimnitz, E., Barnes, P.W. & Phillips R.L. (1984). Geological evidence for 60 meter deep pressure-ridge keels in the Antarctic Ocean. IAHR Symposium, Hamburg, 18 pp.

Rocha-Campos, A.C., Santos, P.R. & Canuto, J.R. (1990). Ice-scouring in Late Paleozoic glacial lake sediments, Paraná Basin, Brazil. Geological Society of America, Abstracts with program, 22 (2), 66.

Rocha-Campos, A.C., dos Santos, P.R. & Canuto, J.R. (This volume). Ice scouring structures in Late Paleozoic rhythmites, Paraná Basin, Brazil.

Savage, N.M. (1972). Soft-sediment glacial grooving of Dwyka age in South Africa. *Journal of Sedimentary Petrology*, **42**, 307–8.

Solheim, A., Russwurm, L., Elverhoi, A. & Nyland Berg, M. (1990). Glacial geomorphic features in the northern Barents Sea: direct evidence for grounded ice and implications for the pattern of deglaciation and late glacial sedimentation. In *Glacimarine environments: processes and sediments*, eds., J.A. Dowdeswell & J.D. Scourse. Geological Society of London Special Publication No. 53, 253–68.

Syvitski, J.P.M. (1991). Towards an understanding of sediment deposition on glaciated continental shelves. *Continental Shelf Research*, **11**, 897–937.

Vaslet, D. (1990). Upper Ordovician glacial deposits in Saudi Arabia. *Episodes*, **13**, 147–61.

Visser, J.N.J. (1985). The Dwyka Formation along the north-western margin of the Karoo Basin in the Cape Province, South Africa. *Transactions of the Geological Society of South Africa*, **88**, 37–48.

Visser, J.N.J. (1989). The Permo-Carboniferous Dwyka Formation of southern Africa: deposition by a predominantly subpolar marine ice sheet. *Palaeogeography, Palaeoclimatology, Palaeoecology*, **70**, 377–91.

Visser, J.N.J. (1990). Glacial bedforms at the base of the Permo-Carboniferous Dwyka Formation along the western margin of the Karoo Basin, South Africa. *Sedimentology*, 37, 231–45.

Visser, J.N.J. & Hall, K.J. (1984). A model for the deposition of the Permo-Carboniferous Kruitfontein boulder pavement and associated beds, Elandsvlei, South Africa. *Transactions of the Geological Society of South Africa*, **87**, 161–8.

Visser, J.N.J. & Loock, J.C. (1988). Sedimentary facies of the Dwyka Formation associated with the Nooitgedacht glacial pavements, Barkly West District. *South African Journal of Geology*, **91**, 38–48.

von Brunn, V. & Marshall, C.G.A. (1989). Glaciated surfaces and the base of the Dwyka Formation near Pietermaritzburg, Natal. *South African Journal of Geology*, **92**, 420–6.

Vorren, T.O., Hald, M., Edvardsen, M. & Odd-Willy Lind-Hansen (1983). Glacigenic sediments and sedimentary environments on continental shelves: general principles with a case study from the Norwegian shelf. In *Glacial deposits in north-west Europe*, ed., J. Ehlers, Groningen: A.A. Balkema, pp. 61–73.

Weber, J.N. (1958). Recent grooving in lake bottom sediments at Great Slave Lake, Northwest Territories. *Journal of Sedimentary Petrology*, **28**, 333–41.

Woodworth-Lynas, C.M.T. (1988). Ice scours in the geological record: why they are not seen. Program with abstracts, Joint Annual Meeting, Geological Association of Canada, Mineralogical Association of Canada, Canadian Society of Petroleum Geologists, Memorial University of Newfoundland, May 23–25, vol. 13, A137.

Woodworth-Lynas, C.M.T. (In press). Ice scour as an indicator of glaciolacustrine environments. In *Glacial environments – processes, sediments and landforms*, ed. J. Menzies, Oxford: Pergamon, Chapter 17.

Woodworth-Lynas, C.M.T., Christian, D., Seidel, M. & Day, T. (1986). Relict iceberg scours on King William Island, N.W.T. In *Ice scour and seabed engineering*, eds., C.F.M. Lewis, D.R. Parrott, P.G. Simpkin & J.T. Buckley, Environmental Studies Revolving Funds Report No. 049, 64–70.

Woodworth-Lynas, C.M.T. & Guigné, J.Y. (1990). Iceberg scours in the geological record: examples from glacial Lake Agassiz. In *Glacimarine environments: processes and sediments*, eds. J.A. Dowdeswell & J.D. Scourse. Geological Society of London Special Publication No. 53, 217–33.

Woodworth-Lynas, C.M.T., Josenhans, H.W., Barrie, J.V., Lewis, C.F.M. & Parrott, D.R. (1991). The physical processes of seabed disturbance during iceberg grounding and scouring. *Continental Shelf Research*, **11**, 939–61.

19 Environmental evolution during the early phase of Late Proterozoic glaciation, Hunan, China

QI RUI ZHANG

Abstract

Glacial rocks of Late Proterozoic age are widely distributed in South China and two glacial epochs are recognized. The epochs are the earlier Dongshanfeng and the later Nantuo, separated by the Datangpo interglacial epoch. The Dongshanfeng epoch is named after the Dongshanfeng Formation found in the Sinian section in the Yangjiaping Area, Shimen County, Hunan province. It was previously thought that the sequence below the lower boundary of the Dongshanfeng was preglacial and was deposited under warm climatic conditions. Recently several possible peri- and proglacial environmental indicators along the lower boundary of the Dongshanfeng Formation and in the top 60 m segment of the underlying Xieshuihe Formation have been discovered. These are ice-keel scoured ridges, dropstones with compressive concave morphology, ice-wedge casts, glacier meltwater-cut grooves, and glaciotectonic thrust structures. The first three are indicative of periglacial environments, and the rest record proglacial conditions.

Based on the stratigraphic order of the above mentioned structures, a depositional model for the Yangjiaping Area during the preceding stage or the early stage of the Dongshanfeng Epoch is proposed. The major controls on deposition were the gradual lowering of temperature and sea-level, the thickening and expanding of an ice-sheet, and the change from a submarine to subaerial setting. Later, due to crustal loading below the ice-sheet, the Yangjiaping Area once again became submarine and accommodated the glaciomarine Dongshanfeng Formation.

The structures described herein are of particular significance because most glacial stratigraphic records fail to show indications of the climatic transition from warm to cold as a result of glacial erosion. The sparse records of the preceding stage of Nanhua Ice Age found in Yangjiaping are, therefore, of special significance.

Introduction

Since the 1970s the Late Proterozoic glaciation in South China, i.e. the Nanhua glaciation (Liu et al., 1980), has been extensively studied, and recent investigations have integrated sedimentology with tectonics. Based on data from Yangjiaping (YJP), Shimen County, Hunan Province, China, evidence of glacial environments and an evolutionary model for the initiation of the Late Proterozoic glaciation are proposed.

Geological setting

YJP is about 90 km to the southwest of the well-known Sinian section of the Yangtze Gorges (YG) section (Fig. 19.1a). The stratigraphic sequence of the Sinian in YJP resembles the ones in YG, but is more complete, for it was located farther out on the ancient shelf. The outline and correlation of the Sinian strata between the two locations are shown in Fig. 19.1b.

From the base upwards, the thickness of Liantuo Formation, the lowest one of Sinian at the type section (YG), is 102 m (Fig. 19.1b). It thins to the northeast and pinches out about 8 km away. It thickens to 300 m about 30 km further south and is 435 m thick in YJP. According to Liu et al. (1980), units equivalent to the Liantuo Formation in YJP comprise the Laoshanya (the lower) the Xieshuihe (the upper) Formations. The sequence is characterized by continental fluvial sedimentary rocks in YG (Liu and Sha, 1963) and tidal flat sediments in YJP.

Nantuo Formation is 64 m thick and is the only glacial record in the YG. Its thickness variations are similar to the underlying Liantuo Formation. To the north, it thins out rapidly and lies directly on the pre-Sinian crystalline basement. The case is different in YJP for the Datangpo Formation (interglacial) and the Dongshanfeng Formation (lower glacial) (11 and 4 m thick, respectively) (Fig. 19.1b), which comprise two additional units of the Nanhua Ice Age (Liu et al., 1980).

The Doushantuo Formation and Dengyeng Formation rest on the glacigenic horizons and are similar in lithology and thickness in YG and YJP. The underlying glacigenic horizon becomes thinner to the north of the YG and finally completely vanishes, allowing the Doushantuo Formation to lie directly on pre-Sinian metamorphic rocks (Sha et al., 1963; Lu et al., 1985). No direct evidence of glacial erosion or substrate deformation has been reported in this region.

Sedimentology of Xieshuihe Formation

Rocks below the Dongshanfeng Formation were previously considered to be pre-Nanhua, and not related to the

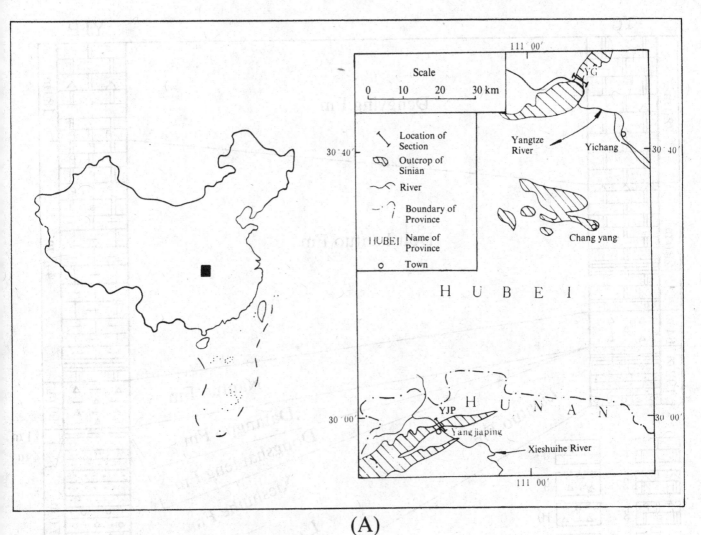

(A)

Fig. 19.1. Location and correlation of the Yangjiaping (YJP) and Yangtze Gorges (YG) Sinian stratigraphic sections. (A) Location of the sections shown in (B).

glaciation. However, new evidence suggests that the top portion of Xieshuihe Formation, just beneath Dongshanfeng Formation (Fig. 19.1b), possesses evidence of peri- and proglacial sedimentation. The thickness of the Xieshuihe Formation is 270 m (Fig. 19.2). The lower half mainly comprises pebbly sandstone and pebble conglomerate, with trough, tabular, and wedge-shaped cross stratifications, and locally, parallel and wavy bedding, and herringbone cross-bedding. These observations suggest an estuarine environment. The upper half mainly comprises coarse, silty, and fine-grained sandstone; claystone is rare and, except in the case mentioned below, no pebbly rocks are found. Besides sparse trough cross-stratifications at the bottom, the predominant sedimentary structures are wave ripple cross-laminations, and wave ripple bedding and/or small current ripple bedding (Reineck and Singh, 1980). The fining upward sequence in the Xieshuihe Formation indicates an increasingly quiet water setting; the environment of deposition is inferred to be tidal flat and lagoon.

A peculiar structure (Fig. 19.3) is found about 50 m below the top

of the Xieshuihe Formation, 6 meters above '–A–' in Fig. 19.2. The structure suggests that it is probably an incomplete record of an ice-keel scoured ridge. For a detailed description and discussion of the structure see Zhang (1992).

In the upper part of the Xieshuihe Formation, quartz and chert pebbles, 0.5 to 4 cm in size, angular to subangular, are occasionally found. These pebbles are usually associated with mud-clasts in greyish-green siltstone (Fig. 19.4). The morphology of mud-clasts is irregular. Quartz pebbles are also found in siltstone with parallel stratification (Fig. 19.5).

The occurrence of mud-clasts and quartz pebbles is generally not in keeping with the fine-grained character of the associated strata. They may have been emplaced by either slumping, ice-rafting, storm, or strong bottom currents. Slumping is not considered likely in the absence of related structures. Storm and bottom currents are possible causes responsible for scour-and-fill structures as shown in Fig. 19.4 but not those structures depicted in Fig. 19.5. The most probable process capable of bringing pebbles to the area was ice-rafting,

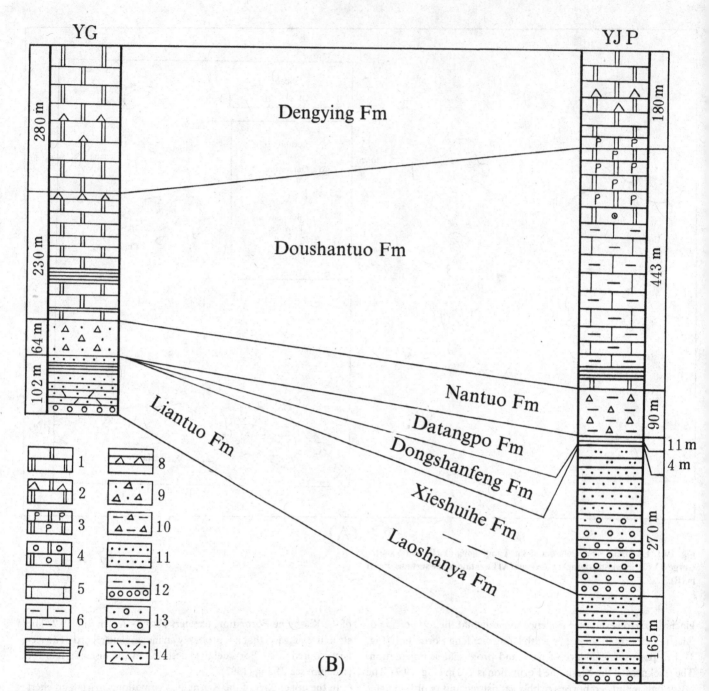

Fig. 19.1 (*cont.*) (B) Correlation of the Sinian between YJP and YG: 1, dolostone; 2, siliceous dolostone; 3, phosphatic dolostone; 4, oolitic dolostone; 5, limestone; 6, argillaceous limestone; 7, shale; 8, chert; 9, diamictite; 10, stratified diamictite; 11, sandstone; 12, muddy sandstone and conglomerate; 13, pebbly sandstone; 14, volcaniclastic rocks.

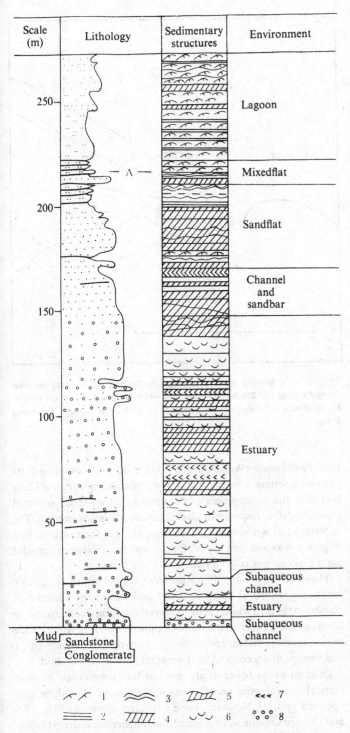

Fig. 19.2. Lithology and sedimentary environments of Xieshuihe Formation. 1, ripple cross-lamination; 2, horizontal bedding; 3, wavy bedding; 4, tabular cross-bedding; 5, planar cross-bedding with wavy boundaries; 6, trough cross-bedding; 7, herringbone cross-bedding; 8-channel lag; and –A–, the stratigraphic position of rock colour changed from reddish to greenish.

Fig. 19.3. A probable ice keel scoured ridge found in YJP section, 6 m above level '–A–' in Fig. 19.2. The hammer is 35 cm long. White spotty floccose layer is apparent.

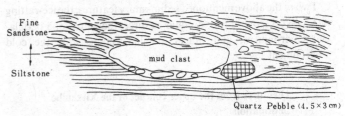

Fig. 19.4. Miniature scour-and-fill structure filled with mud-clasts and quartz pebbles.

Fig. 19.5. Delicately stratified fine sandstone with outsized quartz pebbles. An angular compressive concave pit is seen on the surface of the largest pebble.

which is especially favourable for explaining the outsized quartz pebbles occurring within parallel stratified fine sandstone (Fig. 19.5). An ice-rafting interpretation is in agreement with the cold climate inferred above. The rather frequent scour-and-fill structures suggest that, from time to time in this relatively quiet environment, there were also strong bottom currents.

The preglacial Sinian rocks, i.e. Xieshuihe and Laoshanya Formations, are mostly greyish-purple, purple-red, or reddish in colour; occasionally there are thin greyish-green and light greyish-white beds, suggesting the existence of temporary reducing conditions. This red bed system, more than 400 m thick, was generally deposited under warm climate. The top segment (above '–A–' in Fig. 19.2) of Xieshuihe Formation is, however, exclusively greyish-green in colour. The transition from purple-red to greyish-green is sharp, and is seen within one 135 cm thick layer of siltstone. Generally, the greyish-green colour of sedimentary rocks can be an indication of reducing environmental conditions. In addition, under certain special settings as in YJP, it is attractive to take it as an index of low temperature, because the segment with this colour is just below the level of known glacial deposition.

Two of the above mentioned sedimentary features, the ice-rafting and ice-keel scoured ridges, could be direct indicators of a periglacial environment, and others are in keeping with the cold environment.

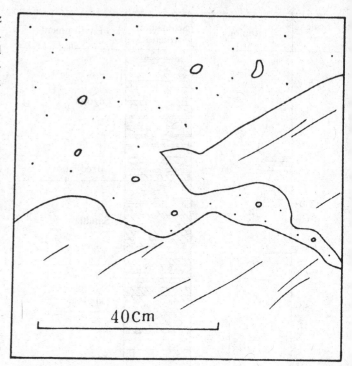

Fig. 19.7. A drawing from a photograph showing the contact between Dongshanfeng and Xieshuihe Formations which appeared in a report written by the Regional Geologic Survey Brigade of the Hunan Geologic Bureau in 1966.

Features from the upper contact of the Xieshuihe Formation

The top of Xieshuihe Formation is greyish-green siltstone and the base of the overlying Dongshanfeng Formation is greyish muddy sandstone, but the contact is not easily located.

The contact, however, shows V-shaped diamictite bodies shown in Fig. 19.6 and Fig. 19.7 which are believed to be ice-wedge casts. A review of Precambrian ice-wedge structures was recently given by Williams (1986). Unfortunately as can be seen from Figs. 19.6 and 19.7, the typical interior structure of the Xieshuihe ice-wedge or sand-wedge casts cannot be seen. Since the outcrop is now unobservable, the problem remains unsolved. The siltstone host rock may resemble the structureless wedge-shaped casts from Scotland's

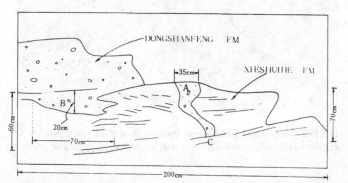

Fig. 19.6. A sketch map of the buried outcrop showing the disconformable contact between Dongshanfeng and Xieshuihe Formations. (Copied by courtesy from the 1967 field survey notebook of Professor Sha Qingan.)

Late Proterozoic (Williams, 1986, table 2), and considering its geological setting, a permafrost origin is possible in YJP. It is likely that with the development of the continental ice-sheet, sea-level gradually fell in response to glacio-eustatic sea-level lowering. The occurrence of wedge-shaped casts suggests that the shoreline had migrated seaward, and the YJP area became a subaerial coastal plain exposed to a cold climate.

The structure B in Fig. 19.6 is probably a fossil record of a glacier meltwater groove. It could be direct evidence of the presence of glaciers in the area which partially draped the permafrost ground. It is reasonable to infer that with continuously lowering temperature the ice-sheet expanded, and the formerly ice-wedge-producing area was then partially covered by the frontal part of the ice-sheet.

Original ice-wedge casts are more or less symmetrical to their vertical central axis, and the meltwater groove should be well openned upward. However, these structures shown in Figs. 19.6 and 19.7 are sinuous and deformed, and furthermore, the extremity of cast A in Fig. 19.6 has been cut off. These deformations may result from glaciotectonic stresses developed below the expanding ice-sheet as it overran the site.

It should be borne in mind that identification of subaerial conditions is dependent upon the correct identification of ice-wedge casts and glacier meltwater grooves. These features could well be subaqueously formed soft-sediment deformation structures. Eyles and Clark (1985) have shown that wedge-shaped casts may form under non-glacially related processes in mudstones in response to

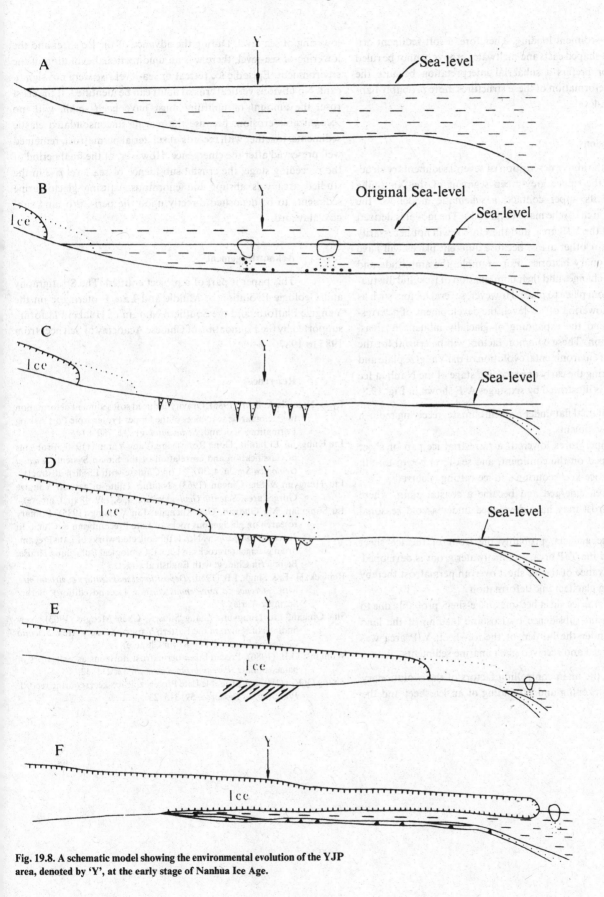

Fig. 19.8. A schematic model showing the environmental evolution of the YJP area, denoted by 'Y', at the early stage of Nanhua Ice Age.

penetrative soft-sediment loading. Therefore, a soft-sediment origin of the wedge-shaped casts and meltwater groove cannot be ruled out. The author prefers a subaerial interpretation because the glaciotectonic deformation of these structures more probably happened above sea-level.

Discussion

From the above description of several sedimentary structures found in the upper grey-green segment of the Xieshuihe Formation and its upper contact, a schematic model for the evolution of the area can be made (Fig. 19.8). The model is derived exclusively from the YJP area, and therefore it will not necessarily be appropriate for other areas, because different places will have different evolutionary histories. For example, the amplitude and time of sea-level changes and their combination of isostatic fluctuation will differ from place to place. However, several factors such as the continuous lowering of sea-level, the development of a terrestrial ice sheet, and the expanding of glacially influenced areas, should be common. These common factors will be crucial for the reconstruction of environmental evolution of the Yangtze plate and related areas during the early or preceding stage of the Nanhua Ice Age. The model is illustrated by six stages A-F shown in Fig. 19.8:

A. Normal tidal flat under a warm climate, receiving reddish clastic sediments

B. As temperatures lowered, a terrestrial ice cap or sheet developed on the continent, and sea-level began to fall; storms, ice-keel scouring and ice-rafting occurred

C. YJP area emerged and became a coastal plain, where permafrost may have developed under a cold seasonal climate

D. Discrete glaciers, probably the front of the ice sheet reached the YJP area, and meltwater grooves developed

E. The advance of the ice sheet overran permafrost thereby creating glacitectonic deformation

F. Finally, the ice sheet became an ice-shelf, probably due to the crustal subsidence by isostatic loading of the land mass under the burden of the ice-sheet; YJP area was submerged and received glaciomarine sediments.

In conclusion, the main controlling factors of the evolutionary model are the thickening and expanding of an ice sheet and the lowering of sea-level. During the advance of the ice sheet and the lowering of sea-level, there was an unidirectional evolution of the environment. Episodic ice retreat or sea-level rise were not significant. No obvious vertical crustal uplift can be identified. If the crust rose, the amount of uplifting must have been small, with no recognizable erosion, because the marine unconsolidated clastic sediments, together with records of subaerial permafrost, remained well preserved after the emergence. However, at the final period of the preceding stage, the crustal subsidence or sea-level rise in the studied area was abrupt and enormous, causing glaciomarine sediments to be deposited directly upon the peri-, pro- and subglacial record.

Acknowledgement

This paper is part of a project entitled 'The Stratigraphy and Geologic Evolution of Middle and Late Proterozoic on the Yangtze Platform and the Southern Margin of Huabei Platform', supported by the Foundation of Chinese Academy of Sciences from 1985 to 1987.

References

Eyles, N. & Clark, B.M. (1985). Gravity-induced soft sediment deformation in glaciomarine sequences of the Upper Proterozoic Port Askaig Formation, Scotland. *Sedimentology*, **32**, 789–814.

Liu Hungyun, Li Jianlin, Dong Rongsheng, Yang Yanjun (1980). Problems of classification and correlation of the Sinian System. *Scientia Geologica Sinica*, **4**, 307–21 (in Chinese with English abstract).

Liu Hungyun & Sha Qingan (1963). Nantuo Tillite in eastern Yangtze Gorges area. *Scientia Geologica Sinica*, **3**, 139–48 (in Chinese).

Lu Songnian, Ma Guogan, Gao Zhengjia, Lin Weixing (1985). Primary research on glacigenous rocks of Late Precambrian in China. In *Precambrian geology*. No. 1, the collected works of Late Precambrian glacigenous rocks in China. Geological Publishing House, Beijing (in Chinese with English abstract).

Reineck, H.-E. & Singh, I.B. (1980). *Depositional sedimentary environments, with reference to terrigenous clastics*, (second edition). Berlin: Springer-Verlag.

Sha Qingan, Liu Hungyun, Zhang Shusen, Chen Meng-e (1963). New findings of Sinian in the Eastern Yangtze Gorges Area. *Scientia Geologica Sinica*, **4**, 177–87 (in Chinese).

Williams, G.E. (1986). Precambrian permafrost horizons as indicators of palaeoclimate. *Precambrian Research*, **32**, 233–42.

Zhang Qi Rui (1992). A probable Late Proterozoic ice-keel scouring record. *Precambrian Research*, **59**, 315–23.

Printed in the United States
By Bookmasters